T0313727

Advances in Longitudinal Survey Methodology

WILEY SERIES IN PROBABILITY AND STATISTICS

Established by WALTER A. SHEWHART AND SAMUEL S. WILKS

The Wiley Series in Probability and Statistics is well established and authoritative. It covers many topics of current research interest in both pure and applied statistics and probability theory. Written by leading statisticians and institutions, the titles span both state-of-the-art developments in the field and classical methods.

Reflecting the wide range of current research in statistics, the series encompasses applied, methodological and theoretical statistics, ranging from applications and new techniques made possible by advances in computerized practice to rigorous treatment of theoretical approaches. This series provides essential and invaluable reading for all statisticians, whether in academia, industry, government, or research.

A complete list of titles in this series can be found at
http://www.wiley.com/go/wsps

Advances in Longitudinal Survey Methodology

Edited by
Peter Lynn
Institute for Social and Economic Research
University of Essex
UK

This edition first published 2021

Registered Offices
John Wiley & Sons, Inc., 111 River Street, Hoboken, NJ 07030, USA
John Wiley & Sons Ltd, The Atrium, Southern Gate, Chichester, West Sussex, PO19 8SQ, UK

Editorial Office
The Atrium, Southern Gate, Chichester, West Sussex, PO19 8SQ, UK

For details of our global editorial offices, customer services, and more information about Wiley products, visit us at www.wiley.com.

Wiley also publishes its books in a variety of electronic formats and by print-on-demand. Some content that appears in standard print versions of this book may not be available in other formats.

Library of Congress Cataloging-in-Publication Data

Names: Lynn, Peter, 1966- editor.
Title: Advances in longitudinal survey methodology / edited by Peter Lynn.
Description: First edition. | Hoboken, NJ : Wiley, [2021] | Series: Wiley series in probability and statistics | Includes index.
Identifiers: LCCN 2020033403 (print) | LCCN 2020033404 (ebook) | ISBN 9781119376934 (hardback) | ISBN 9781119376941 (adobe pdf) | ISBN 9781119376958 (epub)
Subjects: LCSH: Longitudinal method. | Surveys–Methodology.
Classification: LCC BF76.6.L65 A383 2021 (print) | LCC BF76.6.L65 (ebook) | DDC 001.4/33–dc23
LC record available at https://lccn.loc.gov/2020033403
LC ebook record available at https://lccn.loc.gov/2020033404

Cover Design: Wiley
Cover Image: © Peter Lynn

Set in 9.5/12.5pt STIXTwoText by SPi Global, Chennai, India
C9781119376934_040321

Contents

List of Contributors

Zahoor Ahmad
Department of Statistics and Demography
University of Southampton
Southampton, UK

Tarek Al Baghal
Institute for Social and Economic Research
University of Essex
Colchester, UK

Kelsey Beninger
Kantar Public
London, UK

Michaela Benzeval
Institute for Social and Economic Research
University of Essex
Colchester, UK

Nicholas Biddle
Centre for Social Research and Methods
Australian National University
Canberra, Australia

Michael Bosnjak
Leibniz-Institute for Psychology Information (ZPID)
Trier, Germany

and

Department of Psychology
University of Trier
Trier, Germany

Matt Brown
Centre for Longitudinal Studies
UCL Institute of Education
London, UK

Jonathan Burton
Institute for Social and Economic Research
University of Essex
Colchester, UK

Lisa Calderwood
Centre for Longitudinal Studies
UCL Institute of Education
London, UK

Alexandru Cernat
Department of Social Statistics
University of Manchester
Manchester, UK

Yang Cheng
National Agricultural Statistics Service
US Department of Agriculture
Washington, DC, USA

Mick P. Couper
Survey Research Center
University of Michigan
Ann Arbor, MI, USA

Stephanie Eckman
Survey Research Division
RTI International
Washington, DC, USA

Ben Edwards
Centre for Social Research and Methods
Australian National University
Canberra, Australia

Duncan Elliott
Time Series Analysis Division
Office for National Statistics
Newport, UK

Paula Fomby
Survey Research Center
University of Michigan
Ann Arbor, MI, USA

Alessandra Gaia
Department of Sociology and Social Research
University of Milano-Bicocca
Milan, Italy

Emily Gilbert
Centre for Longitudinal Studies
UCL Institute of Education
London, UK

Annette Jäckle
Institute for Social and Economic Research
University of Essex
Colchester, UK

Kristian Kleinke
Department of Clinical Psychology
University of Siegen
Siegen, Germany

Sabine Krieg
Methodology Department
Statistics Netherlands
Heerlen, Netherlands

Martin Kroh
Faculty of Sociology
University of Bielefeld
Bielefeld, Germany

Simon Kühne
Faculty of Sociology
University of Bielefeld
Bielefeld, Germany

Meena Kumari
Institute for Social and Economic Research
University of Essex
Colchester, UK

Carli Lessof
Department of Social Statistics and Demography
University of Southampton
Southampton, UK

Oliver Lipps
Surveys Unit
FORS
Lausanne, Switzerland

Peter Lynn
Institute for Social and Economic Research
University of Essex
Colchester, UK

Colleen A. McClain
Survey Research Center
University of Michigan
Ann Arbor, MI, USA

Katherine A. McGonagle
Survey Research Center
University of Michigan
Ann Arbor, MI, USA

Mary Beth Ofstedal
Survey Research Center
University of Michigan
Ann Arbor, MI, USA

Fiona Pashazadeh
Department of Social Statistics
University of Manchester
Manchester, UK

Darina Peycheva
Centre for Longitudinal Studies
UCL Institute of Education
London, UK

George Ploubidis
Centre for Longitudinal Studies
UCL Institute of Education
London, UK

Jost Reinecke
Faculty of Sociology
University of Bielefeld
Bielefeld, Germany

Caroline Roberts
Institute of Social Sciences
University of Lausanne
Lausanne, Switzerland

Joseph W. Sakshaug
Institute for Employment Research
Nuremberg, Germany

Department of Statistics
Ludwig Maximilian University of Munich
Munich, Germany

and

School of Social Studies
University of Mannheim
Mannheim, Germany

Emanuela Sala
Department of Sociology and Social Research
University of Milano-Bicocca
Milan, Italy

Narayan Sastry
Survey Research Center
University of Michigan
Ann Arbor, MI, USA

Timo Schmid
Institute for Statistics and Economics
Freie Universität Berlin
Berlin, Germany

Eric Slud
Center for Statistical Research and Methodology
US Census Bureau
Suitland, Maryland, USA

and

Mathematics Department
University of Maryland
College Park, Maryland, USA

Paul A. Smith
Department of Social Statistics and Demography
University of Southampton
Southampton, UK

Martin Spiess
Department of Psychological Methods and Statistics
University of Hamburg
Hamburg, Germany

Bella Struminskaya
Department of Methodology and Statistics
Utrecht University
Utrecht, The Netherlands

Yves Thibaudeau
Center for Statistical Research and Methodology
US Census Bureau
Suitland, Maryland, USA

Nikos Tzavidis
Department of Social Statistics and Demography
University of Southampton
Southampton, UK

Jan A. van den Brakel
Methodology Department
Statistics Netherlands
Heerlen, Netherlands

and

School of Business and Economics
Maastricht University
Maastricht, Netherlands

Marieke Voorpostel
Surveys Unit
FORS
Lausanne, Switzerland

Nicole Watson
Melbourne Institute of Applied Economic and Social Research
University of Melbourne
Melbourne, Australia

Erica Wong
Centre for Longitudinal Studies
UCL Institute of Education
London, UK

Li-Chun Zhang
Department of Social Statistics and Demography
University of Southampton
Southampton, UK

Preface

This is a book about methods that are, or could be, used to carry out longitudinal surveys, that is, surveys that collect data repeatedly from the same set of individuals. Longitudinal surveys are unique in many ways and face methodological challenges that are often distinct from those faced by other surveys. The challenges range from how we select the sample of people to study and maintain their participation over time, through how we collect the data, to how we use the data to shed light on population dynamics of various kinds. All are tackled within this book.

Each chapter of this book was presented and discussed at the *Second International Conference on the Methodology of Longitudinal Surveys*, MOLS2, which took place at the University of Essex, UK, 25–27 July 2018. The first *International Conference on the Methodology of Longitudinal Surveys* had taken place 12 years earlier at the same venue and had spawned the eponymous book, published by Wiley in 2009. In the intervening years, much had changed in the world of longitudinal surveys. The advantages of longitudinal data had become more widely understood, leading to a boom in the commissioning of new longitudinal surveys, including many in substantive fields where longitudinal data did not previously exist, and in countries/regions where longitudinal surveys had not previously been carried out. This enthusiasm for longitudinal surveys brought with it a thirst for knowledge about the best ways to design and implement such surveys. Consequently, considerable advances were made in areas such as methods to encourage continued participation, ways of utilising multiple data collection modes, improving the measurement of change, obtaining participant consent to data linkage, linking and analysing administrative data longitudinally, and longitudinal methods for cross-disciplinary research, including bio-social research using biomarkers collected in social surveys.

Many of these advances had been discussed at the biennial *Panel Survey Methods Workshop*, which had itself been spawned by the first MOLS conference. The workshop is smaller and more informal than the MOLS conferences, with presentations focussed on work in progress and ideas under development, and more time set aside for discussion than for presentation. Workshops have taken place at the University of Essex, UK (2008), Mannheim, Germany (2010), Melbourne, Australia (2012), Ann Arbor, Michigan (2014), and Berlin, Germany (2016). Following MOLS2 in 2018, the next *Panel Survey Methods Workshop* was due to have taken place in Lausanne, Switzerland, in 2020, but was postponed due to the COVID-19 pandemic.

The aim of this book is therefore to provide an updated overview of current knowledge and best practice in implementing longitudinal surveys. Some chapters demonstrate how methodological knowledge in areas covered by the earlier book have advanced in the intervening years, while others introduce new topics that have emerged, or become more prominent, since 2006. For example, the collection of biomarker data in longitudinal social surveys was little more than an aspiration in 2006 but has since become a key component of several surveys. This has opened up an array of new research possibilities that have had considerable impact on our understanding of the interacting roles of biological and environmental factors in shaping economic, social, health, and well-being outcomes. The growth in this area is reflected in the inclusion of two chapters (Chapters 2 and 5) addressing issues in the collection of biomarker data. Linkage of longitudinal survey data to individual-level administrative data is another sub-field that was embryonic in 2006, a situation reflected in the contents of the chapter in the earlier book devoted to this topic. Within that chapter, just half a page addressed issues of respondent consent. In this book, three chapters (6, 7, and 8) are devoted to such issues: an indication of the growth in prominence of the need for, and concerns about, respondent consent. A big shift to mixed-mode data collection is reflected in Chapters 9 and 10. This book benefits from its broad and up-to-date coverage of key issues facing longitudinal surveys, but most of all from the wisdom and experience of the chapter authors, whose enthusiasm and diligence has made the book what it is.

MOLS2 was organised and hosted under the auspices of *Understanding Society*: the UK Household Longitudinal Study, which is housed at the Institute for Social and Economic Research (ISER), University of Essex. This followed a proposal to hold the conference which I presented to a meeting of the *Understanding Society* Executive Team in 2016. I am grateful for the foresight of the Executive Team and, particularly, Michaela Benzeval, Principal Investigator of *Understanding Society* for providing support in the organisation of MOLS2. *Understanding Society* is funded by the UK Economic and Social Research Council with co-funding from a consortium of UK Government departments. Special thanks are due to Jolanda James who not only administered all aspects of registration for MOLS2 and organised the meeting rooms and catering but also went far beyond the call of duty in helping participants with a wide range of issues, from travel arrangements and international money transfers to where to find gluten-free pastries on the university campus.

The MOLS2 Scientific Committee consisted of Mick Couper (University of Michigan, USA), Annette Jäckle (ISER, University of Essex, UK), Peter Lynn (Chair: ISER, University of Essex, UK), and Nicole Watson (University of Melbourne, Australia). The Committee drafted conference announcements and calls for papers and carried out the work of reviewing and selecting papers to be presented. We received 43 proposals for monograph papers, from which we accepted 24, of which 4 were subsequently withdrawn or dropped for various reasons, leaving the 20 chapters that we see now in this book.

A subsequent call for contributed papers resulted in 79 submissions, of which 60 were accepted for presentation. The 80 papers were presented over three glorious summer days in which the temperature never fell below 30° centigrade. The conference was a success primarily due to the contributions of the 80 presenters, 29 session chairs, 10 discussants of monograph papers and 180 participants from 28 countries.

A student paper competition was held, with entries judged by a panel consisting of two ISER colleagues, Olena Kaminska and Jonathan Burton. The winner was Ruben Bach (University of Mannheim, Germany) for 'A Methodological Framework for the Analysis of Panel Conditioning Effects'. Two other entries were highly commended: Caroline Vandenplas (University of Leuven, Belgium), Jessica Herzing (University of Lausanne, Switzerland) and Julian Axenfeld (University of Mannheim, Germany) for 'A Method for Optimizing Reminder Procedures During Data Collection of an Online Panel: A Data-Driven Approach' and Nicole James (University of Essex, UK) for 'Tis the Season: Seasonal Effects and Panel Attrition'.

An archived version of the conference website at mols2.org.uk preserves for posterity the conference programme and abstracts of all the paper presented. I am especially grateful to the International Statistical Institute and World Bank Trust Fund for Statistical Capacity Building, whose support funded the attendance of six participants from developing countries. Additional support for MOLS2 was gratefully received from the International Association of Survey Statisticians.

Three workshops took place on the morning of the day before the conference commenced. These were 'The Mechanics of Longitudinal Data Analysis,' presented by Oliver Lipps, 'Introduction to the *Understanding Society* Innovation Panel,' presented by Brendan Read, and 'Introduction to Latent Class Analysis,' presented by Alex Cernat. In the afternoon, we held a 'Longitudinal Studies Showcase,' at which 10 major longitudinal surveys, or suites of surveys, from around the world each presented a 15-minute overview of their survey and current methodological challenges and opportunities, with ample time for questions and discussion. These fringe events greatly enhanced MOLS2.

In addition to the 20 monograph chapters published here, authors of contributed MOLS2 papers were invited to submit their paper for possible publication in a special issue of the journal *Longitudinal and Life Course Studies*. Twenty submissions were received, of which 6 were ultimately accepted for publication and appeared in volume 10, number 4 of the journal (www.ingentaconnect.com/content/bup/llcs/2019/00000010/00000004), guest-edited by Peter Lynn (ISER, University of Essex, UK), Mick Couper (University of Michigan, USA), and Nicole Watson (University of Melbourne, Australia).

There is good reason to believe that longitudinal surveys will continue to provide important insights into an ever-increasing range of important issues facing society. Methodological knowledge will need to keep pace with changing information needs, changing technology, and a changing survey environment.

Colchester, May 2020 **Peter Lynn**

About the Companion Website

This book is accompanied by a companion website:
www.wiley.com/go/lynn/advancesinlongitudinalsurvey

The website includes supplementary material.

1

Refreshment Sampling for Longitudinal Surveys

Nicole Watson[1] *and Peter Lynn*[2]

[1] *Melbourne Institute of Applied Economic and Social Research, University of Melbourne, Melbourne, Australia*
[2] *Institute for Social and Economic Research, University of Essex, Colchester, UK*

1.1 Introduction

As a longitudinal survey matures, there may be a need to consider refreshing the sample by adding a new sample (or indeed several new samples) to the ongoing sample. This may be due to the impacts of non-response or attrition on the size or representativeness of the sample, or a desire to increase the size of the overall sample, target (overrepresent) particular segments of population, or provide coverage of new population entrants.

The type of longitudinal population that the longitudinal survey aims to represent is integral to this discussion of refreshing the sample. Figure 1.1 shows the various elements of the population as it changes from when the sample was originally selected in wave 1 to a later time point, wave *t*. People leave the population via death or otherwise moving out of scope (i.e. by emigrating or ageing out of the population of interest) and others join the population as new births (or by ageing into the population of interest) or otherwise moving in-scope (such as immigrating). The terms 'births' and 'deaths' are used in this chapter to refer to this broader group of population joiners and leavers, respectively. Smith et al. (2009) describe three types of longitudinal populations:

1. A static population defined by the population at the time the first wave is selected (i.e. there are no 'births' but 'deaths' are allowed). This population definition is used in many cohort studies and is shown as the shaded oval in Figure 1.1.[1] Smith et al. (2009) refer to the longitudinal sample selected from

1 An exception appears to be the National Child Development Study which follows a sample of children born in 1958. When the children were aged 7, 11, and 16, an additional sample of

Advances in Longitudinal Survey Methodology, First Edition. Edited by Peter Lynn.

Companion website: www.wiley.com/go/lynn/advancesinlongitudinalsurvey

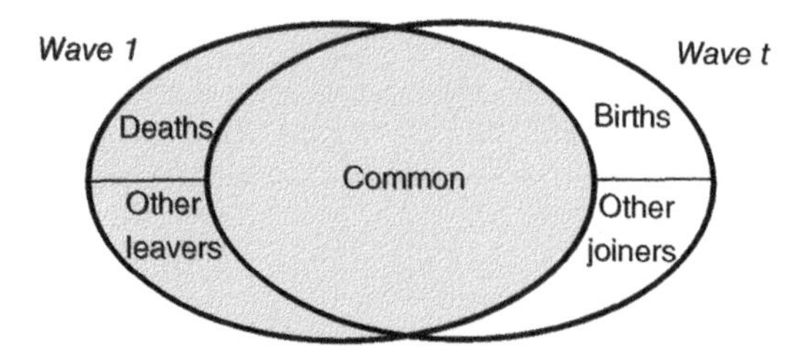

Figure 1.1 Dynamic population, comparing wave 1 and wave t.

this population as a 'fixed panel'. The goal is to study a group of individuals as they progress from a particular event (being born, leaving school, graduating, leaving hospital, receiving a grant, etc.) and examine the development and outcomes following that event. For these populations there are no new population entrants, only new populations (cohorts).

2. A population defined by the *intersection* of the cross-sectional populations at each wave (that is, both 'births' and 'deaths' are excluded). This population definition corresponds to the longitudinal sample typically extracted post hoc from a labour force survey that has a rotating sampling scheme. It is shown as the 'common' section in Figure 1.1 (if $t = 2$).
3. A population defined by the *union* of the cross-sectional populations at each wave (where both 'births' and 'deaths' are allowed). This dynamic population definition is used for most household panel studies and older age panels. It is the combination of the two ovals in Figure 1.1 (if $t = 2$). Smith et al. (2009) refer to the longitudinal sample selected from this population as a 'fixed panel plus "births"'. The goal is to study individuals over time maintaining representativity of the dynamic population so that new entrants to the population can also be studied over time.

Refreshment samples aim to address bias or variance issues (or a combination of both) identified in a longitudinal survey sample. Bias can arise from non-response in the initial wave or attrition from the sample over time affecting the representativity of the sample. Bias can also arise through non-coverage of the population either in the initial sample design or as the population changes dynamically over time due to new 'births'. This latter potential contributor to bias is only cause for concern if there is a desire for the sample to capture these new segments of the population (for example, because they are different in substantial ways to the original population or are of particular policy relevance). Variance is a function of sample size, so can be affected by initial wave non-response and subsequent attrition. Further, new uses of the data may be envisaged after the study began,

children selected from school registers was added at each of these sweeps. These additional children were born outside Britain but were living in Britain by the respective age (Johnson and Brown 2015). Subsequent additions to the sample, however, have not been made.

leading to a desire for an additional sample to obtain more accurate estimates at a finer level (such as more detailed geography) or of particular sub-populations (such as ethnic minorities) to better address policy needs.

The population to which these concerns about the sample relate can vary depending on the concern. Concerns about non-coverage and sample size can relate to (i) the entire initial population; (ii) the entire population of new entrants; or (iii) sub-populations of policy interest, such as ethnic minorities. Concerns about non-response and attrition only apply to (i) and (iii). Depending on the particular combination of concerns being addressed, there is a wide range of potential refreshment sample objectives and resulting sample designs.

We illustrate the various reasons for adding a refreshment sample with a number of examples:

- *Improve representativity of the sample and restore sample size.* The Dutch Longitudinal Internet Study for the Social Sciences (LISS) recruited several additional samples since its inception in 2007. We focus on one additional sample introduced in 2009. This top-up sample had two key objectives: (i) to restore the overall sample size to its original level; and (ii) to address non-response bias identified in the initial wave. This resulted in a sample design, which disproportionately sampled groups with a lower propensity to respond in the initial recruitment in 2007, such as elderly respondents, single-person households, and non-Westerners (Lynn and Lugtig 2017). The remainder of the population was also sampled, but at a lower rate than the groups targeted due to the low initial response propensities, so that the overall responding sample size was brought back to the initial level.
- *Increase the overall sample size.* Some studies have sought to increase the overall sample size beyond the initial level or to return the overall sample to its initial size to reduce sampling variance and make analysis of certain small groups possible. For example, the German Socio-Economic Panel (SOEP), which started in 1984 added large-scale additions to the sample in 2000 and 2011 (Haisken-DeNew and Frick 2005; Kroh et al. 2014) increasing the number of responding households by around 45% and 20%, respectively. Also the Household, Income and Labour Dynamics in Australia (HILDA) Survey, which started in 2001 added a general population-wide sample top-up in 2011, which added about 30% to the size of the responding sample (Watson 2014).
- *Increase the sample size of particular sub-groups.* Boost samples add sample from certain sub-populations of policy interest. The British Household Panel Survey (BHPS), which started in 1991 added boost samples in Scotland and Wales in 1999 to permit more detailed analysis of these countries (Taylor et al. 2010). Further, for a period of five years, the BHPS incorporated a sample of low income households (from 1997 to 2001). The SOEP added a sample

of high-income households in 2002 (Haisken-DeNew and Frick 2005) and second-generation immigrants in 2013 (Kroh et al. 2015). In 2015, the UK Household Longitudinal Study, also known as *Understanding Society*, added a boost sample, known as the Immigrant and Ethnic Minority Boost (IEMB), to the original sample (selected in 2009). This boost sample had two components: (i) people of Indian, Pakistani, Bangladeshi, Black Caribbean, or Black African ethnicities; and (ii) people born outside of the UK, regardless of ethnicity or nationality (Lynn et al. 2018). The first component added additional sample to the initial ethnic minority boost sample selected when the study began, while the second component provided coverage of population entrants (immigrants) since 2009 as well as adding to the sample of pre-2009 immigrants.

- *Increase the sample size and address non-coverage in initial sample design*. The Swiss Household Panel (SHP) added a refreshment sample in 2012 (13 years after the study began) using a different sampling frame from the one used for the original sample and an earlier refreshment sample. The first two samples were drawn from a frame of listed landline and mobile numbers, whereas the 2012 refreshment sample was drawn from a population register that included people with silent telephone numbers and people without telephones. The SHP team estimate the original frame had 92–94% coverage of the resident population, whereas the new frame has close to 100% coverage (Voorpostel et al. 2017). While the main reason for adding the 2012 sample was to increase sample size, the change of frame also helps to address initial non-coverage issues.

Longitudinal studies that aim to be representative of the dynamic population need to consider how to incorporate a sample of 'births' that have joined the population over time. As these studies age, the number of these new entrants grows. Household panel studies, whose samples aim to provide complete coverage of the household population, add to the sample at each wave new births to existing sample members. This mechanism provides a means of sampling the new births in the population (albeit excluding those who are born to recent immigrants). More difficult to sample are immigrants arriving into the population after the study has begun. And these immigrants can be quite different from the original sample (and their descendants). For example, recent immigrants in Australia are much younger than the Australian-born population, and even when age is taken into account, they also are more likely to be married, have lower fertility rates, have higher education levels, have less wealth and savings, settle in capital cities (especially inner-city areas), and work fewer hours per week (Productivity Commission 2016). Further, second-generation immigrants are also likely to have higher education attainment than those with both parents born in Australia. As a result, excluding recent immigrants can have both short-term and long-term consequences on sample representativity. There are a number of ways to address this issue, with the three most common being the following:

1. Add a dedicated immigrant sample that is focused on the immigrants that have arrived since the study began.
2. Add a broader sample of the general immigrant population that will include a portion of recent immigrants but will also add to the sample immigrants who arrived before the study began thus making more detailed analysis of this group possible.
3. Add a sample of the entire population that will reduce the variability of estimates more broadly, but estimates for the recent immigrants will be more variable than for the other approaches due to the smaller sample size for this particular group.

The household panel studies offer various examples of addressing the issue of immigrants arriving after the study began, sometimes combining this particular concern with other reasons for a refreshment sample. The SOEP added, in 1994 and 1995, an immigrant sample of those who had arrived after the study began and then subsequent refreshment samples selected from the general population, which include some recent immigrants (Haisken-DeNew and Frick 2005). The Household, Income, and Labour Dynamics in Australia (HILDA) Survey added a general population-wide sample in 2011, and this was in part due to a desire for a larger sample size, but also to incorporate a representative sample of recent immigrants into the sample. The US Panel Study of Income Dynamics (PSID), which started in 1968, added an immigrant sample in 1997 and again in 2017 (McGonagle et al. 2012; Sastry 2019). An alternative way to meet both cross-sectional and longitudinal requirements is via a rotating panel design like that used in the Canadian Survey of Labour and Income Dynamics (SLID: Latouche et al. 2000) and the European Statistics on Income and Living Conditions Survey (EU-SILC: Wolff et al. 2010). The disadvantage of a rotating panel design is that it restricts longitudinal analysis to a maximum duration, which is a six-year period in the case of SLID and four years in the case of EU-SILC.

In a similar fashion, the older age panel studies, which select a sample from the population over a particular age, need to address new 'births' in their population of interest if there is a desire to represent the cross-sectional population at each wave as well as the longitudinal population from wave 1. For example, the US Health and Retirement Study (HRS) started in 1992 with a sample of people born in 1931–1941 and extended the sample in 1998 to include those who had turned 51 since the study began so that the sample could be representative of the population over the age of 50 in the US. Every six years since then, the study adds a fresh sample of people who have turned 51 since the last sample addition (Sonnega et al. 2014).

In this chapter, we discuss ideal properties of a refreshment sample and provide considerations for sampling frames, screening, sample design, questionnaire design, refreshment frequency, recruitment, data integration, and weighting. We also consider the impact these refreshment samples have on researchers analysing the data. The chapter concludes with suggestions for future work on this topic. We do not consider samples that aim to extend the population coverage of a survey as refreshment samples, such as the 1990 sample extension in the German SOEP to include East Germany or the 2001 BHPS sample extension to include Northern Ireland. These are perhaps better thought of as extension samples and are not considered further here. In the same vein, we also do not cover survey design considerations for new cohort additions to a cohort study as each new cohort is a distinct population.

1.2 Principles

The ideal properties of a refreshment sample include the following aspects:

1. *Known sampling probabilities.* The way in which the refreshment sample is selected should result in clearly defined sampling probabilities. Historical information should also be collected from the respondents to identify if they were part of the population at the times when earlier samples were selected. The type of historical information needed will depend on the nature of the previous samples but should enable identification of overall inclusion probabilities.
2. *High accuracy of frame information.* The sampling frame from which the refreshment sample is selected should have high coverage of the population of interest, low measurement error on the variables used to define the population from which the sample is selected, and, preferably, reasonably low rates of ineligible cases.
3. *Time invariant characteristics define population groups that are oversampled.* If the refreshment sample involves oversampling particular population groups, the characteristics used to define those groups should ideally be constant over time (such as year of birth, country of birth, ethnic origin) rather than characteristics that change over time (such as income or employment status). Using characteristics that change over time is likely to reduce the efficiency of the sample in the longer term.
4. *Minimise overall coverage error.* When adding missing segments of the population, the frequency of refreshment samples and the coverage of the missing segments achieved by the refreshment sample impact on the coverage error for the dynamic population. The frequency of refreshment samples needs to be balanced against the cost of adding these samples.

5. *Maximise sample efficiency*. Depending on the objective of the refreshment sample, the size of the new sample should be chosen to maximise the efficiency of the analysis samples expected to be used by researchers. These analysis samples will typically be a combination of some or all members of the refreshment sample and some or all members of the original/previous samples.

1.3 Sampling

1.3.1 Sampling Frame

Different countries or states will have different frame choices available for selecting a refreshment sample, and indeed organisations within the country may have different access to the various frames. These might include administrative records (such as migration records, health care registrations, electoral rolls, or population registers), address lists, or area-based frames. When using administrative records to select a sample, it is important that these records are of high quality, that is: (i) the frame has high coverage of the desired population groups; (ii) membership of the population groups is defined by variables with minimal error; and (iii) the number of duplicate records is low. Further, the contact details on the frame need to be reasonably up-to-date or tracking of the sampled cases prior to the first interview needs to be possible and effective. Address-based and area-based frames offer high coverage but are also high cost when the coverage required is of only a subset of the total residential population. When sampling a population subset, addresses will need to be approached and screened to identify households or people belonging to the population of interest (see next section). The costs can be reduced by oversampling areas that have a high proportion of cases belonging to the population of interest, as was done in the 2015 IEMB sample in *Understanding Society* (Lynn et al. 2018). The optimal cut-off to define the high-density areas and low-density areas is one that provides the maximum precision of survey estimates for a given budget. For the IEMB sample added to *Understanding Society*, the low-density areas were defined as those that had less than 10% of people from the five ethnic minority groups and less than 12% of people who were born outside of the UK. The high-density areas were then further stratified. In this particular case, the original 2009 *Understanding Society* sample provided the sample for the low-density areas and the 2015 boost sample was only targeted to the higher-density areas. This does, however, mean that the recent immigrant sample (i.e. those immigrants arriving in the UK since the study began) included as part of this boost sample is only selected from those immigrants living in areas with a higher density of ethnic minorities or people born outside the UK.

1.3.2 Screening

Where the refreshment sample cannot be sampled directly from the frame, a screening process is required. This may be simply done by drawing the sample from another survey of the population if this option is available. Variables collected as part of this survey can be used to target particular areas or characteristics of interest for the refreshment sample. For example, the low-income sample included in the BHPS was drawn from the UK European Community Household Panel (Taylor et al. 2010), refreshment samples for the English Longitudinal Study of Ageing are selected from the Health Survey for England (Banks et al. 2016), and the Survey of Labour and Income Dynamics obtained their refreshment samples from the Canadian Labour Force Survey (Latouche et al. 2000). Otherwise, screening can be a large and costly exercise if it must involve original enquiry in the field, especially when the hit-rate is low. For example, if screening to find recent immigrants, it may be necessary to wait quite some time between refreshment samples so the proportion of recent immigrants is high enough for the cost of screening to be acceptable. But a longer wait will also imply greater undercoverage for the survey in the meanwhile. There may be ways to improve the efficiency of the screening, though these tend to come with compromises in other sources of errors. Some examples include:

1. *Sharing screening between two or more surveys.* The Health and Retirement Study and the Panel Study of Income Dynamics shared the cost of screening in 2017. The HRS aimed to find people who have aged into the scope of their survey in the last six years (i.e. those aged 51–56) and the PSID team aimed to find people who have immigrated to the US in the last 20 years (Sastry 2019). A compromise had to be made for recent immigrants who fell in the HRS target age range as the two survey groups agreed that these people should not be asked to participate in both surveys. The PSID team therefore devised an alternative strategy using multiplicity sampling from the relatives of the new immigrants added to the sample to cover immigrants in the 51–56 age range who arrived in the last 20 years.
2. *Using administrative data to target the sample.* In 2013, a refreshment sample was added to the SOEP, which oversampled particular first- and second-generation immigrant groups. The sample was selected from a labour force market register, and the SOEP team estimated that the frame excludes 5–8% of the target population, such as the self-employed, students, and refugees (Kroh et al. 2015). Further, to help improve the coverage and detail of the immigrant information on the register (which is based on recorded citizenship flags and country of origin), onomastic methods were used to classify people into a region of origin based on their first and given names. Tentative generation information was derived from other variables on the register.

The sample was only selected from those people flagged or recoded as first- or second-generation immigrants. Specific migrant groups (in terms of country of origin and generation) were oversampled. A screening exercise was then undertaken to verify the immigration information before proceeding with the interview with those determined to be in-scope.

3. *Using the existing panel to pre-screen residents.* A novel approach to adding an immigrant sample to an ongoing panel (proposed by Lynn 2011) is to have the interviewers return to the original selected addresses where survey interviews are no longer carried out. The people in these dwellings include current non-respondents or people who have moved in after the original sample members have moved out. These households can be approached and screened for immigrants that have arrived after the study began. By not needing to screen the households of current respondents who have not moved since the study began, significant costs can be saved. Further, if the sample addresses were clustered, the cost of screening will be constrained and the new immigrant sample will fit neatly into the existing workloads of the interviewers. However, it does mean re-approaching some non-respondents that may not be desirable. It also means that new dwellings (built or converted to residential use after the study began) are not included unless a separate sample of these is also selected. Though these two approaches (screening original sample addresses and a sample of newly-built addresses) were both originally proposed as part of the design of the IEMB, both were eventually dropped on cost-efficiency grounds, as they would have yielded small numbers of immigrants. As far as we are aware, this screening approach has not been tested.

The process of screening a sample to find cases that match certain criteria can be a source of error. Not enough contact attempts or low co-operation rates may result in eligible cases being missed. This may result in bias if those missed cases are different from the easier-to-find eligible cases. The screening questions may be subject to interpretation, be sensitive, or the person answering them may not know these details about others (e.g. others in their household, or neighbours). This may lead to underreporting and, in some circumstances, overreporting of eligibility, both of which can introduce bias if these reporting errors are not random. As a result, the screening questions need to be as simple and clear as possible. This may mean that certain nuances of eligibility for the refreshment sample may need to be discarded in favour of greater clarity in the questions or, in the situations where the screening questions lead to false positives, more detailed questions could be included in the main interview to further clarify eligibility though this will be at the cost of unnecessary interviews. It is very costly (and therefore difficult) to rectify the problem of false negatives at the screening stage. These sorts of issues highlight the type of trade-offs that need to be made at this stage between coverage errors and cost. Further, if the screening is conducted at a different time to the main interview (as might be the case if the screening is undertaken by telephone and the interviewing is undertaken face-to-face), there is the potential for further non-response at the interview stage. Ideally, the interview would follow immediately after screening, but this may not be possible for cost or logistical reasons.

1.3.3 Sample Design

Sample design considerations for a refreshment sample are primarily those that need to be addressed when designing any new longitudinal survey sample: optimal sample size, stratification, clustering, oversampling, expected eligibility rates, and expected response rates (for a discussion, see Smith et al. 2009).

The main difference in designing the refreshment sample compared to a completely new longitudinal survey is in considering how it will interact with the ongoing sample. Where a screening exercise is included, there may well be a distinction between how the two samples interact in the initial wave, when there is a large sample to screen, and in subsequent waves, when there is a much smaller sample to follow and interview. The efficiency of both phases needs to be considered. If the sample design is addressing issues of non-coverage along with other bias or variance concerns, it would be worth considering what the original sample can offer that the refreshment sample does not need to, as in the case of the IEMB sample added to the *Understanding Society* sample that was mentioned earlier. If the sample is clustered, consideration needs to be given as to how the clusters of the new sample will best fit into the ongoing sample design. For example, the new sample could be selected close to the original sampling points so that efficient field workloads can be formed. This might be done by selecting the new sample from within the same primary sampling units as the original sample, assuming they are reasonably large – this approach was taken for the wave 4 refreshment sample in the *Understanding Society* Innovation Panel (Burton 2012). Having the refreshment sample in the same location as the original sample will save on fieldwork costs though this has to be balanced against the effect it will have on variance. The overlap of the refreshment sample and ongoing sample should be examined to ensure, as far as possible, ongoing sample members are not approached to be part of the refreshment sample.

1.3.4 Questionnaire Design

For a longitudinal survey, the questionnaire used in the initial interview is usually somewhat different from the questionnaires used to interview people who have previously been interviewed. This is because in the first interview, background information is usually collected that either only needs to be collected once (such as country of birth or age left school) or only needs to be updated in subsequent interviews if the information changes (such as education attainment or children born). With the introduction of a refreshment sample, minimal disruption to the longitudinal questionnaire design would suggest the refreshment sample participants be given the same questionnaire routinely provided to new sample members each wave. This questionnaire would collect the relevant historical information together with questions about current or recent activity that is given to people interviewed on a regular basis. If the refreshment sample has been recruited from a specific sub-section of the population (such as immigrants, ethnic minorities, recently come of an age to be interviewed) then there may be a desire to add specific questions to the questionnaire about these circumstances. However, it is important to carefully consider how this information might be used and whether these same questions should also be asked of the ongoing sample – or at least of the equivalent subgroup in the ongoing sample – for comparison purposes.

There is evidence that longer interviews (as indicated by the study material or mentioned by the interviewer) are associated with lower response rates in the cross-sectional literature (e.g. Collins et al. 1988; Roberts et al. 2010; Groves et al. 1999), which is relevant to the first wave of a refreshment sample. Little experimental evidence is available on how much the length of the initial interview in a longitudinal survey affects the re-interview rates. One such study, however, has shown that a 20% decrease (from 31 minutes to 26 minutes) in the length of an interview in wave 1 of the *Understanding Society* Innovation Panel consisting of 2500 respondents had no significant effect on the response rates at subsequent waves (Lynn 2014). Some studies, such as the SOEP, avoid collection of the historical information in refreshment samples in the first wave to help reduce respondent burden and instead collect it in the second wave (Goebel 2012). This means, however, that the individuals interviewed in wave 1 that are not subsequently interviewed are missing this historical information. As the attrition between waves 1 and 2 is almost always greater than that between any other set of consecutive waves, this may represent a considerable loss of information. Another option is to exclude some or all of the rotating content from the initial wave for the refreshment sample, as was done with the HILDA Survey (Watson and Wooden 2013). It is not advisable to drop core content from the questionnaire for the refreshment sample as this will have undesirable flow-on implications for the weights and derived variables.

1.3.5 Frequency

The initial motivations for a refreshment sample may vary in nature: concerns about the sample being too small for particular analyses, concerns with the impact of attrition on the sample size or sample representativity, or concerns about non-coverage of parts of the contemporary population. The magnitude of these concerns will depend on the original sample size, the amount and nature of sample attrition, research consumers' desire for research on new and emerging population sub-groups, and the rate at which new members join the population. Each of these concerns must be tempered by the funding appetite available to support new samples and the availability of the other resources (like interviewers) to undertake the work. Few longitudinal surveys are blessed with funding for more than a few waves, so it is difficult to plan for refreshment samples with any certainty of funding for the longer term.

There are three options for deciding when to incorporate refreshment samples into a longitudinal study:

1. *On a regular, frequent, ongoing basis.* For example, new births are added to the sample each year as occurs in household panels, or an annual sample of recent immigrants arriving could be added from immigration records, or an additional sample of people coming of age to be interviewed could be added every three waves (as occurs with the HRS, which has a biennial interview cycle).
2. *On a regular but relatively infrequent basis.* For example, every 10 years a new refreshment sample could be added to address coverage issues. Or a screening exercise could be undertaken every 10 years or so to find specific sub-groups of the population and add them to the ongoing sample.
3. *On an irregular basis.* Most often, the timing of refreshment samples is driven by external factors as opportunities and needs arise. For example, if the reason for a sample replenishment is to counter the effects of attrition and sample loss, then refreshments might occur more frequently in the early stages of the panel where these issues are the greatest, and less so as the panel matures. Or the sample refreshment may address an emerging population group, such as when the unusually large number of asylum seekers and refugees entering Germany triggered additional samples to be added to the German SOEP in 2016 and 2017 (Kroh et al. 2017).

In balancing the cost of a refreshment sample against the benefit it brings, it is difficult to equate the benefits with a monetary value. A value would need to be placed on the benefits of a reduction in bias or variance that the refreshment sample contributes to the combined estimates. These benefits would also need to be considered across multiple waves. For example, if the choice was between a lower-quality refreshment sample that offers a coverage bias correction each wave and a higher-quality refreshment sample design that brings a more accurate, but sudden, correction every 10 waves, we might value the benefit of a constant correction higher than the sudden correction such that it is enough to counter the effect of a lower-quality sample. Alternatively, if the choice is between a refreshment sample focused on increasing the sample size overall or on a particular target group, a value would need to be placed on a general reduction in variance of multiple longitudinal and cross-sectional estimates compared to those for a specific subgroup of the population. If the analysis for the subgroup of the population was not even possible before adding a refreshment sample (sample size too small), then making this possible might be quite valuable.

1.4 Recruitment

In recruiting the refreshment sample to the study, special consideration will need to be given to the timing of the fieldwork, the mode of recruitment, the interviewers (if interviewers are used), and respondent communications. In terms of the fieldwork, the sample design considerations discussed earlier may have implications for how the fieldwork is to be undertaken. For example, the new sample could be added to existing workloads of interviewers, necessitating an extension to the time they are given to undertake their workloads, or the interviewer workforce could be expanded so that the fieldwork for the new sample can fit within the existing fieldwork timeframe. Both options have implications for the integrity of the data and for costs.

If interviewers are used, deciding on who will recruit the new refreshment sample is an important decision. In many ways, interviewers experienced with a longitudinal study can best explain the study to a potential sample recruit and thus have a better chance of keeping that person involved with the study in the longer term than an interviewer with only cross-sectional survey experience. However, it is important to manage the expectations of the interviewers in terms of the response rates they are likely to achieve in a recruitment exercise as opposed to working in a later wave of a longitudinal survey. When an interviewer is used to getting very high re-interview rates (90–95%), it can be disheartening for them to then work on the initial wave of a refreshment sample where the response rates may be in the range of 50–60%. They also need to be given tools and approaches specifically targeted towards recruiting these new sample members, designed with input from the most proficient interviewers on staff (Groves and McGonagle 2001). Further, for some types of refreshment samples, such as immigrant samples or other targeted sub-samples of the population, using bilingual interviewers (in the case of an immigrant sample), or otherwise tailoring the interviewer to the respondent can help improve response rates. Consideration also needs to be given as to whether the same interviewers be retained in subsequent waves to keep consistency between interviewers and respondents as is usually considered desirable for respondents in a longitudinal study (Watson et al. 2019). Note, however, that experimental evidence on the effect of respondent-interviewer continuity is scarce. One study using an interpenetrated design at wave 2 of the BHPS found no effect of interview continuity on reinterview rates (Campanelli and O'Muircheartaigh 1999). In contrast, a more recent study found some support for interviewer continuity for respondents under 60 years old, however, for older respondents the effects of continuity are dependent on the interviewer age (Lynn et al. 2014).

While in many ways, the recruitment of a refreshment sample will be similar to the initial wave of a longitudinal study, consideration needs to be given to how this new component to the sample is explained to the new sample members. If the study has a respondent website, it is likely to be obvious to the respondent that the study has been going for a long time and it needs to be clear to them why they are only now being invited to participate. Also, when the refreshment sample is targeted towards particular parts of population (such as immigrants), more attention may need to be given to ensure the language used in communicating with the respondents is accessible (though translation, using graphics, etc.).

1.5 Data Integration

When preparing datasets that include both the ongoing sample and the refreshment sample, the sample members need to be flagged as to which sample they belong to. As multiple refreshment samples may be added over the course of the study, this is best done using the same variable, so when the first refreshment sample is added, aim for variable names, labels, values, and value labels that would easily extend to include other sample refreshments. For household panels, there may be some sample members who eventually move in with a person from a different sample, so if household-level sample membership flags were constructed, there would need to be rules governing which sample takes precedence (generally, the earliest sample would take precedence). Indicators of primary sampling unit (cluster) and sampling stratum – necessary to allow unbiased estimation of variances – must be derived in a way that is appropriate for analysis of the combined sample (original plus refreshment). In particular, Primary Sample Unit (PSU) indicators for a refreshment sample should be consistent with those for the original sample if the sample units are selected from the same set of PSUs, but should utilise a distinct range of values if a fresh set of PSUs is selected.

The survey documentation available to users should clearly describe the different samples, how they were selected, any fieldwork differences, response rates, the flag used to delineate the samples, whether (and why) certain questions were or were not asked of the refreshment samples, and which weights incorporate the refreshment samples. Each of these components will help the data users work with the data in an appropriate way.

1.6 Weighting

Longitudinal survey data can be used to make inferences about a variety of populations, as discussed earlier in the introduction. The data can be analysed longitudinally over two or more waves to provide estimates for longitudinal populations. It can also be analysed cross-sectionally each wave to provide estimates about the population at those points in time. Whether the refreshment sample is included with the ongoing sample to make inferences about a particular population will depend on which waves the analyst chooses to include in their analysis and to which population they want to make inferences. In analysis, weights are used to compensate for differential probabilities of selection into the sample and for differential non-response (both in the initial wave and over time). This section discusses the complexities in combining the ongoing and refreshment samples to create suitable weights.

For analysis that includes waves prior to introduction of the refreshment sample, longitudinal weights should continue to be produced (after the introduction of the refreshment sample) for the ongoing sample alone. But for analysis of time periods beginning at or after the first wave of the refreshment sample, depending on the reason for the refreshment sample, separate weights may be provided for the ongoing sample, the refreshment sample, and the combined samples. The combined sample weight is the core weight to provide, and any others may be considered optional extras. Weights for the ongoing sample may be desirable, for example, if some questions were dropped for the refreshment sample. Or the researcher may want to compare the estimates from the refreshment sample to those from the ongoing sample to assess potential biases in the ongoing sample. Weights for the refreshment sample alone may also be useful if it is a population-wide sample or represents an entire subpopulation. Such weights could be used to compare the ongoing sample estimates to those from the combined sample to assess how much of an improvement the refreshment sample has brought to the combined sample in terms of bias and variance reduction. The potential benefits that these additional weights can bring needs to be balanced against the confusion that extra weights can cause for users when attempting to select the correct weight for their particular analysis.

Calculating weights for the combined ongoing and refreshment sample must consider each sample member's multiple chances of selection into the study. For example, a person in the population at wave 1 when the original sample was selected and at wave t when the refreshment sample was selected has two chances of selection, whereas a person joining the population after wave 1 (such as a recent immigrant) only has one chance of selection (i.e. at wave t). Each of these selection probabilities must also be adjusted for the probability of response. For wave 1 selection probabilities, this is the probability of response in wave 1 and subsequent waves to wave t for a continuous balanced panel weight (or in waves 1 and t for a paired wave balanced panel weight). And for refreshment sample selection probabilities, this is the probability of response in the initial wave of the refreshment (i.e. wave t).

There are two ways to create the weights that incorporate sample from both the ongoing sample and the refreshment sample where individuals have multiple chances of selection into the combined sample. One way is to estimate the overall inclusion probability for each individual, which requires estimating the probability of selection and response both in the sample into which they were selected and in the one into which they could have been selected but were not. This can be formalised as follows. Let A denote the ongoing sample and B denote the refreshment sample. For all individuals i in each sample S, identify whether they have left the wave 1 population (through death, emigration, etc.) by the time of wave t (denoted by *a only*), whether they belong to both the wave 1 and wave t populations (denoted by *common*), or whether they are a new entrant to the wave t population since wave 1 (denoted by *b only*). The integrated cross-sectional weight for individual i at time t where the two samples first co-exist is given by:

$$w_{integrated,it} = \begin{cases} 0 & \text{if } i \in S_{A,a\ only} \\ \frac{1}{p_{Ait}+\hat{p}_{Bit}} & \text{if } i \in S_{A,common} \\ \frac{1}{\hat{p}_{Ait}+p_{Bit}} & \text{if } i \in S_{B,common} \\ \frac{1}{p_{Bit}} & \text{if } i \in S_{B,b\ only} \end{cases} \tag{1.1}$$

where p_{Ait} and p_{Bit} are the estimated probabilities of being observed at time t for members of the original sample (sample A) and the refreshment sample (sample B), respectively. Each of these estimated probabilities is the product of the (known) selection probability and the (estimated) conditional probability of response at time t conditional on selection. Equivalently, $\hat{p}_{Ai}$ and $\hat{p}_{Bi}$ are the estimated probabilities of being observed at time t via the sample that individual i was not actually selected in but could have been. Thus, p_{Ait} and p_{Bit} are within-sample estimates and $\hat{p}_{Ai}$ and $\hat{p}_{Bi}$ are out-of-sample estimates. The advantage of this approach is that it gives higher weight to individuals in the refreshment sample that are like those that have attrited from the ongoing sample.

The alternative way to incorporate the refreshment sample with the ongoing sample is to derive weights that have the effect of combining estimates based separately on each sample. However, this approach in its basic form is only appropriate if the refreshment sample covers the entire time t population. If the refreshment sample covers only a subgroup, such as recent immigrants or a geographical sub-region, the combined weighting approach must be modified as shown later in this section. With the basic form of the combined weighting approach, the cross-sectional weight is:

$$w_{combined,it} = \begin{cases} 0 & \text{if } i \in S_{A,a\ only} \\ \theta w_{Ait} & \text{if } i \in S_{A,common} \\ (1-\theta) w_{Bit} & \text{if } i \in S_{B,common} \\ w_{Bit} & \text{if } i \in S_{B,b\ only} \end{cases} \tag{1.2}$$

where $w_{Ait} = \frac{1}{p_{Ait}}$, $w_{Bit} = \frac{1}{p_{Bit}}$, and θ is selected in a justifiable way. There are a number of ways of determining θ such as based on the relative sample size of the two samples, optimising a particular estimate (Skinner and Rao 1996; LaRoche 2007), based on the design effects of the two samples (O'Muircheartaigh and Pedlow 2002), or selecting a convenient value with a plausible range (Spiess and Rendtel 2000).

Watson (2014) evaluates six options for weighting an ongoing sample together with a population-wide refreshment sample in the context of a household panel with a version of the above methods modified for recruitment at the household level and the inclusion of temporary sample members. The various methods were evaluated by considering the variability in the weights, and the bias and root mean square error for a range of cross-sectional estimates. The conclusion from this evaluation study was that the method that performs the best is the one that integrates the samples together (as in Eq. (1.1)) rather than combining the estimates (as in Eq. (1.2)). It is recommended that further analysis of these options be undertaken using other datasets and refreshment sample designs.

If the refreshment sample is of a specific sub-population q (e.g. a geographical region), then if using the integrated samples approach, (1.1) would be adapted as follows:

$$w_{integrated,it} = \begin{cases} 0 & \text{if } i \in S_{A,a\ only} \\ \frac{1}{p_{Ait}} & \text{if } i \in S_{A,common,not\ in\ subpop\ q} \\ \frac{1}{p_{Ait}+\hat{p}_{Bit}} & \text{if } i \in S_{A,common,in\ subpop\ q} \\ \frac{1}{\hat{p}_{Ait}+p_{Bit}} & \text{if } i \in S_{B,common,in\ subpop\ q} \\ \frac{1}{p_{Bit}} & \text{if } i \in S_{B,b\ only,in\ subpop\ q} \end{cases} \tag{1.3}$$

Or if the refreshment sample for the sub-population were incorporated using the combined samples approach (as per Eq. (1.2)), the weight would take the following form:

$$w_{combined,it} = \begin{cases} 0 & \text{if } i \in S_{A,a\ only} \\ w_{Ait} & \text{if } i \in S_{A,common,not\ in\ subpop\ q} \\ \theta w_{Ait} & \text{if } i \in S_{A,common,in\ subpop\ q} \\ (1-\theta) w_{Bit} & \text{if } i \in S_{B,common,in\ subpop\ q} \\ w_{Bit} & \text{if } i \in S_{B,b\ only,in\ subpop\ q} \end{cases} \tag{1.4}$$

where θ is selected in a justifiable way for the sub-population q.

If the refreshment sample solely addresses non-coverage in the ongoing sample (i.e. the individuals in the ongoing sample only have one chance of selection as do those in the refreshment sample) then combining the sample is straightforward. Both (1.3) and (1.4) reduce to the special case where the third and fourth categories are empty:

$$w_{it} = \begin{cases} 0 & \text{if } i \in S_{A,a\ only} \\ w_{Ait} & \text{if } i \in S_{A,common} \\ w_{Bit} & \text{if } i \in S_{B,b\ only} \end{cases} \tag{1.5}$$

In terms of the creation of longitudinal weights incorporating a refreshment sample, the weights for longitudinal populations that begin at the wave the refreshment sample was added or later would use the combined cross-sectional weight (via one of the relevant methods described above) as the starting point, to which adjustments would be made for subsequent non-response.

1.7 Impact on Analysis

Researchers using data that incorporates ongoing and refreshment samples should be aware of the impacts the refreshment samples can have on their analyses. Researchers using a balanced panel will incorporate the refreshment sample only when the balanced panel starts at a point in time that is on or after the wave the refreshment sample is added. Those that use an unbalanced panel will incorporate the refreshment sample if they are using data that cover any of the waves for which refreshment sample data are available. Some researchers will not, therefore, use members of the refreshment sample in their analyses.

Adding a refreshment sample will tend to increase variability in the inclusion probabilities and may also change the nature of the sample clustering. This will cause some analyses to be biased unless the sample design is taken into account in the analyses. Further, the effect of non-response on the ongoing sample and the refreshment sample will be different as they are at different stages of the longitudinal survey process (for example, the ongoing sample may have experienced 10 years of attrition, whereas the refreshment sample may only have experienced one year). For a discussion of the ways in which sample design and non-response can be taken into account, see for example Pfeffermann (2011) and Heeringa et al. (2017).

Even using the sample design information and the best available weights will not correct non-coverage issues that occur over time for surveys that aim to be representative of the dynamic population unless refreshment samples are added every wave and address all components of non-coverage in the ongoing sample. This is because up until the refreshment sample is added, the sample is drifting further from being representative of the current population. This is most evident in the cross-sectional estimates, but will also affect estimates for longitudinal populations starting wave 2 onwards. The HILDA Survey, which added a population-wide refreshment sample 10 years after the study began, provides an example. Recent immigrants who arrived in Australia after the study began only had one chance of selection, so received a relatively high weight to compensate for this under-coverage in the ongoing sample. Cross-sectional estimates from the HILDA Survey (DSS and MI 2018) can be compared to external sources such as the Australian Census (ABS 2007, 2012, 2017) and the Labour Force Survey (ABS 2014).[2] Figure 1.2 shows the proportion of the population aged 15 years and older that are born in Australia, a variable that is highly related to immigration. In a period where the Census and Labour Force Survey shows that this proportion is declining, the HILDA Survey estimates based on the ongoing sample show the proportion increases over time. This is because the HILDA Survey is missing

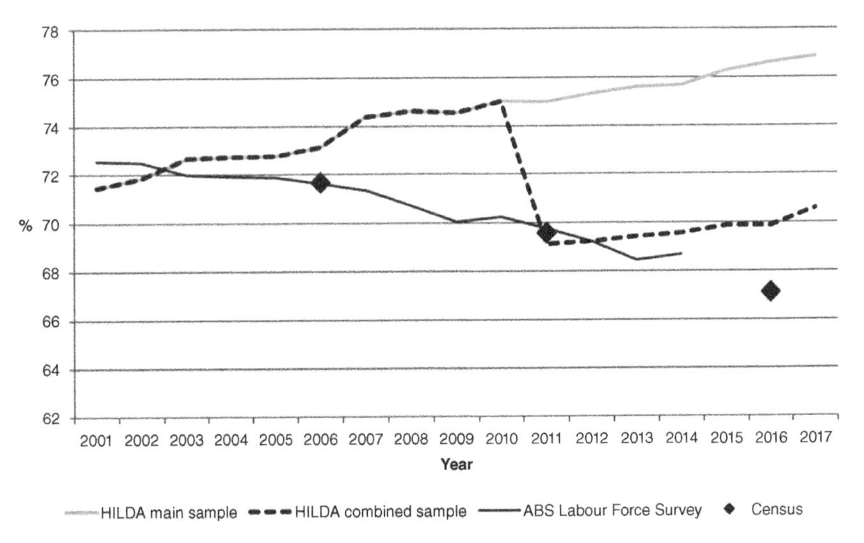

Figure 1.2 Comparison of cross-sectional estimates of the percentage of people aged 15 years and older that are born in Australia.

2 This chapter uses unit record data from the Household, Income and Labour Dynamics in Australia (HILDA) Survey. The HILDA Survey was initiated and is funded by the Australian Government Department of Social Services (DSS) and is managed by the Melbourne Institute of Applied Economic and Social Research (Melbourne Institute). The findings and views reported in this paper, however, are those of the authors and should not be attributed to the Australian Government, DSS, the Melbourne Institute or the University of Essex.

recent immigrants and cannot reflect the changes in the Australian population caused by immigration (and in this case, increasing immigration over the period examined). In 2011, the refreshment sample was added (which includes recent immigrants) and the weighted estimate is corrected and comes close to those from other sources (dashed line). However, without continually addressing this under-coverage of immigrants who move into the population each year, the HILDA estimates for the combined sample once again begin to drift away from those from other sources. Obviously, the proportion born in Australia is highly related to the extent of coverage of recent immigrants and therefore is more affected than other estimates. But, as mentioned earlier in Section 1.1, recent immigrants tend to have quite different characteristics than the Australian-born population and the exclusion of the immigrants has the potential to affect other estimates. Similarly, estimates for different longitudinal populations compared over time can be affected when the refreshment sample corrects the growing level of under-coverage. For example, comparing estimates of change across two years (e.g. the change between 2010 and 2011 compared to the change between 2011 and 2012) can show differences simply because of the addition of the refreshment sample in 2011 and not because there was a real change in the population.

Frequent refreshment samples targeting non-coverage issues would be ideal to avoid large corrections in estimates that can occur with infrequent refreshment samples. The cost and benefits of different approaches need to be considered. Frequent refreshment samples may be of lower-quality due to a poorer quality frame, may have higher overhead, and may include individuals who stay in the population only for a short time (so are costly to include but have low long-term benefit to the panel). On the positive side, frequent refreshments can collect the early experiences of these new individuals as they enter the population.

1.8 Conclusions

Refreshment samples are common in large ongoing longitudinal surveys that aim to be representative of the population or a broad sub-section of it, as is the case with household panel studies and older age panel studies. Often there are numerous refreshment samples, each with different complex designs aiming to meet different survey objectives. For example, the German Socio-Economic Panel has had more than 15 refreshment samples added over the past 25 years. *Understanding Society*, with the incorporation of the British Household Panel Survey, the original Ethnic and Minority Boost sample and the recent IEMB sample, effectively has five refreshment samples and one extension sample.

For groups planning to add a refreshment sample to their ongoing panel, we make a number of general recommendations:

1. If possible, add a sample of new 'births' (population entrants) each wave.
2. Sample from a population register (or linked administrative data with high coverage).
3. If (2) is not possible, then:
   a. screen via another survey;
   b. share screening activities with another survey; or
   c. consider oversampling or targeting high yield areas.

   Note that both options b and c, at least, involve compromises in the sample design.
4. Avoid oversampling based on time-variant characteristics.
5. Ensure you collect information to allow overall (multi-sample) selection probabilities to be calculated or estimated.
6. Construct weights that integrate the ongoing and refreshment samples (via Eqs. (1.1) or (1.3)) for use in analysis.
7. Seek to make the life of the analyst as simple as possible.
8. Put resources into maintaining participation of the ongoing sample (as it is better to avoid the need for refreshment samples due to high attrition).

There are also several areas where we suggest further research is needed. First, to simplify the exposition in this chapter, we have referred to the 'ongoing sample' and the 'refreshment sample' in a way that might imply the ongoing sample is the original sample selected in wave 1 and the refreshment sample is the first such refreshment sample added. In practice, the ongoing sample may consist of the original sample plus any refreshment samples added previously, and for most sections of this chapter, 'ongoing sample' can be interpreted in that way. An exception is the discussion on the construction of the weights that incorporate the ongoing sample and the refreshment sample where the ongoing sample does not include any earlier refreshment samples. If earlier refreshment samples have been added, then some sample members may have more than two chances of selection into the sample (unless the refreshment samples only target new entrants to the population since the last refreshment sample) and each chance of selection needs to be taken account of in the construction of the weights. Although examples exist of how the weights for longitudinal surveys have taken account of multiple refreshment samples (such as Spiess and Rendtel 2000, and Knies 2017), it might be useful to extend the weighting framework presented here in a more generic fashion.

Second, research might usefully assess the optimal frequency for a refreshment sample to address a re-occurring sample issue, such as including new population entrants or restoring the sample to its original sample size, and how this might depend on the type of refreshment sample. To address this, a value needs to be placed on the benefit of reduced bias and increased precision. Also, simplicity for users when there are few refreshments needs to be balanced against the benefits of greater frequency and complexity.

Third, there are a number of research questions regarding the best method to incorporate the sampling and non-response information of the ongoing and refreshment samples in analysis. These relate to clustering, stratification and non-response adjustments for the refreshment sample. Cluster and strata information must be assigned to the refreshment sample in a manner that will enable appropriate variance estimation when the ongoing and refreshment samples are combined in analysis. If the refreshment sample has been selected from the primary sampling units of a previous refreshment sample or the original sample, the same cluster identifiers can be assigned, but in other scenarios new cluster identifiers will need to be used. It is unclear how best to reflect the strata used for the refreshment sample if they overlap, but do not correspond to, those used for the original sample or previous refreshment samples.

We believe that the non-response behaviour of the different samples should be modelled separately (which is a departure from the classic dual-frame approach of Lohr and Rao (2000), which assumes a common non-response model). This is especially important in the early years as their experience with being panel members is quite different – the original sample have experience being part of the panel for much longer. It is less clear whether the non-response behaviour of these samples will be appreciably different in another 10 or 20 years. There may be a point in the evolution of the survey after which it would be better to combine samples to generate estimated response probabilities more accurately rather than keeping them separate. More broadly, little is known about how well different methods to incorporate the sampling and non-response information of the multiple samples work in practice. While some of these issues apply to the ongoing sample alone, the addition of a refreshment sample adds to the complexity (and need for) taking into account the sample design and non-response information.

References

ABS (2012). *2011 Census of Population and Housing – expanded community profile.* Cat. No. 2005.0. Canberra: ABS.

ABS (2014). *Labour force, Australia, detailed – electronic delivery, Sep 2014.* Cat. No. 6291.0.55.001. Canberra: ABS.

ABS (2017). *2016 Census of Population and Housing – general community profile.* Cat. No. 2001.0. Canberra: ABS.

Australian Bureau of Statistics (ABS) (2007). *2006 Census expanded community profile.* Cat. No. 2005.0. Canberra: ABS.

Banks, J., Batty, G.D., Nazroo, J., and Steptoe, A. (2016). The Dynamics of Ageing: Evidence from the English Longitudinal Study of Ageing 2002-2015 (Wave 7). London: The Institute for Fiscal Studies.

Burton, J. (ed.) (2012). Understanding Society Innovation Panel Wave 4: Results from Methodological Experiments. *Understanding Society Working Paper Series* No. 2012–06. Colchester: University of Essex.

Campanelli, P. and O'Muircheartaigh, C. (1999). Interviewers, interviewer continuity, and panel survey nonresponse. *Quality and Quantity* 33: 59–76.

Collins, M., Sykes, W., Wilson, P., and Blackshaw, N. (1988). Nonresponse: the UK experience. In: *Telephone Survey Methodology* (eds. R.M. Groves, P.P. Biemer, L.E. Lyberg, et al.), 213–231. New York: Wiley.

DSS & MI (2018). The Household, Income and Labour Dynamics in Australia (HILDA) Survey, General Release 16 (Waves 1-16), doi: https://doi.org/10.4225/87/VHRTR5, ADA Dataverse, V3.

Goebel, J. (2012). Biography and life history data in the German Socio Economic Panel (SOEP, v28, 1984–2011). In: *DIW Data Documentation*, vol. 67. Berlin: DIW.

Groves, R.M. and McGonagle, K.A. (2001). A theory-guided interviewer training protocol regarding survey participation. *Journal of Official Statistics* 17 (2): 249–265.

Groves, R.M., Singer, E., Corning, A.D., and Bowers, A. (1999). A laboratory approach to measuring the effects on survey participation of interview length, incentives, differential incentives, and refusal conversion. *Journal of Official Statistics* 15 (2): 251–268.

Haisken-DeNew, J.P. and Frick, J.R. (2005). *Desktop Companion to the German Socio-Economic Panel.* Version 8.0. Berlin: DIW.

Heeringa, S.G., West, B.T., and Berglund, P.A. (2017). *Applied Survey Data Analysis*, 2e. Boca Raton: CRC Press.

Johnson, J. and Brown, M. (2015). *National Child Development Study: User Guide to the Response and Deaths Datasets.* London: University College London.

Knies, G. (ed.) (2017). *Understanding Society: The UK Household Longitudinal Study Waves 1–7 User Guide.* Colchester: University of Essex.

Kroh, M., Käppner, K., and Kühne, S. (2014). Sampling, nonresponse and weighting in the 2011 and 2012 refreshment samples J and K of the Socio-economic Panel. In: *SOEP Survey Papers Series C*, 260. Berlin: DIW.

Kroh, M., Kühne, S., Goebel, J., and Preu, F. (2015). The 2013 IAB-SOEP migration sample (M1): Sampling design and weighting adjustment. In: *SOEP Survey Papers Series C*, 271. Berlin: DIW.

Kroh, M., Kühne, S., Jacobsen, J. et al. (2017). Sampling, nonresponse, and integrated weighting of the 2016 IAB-BAMF-SOEP Survey of Refugees (M3/M4) – revised version. In: *SOEP Survey Papers Series C*, 477. Berlin: DIW.

LaRoche, S. (2007). Longitudinal and cross-sectional weighting of the Survey of Labour and Income Dynamics. *Income Research Paper Series*, 75F0002MIE(007).

Latouche, M., Dufour, J., and Merkouris, T. (2000). Cross-sectional weighting: Combining two or more panels. In: *75F0002MIE – 00006.* Ottawa: Statistics Canada.

Lohr, S. and Rao, J.N.K. (2000). Inference from dual frame surveys. *Journal of the American Statistical Association* 95 (449): 271–280.

Lynn, P. (2011). Maintaining cross-sectional representativeness in a longitudinal general population survey. In: *Understanding Society Working Paper Series 2011–04*. Colchester: University of Essex.

Lynn, P. (2014). Longer interviews may not affect subsequent survey participation propensity. *Public Opinion Quarterly* 78 (2): 500–509. https://doi.org/10.1093/poq/nfu015.

Lynn, P. and Lugtig, P.J. (2017). Total survey error for longitudinal surveys. In: *Total Survey Error in Practice* (eds. P.P. Biemer, E.D. de Leeuw, S. Eckman, et al.), 279–298. Hoboken, NJ: Wiley.

Lynn, P., Kaminska, O., and Goldstein, H. (2014). Panel attrition: how important is interviewer continuity? *Journal of Official Statistics* 30 (3): 443–457. https://doi.org/10.2478/jos-2014-0028.

Lynn, P., Nandi, A., Parutis, V., and Platt, L. (2018). Design and implementation of a high quality probability sample of immigrants and ethnic minorities: lessons learnt. *Demographic Research* 18: 513–548.

McGonagle, K.A., Schoeni, R.F., Sastry, N., and Freedman, V.A. (2012). The panel study of income dynamics: overview, recent innovations, and potential for life course research. *Longitudinal and Life Course Studies* 3: 268–284. https://dx.doi.org/10.14301/llcs.v3i2.188.

O'Muircheartaigh, C. and Pedlow, S. (2002). Combining samples vs. cumulating cases: a comparison of two weighting strategies in NLSY97. In: *American Statistical Association Proceedings of the Joint Statistical Meetings*, 2557–2562. American Statistical Association.

Pfeffermann, D. (2011). Modelling of complex survey data: why model? Why is it a problem? How can we approach it? *Survey Methodology* 37 (2): 115–136.

Productivity Commission (2016). *Migrant intake into Australia*. In: *Productivity Commission Inquiry Report No. 77*. Canberra: Productivity Commission.

Roberts, C., Eva, G., Allum, N., and Lynn, P. (2010). Data quality in telephone surveys and the effect of questionnaire length: a cross-national experiment. In: *Institute for Social and Economic Research Working Paper 2010-36*. Colchester: University of Essex.

Sastry, N. (2019). Creating a new nationally representative sample of immigrants to the United States: Design and results from the 2017 PSID new immigrant refresher. Paper presented at the Annual Meeting of the Population Association of America, 10–13 April 2019, Austin, TX, United States.

Skinner, C.J. and Rao, J.N.K. (1996). Estimation in dual frame surveys with complex designs. *Journal of the American Statistical Association* 91 (433): 349–356.

Smith, P., Lynn, P., and Elliot, D. (2009). Sample design for longitudinal surveys. In: *Methodology of Longitudinal Surveys* (ed. P. Lynn), 21–33. Chichester: Wiley.

Sonnega, A., Faul, J.D., Ofstedal, M.B. et al. (2014). Cohort profile: the health and retirement study (HRS). *International Journal of Epidemiology* 43 (2): 576–585. https://doi.org/10.1093/ije/dyu067.

Spiess, M. and Rendtel, U. (2000). Combining an ongoing panel with a new cross-sectional sample. In: *DIW Discussion Paper Series*, 198. Berlin: DIW.

Taylor, M.F., Brice, J., Buck, N., and Prentice-Lane, E. (2010). *British Household Panel Survey: User Manual Volume A: Introduction, Technical Report and Appendices*. Colchester: University of Essex.

Voorpostel, M., Tillmann, R., Lebert, F. et al. (2017). *Swiss Household Panel user guide (1999-2016)*. Lausanne: FORS.

Watson, N. (2014). Evaluation of weighting methods to integrate a top-up sample with an ongoing longitudinal sample. *Survey Research Methods* 8: 195–208. https://dx.doi.org/10.18148/srm/2014.v8i3.5818.

Watson, N. and Wooden, M. (2013). Adding a top-up sample to the household, income and labour dynamics in Australia survey. *The Australian Economic Review* 46 (4): 489–498.

Watson, N., Leissou, E., Guyer, H., and Wooden, M. (2019). Best practices for panel maintenance and retention. In: *Advances in Comparative Survey Methods* (eds. T.P. Johnson, B. Pennell, I.A.L. Stoop and B. Dorer), 597–622. New York: Wiley.

Wolff, P., Montaigne, F., and González, G.R. (2010). Investing in statistics: EU-SILC. In: *Income and Living Conditions in Europe* (eds. A.B. Atkinson and E. Marlier), 37–55. Luxembourg: Publications Office of the European Union.

2

Collecting Biomarker Data in Longitudinal Surveys

Meena Kumari and Michaela Benzeval

Institute for Social and Economic Research, University of Essex, Colchester, UK

2.1 Introduction

This chapter provides an overview and summary of methodological issues in collecting biomarker and other biological data in longitudinal surveys. To motivate the chapter, we begin with a discussion of why it is particularly valuable to collect biomarker data together with social and economic information within longitudinal surveys. Biomarkers help us measure different aspects of health earlier and with more precision than survey questions allow, and identify the underlying biological pathways to understand how the 'social gets under the skin'. We will then focus on methodological issues in biomarker collection where the longitudinal survey context makes a difference. In the main these are similar challenges to those faced by all longitudinal surveys, but with some added complexities caused by the types of measures employed. Key challenges include: consistency and relevance of measures over time and future proofing the study – this is mainly focused on the measures that are taken within data collections but also issues around subsequently storing samples for 'unknown future use'; possible panel conditioning effects from repeating measures but also from the results of medical tests being fed back to participants; and, the choice of wave, sample and repetition period for biomarker collection to maximise uptake and value but minimise burden and possible subsequent attrition. The chapter concludes with a brief discussion of likely future methodological advances in biomarker collection in longitudinal surveys and of relevant knowledge gaps.

Understanding the interaction between people's social and economic circumstances and their health across the life span is essential to develop policies not only to improve the nation's health but also its social and economic capacities. While much research has been produced in different disciplines to examine

Advances in Longitudinal Survey Methodology, First Edition. Edited by Peter Lynn.

Companion website: www.wiley.com/go/lynn/advancesinlongitudinalsurvey

these issues, social science research often treats health as a unitary concept while clinical or biomedical studies generally control for a single measure of social position (Herd et al. 2007). However, to investigate these complex links, studies are required that bring together rich data on people's social and economic lives with accurate measures of health and health functioning – 'biomarkers' such as blood pressure or cholesterol and data on underlying biological processes. While there is a substantial body of research in the biomedical sciences on the development and calibration of specific biomarkers for measuring health and disease, in social research, where these measures are increasingly being employed, a broader perspective that considers total survey error (Groves et al. 2009) is required in reviewing their effectiveness.

In this chapter, we will first explain biomarkers, the aspects of health they measure and the value to research of including them in longitudinal surveys. We then focus on key methodological issues associated with measuring biomarkers, and the particular challenges of doing this in longitudinal surveys. Finally, we look to the future: how can longitudinal studies sustain repeating biomarker measurement as surveys move online and to new technologies to reduce cost and burden?

2.2 What Are Biomarkers, and Why Are They of Value?

The National Institute for Health has defined biomarker as '*a characteristic that is objectively measured and evaluated as an indicator of normal biological processes, pathogenic processes, or pharmacologic responses to a therapeutic intervention*' (Biomarkers definitions working group 1998). Clinical and basic biologists think of biomarkers as factors such as potential prognostic biological markers of disease. In the context of social surveys, the term *biomarker* is often used to encompass a variety of biological data from clinical risk factors for disease to factors of relevance to population health and well-being. Occasionally, the latter can be termed *biomeasures,* which include physiological measurements such as blood pressure but also wider function such as cognitive function (with 'biomarker' reserved for blood based markers). Technological advances have enabled a new set of '-omics' data, which may also be considered biomarkers. *Omics* is an umbrella term that encompasses the measurement of thousands of markers at once (Hoefer et al. 2015), for example, measurement of thousands of entities; proteins is 'proteomics', metabolic markers 'metabolomics', while genetic markers and epigenetic markers are referred to as genomics. With genomic data, one measure may or may not be a meaningful marker of disease progress, for example, genes for red hair (e.g. the melanocortin 1 receptor gene (Valverde et al. 1995)) are biological characteristics but not markers of disease, other markers such as BRCA genes for breast cancer indicate the risk of a disease (Toland et al. 2018). Increasingly biologists are creating composite indices from genomic and more recently, proteomic data that indicate the risk or progression of diseases, which brings them closer in meaning to biomarkers. Here we will use *biomarkers* as an encompassing term that includes blood-based markers of biological processes, physiological measures such as grip strength and the more recent novel -omics measures. The breadth of possible measures has the potential to provide a detailed understanding of the biological pathways that may be operating in different environments.

Biomarkers are important in population studies in three key ways. First, they provide objective, earlier and precise measurements of health and functioning. Biomarkers can measure a wide range of specific diseases and biological processes, and also act as markers for the development of disease enabling the examination of precursors of conditions and their correlates. Second, these measures help our understanding of the biological mechanisms and pathways and hence causality between psychosocial and economic factors and health and well-being. Third, genetics data are increasingly being measured in social surveys to investigate the interaction between biology and environment in health and social outcomes.

2.2.1 Detailed Measurements of Ill Health

Biomarkers provide measures of health that are objective, precise and can capture development of disease at an early stage. Population studies and social surveys include biomarkers that fall into two broad groups. The first group is related to specific disease outcomes and include measures such as blood cholesterol, which is a cardiovascular risk factor, lung function measures such as forced expiratory volume for the identification of obstructive pulmonary disease and glycated haemoglobin or blood glucose for the assessment of diabetes. Including such measures in longitudinal surveys enables us to identify the prevalence of diseases in different sub-groups of the population and how this changes as people age and over time. For example, in *Understanding Society*, a study that includes participants in the entire adult age range, over half of all participants (58%) have clinically raised levels of total cholesterol (Benzeval et al. 2014) representing substantial increased population level risk for future cardiovascular disease. This is evidence of the so-called 'clinical iceberg' as it provides evidence of potential future levels of disease. Combined with other information from the survey such as reporting of diagnosed diseases or medications taken, biomarkers can help to identify unmet need. In *Understanding Society,* it is apparent that of those participants classified as having diabetes, a proportion have undiagnosed diabetes, identified with the biomarker HbA1c compared to reports of doctor diagnosed diabetes and analysis of medications taken. In the *Understanding Society* dataset, where 4% of participants are classified as with diabetes, 10% of these participants are not diagnosed with the condition. Analysis suggests that levels of undiagnosed diabetes do not vary by socioeconomic position, but there was evidence of social differences in poorly managed diabetes such that participants who know they have diabetes with more educational attainment are more likely to maintain HbA1c level below the target than those without qualifications. This is important as uncontrolled diabetes is associated with a wide variety of poor outcomes, including sight loss and amputation.

2.2.2 Biological Pathways

The second set of markers that are included in social surveys provide a wider set of measures than risk factors for disease but help the understanding of biological pathways to disease. Of particular interest to social researchers is the mechanisms by which the environment 'gets under the skin'. Here there are several broad pathways. The physical environment, for example pollution or occupational toxin exposure may involve pathways that may reflect changes in lung function (Rückerl et al. 2011) and subsequent inflammation (Thompson et al. 2010). Stressful exposures due to social and economic experiences manifest themselves in different biological processes, which with repeated experiences, can lead to long-term biological damage. The biological processes associated with stress include activation of both the sympathetic/parasympathetic system and the hypothalamic-pituitary-adrenal axis to effect changes in adrenaline/noradrenaline and cortisol, respectively (Adam and Kumari 2009). There are a number of broad pathways downstream from these processes. For example, chronic inflammation, which has theoretical links to chronic stress (Yudkin et al. 2000), is particularly important in pointing to pathways that may relate social stress to poor health and disease (Tabassum et al. 2008). As well as measures that may be directly associated with social stress, researchers are also investigating ways of measuring the cumulative wear and tear on a body due to stress. A range of measures have been proposed here, for example allostatic load (Seeman et al. 2001). Studies have also tried to capture the wear and tear associated with ageing, for example telomeres (Robertson et al. 2013) and more recently the 'epigenetic clock' has been proposed as a measure of biological ageing (Marioni et al. 2016). By adding DNA methylation to *Understanding Society,* we have been able to understand the association of social position across the life course with measures of biological age (epigenetic age). DNA methylation serves to modulate the extent to which genes function to make their gene product. Participants can be characterised as being biologically older than chronological age where epigenetic age is greater than chronological age (accelerated age), which is associated with social position (Hughes et al. 2018).

With these advantages in employing biomarkers more generally, the value of biomarkers in longitudinal studies becomes apparent. In life course research, earlier social and economic experiences impact directly on biomarkers assessed at later time points. Similarly, health in childhood (ideally captured by biomarkers, although this is not normative due to the difficulties in collection – see Section 2.3.3 below) may track into later life and/or influence later life social and economic circumstances through engagement with education (Boyd et al. 2012). By understanding, through biomarkers, what aspects of physiological functioning or disease risk are most important for different kinds of subsequent ill health or are most influenced by earlier life circumstances, researchers can identify critical periods and the potential for a variety of interventions in terms of social and economic circumstances, life stage, and disease development. Biological changes may in themselves represent development, and have been used to identify key transitions such as puberty and hence investigate precursors or consequences of early and late puberty.

In adulthood, the repeated measurement of biomarkers in population studies is rare, but of considerable value as it allows researchers to study the development of disease processes as people age and to identify the temporal ordering of the interplay between different biological processes and between biological changes and social and environmental changes. For example, evidence suggests that inflammatory marker levels begin to change some 10 years prior to diagnosis of disease (Tabák et al. 2010). Similarly, measurement of biomarkers can illuminate both long- and short-term processes in disease progression. Thus, body mass index (BMI) is positively associated with dementia when measured 20 years or more before diagnosis, not associated with dementia when measured between 10 and 20 years before diagnosis but positively associated with dementia when measured 10 years before diagnosis (Kivimäki et al. 2018). Also, in the Whitehall II study, repeatedly collected biomarker data were used to examine the temporal relationship between depressive symptoms and both C-reactive protein and IL-6 (Gimeno et al. 2009), which were measured at two time points 11 years apart. It was found that baseline levels of both C-reactive protein and IL-6 predicted the development of depressive symptoms at follow-up, independently of baseline depressive symptoms, age, gender, and ethnicity. In contrast, depression at baseline did not predict C-reactive protein or IL-6 at follow-up, suggesting that inflammation is a predictor of depressive mood, and not vice versa.

Including biomarkers in longitudinal studies not only enables us to understand precursors to disease progression in different population groups but also enables us to identify short- and long-term subsequent comorbidities and disabling processes. For example, Sacker et al. (2008) discusses a debate about the direction of causality between heart disease (CHD) and depression: Does the diagnosis and restricted lifestyle associated with having a heart condition lead to depression, or does depression create biochemical changes in the body and altered behaviours that lead to heart disease? Measuring biomarkers for CHD over time in longitudinal studies, alongside survey measures of depression, enables us to unravel such associations to get a better sense of causality and hence how best to prevent disease and/or treat the conditions when comorbid. Identifying at-risk populations is an important public health imperative. Research in *Understanding Society* has shown that biomarkers are better predictors of subsequent disability and hence need for services than self-report measures (Davillas and Pudney 2018). If biomarkers are better discriminators of who becomes disabled by different conditions, this may contribute understanding of how to prevent disability despite disease.

2.2.3 Genetics in Longitudinal Studies

Technological advances have meant that it has become possible to measure a large number of genetic variations between individuals, so-called single nucleotide polymorphisms (SNPs – pronounced Snips) in large scale. Thus, recently a number of social and population surveys have supplemented their datasets with the addition of genetic markers. This includes longitudinal studies such as the National Child Development Study (NCDS), *Understanding Society*, English Longitudinal study of Ageing (ELSA), and Whitehall II.

To date, SNP data have mainly been used for cross-sectional association analyses to examine whether genetic variations between people are associated with outcomes of interest (Pearson and Manolio 2008). However, including genetics in longitudinal studies enables research to investigate not just associations with diseases at specific point in time, but with the persistence or progression of diseases over time and at different life stages. Research of this kind is still in its infancy, but an example of longitudinal research using genetic markers concerns the SNPs most strongly associated with BMI at a population level. This was examined in the 1946 birth cohort, and suggests the associations of genotype with BMI measured early in life and remain constant until measured in middle age (Hardy et al. 2010). These analyses have recently been repeated with a genetic score that summed 97 genetic markers associated with BMI and described an association that is persistent in middle age but declined as people aged into older age groups (Song et al. 2018). Longitudinal data can also help examine whether genetics are implicated in how people respond to interventions. For example, genetic markers associated with BMI are correlated with small changes in BMI following such an intervention (Celis-Morales et al. 2017). Investigation of associations between genetic markers and outcomes of interest in longitudinal studies is limited by data availability as effect sizes are generally small and require large numbers, often facilitated by cross-study analyses with common measures, which are not widely available at multiple time points. Nevertheless, such data should be aspired to as they have the potential to provide much greater insights into the genetic contribution and pathways to the development of health conditions.

2.3 Approaches to Collecting Biomarker Data in Longitudinal Studies

In a longitudinal setting, study design needs to establish and initiate protocols that ensure sufficient information of interest at onset of the study but not be so onerous that participants across the social spectrum are deterred from current and future participation. At the same time the measures need to be future proofed so that repetition of them to investigate change is viable and valuable. This requires careful consideration of the methods used to collect these data and managing the participant burden associated with sample collection. Further, biomarker collection is a relatively expensive form of data collection and the costs and benefits of further re-collection should be considered.

Studies that incorporate biomarker collection into their protocols generally use one of two approaches. The first approach involves a centralised collection in which participants are invited to a clinic for physiological measurements and sample collection. This enables the use of protocols that are considered 'gold standard' in the collection of biomarkers. Medical studies and funders place a high level of importance on the specificity and validity of individual measures, which is obviously necessary in individual clinical settings, but perhaps over demanding for observational studies that need to rank and identify broadly at risk groups rather than diagnose an individual's disease. Studies that have adopted this method are generally funded by medical funders, and include the Whitehall II study, the Avon Longitudinal Study of Parents and Children (ALSPAC), Southampton Women's Study, and UK Biobank. The second common approach is to collect biomarker data from participants in home settings or other participant-convenient settings such as schools or workplaces. Here, protocols are modified to enable sample collection in the participants' environment, usually by a nurse, but increasingly also by social interviewers (McFall et al. 2014). This approach may serve to promote representativeness by enabling participation by hard to reach groups such as children and the elderly. However, it is not always possible to adhere to gold standard collection and processing protocols in such settings. Studies that have adopted this approach prioritise national representativeness in their design and include the ELSA, NCDS, and *Understanding Society*, all mainly funded by social science funders. Thus it is apparent there may be a trade-off between gold standard clinically relevant measures, which require clinical settings but do not necessarily reach the full population spread or achieving high response rates and/or a better population spread with measures that meet research needs rather than being accurate clinical measures.

A third approach to sample collection not traditionally used to collect biomarker data is based on remote methods, either by providing participants with instructions through the post or by electronic or other means. This method is likely to be more cost efficient but development is in its early stages. An example of a successful implementation of web delivered blood spot collection is Food4me (Celis-Morales et al. 2015), which was a pan-European study examining nutritional biomarkers selected from a (low response) web survey. By contrast, blood spot collection was possible in only 12 of the 20 countries in the Survey of Health, Ageing, and Retirement (SHARE) due to legal and ethical issues (Schmidutz 2016). How successful remote methods would be in a representative general population study is an open question.

Wearable technology such as accelerometers to assess physical activity and sleep behaviours have been used on small or selected samples to compare subjective and objective reports of behaviour (Scholes et al. 2014; Hassani et al. 2014). Data collection through such wearable devices could, in principle, be combined with any of the three main approaches outlined above.

Whatever kind of setting biomarkers are collected in, doing this longitudinally poses a range of methodological challenges. Broadly, these challenges are analogous to those faced by any longitudinal measurement, but with some additional complexities introduced by the nature of the measures. Such challenges include:

- Consistency and relevance of measures over time
- Panel conditioning and potential feedback
- Choices of when and who to ask for sensitive and invasive data
- Cost

2.3.1 Consistency and Relevance of Measures Over Time

If repeated measurement of biomarkers in a longitudinal survey is to be used to infer change, then the measurement properties of the biomarker should remain the same. Ideally, biomarkers should be measured using the same protocol, equipment, and laboratories throughout the lifetime of the study. For studies such as ELSA, the Whitehall II study and National Study of Health and Development (NSHD) that have been collecting biomarkers for up to 30 years, this is clearly impossible for a number of reasons. These include changes in equipment. For example, in the last 30 years the routine measurement of blood pressure has changed from using manual zero sphygmomanometry to a variety of automated blood pressure devices. Systolic blood pressure readings tend to be higher using automated devices than sphygmomanometry and readings should be adjusted in order to avoid biasing analyses of change where methods may have changed over time (Wills et al. 2011). Even without a change in practice such as this, equipment may break and require replacement with newer models not necessarily being consistent with their older versions.

Blood samples collected in a survey must be sent to a laboratory to measure blood-based biomarkers, but laboratory practice and even laboratories can change over time. A number of biomarkers included in social surveys are 'clinical' in that sense that they are widely measured and used in management of health in clinical settings. In the UK, these biomarkers, for example total cholesterol are measured in National Health Service (NHS) laboratories using internationally recognised and cross-validated laboratory assays that enable comparison over time and between places. This has not always been the case for all biomarkers that are collected in surveys. For example HbA1c, a biomarker used to categorise type 2 diabetes, has not always had a well-validated laboratory assay and international standards were agreed only in 2007 (Consensus committee 2007). Over the lifetime of a study, laboratories can change protocols or laboratories can close. For the Health and Retirement Study (HRS), this occurred within a wave on one occasion (Crimmins et al. 2015). In such cases, it is useful to be able to cross – calibrate analytes across laboratories. It is not always possible to do this. An example of this was the change in platforms available to assess DNA methylation across the genome from a chip that measured 450 K sites to 850 K sites in 2017, some of which were not overlapping and therefore making it difficult to cross calibrate. Analysis of clinical analytes should be least susceptible to laboratory differences as measurement assays should be standardised across place and time; however, where analytes are not routinely measured or where protocols are not universal, laboratory differences may become more evident and need addressing, especially in the context of cross study analyses (Ruiz et al. 2017).

Further, the importance or salience of some biomarkers have changed over time. Thus, when the Whitehall II study began in 1985–1989, inflammatory markers were not identified as important to cardiovascular disease or social differences in health. These markers were first proposed to be important in this regard some 12 years after the study began (Yudkin et al. 2000; Willerson and Ridker 2004) and the biomarkers CRP and IL-6 measured in 2004. By this time, the study was in wave 7 (2002–2004). However, in this case it was also possible to measure these markers retrospectively as well because the blood samples had been stored appropriately and managed such that they could be used for these measures.

As outlined above, it is not possible to foresee all possible research uses at the outset of a study. One benefit of blood (and other tissue) samples is that they can be taken and with consents from participants stored for future unknown use. This enables new biomarkers to be analysed for past waves in response to new research imperatives. But this practice also creates challenges for studies. First, over time ethical perspectives change. For example, there is an increasing emphasis on feeding back clinically relevant genetics results to participants. However, many studies when they requested blood for storage for unknown analyses did so on the explicit understanding that results were not to be fed back. This was the basis for the informed consent to store the blood. Some studies are now considering returning to their full sample to obtain consent from participants for the return of clinical information.

Secondly, the impact on many biomarkers of the storage of samples for long-time horizons is unknown. We do not know, when frozen, how samples might degrade over what time periods or if degradation affects different analytes in the same way. Thus, there is a difficult balance in managing blood samples between using them to make effective use of a valuable resource and keeping them for future discoveries when doing so for too long may make them unusable.

Thirdly, in the social sciences such stored samples are made available, and funders encourage, external researchers to apply to use them. However, even this is costly, and few researchers want to do so when they are then reliant on secondary survey data they have not designed for their specific research questions. It is not sensible for samples to be defrosted for a single research use as defrosting and refreezing samples degrades them faster, but waiting for multiple users to build up the volume required to make defrosting the samples efficient is likely to result in lost research opportunities. As a result, many studies still have significant levels of stored tissues after 10 years, when samples are known to begin to degrade.

2.3.2 Panel Conditioning and Feedback

Feedback of potentially clinically relevant results is ethically responsible and perceived as a mechanism to aid and maintain response rates in studies. There are two broad kinds of feedback, planned feedback that respondents consent to in advance and feedback of unexpected 'incidental findings'.

In relation to the first kind of feedback, qualitative studies of participants in Whitehall II and Twenty-07 suggests that participants value feedback as part of their participation in their studies (Lorimer et al. 2011; Mein et al. 2012). Studies may decide to feedback all measurements to participants or only those that are clinically meaningful. If feedback is offered the timeliness of the feedback needs to be considered or managed by the study. A number of factors need to be considered in these decisions. For example, the Whitehall II study conducts an electrocardiogram during the participants' clinic visit and these are reviewed by a clinician each day and where clinically meaningful findings are fed back to participants immediately or as soon as possible. Other results, such as cholesterol, are fed back within a specified amount of time, usually one month. This level of feedback is expensive as immediate measurement and feedback services are more expensive to conduct than 'batch' measurements made intermittently. By contrast, a lack of feedback may have the effect of alienating participants from the study and thus impacting response rates in subsequent waves of data collection. Evidence from *Understanding Society* suggests that the decision not to feedback results to participants may have been important in the slightly lower than expected response rate to blood sample collection in the biomarker wave. For example, in those that agreed to a nurse visit, the proportion of those that agreed to a blood samples was 75%, which is lower than the equivalent response rate in ELSA (response rate over 90% of eligible participants that agreed to a nurse visit), a study which did feedback results to participants (Banks et al. 2006). Further, when we examined the reasons to refuse to blood sample collection (own analysis), 12% of participants cited lack of feedback of results.

In addition to designed feedback, a further area for ethical consideration is that of incidental findings, which are discovered in the course of research but beyond planned feedback or the original aims of the study. Studies may need to develop appropriate protocols for handling incidental findings, for example, it is currently not normative to feed back information from genetic measurements in the research setting. This would require development of appropriate staffing and feedback protocols to ensure adequate duty of care to participants. In the case of genetic feedback, in the clinical setting, counselling and other support is typically provided. Similarly, there has been a move towards imaging in large-scale studies (Gibson et al. 2017; Filippini et al. 2014) with no consensus on protocols to deal with incidental findings. Incidental findings may therefore place a much greater burden on studies in the future in terms of review and appropriate feedback.

Feedback of results, planned or incidental, has a potential consequence in a longitudinal setting of influencing behaviour and/or outcomes (see Chapter 12) and may therefore affect subsequent data analysis in the study. Findings from the HRS, on feedback of high-risk levels of blood pressure and HbA1c to participants suggest that health behaviours are impacted with reductions in weight and increases in self-reported physical activity (Edwards 2013). In a qualitative follow up of feeding back BMI, body fat, BP, cholesterol, and HBa1C, respondents in the Twenty-07 Study reported planning and/or carrying out improvements in their lifestyle (Lorimer et al. 2011), but subsequent measurement was not conducted to assess whether this persisted or altered measures of interest subsequently.

2.3.3 Choices of When and Who to Ask for Sensitive or Invasive Measures

A key decision in establishing a longitudinal study that includes biomarker measurement is when to first ask respondents to undertake such data collection; studies need to balance the value of early measurement with the benefits of establishing the study and effective engagement with respondents before requesting what might be considered burdensome data collection.

Studies that begin as 'biosocial' surveys will often have explicit hypotheses necessitating the collection of these data from the start. The focus is often salient to respondents (e.g. studies of a specific condition they have experienced), creating considerable study loyalty from the start. This can mean that studies can begin in an intensive way that can represent a large participant burden, but which respondents accept and value given the study purpose. In broader social studies, the issue of participant burden tends to make researchers reluctant to include the collection of biological data in the initial contact with the participant. For example, the ELSA alternated 'clinic' and 'non-clinic' waves of data collection, but its first wave was a 'non-clinic wave' in which participants were interviewed only. Once participants were established as members of the study, collection of a longer and more detailed set of data was included in the second wave of data collection. *Understanding Society* was another study that used this model of data collection, with wave 1 an interview wave and waves 2 or 3 used for biomarker collection. Examples of studies that did not use this method of 'easing in' participants include the Whitehall II study and UK Biobank. The Whitehall II study first wave of data collection was a limited biomarker collection, conducted in an occupational setting, and the themes and aims of the collection were salient and explicitly stated to the participants (the participants know the study as the 'stress and health' study). The Irish Longitudinal Study on Ageing (TILDA) also did not 'ease in', but again, the aims of the project were described to the participants as being to 'make Ireland the best place in the world to grow old' and improving health was explicitly stated as an aim (Kenny et al. 2010) and extensive efforts were made to engage and create loyalty to the study. By contrast, the UK Biobank is a more generalised study with a wider range of objectives, which did not initially engage in activities to create loyalty with participants and data collection was quite burdensome, which may have resulted in the lower response rates obtained (Fry et al. 2017).

A number of longitudinal studies have supplemented their data collection with the addition of biomarkers at later stages when respondents are more loyal and – in the case of the age cohort surveys such as the UK birth cohort studies – health conditions more prevalent and hence salient. Examples include the NCDS (1958 birth cohort), when biomarkers were collected at age 44–45 and National Study of Health and Development (NSHD), also called the 1946 cohort, (Wadsworth et al. 2006), when participants were aged 53 years and 60–64 years. Here, collection of biomarker data was made to capitalise on the life course data available in the studies. As a result of the earlier waves of data collection, participant characteristics and adherence to the study were well established. However, collection of new types of data in an established study in which the participants may have expectations of burden and protocols requires a certain amount of participant management. It is assumed that participants identify as members of the study but there is a concern that collection of biomarker data is a departure, as it might be not considered to be consistent to the aims of the study. Evidence from *Understanding Society* suggests there was a slight temporary decline in response rates among those asked to participate in the biomarker data collection compared to those who were not invited to participate (see Chapter 5 of this book, Pashazadeh et al. 2021).

Studies can use previously collected information to inform the design of the biomarker information collected: for example in *Understanding Society*, epigenetic data in the form of DNA methylation was measured in samples taken from a subset of participants with the most complete data in previous years, while NCDS targeted inclusion into the biomarker wave to those that had responded to the previous wave of data collection in the survey. The richness of previously collected data enables researchers to account for sample selection in analyses.

In the UK, social surveys that collect biomarker data have often used a protocol that involves an interview followed by 'clinic' or nurse visit. This two stage process has tended to impact the response rate for the nurse visit as it is a second visit following an interview, which provides an opportunity for participants to refuse. However, it ensures a high response for the core business of the study and keeps costs lower by only using nurse interviewers, who are more expensive (McFall et al. 2014) for specialist parts of the study. The loss of participants associated with this approach is evident in ELSA, when the nurse visit occurs shortly following the interview with 80% of the participants that agree to an interview going on to agree to a nurse visit, and in *Understanding Society*, when the nurse visit occurred some months following the interview (58% response rate). In NCDS, the biomedical wave was treated as a new wave of data collection, conducted entirely by the nurse (78% response rate) but the sample invited had taken part in a wave two years earlier. Once participants agree to a nurse visit, there are differential consent rates to individual measures and a number of factors will influence this. Response rates have been managed in a number of studies by using a mixed-mode data collection. Thus, the Whitehall II study, which was originally an occupational cohort that invited participants to a clinic, began to offer 'home visits' in later waves of the study (Marmot & Brunner 2005). This had implications for protocols and analytes that could be measured and carried out in both settings and required an examination of the impact of the change in mode (Ruiz et al. in prep). This is important for biosocial studies as participants who choose home visits tend to be less healthy and wealthy than participants who choose to travel to clinic. Thus, in a longitudinal setting, the different participation and consent rates within and between waves can lead to complex patterns of data availability, making weighting and analysis further complicated by the patterns of drop out associated with the nurse visit. Further to this, there is item loss of data due to missingness associated with measurement and sample collection within the clinical visit.

One debate in social surveys that incorporate biomarker data collection is the appropriate interval between data collections to capture change. In ELSA, as noted above, this is done at alternate waves (every four years) which may reflect the faster change in health at older ages. However, in a survey such as *Understanding Society* which covers the full age span, the appropriate interval is more problematic. Indeed, it would almost certainly be optimal to collect biomarkers at different intervals for different life stages, but managing this in a panel study would be complex.

There has been a move in some observational studies to embed interventions and other types of studies based on prior knowledge of biomarkers in participants where trials target individuals with specific phenotypes or attributes (Newman et al. 2016). This approach has the benefit of accruing data from pre-existing, well-characterized, and engaged populations that are already under active follow-up. However, some observational studies may not offer sufficient sample size, particularly if effect sizes are small and/or there are multiple inclusion/exclusion criteria. Furthermore, any intervention study needs to be important enough for embedding to be allowed to occur, given possible impacts on the main study.

In addition to intervention studies targeting individuals with specific phenotypes, studies also conduct targeted sub-studies for detailed measurements where measurements might be expensive or difficult to conduct. For example, Whitehall II ran a nested case control study in people with and without metabolic syndrome to measure atherosclerosis, using ultrasound in a clinic (Brunner et al. 2002). An extension of this is the move to conduct a version of this in the form of 're-call by genotype' studies (Corbin et al. 2018). Established studies may need to carefully manage participation in this type of recall study as the basis of engagement of participants in the study has changed from the original stated aims of the study.

2.3.4 Cost

The addition of biomarkers can add significantly to the value of studies, and evidence suggests that the breadth of users of the survey is widened. However, the addition of biomarker data to social surveys requires a substantial investment into those studies made up of the additional costs of data collection (extra time/interview, more costly specialist interviewers, equipment and consumables, medical expertise for clinical issues that arise, etc.), the cost of laboratory measurement of the biomarkers, support of data analysis, and management and data sharing. Adding these costs regularly on top of the standard costs of running a longitudinal study can be prohibitive. In a longitudinal setting, measurement of biomarkers can be managed, as these data can be collected periodically or on specific sub-sections of the sample. However, the lack of 'longitudinal biomarker' data in social surveys, which is needed for understanding and policy development, provides evidence that cost may be prohibitive to repeated collections. Development work is required to understand how to modify protocols to enable these collections, for example the collection of blood spots rather than phlebotomy, while maintaining longitudinal comparisons to enable analysis of the social environment and change in biomarker levels.

2.4 The Future

As the trend in many countries is for surveys to move away from face to face interviews, to reduce cost and be less burdensome to respondents, whether it is viable to continue to collect biomarker data regularly in clinical settings or on large population samples via home visits is questionable. In particular, many longitudinal surveys are moving to mixed-mode designs that include online data collection (see Chapter 9). This raises questions about how biomarkers could be collected in such a context, while maintaining comparable measurement properties between modes and over time. In *Understanding Society*, a series of experiments has been carried out, starting in the 2019 wave of the innovation panel (IP, the so-called health IP), to identify appropriate and cost effective ways of collecting high-quality biosocial data in three different modes (traditional nurse interview, by social survey interviewers and by participants themselves) for a fully representative population sample. Results are not yet available at the time of going to press. The experiments draw on new technologies and laboratory techniques, which mean that a much wider range of bio-measures can be collected from minimally invasive and less specialised protocols. The IP experiments cover three approaches:

1. New ways of collecting traditional measures in less specialised ways (e.g. blood spots, hair samples);
2. Self-report or self-administration for measures traditionally collected by nurses (e.g. height and weight, blood pressure);
3. Exploring new biological science in population surveys (e.g. microbiome).

A key challenge in adopting new protocols for measuring standard biomarkers is how they compare to more conventional collection methods in a number of ways, including measurement properties, study participation, and total cost of collection. For longitudinal surveys, consistency in measurement over time is vital, and changes in mode of collection can obviously impact this. For example, in waves 2 and 3 of *Understanding Society*, biomarker data were collected by a nurse in the participants' home. A move to self-collection of blood pressure or blood samples requires careful evaluation to ensure that new methods can provide comparable data over time.

In the *Understanding Society* IP, blood spot collections are being compared in terms of their acceptability and cost and the resulting validity of analytes with venous blood collected by nurses in the home. Similarly, a key stress hormone – cortisol – has been difficult to collect in social surveys because of the demanding protocol. It requires consistent repeat measures throughout the day (Adam and Kumari 2009) and, as a result, is relatively difficult to administer and burdensome for participants and has not been adopted in some studies for these reasons (Halpern et al. 2012). Recently, a new method to measure cortisol in hair samples has been developed, and samples collected in the clinic in the Whitehall II study (Abell et al. 2016). Reports suggest that hair samples can be collected remotely; however, these studies have been conducted using small convenience samples (Ouellet-Morin et al. 2016), and the IP represents an opportunity to test collection protocols in the context of a large general population probability-based survey.

While respondents have often been asked to self-report height and weight data, and we assume this can be biased (Gunnell et al. 2000; Uhrig 2012), the IP experiments involve testing what sorts of bias might occur by both measuring and asking for height and weight data within the same interview. More challengingly, the study plans to ask respondents to measure their own blood pressure (BP), making use of the easy availability of BP machines in pharmacies, GP clinics, and at home. We will encourage respondents to measure and report their own BP, but also measure it in the interview modes to compare results.

Gut microbiota has emerged as a scientifically innovative measure with exciting possibilities for understanding human biological processes and their interaction with the environment, but collecting it provides challenges in the context of longitudinal studies. Evidence from basic science suggests that gut bacteria may be important for a number of diseases and conditions, and this has been supported by research into the bacterial component of the human microbiome (Gareau 2014; Moreno-Indias et al. 2014; Kelly et al. 2015). It is proposed that the microbiome will provide insight into biosocial processes (https://commonfund.nih.gov/hmp/index). However, it is unclear how or whether this is relevant to population-level health, as the literature on health-microbiota associations has not appreciated the complexity of interactions that should be accounted for when dealing with large observational datasets and fails to take an epidemiological perspective (Hanage 2014). Recently, there has been a call for the introduction of microbiota collection to large, richly characterised studies in order to understand the full complexity of previously described associations (Lynch and Pedersen 2016). Further, while scientifically innovative and exciting, in the context of a longitudinal study there are many potential pitfalls to negotiate as there is a strong potential to compromise participant response. This is because measurement of the gut microbiota requires the collection of stool samples, which may or may not be a barrier – both in terms of training for sample collection and respondent participation. Sample collection and storage are difficult, and it is currently unclear whether 'future proofing' is possible. Additionally, measurement of microbes is expensive. Each of these areas must be addressed for large scale and in the context of a longitudinal survey.

It is important to develop evidence-based recommendations for the collection of biomarkers considering the 'total survey error framework', which examines both errors based on who responds and hence how representative the findings are to the population of inference and how valid the data are for measuring the theoretical concepts of interest (Lynn and Lugtig 2017). Adopting this approach in the health IP is much broader than traditional medical methodological research, which focuses on the validity of specific measures and will create not only new understandings of how to collect biomarkers in population studies but, through the collaboration of survey methodologists and biologists, develop a new approach to evaluating their effectiveness.

The adoption of state-of-the-art biology into well-conducted and deeply phenotyped social studies enables researchers to build on solid social science to understand what is happening inside the body, what is happening outside the body, and the processes that link them. Collection of data using new methods, such as genetic sequencing or the development of -omics technologies and platforms in these surveys in the rapidly social environments of today, have the potential to illuminate old questions or open new lines of inquiry to help the understanding the complex interactions at play. This integration of biological information with measures of social environments will generate unique insights and provide new opportunities for discovery and the development of policy relevant findings.

References

Abell, J., Stalder, T., Ferrie, J. et al. (2016). Assessing cortisol from hair samples in a large observational cohort: the Whitehall II study. *Psychoneuroendocrinology* 73: 148–156.

Adam, E. and Kumari, M. (2009). Assessing salivary cortisol in large-scale, epidemiological research. *Psychoneuroendocrinology* 34: 1423–1436.

Banks, J., Breeze, E., Lessof, C., and Nazroo, J. (2006). *Retirement, Health and Relationships of the Older Population in England: ELSA 2004 (Wave 2).* London: IFS.

Benzeval, M., Davillas, A., Kumari, M., and Lynn, P. (2014). *Understanding Society: UK Household Longitudinal Study: Biomarker User Guide and Glossary*. Colchester: University of Essex.

Biomarkers definitions working group (1998). *Biomarkers and Surrogate Endpoints: Preferred Definitions and Conceptual Framework*. Bethesda: NIH.

Boyd, A., Golding, J., Macleod, J. et al. (2012). Cohort profile: the 'children of the 90s'—the index offspring of the Avon longitudinal study of parents and children. *International Journal of Epidemiology* 42: 1–17.

Brunner, E., Hemingway, H., Walker, B. et al. (2002). Adrenocortical, autonomic, and inflammatory causes of the metabolic syndrome: nested case-control study. *Circulation* 106: 2659–2665.

Celis-Morales, C., Livingstone, K., Marsaux, C. et al. (2015). Design and baseline characteristics of the Food4Me study: a web-based randomised controlled trial of personalised nutrition in seven European countries. *Genes & Nutrition* 10: 450.

Celis-Morales, C., Marsaux, C., Livingstone, K. et al. (2017). Can genetic-based advice help you lose weight? Findings from the Food4Me European randomized controlled trial. *The American Journal of Clinical Nutrition* 105: 1204–1213.

Consensus committee (2007). Consensus statement on the worldwide standardization of the hemoglobin A1C measurement: the American Diabetes Association, European Association for the Study of Diabetes, International Federation of Clinical Chemistry and Laboratory Medicine, and the Inte. *Diabetes Care* 30: 2399–2400.

Corbin, L., Tan, V., Hughes, D. et al. (2018). Formalising recall by genotype as an efficient approach to detailed phenotyping and causal inference. *Nature Communications* 9: 711.

Crimmins, E., Faul, J., Kim, J., and Weir, D. (2015). *Documentation of Biomarkers in the 2010 and 2012 Health and Retirement Study*. Ann Arbor, MI: Survey Research Center University of Michigan.

Davillas, A., & Pudney, S. (2018). Biomarkers as precursors of disability. *ISER Working Paper Series.*

Edwards, R. (2013). If My Blood Pressure Is High, Do I Take It To Heart? Behavioral Impacts of Biomarker Collection in the Health and Retirement Stud. *NBER Working Paper Series.*

Filippini, N., Zsoldos, E., Haapakoski, R. et al. (2014). Study protocol: the Whitehall II imaging sub-study. *BMC Psychiatry* 14: 159.

Fry, A., Littlejohns, T., Sudlow, C. et al. (2017). Comparison of Sociodemographic and health-related characteristics of UK biobank participants with those of the general population. *American Journal of Epidemiology* 186: 1026–1034.

Gareau, M. (2014). Microbiota-gut-brain Axis and cognitive function. In: *Microbial Endocrinology: The Microbiota-Gut-Brain Axis in Health and Disease* (eds. M. Lyte and J. Cryan). New York, NY: Springer.

Gibson, L., Littlejohns, T., Adamska, L. et al. (2017). Impact of detecting potentially serious incidental findings during multi-modal imaging. *Wellcome Open Research* 2: 114.

Gimeno, D., Kivimäki, M., Brunner, E. et al. (2009). Associations of C-reactive protein and interleukin-6 with cognitive symptoms of depression: 12-year follow-up of the Whitehall II study. *Psychological Medicine* 39: 413–423.

Groves, R., Fowler, F., Couper, M. et al. (2009). *Survey Methodology, 2nd* edition. Hoboken, NJ: Wiley.

Gunnell, D., Berney, L., Holland, P. et al. (2000). How accurately are height, weight and leg length reported by the elderly, and how closely are they related to measurements recorded in childhood? *International Journal of Epidemiology* 29: 456–464.

Halpern, C., Whitsel, E., Wagner, B., and Harris, K. (2012). Challenges of measuring diurnal cortisol concentrations in a large population-based field study. *Psychoneuroendocrinology* 37: 499–508.

Hanage, W. (2014). Microbiome science needs a healthy dose of scepticism. *Nature* 512: 247–249.

Hardy, R., Wills, A., Wong, A. et al. (2010). Life course variations in the associations between FTO and MC4R gene variants and body size. *Human Molecular Genetics* 19: 545–552.

Hassani, M., Kivimaki, M., Elbaz, A. et al. (2014). Non-consent to a wrist-worn accelerometer in older adults: the role of socio-demographic, behavioural and health factors. *PLoS One* 9: e110816.

Herd, P., Goesling, B., and House, J. (2007). Socioeconomic position and health: the differential effects of education versus income on the onset versus progression of health problems. *Journal of Health and Social Behavior* 48: 223–223.

Hoefer, I., Steffens, S., Ala-Korpela, M. et al. (2015). Novel methodologies for biomarker discovery in atherosclerosis. *European Heart Journal* 36: 2635–2642.

Hughes, A., Smart, M., Gorrie-Stone, T. et al. (2018). Socioeconomic position and DNA methylation age acceleration across the life course. *American Journal of Epidemiology* 187: 2346–2354.

Kelly, J., Kennedy, P., Cryan, J. et al. (2015). Breaking down the barriers: the gut microbiome, intestinal permeability and stress-related psychiatric disorders. *Frontiers in Cellular Neuroscience* 9: 392.

Kenny, R., Whelan, B., Cronin, H. et al. (2010). *The Design of the Irish Longitudinal Study on Ageing*. Dublin: Trinity College.

Kivimäki, M., Luukkonen, R., Batty, G. et al. (2018). Body mass index and risk of dementia: analysis of individual-level data from 1.3 million individuals. *Alzheimer's & Dementia* 14: 601–609.

Lorimer, K., Gray, C., Hunt, K. et al. (2011). Response to written feedback of clinical data within a longitudinal study: a qualitative study exploring the ethical implications. *BMC Medical Research Methodology* 11: 10.

Lynch, S. and Pedersen, O. (2016). The human intestinal microbiome in health and disease. *The New England Journal of Medicine* 375: 2369–2379.

Lynn, P. and Lugtig, P. (2017). Total survey error for longitudinal surveys. In: *Total Survey Error in Practice* (eds. Biemer, de Leeuw, Eckman, et al.). Hoboken, NJ: Wiley.

Marioni, R., Harris, S., Shah, S. et al. (2016). The epigenetic clock and telomere length are independently associated with chronological age and mortality. *International Journal of Epidemiology* 45: 424–432.

Marmot, M. and Brunner, E. (2005). Cohort profile: the Whitehall II study. *International Journal of Epidemiology* 34: 251–256.

McFall, S., Conolly, A., and Burton, J. (2014). Collecting biomarkers using trained interviewers. Lessons learned from a pilot study. *Survey Research Methods* 8: 57–65.

Mein, G., Seale, C., Rice, H. et al. (2012). Altruism and participation in longitudinal health research? Insights from the Whitehall II study. *Social Science & Medicine* 75: 2345–2352.

Moreno-Indias, I., Cardona, F., Tinahones, F., and Queipo-Ortuño, M. (2014). Impact of the gut microbiota on the development of obesity and type 2 diabetes mellitus. *Frontiers in Microbiology* 5: 190.

Newman, A., Avilés-Santa, M., Anderson, G. et al. (2016). Embedding clinical interventions into observational studies. *Contemporary Clinical Trials* 46: 100–105.

Ouellet-Morin, I., Laurin, M., Robitaille, M. et al. (2016). Validation of an adapted procedure to collect hair for cortisol determination in adolescents. *Psychoneuroendocrinology* 70: 58–62.

Pashazadeh, F., Cernat, A. and Sakshaug, J.W. (2021). The effects of biological data collection in longitudinal surveys on subsequent wave cooperation. In: *Advances in Longitudinal Survey Methodology* (ed. P. Lynn). Chichester: Wiley.

Pearson, T. and Manolio, T. (2008). How to interpret a genome-wide association study. *JAMA* 299: 1335–1344.

Robertson, T., Batty, G., Der, G. et al. (2013). Is socioeconomic status associated with telomere length? Systematic review and meta-analysis. *Epidemiological Reviews* 35: 98–111.

Rückerl, R., Schneider, A., Breitner, S. et al. (2011). Health effects of particulate air pollution: a review of epidemiological evidence. *Inhalation Toxicology* 23: 555–592.

Ruiz, M., Benzeval, M., and Kumari, M. (2017). *A Guide to the Biomarker Data in the CLOSER Studies. A Catalogue Across Cohort and Longitudinal Studies*. Colchester: University of Essex.

Ruiz, M., Shipley, M., & Kumari, M. (in preparation), Comparing the nurse collection of blood samples in the home and clinic.

Sacker, A., Head, J., and Bartley, M. (2008). Impact of coronary heart disease on health functioning in an aging population: are there differences according to socioeconomic position? *Psychosomatic Medicine* 70: 133–140.

Schmidutz, D. (2016). *Synopsis of policy rules for collecting biomarkers in social surveys*. MEA (MPG)-SHARE-ERIC.

Scholes, S., Coombs, N., Pedisic, Z. et al. (2014). Age- and sex-specific criterion validity of the health survey for England physical activity and sedentary behavior assessment questionnaire as compared with accelerometry. *American Journal of Epidemiology* 179: 1493–1502.

Seeman, T., McEwen, B., Rowe, J., and Singer, B. (2001). Allostatic load as a marker of cumulative biological risk: MacArthur studies of successful aging. *Proceedings of the National Academy of Sciences of the United States of America* 98: 4770–4775.

Song, M., Zheng, Y., Qi, L. et al. (2018). Longitudinal analysis of genetic susceptibility and BMI throughout adult life. *Diabetes* 67: 248–255.

Tabák, A., Kivimäki, M., Brunner, E. et al. (2010). Changes in C-reactive protein levels before type 2 diabetes and cardiovascular death: the Whitehall II study. *European Journal of Endocrinology* 163: 89–95.

Tabassum, F., Kumari, M., Rumley, A. et al. (2008). Effects of socioeconomic position on inflammatory and hemostatic markers: a life-course analysis in the 1958 British birth cohort. *American Journal of Epidemiology* 167: 1332–1341.

Thompson, A., Zanobetti, A., Silverman, F. et al. (2010). Baseline repeated measures from controlled human exposure studies: associations between ambient air pollution exposure and the systemic inflammatory biomarkers IL-6 and fibrinogen. *Environmental Health Perspectives* 118: 120–124.

Toland, A., Brody, L., and Committee., B. S (2018). Lessons learned from two decades of BRCA1 and BRCA2 genetic testing: the evolution of data sharing and variant classification. *Genetics in Medicine*.

Uhrig, S. (2012). Understanding panel conditioning: an examination of social desirability bias in self reported height and weight in panel surveys using experimental data. *Longitudinal and Life Course Studies* 3: 120–136.

Valverde, P., Healy, E., Jackson, I. et al. (1995). Variants of the melanocyte-stimulating hormone receptor gene are associated with red hair and fair skin in humans. *Nature Genetics* 11: 328–330.

Wadsworth, M., Kuh, D., Richards, M., and Hardy, R. (2006). Cohort profile: the 1946 national birth cohort (MRC National Survey of health and development). *International Journal of Epidemiology* 35: 49–54.

Willerson, J. and Ridker, P. (2004). Inflammation as a cardiovascular risk factor. *Circulation* 109: II2–II10.

Wills, A., Lawlor, D., Matthews, F. et al. (2011). Life course trajectories of systolic blood pressure using longitudinal data from eight UK cohorts. *PLoS Medicine* 8: e1000440.

Yudkin, J., Kumari, M., Humphries, S., and Mohamed-Ali, V. (2000). Inflammation, obesity, stress and coronary heart disease: is Interleukin-6 the link? *Atherosclerosis* 148: 209–214.

3

Innovations in Participant Engagement and Tracking in Longitudinal Surveys

Lisa Calderwood, Matt Brown, Emily Gilbert and Erica Wong

Centre for Longitudinal Studies, UCL Institute of Education, London, UK

3.1 Introduction and Background

Locating sample members who move and keeping them engaged over time are challenges unique to longitudinal surveys. Non-location of sample members who move between waves is a major reason for non-response. In their conceptual framework for non-response in longitudinal surveys, Lepkowski and Couper (2002) argued that non-location should be treated as a distinct part of the response process that should be studied separately. The reasons why sample members are not located may be different from non-contact and non-cooperation, and non-location can therefore have a different impact on non-response bias. As residential moves are often associated with other life events such as changes in employment and partnership status, failure to track people who move may lead to underestimation of change, a key aim of longitudinal surveys.

Couper and Ofstedal (2009) presented new evidence about tracking methods used on two major longitudinal surveys in the US. They noted that although many large-scale longitudinal surveys have high rates of tracking success, they devote considerable resources to this and there is relatively little research about the effectiveness of tracking procedures, particularly in relation to cost-effectiveness. This led to a growth of research in this area, much of it focusing on the design of between-wave mailings, which is one of the most commonly used tracking methods (e.g. McGonagle, Couper, and Schoeni 2011; McGonagle, Schoeni, and Couper 2013; Fumagalli, Laurie, and Lynn 2013; Calderwood 2014).

However, there has been relatively little research on the use of newer, more innovative methods of participant engagement and tracking, particularly on large-scale longitudinal surveys. This chapter will provide a review of the literature and current practice regarding the use of innovative methods of tracking, and provide new

Advances in Longitudinal Survey Methodology, First Edition. Edited by Peter Lynn.

Companion website: www.wiley.com/go/lynn/advancesinlongitudinalsurvey

evidence regarding the use of the internet and social media for tracking and participant engagement and the use of administrative data for tracking in the three major longitudinal surveys in the UK: 1970 British Cohort Study (BCS70), Next Steps (previously known as the Longitudinal Study of Young People in England), and the Millennium Cohort Study (MCS). These are large-scale cohort studies following generations born in 1970, 1989/1990, and 2000/2001 respectively.

3.2 Literature Review

Various typologies or categorisations of tracking procedures have been proposed in the literature. The main distinction made is between prospective (or forward) tracking, which involves attempts to prevent loss of contact with sample members due to mobility or mitigate the impact of this, for example by collecting alternative contact details or sending between-wave mailings to sample members, and retrospective tracking, which involves trying to find sample members when it is discovered that they have moved and where they have not provided their new address (Burgess 1989; Laurie, Smith, and Scott 1999). Couper and Ofstedal (2009) extend this typology to further distinguish between field (or ground) tracking, which is usually done by interviewers, and centralised tracking, done in the office remotely by a central team of specialists. A useful distinction can also be made between tracking which must be done on an individual basis for each sample member (case-level tracking), which is very resource intensive, and tracking which can be done in an automated manner for large numbers of sample members at the same time (batch-level tracking), which may be more cost-effective. We conceptualise participant engagement as a form of prospective tracking, as its main purpose is to keep sample members engaged with and interested in the survey over time with the dual aim of encouraging them to keep the study updated with their contact information, particularly if they move address, and also to encourage them to continue to participate.

The focus of this chapter is on innovative methods of participant engagement and tracking. We include those which take advantage of technological advances in digital and mobile technologies, such as websites, mobile phones, and social media. We also include the use of administrative data for tracking, specifically where this requires applying to specific government departments and agencies rather than using publicly available administrative records such as electoral roles, telephone directories, and motor vehicle registration. This is a relatively new area of tracking, in part because technological advances have led to greater centralisation and digitisation of these records and have made them easier to access.

There is a modest and growing literature on the use of social media and the internet, primarily for retrospective tracking. A few papers also explore the use of these methods for participant engagement and recruitment. Almost all of this literature is based on relatively small-scale studies involving longitudinal follow-up for the purposes of evaluating intervention programmes for particular at-risk populations such as drug users, or for exploring outcomes for students after they have left particular educational programmes. As a result, these studies involve particular special population groups, rather than general population samples, which may limit their generalizability and applicability to other contexts. However, their findings are informative, and we review them in this section. In relation to at-risk populations, one of the motivations for exploring the use of the social media for tracking discussed in much of the literature is that although such population groups tend to have relatively unsettled lives and high mobility, they also have high rates of access to digital technology either through mobile phones or the use of internet terminals in public places such as libraries. Regardless of the population group, social media accounts provide a stable point of contact for an individual which tends not to change over time. This makes them very attractive from the point of view of tracking in longitudinal surveys. However, population coverage is clearly one of the main drawbacks of using social media for tracking, as not everyone has social media accounts and particular population groups are more likely to have them than others. Coverage rates for many forms of social media have been increasing over time for all population groups, though they are at their highest for younger age groups. In the UK in 2016, 76% of internet users had a social media account, but when broken down by age, disparity becomes apparent. Ninety six percent of 18- to 24-year-olds had a social media profile, but this dropped off steadily as age increased, to 41% for those aged 75+ (Ofcom 2017). In the US in 2016, the same trend emerges. Sixty nine percent of adults have at least one social media profile, but those aged 18–29 are far more likely to have one than those aged 65+, the proportions standing at 86% and 34% respectively (Pew Research Center 2017). For this reason, the research in this area has tended to focus on surveys of younger people, and in particular students.

Although there are many different social media platforms, most of the examples in the literature discuss the use of Facebook. This in part because it is the most widely used forum but also because, unlike many other forms of social media (e.g. Twitter and Instagram), it requires users, when they agree to the Facebook terms of service, to use their own name in their personal profiles and to create only one personal account per person. Facebook also provides the opportunity for users to upload fairly extensive biographical information (e.g. places of work and study), and to make available other contact information. This makes it more useful than many other platforms for tracking. We identified 12 articles on the use of Facebook for retrospective tracking in the research literature and have summarised their findings below. All of the articles are based on studies in the USA, UK, or Canada. Most of these studies used a Facebook tracking protocol involving searching for individuals by name, often using other information such as gender, date of birth and school or university attended for additional verification that the correct individual had been located. The individual was then sent a private message on Facebook (which is not visible to anyone else) from the study asking them to get in touch. A small number of studies also sent 'Friend' requests to the individuals. Most of the articles report success rates for Facebook tracking. Only a few of them provide information about the characteristics of those located on Facebook compared with other methods. Three of the articles (Daniel et al. 2011; Marsh and Bishop 2014; Mitchell et al. 2015) report the use of Facebook for retrospective tracking but do not provide information on success rates.

Five of the papers are about longitudinal follow-ups of at-risk groups. Bolanos et al. (2012) used Facebook to try to re-contact adult methamphetamine users for a follow-up eight years after the original intervention study. They successfully located 48 (9%) of the 551 target sample, and those located were more likely to be younger, female, to have moved out of state and have higher rates of anxiety and cognitive problems compared with those not located on Facebook. Masson et al. (2011) used Facebook, MySpace, and Friends Reunited to try to locate 117 individuals aged 18–30 who had received an intervention in childhood due to behavioural problems for a follow-up around a decade later. They found that Facebook was much more effective than the other social media sites and that they were able to locate 34 (29%) of the sample members and 35 (30%) of the families of the sample members. Overall, 81 cases were located, and of those almost a third ($n = 31$) were found solely on social networks, almost all on Facebook. Mychasiuk and Benzies (2011) used Facebook for a follow-up of a preschool programme of at-risk families, and were able to locate 19 families, reducing their overall attrition rate by 16%. Nwadiuko et al. (2011) used Facebook and MySpace to try to locate 151 people aged 20/21 for a follow-up of a sample of children who had experienced child abuse and neglect. They located 35 (23%) on Facebook, of whom 7 responded, which reduced their overall sample attrition by 4.6%. Rhodes and Marks (2011) attempted to locate 919 mothers for a follow-up three years after baseline interview in an experimental study. They located 294 (32%) mothers on Facebook, of whom 91 (31% of those located) went on to be interviewed. This represented 4% of all those interviewed. Those located on Facebook were more likely to be American Indian or Hispanic, have not attended college, be single and be over 30. Schneider et al. (2015) used Facebook to trace young people aged 15–19 for a follow-up two years after an intervention for youth in foster care. They searched for 31 individuals, resulting in 20 interviews.

A further two articles used Facebook for tracking student populations. Amerson (2011) successfully located 18 out of 22 (82%) ex-nursing students, all aged under 24, on Facebook who all responded to a private message from the study. Jones et al. (2012) used Facebook to try to locate 175 girls, originally recruited as eighth graders, for a follow-up three years later. Of these, they located 78 (45%) girls on Facebook of whom 68 (87% of those located) accepted a friend request and 43 (63% of those who accepted) provided consent to take part in the follow-up. They conclude that although the participation rate amongst those found on Facebook was lower than those found through other methods (84%), overall 6% of the target sample for the follow-up were found on Facebook, which increased their response rate for the follow-up to 81% from 75%. They report that those who were found on Facebook had lower body mass index (BMI) and body-fat and lower physical activity in eighth grade than those found through traditional methods.

Only one of the articles we identified (Meehan et al. 2009) was based on a large-scale population-based survey: the National Longitudinal Study of Adolescent Health in the US. They report searching for 2450 adults aged 24–32 on MySpace and Facebook as part of the fourth wave of the study. They sent private messages to 359 individuals (15%), and received about 40 responses (11% of those messaged).

Given the diversity of different types of study, populations, and interval between waves, it difficult to draw firm conclusions from this literature or to generalise to large-scale population-based surveys. However, it does appear that tracking using Facebook can be an effective additional method for those who cannot be found through more traditional methods, in particular for younger age groups. However, all but one of the articles we identified involved relatively small samples, and it is likely that as this is a time-consuming case-level tracking approach, the high degree of effort per case is one of the reasons why it does not appear that these methods are more widely used in large-scale longitudinal surveys.

In relation to the use of the internet more widely for retrospective tracking, most of the literature in this area discusses the use of web-based directories or databases of public records such as electoral and phone records, which is outside the scope of our chapter. Our focus is on internet tracking involving using search engine technologies and entering names of individuals lost to follow-up. For most of the articles from the literature, it isn't clear if this is a component of the 'web-based directories or databases' that they are referring to. One specific study, Masson et al. (2011), reports on the use of Google searches alongside social media, and managed to locate 5% of sample members and 1% of their families through this method. Tracking using internet searches is also time-consuming, and needs to be done on a case-by-case basis. In practice, these methods are not easily separable from each other and from other public records, as internet searches generate social media hits, social media profiles can provide additional information for further internet searching, and both social media and internet tracking require extensive cross-referencing of other 'traditional' sources (e.g. searching on both the internet and social media), and cross-referencing with contact information held by the study and information found on publicly available directories such as phone and electoral records.

In relation to administrative records which require specific applications to the data holders, we only found one study, Haggerty et al. (2008) which reports on this. They located 13% of parents and 12% of children through records held by the state's Department of Social and Health Services and Department of Corrections.

In relation to the use of social media for participant engagement, we identified four articles in this area. Three of them are based on student populations, and involved asking survey participants to 'like' a Facebook page (Berry and Bass 2012; McGinley et al. 2015; Sheffield and Kimme Hea 2016).

3.3 Current Practice

There is wide variety in the level and type of participant engagement and participant tracking activities that different longitudinal studies around the world carry out. Some of the key factors related to variation between studies are the geographic scope of the study, and in particular the difference between local area and national studies, the disciplinary tradition of the study, with differences between social science and biomedical science studies and, of course, country differences related in particular to whether population registers were available, and also privacy laws (Calderwood 2012; Park, Calderwood, and Wong, 2019).

In order to gather information about innovative methods of participant engagement and tracking used by longitudinal studies, we carried out a short online survey, inviting 143 studies from across the world to participate, through a direct email to the study director or other relevant contact. The email contained a link to the online questionnaire. The studies covered a broad range of disciplines, from social science through to biomedical, as well as those operating on both local and national scales. We received responses from 48 (see Appendix 3.A for list of studies that responded). The four cohort studies run by the Centre for Longitudinal Studies are not included in the results.

A wide variety of different methods of participant engagement were reported across the studies who responded to the survey. The vast majority (94%) reported using some form of offline/non-electronic engagement method, such as letters, leaflets, postcards, or birthday cards. Around one-third (35%) held social events for participants, ran participant advisory groups, or carried out other forms of consultation or research with their study members.

Regarding more innovative methods, messaging participants electronically is still less widespread than postal methods, with 62% of studies saying they did this, and also all of these using email for electronic messaging. Text messages were sent by 27% of studies. Both emails and texts were most commonly sent twice a year or less frequently, though two studies send both text messages and emails once a week. In relation to the effectiveness of this method, a few studies suggested they felt anecdotally it was helpful, as they believed participants wanted to be contacted about the study in this way, though two studies mentioned that although email was a more cost-effective and quick way for them to keep in touch with participants, they have encountered a number of problems with implementing it, such as 'bounce-backs', messages going to 'junk' folders, and concern from some study members about the legitimacy of the email. None of the studies had more formally evaluated the use of this method.

Participant websites were reasonably common, with 66% of studies reporting having one. The frequency with which websites were updated varied vastly across studies, from more than weekly to less than annually, although the most common response was quarterly. None of the studies surveyed reported on the success of their website, although a small number mentioned they felt the website was a cost-effective resource to maintain.

The use of social media for participant engagement was less common, with 47% saying they used some form of social media to engage study members. Of those studies who used social media, 64% used Facebook and 59% used Twitter. Many studies used one or the other, with only 2% reporting using both. Instagram, YouTube, and LinkedIn were used by a small number of studies.

The use of social media was generally reported to be carried out in one of two ways. Social media can either be used in a more anonymised way, where measures are taken to limit what followers of a social media page can see about other followers, and where followers are limited in how they can interact with other followers and the page itself. Alternatively, a social media page can be more 'open', allowing users to interact with others on the page, and the page itself. Seventy percent of studies with social media pages took the first, more anonymised route, saying they either did not allow or did not encourage participants to interact with one another through the platform.

Social media accounts were mainly used as a way to provide regular updates to participants, such as findings and news reports relevant to the study, as well as information about each sweep of the survey as it was happening. None of the studies had carried out any formal evaluation of the success or cost-effectiveness of using social media to engage with participants.

In relation to participant tracing, again it is clear that there are a wide variety of methods used for tracking participants between survey waves. The most commonly used of these was administrative data sources to locate study members, with 59% of studies doing this. Fifty-two percent of studies report supplying participants with change of address cards, or sending them the latest contact details the study holds about them and asking them to confirm. Closely following these methods in terms of popularity was sending study members between-wave mailings, with 50% of studies using this method, and having a website form available for participants to proactively update their contact details, at 46%. Less common was using web-based tracking (28%) and the use of specialist software (11%) to locate sample. Additionally, 9% of studies reported using none of these methods.

A wide variety of different administrative data sources were used by the studies surveyed to track their participants. These include national population registers (where available), electoral registers, postal databases, health, prison, and education records. Some studies suggested they had little success tracking study members using particular administrative data sources, either due to poor-quality data provided by the source or because the information may be out of date (e.g. when using education records for participants who had left education). Conversely, other studies had found administrative data sources helpful to trace participants they had lost contact with.

Studies who reported carrying out web-based tracking generally used two methods: search engines and directories (such as address or telephone directories available online either free or for a fee). The use of specialist software was more uncommon, and mixed success was reported with this method. Some studies reported moderate success in locating untraced study members this way, while others highlighted frequently receiving incorrect information, and the fact they've found using specialist software to be a very expensive method of tracing.

When asked specifically about the use of social media platforms to track participants, 78% reported not using social media at all for the purposes of tracking. The remaining 22% said they used Facebook, with two of the studies using Facebook also using LinkedIn.

Many studies using Facebook to trace felt they had at least moderate success locating study members this way, and generally used this method after other tracing methods had failed. However, the nature of Facebook terms and conditions which restricts methods of contact and ethical concerns were mentioned as a barrier to using this method. One study mentioned asking for consent, as part of a survey, to collect information about a study member's Facebook account in order to use it for tracking in the future. They found this to be most successful for their parent respondents, rather than their teenagers. A general barrier to using social media mentioned by a few studies was legislation surrounding privacy, as well as gaining ethical approval to carry out activities using social media.

3.4 New Evidence on Internet and Social Media for Participant Engagement

3.4.1 Background

In recent years, MCS and Next Steps have expanded their use of the web and social media for participant engagement. The MCS study members were born in 2000–2001, making them 14 years old at the time we started using the internet and social media as an engagement tool, and Next Steps participants, born in 1989–1990, were 24, meaning they belonged to age groups with high rates of social media and internet use. Ofcom reports that in 2014, 71% of 12- to 15-year-olds had a social media profile, with Facebook being the most popular by far – 96% of those with a profile had Facebook (Ofcom 2014). Of 16- to 24-year-olds, 93% who used the internet in 2014 also had at least one social media profile (Ofcom 2015).

The decision to use social media for participant engagement was also informed by audience research which revealed an interest in receiving more regular updates from the study and an interest in receiving information via Facebook for MCS cohort members and interest in online communications from those who were of a similar age to the Next Steps cohort.

Both studies have their own participant-facing websites, which contain information about the study, research findings, FAQs, copies of participant materials, and a contact form that study members can use to get in touch with the study and update their contact details.

In 2014, we set up a Facebook and Twitter account for each study. We run social media campaigns three times a year, posting twice a day for seven days, as a way of disseminating findings and other information about the study. In addition, we post regular updates if a new piece of research is released, if a survey is ongoing and we reach a milestone (for example, 1000 interviews achieved), and also on national holidays or special days of interest.

Much consideration was given to how the social media channels were set up, and a risk assessment was carried out. It was important that the identity and anonymity of study members was protected as far as possible, and while we wanted the social media channels to allow us to communicate to study members and drive traffic to the study websites, we wanted to discourage participants from interacting with the accounts, and in particular we wanted to discourage interaction between study members as far as possible.

For these reasons, the Facebook and Twitter profiles were set to 'private', so potential followers of these channels had to be approved by someone in the team before they were able to view content on the profiles. On the Facebook pages, we disabled the 'wall' function, so study members cannot write a message to the study that was viewable by other page followers. Additionally, we posted guidance on the participant websites about how to stay safe online, which is also posted regularly on the Facebook and Twitter pages. The social media profiles are monitored on a daily basis, and comments that share or request personal information, or contain inappropriate content, are hidden. Additionally, inappropriate content is reported to the site immediately.

3.4.2 Findings

3.4.2.1 MCS

By the middle of 2018, around three years after the participant Facebook and Twitter accounts were set up, we had gained 462 Facebook followers and 167 Twitter followers. Facebook followers are those who have liked the study page. The study had around 12 000 families actively participating at the last wave in 2015, so this is a very low proportion. Analysis of the demographics of Facebook followers suggest that while most people fit with the correct age to be study members, a large number who have liked the page are of the age we would expect study members' parents to be. Additionally, we know that a number of interviewers who work on the study have also liked the Facebook page. The vast majority of people liking the page did so within the first year of it being set up (391 of the 462), with few additional followers in subsequent years. The trend is even flatter for Twitter followers.

In terms of what makes a popular post for MCS study members, the content falls into three categories: (i) general posts highlighting the importance and uniqueness of MCS; (ii) specific information about findings and impact; and (iii) posts about national holidays or events. The top five Facebook posts, measured using the unique number of people who interacted with a post (through likes, shares, comments, or clicking through to a link included in the post), can be seen in Table 3.1.

The figures for Twitter (not shown) reveal a similar pattern in terms of preferred content; posts highlighting the uniqueness of the study, information about findings and impact, and posts on national holidays make the most popular content. However, the number of engagements with each Twitter post is considerably lower.

Turning to the website, the total number of website page views was around 20 000 in 2015, 9000 in 2016, and 8000 in 2017. Note that 2015 was a survey year, which may explain why the views are significantly higher. This is supported when we look at the most popular individual pages each year. In 2015, the homepage received 4651 hits and the page about the current survey was visited 1597 times. In 2016, a non-survey year, the most popular pages were a news report about findings related to happiness (1539) and the homepage (1388). In 2017 (again, a non-survey year), the two most visited pages were the homepage (2018) and a page about the next survey, due to take place in 2018 (394).

Table 3.1 Five most popular MCS Facebook posts.

Post	Unique interactions
Evidence from CNC has been used to inform tonight's (6 March 2017) episode of BBC Panorama, titled 'Sleepless Britain'. It looks at sleep problems among children, and the producers have drawn on data from a series of sleep questions we asked you during the Age 14 Survey. Watch it tonight on BBC One at 20 : 30 or catch it later on BBC iPlayer. Remember CNC is also known as the Millennium Cohort Study so listen out for a potential name-drop!	134
CNC is one of the biggest and most important studies of the generation born in 2000–2001 anywhere in the world.	63
Child of the New Century is one of Britain's 'scientific crown jewels' – no other country in the world has a history of tracking the lives of its people over such a long period. CNC has inspired similar studies in other countries including Ireland, France, New Zealand, and Japan. Find out more about the history of CNC and Britain's other birth cohort studies.	58
Happy #Easter to everyone in Child of the New Century celebrating today!	46
Hola CNC! Researchers from Uruguay visited our team earlier this month to learn about the study's contributions to UK science and society.	39

The vast majority of traffic to the website comes via organic searches, or by users typing the URL directly. However, social media has become an increasing source of website visits since 2015. In 2015, 5.1% of visits came via the study Facebook or Twitter page, rising to 6.9% in 2016 and 9.8% in 2017.

3.4.2.2 Next Steps

By the middle of 2018, the Next Steps Facebook page had 86 followers, and the Twitter account had 47 followers, three years since being set up. The study had almost 8000 cohort members actively participating at the last wave in 2015–2016, so this is a very low proportion. Of those who have liked the Facebook page, the vast majority are in the right age category to be study members, and like MCS, the vast majority of those liking the Facebook page did so within the first year or it being set-up.

There is far less interaction with the Next Steps page compared with the MCS one, with the most popular post only securing six interactions (compared with 134 for MCS). The Twitter page for Next Steps performs marginally better than Facebook, with the most popular tweet being interacted with (either favourited, commented on, or a link clicked on) 12 times. Given the very low level of engagement, we have not included findings on the popularity of different posts.

Looking at the website for study members, a pattern similar to that of MCS emerges. The Age 25 Survey took place from August 2015–September 2016, and consequently the number of page views on the website were highest in 2015 (around 12 000) and 2016 (around 10 000), tailing off in 2017 (around 4000). In 2015, the most popular pages were the homepage (3186 views) and the 'Contact Us' page (1471). In 2016 the homepage remained the most popular (4504), and the two Age 25 Survey pages came in second (2464 combined views). In 2017, a non-survey year, the homepage followed by the 'Contact Us' page were again the most popular, with 1460 and 323 views, respectively.

As for MCS, the most common sources of visits to the website were organic searches and direct visits, with social media playing a small role. In 2015 and 2016, 2.6% of visitors came to the website via the study's Facebook or Twitter page, rising to 4.6% in 2017.

3.4.3 Summary and Conclusions

The internet and social media have been used with some success for engaging study members in two different cohorts. It is clear that the website is an important and useful method to engage participants, with high levels of traffic, especially in survey years. However, the level of engagement on social media is low for both studies, particularly Next Steps. This is despite the fact the study members in the two cohorts are digital natives. It is worth noting that Next Steps as a study had been dormant for a number of years, without any contact with participants, before the age 25 survey, which may also be a factor in the very low level of engagement.

For the future, we are considering the use of other social media platforms for the two cohorts, such as Instagram, and investigating ways of encouraging more study members to follow us on social media, such as through email newsletters. However, given that social media for engagement takes a lot of staff time and effort, our findings to date suggest that this may not be a cost-effective method of engagement.

3.5 New Evidence on Internet and Social Media for Tracking

3.5.1 Background

This section reports findings from a small-scale pilot study using the internet and social media for retrospective tracking on Next Steps and BCS70. As noted above, the Next Steps cohort members, born in 1989/1990, are from a generation who are very likely to have social media profiles, particularly Facebook. The BCS70 generation are slightly older, though a reasonably high proportion of this age group also have social media profiles. Ofcom report that in 2015, 74% of this age group had at least one social media profile, with Facebook again being by far the most popular (Ofcom 2016). We describe the tracking protocols used and describe success rates, in relation to the proportion of study members who were located. We selected a sample of 50 study members for each study, who could not be located through other methods, for this pilot. For BCS70, this comprised 29 cases who were long-term untraced and had not been issued for the current survey wave, and 21 cases who has been issued for the current survey wave but could not be located. For Next Steps, the 50 study members selected were untraced movers at the most recent survey wave in 2015/2016. The study members for the pilot were selected proportionately from a list of all study members fulfilling these criteria, which was sorted by status code and/or survey outcome code (which classified the type of mover case) and by status confirmed date (which is a proxy for length of time the case has been identified as a mover).

The primary social media channel used for tracking was Facebook. However, in contrast to many of the studies reported in the literature, we chose not to attempt to make contact with participants through Facebook. The main reason for this was that we judged that a direct contact with participants, through an institutional account or a duplicate personal account, would constitute a violation of Facebook's Terms and Conditions which does not permit institutional accounts to use private messaging and also does not permit individuals to have more than one account. It was considered inappropriate to ask tracking staff to send messages from their personal accounts to conduct tracking.

Instead, we first attempted to locate the Facebook profile of the sample member using name and other information such as school and/or university attended. Cross-verification with other sources was also carried out where needed. Secondly, the information found on Facebook profiles was used to find contact details such as personal webpages, telephone numbers, and email addresses. Internet searches were used alongside this, and information from Facebook and other sources was cross-referenced with information from other publicly available databases such as residential listings, electoral registers, and telephone directory searches. Furthermore, information from family members and spouses listed on Facebook was also used for tracking the study member.

In Facebook, the date of birth, hometown, secondary school, and names of family members were used to check that the Facebook profile matched with the cohort member data held in our records. In some cases, it was possible to identify a change of surname for women due to marriage on Facebook, or the reverse – that is, other search methods revealed a change of surname, which then led to a Facebook search using the married name.

Due to ethical concerns, we only used contact information found on the internet or social media for the study member themselves to contact them. We did not use information to contact other friends and family members, unless they had themselves participated in the study previously. We report on the proportion of the sample for whom an online profile was identified using these methods, and are not able to say whether this contact information proved to be accurate or led to an interview.

3.5.2 Findings

The results of the pilot study are shown in Table 3.2. For both studies, a remarkably high degree of success was obtained through social media and internet tracking.

Table 3.2 Tracking using Facebook and internet on next steps and BCS70.

	Next steps	BCS70
	N (%)	N (%)
Located using Facebook	**30 (60%)**	**17 (35%)**
Of those located using Facebook:		
New contact information obtained for sample member	*18 (60%)*	*10 (59%)*
Sample member ineligible (died or emigrated)	*2 (7%)*	*1 (6%)*
Other information e.g. *contact information for someone else or other tracking lead*	*5 (17%)*	*2 (12%)*
No information or tracking leads	*5 (17%)*	*4 (24%)*
Located using Internet searches:	**22 (44%)**	**18 (36%)**
Of those located using Internet:		
New contact information obtained for sample member	*14 (64%)*	*12 (67%)*
Sample member died or emigrated	*2 (9%)*	*5 (28%)*
Other information e.g. *contact information for someone else or other tracking lead*	*3 (14%)*	*1 (6%)*
No information or tracking leads	*3 (14%)*	*0 (0%)*
Located using Facebook and/or Internet searches:	**35 (70%)**	**29 (58%)**
Base (N)	50	50

Seven in 10 Next Steps study members and almost 6 in 10 BCS70 study members were located through these methods. The differences in the overall success rates for the different studies is driven primarily by social media tracking which was more successful for the younger cohort; 60% of Next Steps cohort members were located using Facebook, whereas for BCS70 it was 35%. Similar proportions of cohort members from both studies were located through internet searches: 44% for Next Steps and 36% for BCS70.

Of those who were located through Facebook, new contact information was found for 60% of Next Steps sample members and 59% of BCS70 sample members. Some kind of additional information or tracking leads were found for a higher proportion of cases; 77% for Next Steps and 71% for BCS70.

Of those who were located through Internet searches, new contact information was found for 64% of Next Steps sample members and 67% of BCS70 sample members. Some kind of additional information or tracking leads were found for a higher proportion of cases: 78% for Next Steps and 73% for BCS70.

As discussed earlier, this method of tracking relies heavily on cross-referencing between both Facebook and internet searches, and with other more 'traditional sources' of information. This was particularly true for Next Steps. For example, of the 30 sample members located through Facebook, electoral, phone, and postal directories were consulted for 17 cases. For a small proportion of cases, other Facebook profiles, such as friends or parents, were also consulted as part of the tracking.

We looked at how those who were located through Facebook and web tracking differed from those who were not on a few key demographic characteristics; sex, social class, ethnic group, as well as wave of last interview. The only statistically significant difference found was in relation to the sex of the sample member for internet tracking, and interestingly the direction of association was different for the two studies. For Next Steps, women were more likely to be found through internet searches than men, whereas for BCS70, women were less likely to be located using this method. However, small sample sizes mean that some caution should be exercised in the interpretation of these findings, including the absence of significant associations with other variables.

Social media and web tracking were resource intensive, as it was done on a case-by-case basis. The length of time taken was approximately 26 minutes per case for Next Steps and 20 minutes per case for BCS70.

3.5.3 Summary and Conclusions

Overall, the level of tracking success through these methods is relatively high, indicating that it may be a promising method for other large-scale longitudinal surveys, particularly where study populations have high rates of social media coverage. Although it is quite resource intensive, it is often the case that most large-scale surveys already routinely carry out office-based or centralised tracking using more traditional methods on a case-by-case basis, so it may be that adding tracking through social media and internet may not be more burdensome then other case-level tracking methods. However, this method does require high levels of proficiency with Facebook and internet searching, as well as detective skills such as the ability to identify links between different pieces of information and to follow leads.

3.6 New Evidence on Administrative Data for Tracking

3.6.1 Background

In this section we present evidence on using administrative data to track participants in the Age 11 follow-up of the UK MCS in 2012, the Age 25 follow-up of Next Steps in 2015, and the Age 42 follow-up of BCS70 in 2012. Education records were used to track participants in MCS and Next Steps, allowing us to compare the effectiveness of using these data to track school age children and young adults. Health records were used to track participants in all three studies so in addition to the two younger generations we can also examine the effectiveness of using these data to track those in their forties.

In MCS and BCS70, administrative data tracking was used retrospectively to attempt to acquire new addresses for untraced study members, whereas in Next Steps it was used prospectively to attempt to acquire new addresses for all cases.

The National Pupil Database (NPD) was used for tracking in both MCS and Next Steps. This is a major source of administrative data on education maintained by the Department for Education (DfE) in England. All state primary and secondary schools are required to update the home addresses of all pupils on a termly basis as part of a termly Schools Census, which means that new addresses from this source obtained concurrently should be very up-to-date. In Next Steps, Individualised Learner Records (ILR) data, also held by DfE, was also used. This contains addresses for those in further education and vocational training between the ages of 16 and 19. These data sources only cover England, so for MCS, which is a UK-wide study, we exclude cases in Scotland, Northern Ireland, and Wales. Next Steps is England only.

Due to the age of the cohort, in MCS, almost all families would be expected to have a current address in NPD, whereas for Next Steps the participants had left school and vocational education (if they entered it) many years ago. Thus, any addresses from these sources may not be current. However, as our aim was to obtain updated addresses for everyone who had ever taken part in the study including those who last took part during their teenage years, these addresses may have been more up-to-date than those we held which primarily came from the last interview.

For MCS, two matching exercises to NPD were conducted, one before fieldwork and a second towards the end of fieldwork, and these are combined in our analysis. In Next Steps, the matching exercises to NPD and ILR were conducted by DfE, and we are not able to disentangle the different sources.

The National Health Service (NHS) Central Register, which holds the address that individuals provide to their GP practice, was used for tracking in all three studies. The NHS Central Register covers England and Wales only. BCS70 and MCS study members not known to have previously lived in England and Wales are excluded from our analysis.

For both education and health records deterministic matching was used, and the variables used for matching were name, sex, date of birth, previous address, and, for education, details of schools attended. Successful matching required agreement on some, not all of the variables and a number of different combinations of these variables constituted a valid match.

In Next Steps, this tracking was conducted in advance of fieldwork, whilst in MCS and BCS70, it was completed midway through fieldwork. For both education and health data tracking, addresses were the only contact details obtained from administrative data; no telephone numbers or email addresses could be obtained via these sources.

3.6.2 Findings

We present the results from two operational phases. The first 'pre-issue' phase relates to the matching process, the evaluation of addresses obtained and the decision about whether to issue the address to the field. The second 'post-issue' phase relates to the success of using the newly provided addresses during fieldwork. Table 3.3 provides the results of the pre-issue phase.

For all studies and both sources, the match rate (i.e. the proportion of cases sent who were successfully matched to the correct person in the administrative data), is high (Row B). In both MCS and Next Steps the health records match rate was higher than the education records match rate. The lowest match rate (85%) was obtained in BCS70 health records. For some individuals, it was clear that they had been matched to the wrong person, whereas for others this was less clear and required a case-by-case judgement.

Row C shows whether addresses were provided. In MCS and Next Steps, for both education and health records, addresses were provided for almost all matched cases. In BCS70, a significant number of cases were matched but no address was provided. In the main these were individuals who had either moved

Table 3.3 Pre-issue phase: whether cases sent for matching, whether matched, whether address provided, whether new address provided; by study and data source.

		MCS 11 Educ.	MCS 11 Health	BCS 42 Health	NS 25 Educ.	NS 25 Health
A	Sent for matching	1275	1372	4531	15 629	15 620
B	Person matched	1125	1345	3854	14 273	14 848
		88.2%	*98.0%*	*85.1%*	*91.3%*	*95.1%*
C	Address provided	1124	1300	3202	14 273	14 277
		88.2%	*94.8%*	*70.7%*	*91.3%*	*91.4%*
D	New address provided	800	490	2345	3367	2776
		62.8%	*35.7%*	*51.8%*	*21.5%*	*17.8%*
E	Issued	784	235	1915	2779	2776
		61.5%	*17.1%*	*42.3%*	*17.8%*	*17.8%*
	New issue	*133*	*43*	*1376*	—	—
		10.4%	*3.1%*	*30.4%*	—	—
	Update to issued case	*651*	*192*	*539*	—	—
		51.1%	*14.0%*	*11.9%*	—	—

out of England/Wales, died, or cancelled their GP registration (most likely due to moving) and not re-registered.

Row D shows whether addresses from administrative data were 'new', by which we mean it was different to the addresses already held for the study member. Clearly, obtaining a new address is a key measure of success. However, the interpretation of this varies by study, depending on when the tracking exercise took place, how the different exercises interacted with each other and the selection of cases, which means that it is difficult to make comparisons between the studies and data sources. For MCS, as education tracking was conducted before health tracking, if addresses from health records matched those from education records they were not counted as new. This explains, at least in part, why the proportion of cases with new addresses from health records (36%) is considerably lower than from education records (63%). In BCS70, new addresses were obtained from health records for just over half (52%) of the cases sent for matching. In Next Steps, the proportion of cases with new address is much lower as tracking was attempted for all cases, rather than only untraced movers. In many cases, the addresses obtained from administrative data were dated earlier than the addresses collected at the last interview and as such were not new addresses. Additionally, as with MCS, health data tracking took place after education data tracking, so addresses were not considered new if they matched the address obtained from education records, which in part explains why the proportion of cases with new addresses from health records (18%) is lower than education records (22%).

The final row in Table 3.3 (Row E) shows whether new addresses were issued to interviewers. In MCS, almost all new addresses obtained from education data were issued. In both MCS and BCS70, health data tracking was completed after fieldwork had begun meaning a number of cases for whom new addresses had been obtained had by this time already been located so the new address was not required. In Next Steps, all new addresses obtained from health records were issued, and where new addresses were obtained from both health and education records, addresses from health records were prioritised on the basis that they had been more recently confirmed so a lower proportion of education addresses were issued. Row E also shows, for MCS and BCS70, how issued addresses were split between newly issued cases, which would not otherwise have been attempted, and updated addresses for cases that were already issued. In BCS70 in particular, health data tracking resulted in a very significant boost to the issued sample and in many cases these were individuals who had not participated in a very long time. This split is not shown for Next Steps, as all cases were issued to interviewers at the outset. Table 3.4 provides the results of the post-issue phase.

Table 3.4 Post-issue phase: whether located, whether interviewed, and whether interviewed at address obtained from administrative data.

		MCS 11 Educ.	**MCS 11 Health**	**BCS 42 Health**	**NS 25 Educ.**	**NS 25 Health**
A	Sent for matching	1275	1372	4531	15 629	15 620
B	Issued	784	235	1915	2779	2776
C	Whether located	628	152	1333	1557	1882
	Percentage of A (%)	*49.3*	*11.1*	*29.4*	*10.0*	*12.1*
	Percentage of B (%)	*80.1*	*64.7*	*69.6*	*56.0*	*67.8*
D	Whether interviewed	479	104	771	837	1394
	Percentage of A (%)	*37.6*	*7.6*	*17.0*	*5.4*	*8.9*
	Percentage of B (%)	*61.1*	*44.3*	*40.3*	*30.1*	*50.2*
E	Whether interviewed at administrative data address	405	94	718	485	855
	Percentage of A (%)	*31.8*	*6.9*	*15.9*	*3.1*	*5.5*
	Percentage of B (%)	*51.7*	*40.0*	*37.5*	*17.5*	*30.8*
	Percentage of D (%)	*84.6*	*90.4*	*93.1*	*57.9*	*61.3*

The location rate (i.e. the proportion of study members who were found) is shown in Row C. In order to compare across studies, we consider the location rate for cases issued to the field. Overall, location rates are high for all studies and for all data sources. Education records proved highly successful for locating the school-aged MCS participants. Eighty percent of those for whom a new address was supplied were found. In Next Steps, where at age 25 participants were unlikely to still be in education, the education records location rate was unsurprisingly lower at 56%. Across all three studies, the location rates amongst those where a new address was obtained from health records were between 65% and 70%.

It should be noted that our data do not allow us to know with certainty that study members were located at the addresses obtained from administrative records, although it is reasonable to assume that for the majority this would have been the case, and where it was not, the new address may well have provided a new lead for in-field tracing. For BCS and MCS, where participants had all previously been untraced, it is likely to be the case that very few of these study members would have been found were it not for the newly supplied addresses.

Row D shows whether the study members were interviewed as part of the relevant survey wave. It shows that large numbers of interviews were achieved with study members traced through administrative data: almost 600 for MCS, almost 800 for BCS70, and around 2300 for Next Steps. The proportion of issued cases successfully interviewed is difficult to compare across studies, as this interacts with other factors that vary between studies, such as the length of time since last interview and overall response rate to the survey wave.

Row E shows whether interviews were conducted at the addresses obtained from administrative data. The vast majority of MCS and BCS70 interviews were conducted at the addresses obtained from administrative data, meaning that the number of additional interviews that were achieved directly due to this administrative data tracking is around 500 for MCS and 700 for BCS70. For Next Steps, the proportion of interviews that took place at the administrative data address is much lower. This is in part because the tracking was carried out prospectively rather than retrospectively, and it also is likely to be the case that the addresses obtained from administrative data were parental addresses, at which the study member was no longer residing, though this interim address may have led to successful location of the study member at a different address. In total, around 1300 interviews were carried out at addresses obtained from the administrative data.

We also looked at whether the proportion of cases for whom the administrative data led to an interview varied by the time since participants were last interviewed, ethnicity, and parents' social class at birth. In general, the differences by ethnicity and social class were not large and inconsistent between the studies and data sources. The interpretation of findings in relation to time since last interviewed is difficult, as this is also likely to be a factor in participants' willingness to be interviewed. However as noted earlier, for BCS70, a substantial number of interviews – around 600 – were obtained with study members who had not taken part for over 10 years.

3.6.3 Summary and Conclusions

Overall, it is clear that administrative data tracking using both education and health records was successful in all three studies. Significant numbers of interviews were obtained with study members at a new address, which was obtained through administrative data: 500 for MCS, 700 for BCS70, and 1300 for Next Steps. Administrative data tracking may also have been an important factor in obtaining a further 100 interviews for MCS and BCS70, and a further 1000 interviews for Next Steps. In BCS70, the total achieved sample in the Age 42 survey was higher than the two previous waves (at ages 34 and 38), which was primarily due to the administrative data tracking exercise. In Next Steps, the administrative data tracking was an important part of the overall success of the Age 25 survey, which involved re-starting a study that had been dormant for five years and for which we attempted to re-contact all study members who had ever taken part, which was 5–10 years ago.

Given the differences between the studies in terms of how and when the tracking exercises were conducted, how they interacted with each other, and whether they were prospective or retrospective, it is difficult to compare across studies and across the different data sources. However, we are able to identify some patterns and draw some tentative conclusions. One important factor in the success of administrative data tracking is clearly whether the address held in the administrative data is up-to-date. As noted earlier, the addresses in the NPD are updated termly, and hence, this administrative data source is able to provide current addresses for study populations, like MCS, of school-aged children. The use of this data source for tracking on MCS at age 11 was very effective, with new addresses obtained for 63% of cases sent for matching, and 80% of cases for whom a new address from these records was issued being located. Education records, ILR as well as NPD, were also used for Next Steps, though prospectively rather than retrospectively and for a study population aged 25, so for whom it was known that the address in these records was likely to be less current then for MCS, as these education records stop at age 19. It is therefore unsurprising that education records were less effective for Next Steps than MCS, with just over half of cases (56%) for whom a new address was issued being located. In relation to health records, the differences between studies in the proportion located were overall not very large – 65% for MCS, 70% for BCS70, and 68% for Next Steps – and the proportions located at these addresses was high, showing that this is a useful source of new addresses for all age groups. For Next Steps, health records were more successful than education records, which is unsurprising, as these are likely to be more up-to-date for this age group, whereas for MCS the reverse was true.

3.7 Conclusion

Overall, the conclusions drawn from this work are that both tracking using the internet and social media and tracking using administrative data can be effective in large-scale longitudinal studies. In relation to participant engagement, we found that while participant websites are well used, social media has not been a very effective tool for engagement. Given the relatively high-levels of staff resources required to run the social media accounts, the lack of engagement is also a concern in relation to cost-effectiveness. Although internet and social media tracking needs to be done on a case-by-case basis and it is resource intensive, our findings suggest this may be a promising method for other large-scale longitudinal studies whose populations have high levels of web and social media coverage, and we note that such tracking may not be more resource intensive than other case-by-case tracking carried out on longitudinal surveys. However, we have yet to explore whether this tracking leads to successful location and interviewing in the field, which is clearly an important area for future research and evaluation. In relation to administrative data, we have provided evidence of the effectiveness of this method using data from education records and health records, and for three different age groups. As this method involves batch-level tracking, it is also relatively more cost-effective than those methods that require case-by-case tracking, though considerable staff resources are required to process the data and to make the applications to data holders.

There are also ethical and privacy considerations about the use of some of these methods. In particular, the issue of whether participant consent should be explicitly sought for these forms of tracking is also one that needs to be carefully considered, particularly given the recent changes in relation to data protection legislation. Moreover, the availability of some of these methods is also subject to change and outside the control of researchers (e.g. Facebook) may change its terms and conditions and/or administrative data holders may become more restrictive in their release of data for tracking. Finally, we have relatively little information about the acceptability of these methods to study participants, and we should note that although study members may make use of a variety of different media and communication tools in their personal lives, it does not necessarily follow that they are willing to use them to engage with our studies.

More generally, within the context of increased use digital technologies by sample members and of increased use of the web for data collection, it is important for longitudinal surveys to develop effective tracking methods that do not rely on field tracking by interviewers, and to learn how to use such technologies effectively to keep participants engaged between waves. Keeping addresses updated may become less important in the future than updating other types of contact information. Improving our ability to contact and track sample members through social media is a key area for further experimentation and research.

Acknowledgements

We would like to thank Robert Browne and Tony Ball, Database Managers at Centre for Longitudinal Studies, for their work in preparing the data for the section on administrative data tracking.

References

Amerson, R. (2011). Facebook: a tool for nursing education research. *Journal of Nursing Education* 50 (7): 414–416.

Berry, D.M. and Bass, C.P. (2012). Successfully recruiting, surveying, and retaining college students: a description of methods for the risk, religiosity, and emerging adulthood study. *Research in Nursing and Health* 35 (6): 659–670.

Bolanos, F., Herbeck, D., Christou, D. et al. (2012). Using Facebook to maximize follow-up response rates in a longitudinal study of adults who use methamphetamine. *Substance Abuse: Research and Treatment* 6: 1–11.

Burgess, R.D. (1989). Major issues and implications of tracing survey respondents. In: *Panel Surveys* (eds. D. Kasprzyk et al.), 52–75. New York: Wiley.

Calderwood, L. (2014). Improving between-wave mailings on longitudinal surveys: a randomised experiment on the UK Millennium Cohort Study. *Survey Research Methods* 8 (2): 99–108.

Calderwood, L. (2012). Tracking sample members in longitudinal studies. *Survey Practice* 5 (4).

Couper, M.P. and Ofstedal, M.B. (2009). Keeping in contact with mobile sample members. In: *Methodology of Longitudinal Surveys* (ed. P. Lynn), 183–203. Chichester: Wiley.

Daniel, C.S., Brooks, C.M., and Waterbor, J.W. (2011). Approaches for longitudinally tracking graduates of NCI-funded short-term cancer research training programs. *Journal of Cancer Education* 26 (1): 58–63.

Fumagalli, L., Laurie, H., and Lynn, P. (2013). Experiments with methods to reduce attrition in longitudinal surveys. *Journal of the Royal Statistical Society Series A* 176 (2): 499–519.

Haggerty, K.P., Fleming, C.P., Catalano, R.F. et al. (2008). Ten years later: locating and interviewing children of drug abusers. *Evaluation and Program Planning* 31: 1–9.

Jones, L., Saksvig, B.I., Grieser, M. et al. (2012). Recruiting adolescent girls into a follow-up study: benefits of using a social networking website. *Contemporary Clinical Trials* 33 (2): 268–273.

Laurie, H., Smith, R., and Scott, L. (1999). Strategies for reducing nonresponse in a longitudinal panel survey. *Journal of Official Statistics* 15 (2): 269–282.

Lepkowski, J.M. and Couper, M.P. (2002). Nonresponse in the second wave of longitudinal household surveys. In: *Survey Nonresponse* (eds. R.M. Groves et al.), 259–272. New York: Wiley.

Marsh, J. and Bishop, J.C. (2014). Challenges in the use of social networking sites to trace potential research participants. *International Journal of Research and Method in Education* 37 (2): 113–124.

Masson, H., Balfe, M., Hackett, S. et al. (2011). Lost without a trace? Social networking and social research with a hard-to-reach population. *British Journal of Social Work* 43 (1): 24–40.

McGinley, S.P., Zhang, L., Hanks, L.E. et al. (2015). Reducing longitudinal attrition through Facebook. *Journal of Hospitality Marketing and Management* 24 (8): 894–900.

McGonagle, K.A., Couper, M.P., and Schoeni, R.F. (2011). Keeping track of panel members: an experimental test of a between-wave contact strategy. *Journal of Official Statistics* 27 (2): 319–338.

McGonagle, K.A., Schoeni, R.F., and Couper, M.P. (2013). The effects of a between-wave incentive experiment on contact update and production outcomes in a panel study. *Journal of Official Statistics* 29 (2): 261–276.

Meehan, A., Saleska, E., Hinsdale-Shouse, M. et al. (2009). The challenges of locating young adults for a longitudinal study: Improved tracing strategies implemented for the National Longitudinal Study of Adolescent Health, Wave IV. Presentation at the American Association for Public Opinion Research in Hollywood, FL (17 May 2009).

Mitchell, S.G., Schwartz, R.P., Alvanzo, A.A.H. et al. (2015). The use of technology in participant tracking and study retention: lessons learned from a clinical trials network study. *Substance Abuse* 36: 420–426.

Mychasiuk, R. and Benzies, K. (2011). Facebook: an effective tool for participant retention in longitudinal research. *Child Care, Health and Development* 38: 753–756.

Nwadiuko, J., Isbell, P., Zolor, A.J. et al. (2011). Using social networking sites in subject tracing. *Field Methods* 23 (1): 77–85.

Ofcom (2014). Children and Parents: Media Use and Attitudes Report. Research Document. www.ofcom.org.uk/__data/assets/pdf_file/0027/76266/childrens_2014_report.pdf (accessed 22 January 2018).

Ofcom (2015). Adults' media use and attitudes. Research Document. www.ofcom.org.uk/__data/assets/pdf_file/0014/82112/2015_adults_media_use_and_attitudes_report.pdf (accessed 22 January 2018).

Ofcom (2016). Adults' media use and attitudes. Research Document. www.ofcom.org.uk/__data/assets/pdf_file/0026/80828/2016-adults-media-use-and-attitudes.pdf (accessed 22 January 2018).

Ofcom (2017). Adults' media use and attitudes. Research Document. www.ofcom.org.uk/__data/assets/pdf_file/0020/102755/adults-media-use-attitudes-2017.pdf (accessed 22 January 2018).

Park, A., Calderwood, L., and Wong, E. (2019). Participant engagement: current practice, opportunities and challenges. *Social Research Practice* 7: 4–14.

Pew Research Center (2017). Social Media Fact Sheet. http://www.pewinternet.org/fact-sheet/social-media (accessed 22 January 2018).

Rhodes, B.B. and Marks, E.L. (2011). Using Facebook to locate sample members. *Survey Practice* 4 (5).

Schneider, S.J., Burke-Garcia, A., and Thomas, G. (2015). Facebook as a tool for respondent tracing. *Survey Practice* 8 (20) https://www.surveypractice.org/article/2850-facebook-as-a-tool-for-respondent-tracing.

Sheffield, J.P. and Kimme Hea, A.C. (2016). Leveraging the methodological affordance of Facebook: a model of social networking strategy in longitudinal writing research. *Composition Forum* 33 (Spring 2016).

3.A List of Studies that Responded to the Survey

ABCD Study
Aberdeen Children of the Fifties
Avon Longitudinal Study of Parents and Children
Born in Bradford
BRIGHTLIGHT
Cohort 5–9 yr: Inflammation and Metabolic Abnormalities in Pollutant-exposed Children
CRELES
Early Language in Victoria Study
ELIPSS Panel (Longitudinal Internet Study for Social Sciences)
English Longitudinal Study of Ageing
EPIC-Oxford
Evaluation Through Follow Up (ETF)
Fragile Families and Child Wellbeing
GASPII
Generation R
Growing Up in Ireland
Growing Up in New Zealand
Growing Up in Scotland
Health and Retirement Study
Hertfordshire Cohort Study
Household, Income, and Labour Dynamics in Australia Survey
Lifeways Cross-Generation Cohort Study
LISS panel
Longitudinal Study for Young People in England (second cohort)
Longitudinal Surveys of Australian Youth
MRC National Survey of Health and Development
MUBICOS
National Educational Panel Study
National Longitudinal Survey of Youth
New Zealand Health, Work and Retirement Study
Pairfam
Panel Study Labour Market and Social Security
Panel Study of Income Dynamics

Piccolipiù
Polish Mother and Child Cohort (REPRO_PL)
PRIDE Study
PROBIT
Survey of Health, Ageing, and Retirement in Europe
Southampton Women's Survey
Swiss Household Panel
The Irish Longitudinal Study on Ageing (TILDA)
The Longitudinal Study of Australian Children
The OCTO Twin Study
UK Biobank
Understanding Society: The UK Household Longitudinal Study
West of Scotland Twenty 07 Study

4

Effects on Panel Attrition and Fieldwork Outcomes from Selection for a Supplemental Study: Evidence from the Panel Study of Income Dynamics

Narayan Sastry, Paula Fomby and Katherine A. McGonagle

Survey Research Center, University of Michigan, Ann Arbor, MI, USA

4.1 Introduction

A key issue for panel surveys is the relationship between changes in respondent burden and resistance or attrition in future waves. In particular, does asking respondents to participate in longer or more frequent interviews, additional study components, or more demanding activities have negative effects on panel attrition, fieldwork effort, and survey costs? Of course, any negative effects may be balanced by the benefits of collecting valuable additional data, and hence may be worth pursuing. Among the potentially most burdensome activities for panel respondents is participation in a supplemental study, which may require a separate interview, interaction with other family members (such as children or a spouse), a home visit or other new mode, or providing new types of data (such as biological samples). Respondents may perceive the burden to be larger if the topic, content, and nature of the supplemental study diverge significantly from that of the main study. On the other hand, some respondents may enjoy the opportunity to participate in a supplement – or, at least, may appreciate the additional incentives.

An understanding of the effects of a supplemental study invitation on panel survey outcomes can be useful in several ways. First, knowledge of the potential effects of supplemental studies permits an informed assessment of the trade-off between the benefits and costs of these supplemental studies and can help guide decision-making about whether to launch a supplemental study. Second, comparing outcomes for cases that participated in a supplemental study with those for cases that were not invited may provide some insight into the effect of participating, though such evidence would only be suggestive because of the endogenous nature of the decision to participate among those invited to the supplemental study.

Advances in Longitudinal Survey Methodology, First Edition. Edited by Peter Lynn.

Companion website: www.wiley.com/go/lynn/advancesinlongitudinalsurvey

A major challenge in analysing the effects of supplemental studies on panel outcomes is that the offer to participate is rarely randomised to provide a clear comparison group (however, see Chapter 5 in this book for an exception). Although this presents a challenge for our analysis, we use several complementary approaches to create appropriate comparison groups to those selected for the supplement.

This study uses data from multiple waves of the Panel Study of Income Dynamics (PSID) from 1997 to 2015 to examine the effects on attrition and on various other measures of respondent cooperation of being invited to take part in a major supplemental study to PSID, namely the 1997 PSID Child Development Supplement (CDS). In the next two sections we describe our conceptual framework and previous research. We then describe the data and methods. We present our results next, and then end with our conclusions.

4.2 Conceptual Framework

Groves and Couper (1998) describe a generalised model of survey participation that focuses on the initial decision to participate in a survey but is nevertheless relevant to the process of deciding whether to participate in each wave of a panel study. The choice to participate in a survey interview is characterised by Groves and Couper (1998) as being based on heuristic decision-making, rather than by deep, thoughtful consideration of the pros and cons of participation. The heuristics include *reciprocation* (related to the heuristic of social exchange); *authority* (counterbalanced by social isolation, whereby people who feel socially excluded may have less regard for authority); *consistency* (doing what you did before); *scarcity* (perceiving participation as a rare opportunity); *social validation* (being more likely to participate if you think others like you are also participating); and *liking* (connecting with the interviewer). In addition, salience, relevance, and interest in the study topic are likely to influence participation. Each of these factors is likely to be relevant to: (i) the decision to continue to participate in subsequent waves of PSID after having been asked to participate in CDS; (ii) the level of effort required by interviewers to complete the interviews with respondents; and (iii) respondents' behaviour in response to the interview request even if they ultimately participate in the interview.

Guided by these heuristics, Groves and Couper (1998) develop three hypotheses. The first is that people are less likely to participate in a survey when opportunity costs are higher. In our analysis, we control for the effects of variables that reflected higher opportunity costs associated with participating in CDS and subsequent waves of PSID, such as larger family size, higher income, and being unmarried. The second hypothesis, based on social exchange, suggests that an equitable relationship between respondents and the study sponsor or representative is likely to lead to higher response rates; in addition, social exchange is influenced by respondent incentive payments. In our analysis, we expect that families of low socioeconomic status are more likely to value the social exchange of participating in CDS and, therefore, have a greater likelihood of participating in future waves of PSID. The third hypothesis is that individuals who are socially isolated are less likely to be persuaded to participate in a study out of a sense of obligation, duty, or belonging. We use several respondent characteristics as indicators of social isolation, including being unpartnered and not having children.

The leverage-saliency theory (Groves et al. 2000), which extends Groves and Couper's original conceptual framework, suggests that survey design features have different leverage on the decision to cooperate for different individuals. For instance, respondents' interest in the survey topic might increase their likelihood of participation. One example is provided by Barber et al. (2016), who found that respondents in a longitudinal study using weekly web surveys who experienced the behaviours measured by the study maintained higher participation levels than respondents who did not experience those behaviours. We consider the effects of survey-related variables that reveal otherwise-hidden propensity to participate. These variables indicate whether a person was previously a non-respondent and include the sampling weight as an (inverse) indicator of cumulative attrition in the past of individuals with similar characteristics. More generally, previous research shows that characteristics of respondents affect subsequent survey behaviour and outcomes in panel studies (e.g. Fitzgerald et al. 1998; Groves 2006; Lugtig 2014; Lugtig et al. 2014).

The relevance of the extended Groves-Couper framework for our analysis is reflected in the crucial need for us to adequately control for these various types of respondent characteristics when comparing subsequent panel outcomes between individuals who were and were not asked to participate in the supplemental study. To the extent that the invitation to participate – which was not randomised – reflects individuals' behaviour and characteristics, it is important that our analysis adjusts for or removes the effects of these factors. We use regression analysis and inverse probability of treatment weights to control for these factors in our analysis, and also exploit a discontinuity in selection based on children's ages.

Once we incorporate an appropriate set of controls for selection into the supplemental study, we hypothesise that the invitation to participate in the supplemental study will lead to higher attrition and, among those who do not attrit but instead continue in the study, with greater fieldwork effort needed to contact, track, and persuade respondents to participate in panel interviews in subsequent waves. We conceptualise these worse subsequent fieldwork outcomes to be the result of the higher burden associated with completing interviews for the supplemental study, especially when we refine our comparison group to observationally similar panel respondents who just missed being eligible for participation in the supplemental study. The magnitude of the negative effect of participation is difficult to predict because respondents often enjoy the interviews and receive a significant financial incentive for participating. We also conceptualise a negative effect of the invitation that operates regardless of the burden of the supplemental study, which arises because the supplemental study focuses on a different topic – children's development – than the focus on family economics of the main study in which respondents originally agreed to participate. Such a switch in topic could lead to respondents reassessing their decision about participating in the original study and deciding to end their participation.

4.3 Previous Research

Although a number of studies have examined the effects of survey experiences on subsequent participation in the context of panel studies, there are only two studies of which we are aware (aside from Pashazadeh et al. in Chapter 5 of this book) that have considered whether participation in between-wave supplemental surveys affects participation in an ongoing panel study. First, Ofstedal and Couper (2008) examined the impact of supplemental requests on panel non-response in the US Health and Retirement Study (HRS). The overall findings were that most supplemental requests had no significant effects on subsequent panel attrition in HRS. Supplements with high topic relevance had positive effects on subsequent panel retention (an internet survey and diabetes mail survey).

Second, Deeg et al. (2002) assessed whether differential inclusion in a variety of supplemental studies affected participation in subsequent waves of the Longitudinal Aging Study Amsterdam (LASA) among a sample of adults aged 55–85 years at baseline in 1992 ($N = 3805$). Supplemental study topics included health, social networks, widowhood, and depression. The authors' main conclusion is that the risk of attrition from the main panel study is increased by approaching respondents to participate in a supplemental study. Further, the authors note that respondent burden of the supplement (as measured by questionnaire length, effort, and subject matter) was unrelated to subsequent attrition.

In a related study, Kantorowitz (1998) reported that in the Israeli Labour Force Survey conducting either 'easy' or 'heavy' supplements, in contrast to having no supplement, was not associated with higher levels of non-response in the next cycle. Phillips et al. (2005) tested two conditions to assess whether a request to complete a supplemental questionnaire influenced the likelihood that a respondent would complete the primary questionnaire. In the first condition, respondents were asked to complete the secondary questionnaire depending on their responses to items in the primary questionnaire. In the second condition, which was intended to impose respondent burden, respondents were asked to complete the supplemental questionnaire unconditionally. Those in the latter group were less likely to complete either questionnaire, but among those who did, response rates on the supplemental questionnaire were higher compared to the screener condition. McCarthy et al. (2006) considered whether frequency of contact and cumulative interview length affected the likelihood of participating in agricultural surveys sponsored by the US Department of Agriculture's National Agricultural Statistics Service. In this case, the agency fields frequent surveys on a variety of topics in a relatively small population. As a result, individual agricultural operations have a relatively high probability of being selected into multiple, independent surveys over time. Frequency of contact and cumulative interview length over a three-year period were not routinely associated with refusal to participate in subsequent interviews.

There is related research that considers whether interview length and other measures of respondent burden affect subsequent panel attrition (e.g. Lynn 2014; Hart et al. 2005; Phillips et al. 2005; Porter et al. 2004; Rolstad et al. 2011). For instance, Sinibaldi and Karlsson (2016) identified individuals selected into more than one sample for all Statistics Iceland general population household surveys over a 12-year period to examine whether the decision to participate in a second survey is influenced by the amount of time since the first survey. The results show a weak linear positive effect of length of time since the first survey on participation in the second survey, but this is explained by both demographic characteristics of respondents and survey indicators. Overall, these studies present a mixed set of findings suggesting that the survey burden generally does not affect subsequent attrition but may increase it under some limited circumstances.

In summary, previous research suggests that an invitation to participate in a supplemental study may cause respondents to recalibrate the perceived costs to participating in a panel study. Those who determine that the cost of participating in the supplement is too great, or who were already considering ending their participation in the panel study, will withdraw at the point of being invited to the supplementary study, while those who do participate will remain committed to future cycles of the supplementary study as well as the panel study in which it is embedded. Although not reviewed here, salience of the study topic and appeals to respondents' unique value also may help to retain respondents selected for supplemental studies.

4.4 Data and Methods

The PSID is the world's longest-running household panel survey. It began in 1968 and collects nationally representative data for the United States through interviews conducted annually through 1997 and biennially thereafter

(McGonagle et al. 2012). One adult is interviewed in each household; respondents report information about themselves, their spouse/partner, and all other family members. PSID has achieved response rates of 95–98% for the continuing panel in most years. PSID has a number of supplemental studies, which began in 1997 with the original CDS. CDS collected information on up to two randomly selected children ages 0–12 years and their caregivers in 2380 PSID families, including detailed information on health, skills, behaviour, time use, parenting and the home environment, and many related topics (McGonagle and Sastry 2015). Interviews and assessments with children were conducted during in-home visits, and with caregivers using in-person visits and the telephone. Two additional rounds of CDS interviewers were conducted in 2002 and 2007.

Our analysis focuses on the effects of selection for Wave I of the original CDS in 1997 on PSID outcomes in subsequent years. We analyse outcomes for a focal PSID adult sample member in each PSID family who is either the household head or the head's spouse. Most PSID households have just a single such person; however, in the small number of households in which there are two such individuals, we select one of these individuals at random as our focal PSID adult sample member.

The treatment of interest is whether the PSID adult sample member lived in a family unit in which one or more children were selected to participate in the first wave of CDS in 1997. The main treatment indicator does not distinguish between whether or not the family participated in CDS, and hence represents an intent-to-treat (ITT) indicator. An ITT analysis is appropriate for addressing our first research question about whether the *invitation* to participate in the supplemental study is an important determinant of panel attrition. Because there is higher panel attrition among non-compliers (i.e. CDS non-respondents), the ITT approach provides an upper bound on the effects of treatment on the treated. In order to examine the effects of actual participation, our second research question, we need to contrast subsequent panel outcomes of these two groups (CDS respondents and non-respondents) with each other and with the control group – while acknowledging that the decision to respond in CDS is endogenous. We cannot directly assess the effects of CDS participation on subsequent attrition without considering the non-compliers, which do not belong with the control group (because they received the invitation to participate) and, at the same time, should not be omitted from the analysis (because they were present at baseline and hence were eligible for the treatment). We use an ITT approach that distinguishes between compliers and non-compliers as a way to gain insights into the effects of actual participation in the absence of an appropriate causal analysis approach to adequately control for the participation decision. Note that we do not separately examine the effects of subsequent CDS-related participation, which includes either one or two additional waves of CDS in 2002 and 2007 and participation in the biennial PSID Transition into Adulthood Supplement (TAS) from 2005 onwards.

The main outcome of interest for this analysis is the post-1997 attrition of the focal PSID adult sample member. PSID classifies a household as a permanent panel refusal if the household is non-response for two consecutive waves. We treat such cases as having attrited at the first wave in which they do not respond. In addition, a single clear and explicit request by a respondent to be removed from the ongoing sample can also lead to a permanent panel refusal. A PSID adult sample member can leave the study through death and can miss a wave through being declared ineligible following institutionalization or moving into another panel household. Observations are censored at the time of death for deceased sample members and are omitted for waves in which sample members were classified as ineligible.

A second set of outcomes we examine are five indicators of fieldwork difficulty associated with completing interviews in each wave for continuing sample members. These indicators capture the effort required at various fieldwork stages to contact and initiate or to complete an interview (the number of telephone calls to complete an interview and whether a face-to-face visit was necessary), respondent cooperation with the interview request (any resistance to completing an interview and any interview suspension that requires the interview to be completed in two or more calls for any reason), and respondent residential mobility (whether any tracking was required in order to find a sample member and conduct an interview).

Tables 4.1 and 4.2 present the outcome data for our analysis. We begin with 6308 observations in 1997, as shown in Table 4.1, which comprise all households that completed a PSID interview in that year. The attrition rate averaged 3.5% per wave over the subsequent eight biennial waves while ineligibility rates averaged 0.5% per wave. Table 4.1 shows that we have a total of almost 44 000 person-wave observations for our attrition analysis. Table 4.2 shows that the average number of telephone calls needed to complete an interview increased dramatically from 1997 to 2015, almost tripling between 1997 (when an average of 6.0 calls were needed to complete an interview) and 2015 (when 16.2 calls were needed). The fraction of the sample receiving face-to-face visits or exhibiting any resistance both followed a U-shaped pattern of decline and then increase, ending the period at a similar level as at the beginning. Interview suspensions followed an inverse U-shaped pattern, as did tracking rates.

Summary statistics for the covariates used in our analysis are presented in Table 4.3. The main independent/treatment variable is the family's CDS status in 1997. Approximately 40% of PSID families had a sample member eligible for CDS, and 87% of these families participated in the CDS survey.

The next set of covariates in Table 4.3 describes various sample characteristics. The PSID sample has several sources: the original 1968 PSID sample came from

Table 4.1 PSID sample observations by biennial wave for 1997–2013 among the 1997 baseline sample.

Year	Interviewed	Attrited	Ineligible	Died	Total
1997	6308	–	–		6308
	100.0%	–	–		100.0%
1999	5985	249	29	45	6308
	94.9%	3.9%	0.5%	0.7%	100.0%
2001	5770	174	20	50	6014
	95.9%	2.9%	0.3%	0.8%	100.0%
2003	5560	161	36	33	5790
	96.0%	2.8%	0.6%	0.6%	100.0%
2005	5382	156	30	28	5596
	96.2%	2.8%	0.5%	0.5%	100.0%
2007	5145	190	31	46	5412
	95.1%	3.5%	0.6%	0.8%	100.0%
2009	4940	174	28	34	5176
	95.4%	3.4%	0.5%	0.7%	100.0%
2011	4696	204	29	39	4968
	94.5%	4.1%	0.6%	0.8%	100.0%
2013	4463	210	24	28	4725
	94.5%	4.4%	0.5%	0.6%	100.0%
Total	41941	1518	227	303	43989
	95.3%	3.5%	0.5%	0.7%	100.0%

Note: The first column includes all interviewed cases. Cases were coded as attrited in the first wave they were non-response, with attrition occurring after two successive waves as non-response. Ineligible cases comprise respondents who were institutionalised or joined another sample family unit as a non-head/non-spouse. The last column is the row sum and is also equal to the sum of cases from the previous wave that were either interviewed or ineligible.

the Survey of Economic Opportunity (SEO sample) or a nationally representative sample frame maintained by the University of Michigan's Survey Research Center (SRC sample). The CDS sample also included new immigrants that were added to PSID in 1997. Approximately one-quarter of the 1997 PSID sample were from the SEO sample, two-thirds were from the SRC sample, and the remaining 7% were from the new immigrant refresher. The final sample-related variables describe whether the focal respondent did not respond to any prior wave of PSID, whether the family unit had recently split from another PSID family to form a new family

Table 4.2 Summary statistics for PSID fieldwork outcomes.

Year	Continuing sample	Number of calls	Any face-to-face	Any resistance	Interview suspension	Any tracking
1997	6308	6.0	–	–	–	–
1999	5985	7.0	–	–	–	–
2001	5770	7.3	–	–	–	–
2003	5560	9.2	11.6%	4.0%	–	10.6%
2005	5382	9.4	11.6%	3.3%	–	12.7%
2007	5145	9.5	8.4%	3.0%	–	16.3%
2009	4940	11.0	8.9%	3.0%	22.9%	16.8%
2011	4696	11.3	3.4%	3.1%	39.4%	15.7%
2013	4463	11.9	5.3%	3.4%	29.9%	18.3%
2015	4145	16.2	9.2%	5.0%	29.5%	8.7%

Note: Number of calls is the count of telephone calls made to respondents. No data were collected prior to 2003 for face-to-face visits, resistance, or tracking and prior to 2009 for interview suspensions

unit, the type of respondent who was interviewed in 1997 (most often the head or the spouse, but occasionally the head's partner or another individual), and the family's PSID sample weight.

The next set of variables describes demographic and socioeconomic characteristics of the focal respondents and their families in 1997, including the person's age, sex, race or ethnicity, marital status, education, income, rural–urban place of residence, and whether the family had moved since the last interview. The average age in 1997 was 44 years, and well over half of focal respondents were female. Due to the presence of the SEO sample, PSID includes an oversample of poor families and African Americans – with the latter group accounting for one-in-three sample members. Nearly two-fifths of families were headed by an unpartnered individual, with the remainder either married (55%) or cohabiting (6%). The average education level was 12.8 years of schooling, and one-quarter of families resided in rural areas. One-in-five families had moved since the previous interview. Not shown in the table (or included in our analysis) is the fact that just under half of PSID families in 1997 had children aged 0–17 years.

To describe and analyse the effects of CDS on sample attrition in PSID, we use survival curves and univariate and multivariate discrete time hazard models. Survival analysis is the unifying approach, which allows us to examine if a respondent attrited ($y = 1$) or did not attrit ($y = 0$) from PSID in each wave from 1999 to 2013,

Table 4.3 PSID 1997 baseline sample characteristics.

Variable	Percent or mean (std. dev.)
1997 CDS-I status	
Not selected	60.8%
Response	34.1%
Non-response	5.2%
PSID sample source	
SEO	24.8%
SRC	68.2%
Immigrant	7.0%
Nonresponse prior to 1997	
No	92.2%
Yes	7.8%
Family unit split off in 1997	
No	97.1%
Yes	2.9%
Respondent in 1997	
Head	68.3%
Spouse	28.4%
Partner	1.8%
Other family member	0.7%
Proxy, not family member	0.8%
Family sample weight	24.91 (17.39)
Age (years)	44.02 (16.29)
Sex of Head	
Female	58.1%
Male	41.9%
Race/Ethnicity of Head	
White	61.0%
Black	29.4%
American Indian	0.5%
Asian	1.7%
Hispanic	4.5%
Other	2.9%

(Continued)

Table 4.3 (Continued)

Variable	Percent or mean (std. dev.)
Marital status of head	
Unpartnered	38.5%
Married	55.4%
Cohabiting	6.1%
Education of head (years)	12.76 (2.67)
Income-to-needs ratio 1996	3.65 (4.10)
Place of residence in 1997	
Rural	26.0%
Urban	74.0%
Family moved since last interview	
No	80.6%
Yes	19.4%
Observations	6308

while appropriately accounting for observations that were censored due to reaching the end of the observation period or missing a wave due to being ineligible.

To describe and analyse the fieldwork outcomes, we use linear regression panel models for the number of telephone calls to complete an interview in each wave and logistic regression panel models for the remaining binary dependent variables ($y = 1$ if a respondent had a face-to-face visit, any resistance, an interview suspension, or any tracking in each wave; $y = 0$ otherwise).

The key methodological challenge is to control for differences between families that were and were not selected for CDS, which is complicated by the fact that all families with a child in the target age range of 0–12 years were selected for CDS. The controls amount to achieving balance between families with and without children in a particular age range, and hence represent a family-level model for childbearing and, in particular, the timing of childbearing. We use three complementary approaches to address this issue. First, we use regression analysis to control for the observed covariates listed in Table 4.3. Second, we use inverse probability of treatment weights, based on a propensity score for treatment – i.e. being selected for CDS – that is estimated using logistic regression to model whether a household was eligible ($y = 1$) or was not eligible ($y = 0$) for participation in CDS. We estimate a propensity model for the full sample, as well as separate models for comparisons based on sub-samples described below. Third, we stratify the

CDS and non-CDS samples and compare outcomes among sub-samples that are substantially more similar to each other based on the ages of their children, essentially providing a form of discontinuity analysis. In particular, we narrow the comparisons to focus on observationally-similar families who just missed eligibility for CDS participation due to: (i) their oldest children being ages 0–2 years in 1999 and hence born just after the youngest children selected for CDS and (ii) their youngest children being ages 13–17 years in 1997 and thus born just before the oldest children selected for CDS. We compare these two non-CDS sub-samples with CDS families having the youngest and oldest CDS children, respectively, as well as with all CDS families. We apply each of these three methods on their own and in combination.

Four logistic regression models of selection for CDS were estimated in order to construct the inverse probability of treatment weights for our main analysis. The first is estimated on the full PSID sample. The second model is restricted to families with children (all CDS families as well as non-CDS families whose youngest child was aged 13–17 years in 1997 and non-CDS families whose oldest child was aged 0–2 years in 1999, with a further restriction for all groups to those who responded in 1999). The third model is estimated on families with younger children (CDS families with a child aged 0–2 years in 1997 and who responded in 1999 and non-CDS families whose oldest child was aged 0–2 years in 1999), and the fourth on families with older children (CDS families with a child aged 10–12 years in 1997 and non-CDS families whose youngest child was aged 13–17 years in 1997). In all cases, the models fit the data well and reveal several covariates that consistently predict selection for CDS, including being a split-off family, having a younger household head, Hispanic ethnicity, being married, having higher family income, and living in an urban area. The inverse probability of treatment weights was constructed from the propensity score, which is the predicted probability from these models. Full results of the models are presented in Part A of the online supplementary material. The predicted probabilities of selection for CDS of those sample members who were selected and those who were not selected for CDS in each of the four sub-samples corresponding to the four different logistic regression models are compared in Figure A.1. Overall, the results show that there are indeed PSID sample members who, based on observed covariates, appear to have had a reasonably high likelihood of being selected for CDS, even though they were, in fact, not selected. The least overlap between the CDS and non-CDS samples is observed in the first model, based on the full PSID sample, which reflects the inclusion of a large number of families that had no children eligible for CDS and little likelihood of having such children due to their age, marital status, and other basic demographic factors. However, these families would receive low weights, and hence have only small influence on the results. The other three panels in Figure A.1 show considerable overlap between families that were and were not selected for CDS, and, in particular, a significant proportion of cases that were not selected for CDS but that had a likelihood of greater than 50% of being selected. Overall, the results suggest that the inverse probability of treatment weights should perform reasonably well in improving the comparability of those who were and were not invited to participate in CDS.

4.5 Results

Our findings are presented in Tables 4.4 and 4.5 and Figures 4.1 and 4.2. Figure 4.1 shows observed group differences in attrition over the 16-year observation period. The panel in the top left compares all CDS families with all non-CDS families, and shows that families selected for CDS are substantially less likely to attrit from PSID over the observation period. This result is affected by compositional differences between the two groups, which in the top-right panel we partially control for by restricting attention to non-CDS families with either a slightly younger or slightly older child. The top-right panel reveals that respondents in non-CDS families with a child were slightly less likely to attrit from PSID than CDS families. However, the lines are close together, and the 95% confidence intervals indicate that the difference in attrition is not statistically significant. The bottom two panels of Figure 4.1 present comparisons focusing separately on younger children (bottom left) and older children (bottom right). In both cases, individuals from families not selected for CDS were less likely to attrit from PSID. The differences are not statistically significant for either group, although the difference appears larger among families with the youngest children.

The results from the figure are replicated in the first row of Table 4.4, which shows the estimated effect of CDS selection on subsequent attrition from PSID without controlling for any observed covariates. Column 1 compares all 6308 individuals based on CDS vs. non-CDS status, and indicates that likelihood of attriting from PSID is $(1 - 0.502 = 0.498)$, or approximately 50% lower for individuals whose families were selected for CDS. This result is statistically significant at the 0.001 level. Column 2 restricts the comparison to all CDS families and non-CDS families with a slightly younger or older child, and reveals that CDS selection was associated with marginally higher attrition – although this effect is not statistically significant. The same finding emerges for the remaining two comparisons in Row 1: in Column 3 between CDS families with children aged 0–2 years in 1997 and non-CDS families with children aged 0–2 years in 1999 and in Column 4 between CDS families in 1997 with children aged 10–12 years and non-CDS families with children aged 13–17 years. Sample sizes are much smaller for the more focused comparisons in Columns 3 and 4, but the point estimates for the CDS effect are close to unity.

Table 4.4 Discrete time hazard model regression results for the effects of CDS selection on subsequent PSID attrition, 1999–2013.

	Comparison			
	(1)	(2)	(3)	(4)
Model 1 (no covariates)				
CDS selection	0.502*** (0.030)	1.042 (0.120)	1.120 (0.282)	1.022 (0.202)
Model χ^2 (7 df)	187.48***	29.89***	16.01*	9.64
Model 2 (all covariates)				
CDS selection	0.949 (0.116)	1.043 (0.158)	2.148# (0.938)	0.824 (0.199)
Model χ^2 (37 df)	1218.91***	149.94***	60.23**	77.90***
Model 3 (inverse probability of treatment weights)				
CDS selection	0.863 (0.242)	0.956 (0.169)	1.077 (0.308)	0.984 (0.222)
Model χ^2 (7 df)	125.01***	20.68**	22.43**	6.098
Model 4 (inverse probability of treatment weights and all covariates)				
CDS selection	0. 919 (0.186)	0.989 (0.171)	1.954 (0.903)	0.921 (0.251)
Model χ^2 (37 df)	535.87***	148.77***	100.37***	81.86***

(continued)

Table 4.4 (Continued)

	Comparison			
	(1)	(2)	(3)	(4)
Model 5 (inverse probability of treatment weights and all covariates)				
CDS response	0.628# (0.156)	0.822 (0.145)	1.861 (0.857)	0.698 (0.204)
CDS non-response	4.005*** (1.438)	2.235*** (0.442)	2.420 (1.374)	3.756** (1.907)
Model χ^2 (38 df)	528.68***	223.90***	104.40***	99.56***
Comparison groups				
Non-CDS	All non-CDS FUs	Non-CDS FUs with children just younger and older	Non-CDS FUs with children just younger (age 0–2 years)	Non-CDS FUs with children just older (age 13–17 years)
CDS	All CDS FUs	All CDS FUs	CDS FUs with youngest children (age 0–2 years)	CDS FUs with oldest children (age 10–12 years)
Person-periods of observation				
Non-CDS	25 392	4410	969	3441
CDS	17 921	17 921	2711	1893
Total	43 313	22 331	3680	5334

Note: Parameters are relative risks; standard errors in parentheses; *** $p < 0.001$; ** $p < 0.01$; * $p < 0.05$; # $p < 0.10$; FU: 'family unit.' Model 1 covariates are survey year and CDS selection; Models 2–5 add the covariates listed in Table 4.2 plus squared terms for age, sample weight, years of education, and income-to-needs ratio. The inverse probability of treatment weights are derived from the predicted probabilities of logistic regression models of participation in CDS-I with the same set of covariates as used in Models 2–5; separate logistic regression models were estimated and separate inverse probability of treatment weights were constructed for the different comparison groups (see the online supplementary material).

Table 4.5 Panel data regression model results for the effects of CDS selection on PSID fieldwork outcomes.

	Comparison			
	(1)	(2)	(3)	(4)
A. Number of Telephone Calls to Complete a PSID Interview, 1999–2015				
Model 1 (no covariates)				
CDS selection	2.968*** (0.128)	1.878*** (0.215)	−0.041 (0.576)	1.201*** (0.343)
Model 2 (inverse probability of treatment weights and all covariates)				
CDS selection	0.041 (0.330)	0.351 (0.376)	1.099 (0.869)	0.212 (0.399)
B. Any Face-to-Face Visits to Complete PSID Interview, 2003–2015				
Model 1 (no covariates)				
CDS selection	1.562*** (0.080)	1.349*** (0.118)	0.943 (0.153)	1.113 (0.181)
Model 2 (inverse probability of treatment weights and all covariates)				
CDS selection	0.870 (0.121)	1.189 (0.157)	1.223 (0.381)	0.947 (0.199)
C. Any Resistance to Completing a PSID Interview, 2003–2015				
Model 1 (no covariates)				
CDS selection	1.364*** (0.080)	1.311** (0.134)	0.919 (0.173)	0.971 (0.191)
Model 2 (inverse probability of treatment weights and all covariates)				
CDS selection	1.319# (0.209)	1.344# (0.216)	1.052 (0.368)	0.753 (0.163)

(continued)

Table 4.5 (Continued)

	Comparison			
	(1)	(2)	(3)	(4)
D. Any Interview Suspension when Completing a PSID Interview, 2009–2015				
Model 1 (no covariates)				
CDS selection	1.384*** (0.051)	1.241*** (0.076)	0.811# (0.097)	1.214 (0.132)
Model 2 (inverse probability of treatment weights and all covariates)				
CDS selection	1.465*** (0.159)	1.382*** (0.138)	0.901 (0.189)	1.054 (0.145)
E. Any Tracking to Complete a PSID Interview, 2003–2015				
Model 1 (no covariates)				
CDS selection	1.552*** (0.048)	1.586*** (0.089)	1.424** (0.156)	1.355** (0.129)
Model 2 (inverse probability of treatment weights and all covariates)				
CDS selection	1.034 (0.095)	0.963 (0.094)	1.003 (0.186)	1.007 (0.124)
Comparison groups				
Non-CDS	All non-CDS FUs	Non-CDS FUs with children just younger and older	Non-CDS FUs with children just younger (age 0–2 years)	Non-CDS FUs with children just older (age 13–17 years)
CDS	All CDS FUs	All CDS FUs	CDS FUs with youngest children (age 0–2 years)	CDS FUs with oldest children (age 10–12 years)

Note: Linear regression coefficients reported in Panel A and odds ratios reported in Panels B–E; standard errors in parentheses. *** $p < 0.001$; ** $p < 0.01$; * $p < 0.05$; # $p < 0.10$. Covariates for Model 1 are survey year and CDS selection; for Model 2 are survey year, CDS selection, and the variables listed in Table 4.2 plus squared terms for age, sample weight, years of education, and income-to-needs ratio. See text for description of the inverse probability of treatment weights. Person-period observations for Comparisons 1–6 are, respectively, 52 355, 27 313, 4624, and 6506 for Panel A; 34 382, 18 402, 3107, and 4378 for Panels B, C and E; and 18 307, 10 039, 1681, and 2385 for Panel D.

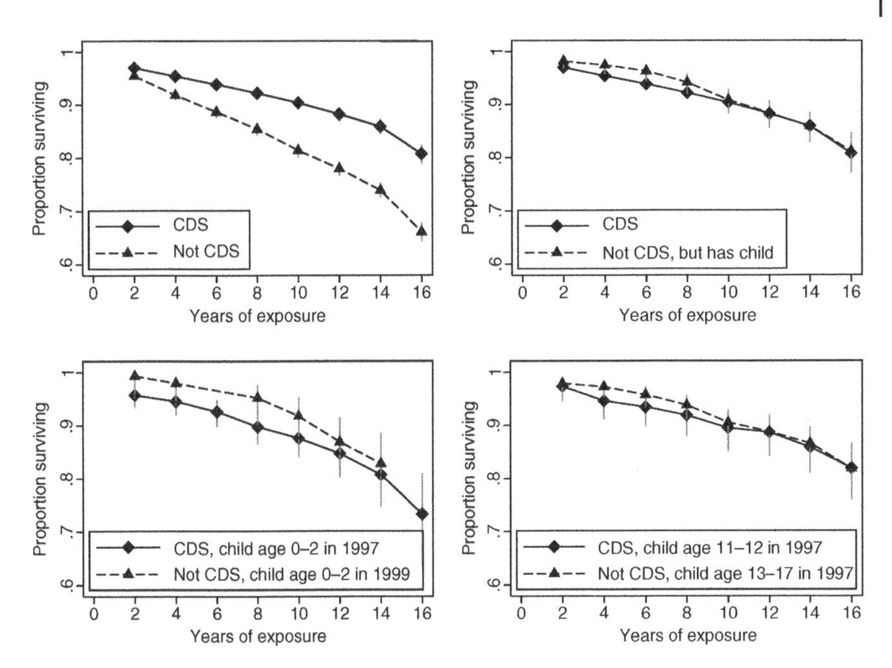

Figure 4.1 Observed trends in PSID response by CDS selection status for full sample and sub-samples with children, 1997–2013. Note: Clockwise from top left: full sample in 1997; family units with children aged 0–17 years in 1997 or 0–2 years in 1999; family units with children aged 11–17 in 1997; family units with children aged 0–2 years in 1997 or 1999. Observed trends are shown, without any covariate controls or weights.

The results for Model 2 in Table 4.4 include covariate controls for observed respondent characteristics. The results in the first column change substantially, with the apparent lower attrition for individuals from CDS families being accounted for entirely by observed characteristics and rendering statistically insignificant the difference in their likelihood of attrition. The results in Columns 2–4 continue to show no statistically significant differences in the effects of CDS participation in subsequent attrition – with the exception of the results in Column 3 which suggest that, compared to individuals in families with children just younger than the youngest children in CDS, the likelihood of attriting from PSID is 2.15 times higher, an effect that is statistically significant at the 0.10 level.

The results for Model 3 in Table 4.4 use the alternative approach of inverse probability of treatment weighting to control for the likelihood of CDS selection. This approach provides a near-uniform set of findings suggesting that CDS selection leads to no change in attrition from PSID in subsequent years. Model 4 in Table 4.4 extends these results to include covariate controls in addition to inverse probability of treatment weights, providing us with 'doubly-robust' estimates of the effects

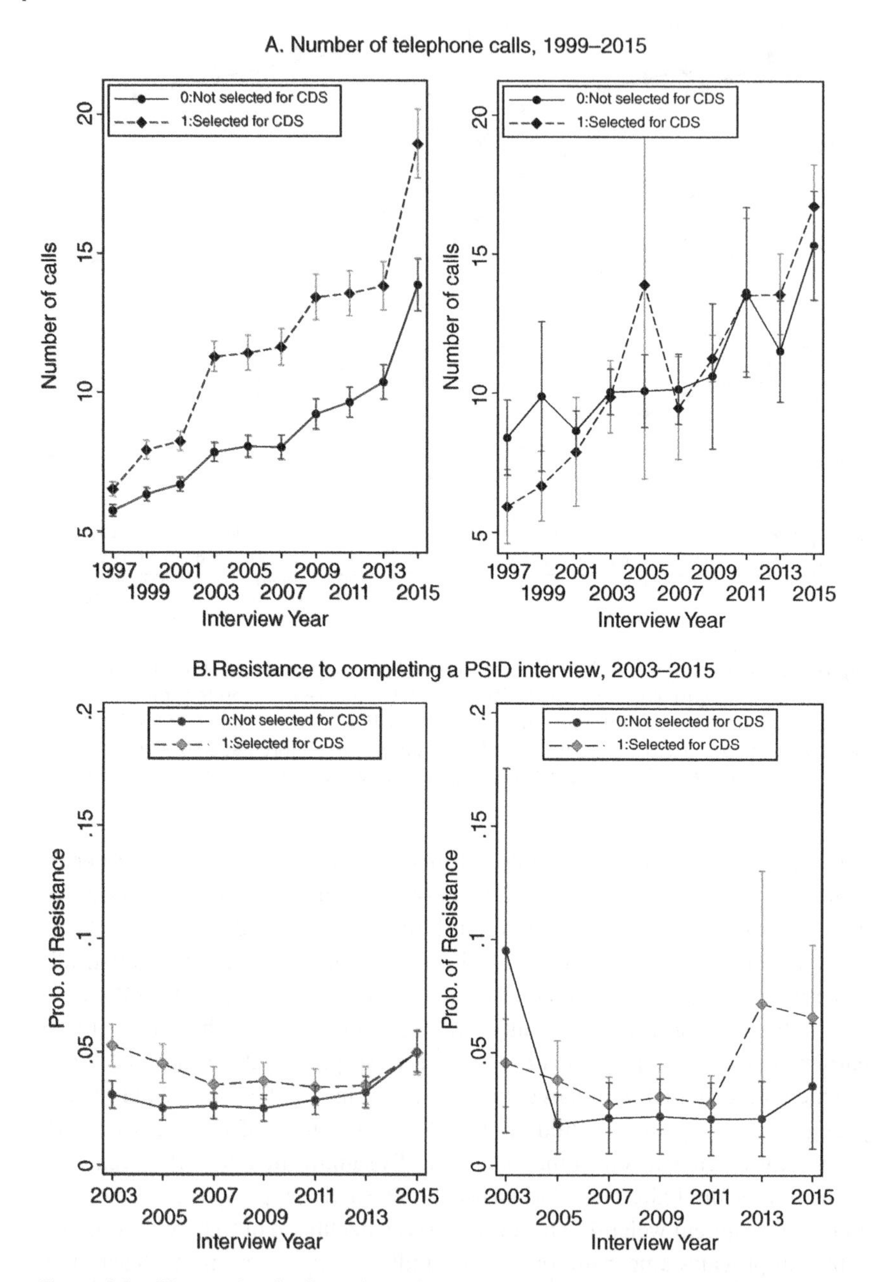

Figure 4.2 Observed and adjusted trends in PSID fieldwork outcomes by CDS selection. A. Number of telephone calls, 1999–2015. B. Resistance to completing a PSID interview, 2003–2015. C. Tracking to complete a PSID interview for full PSID sample, 2003–2015.

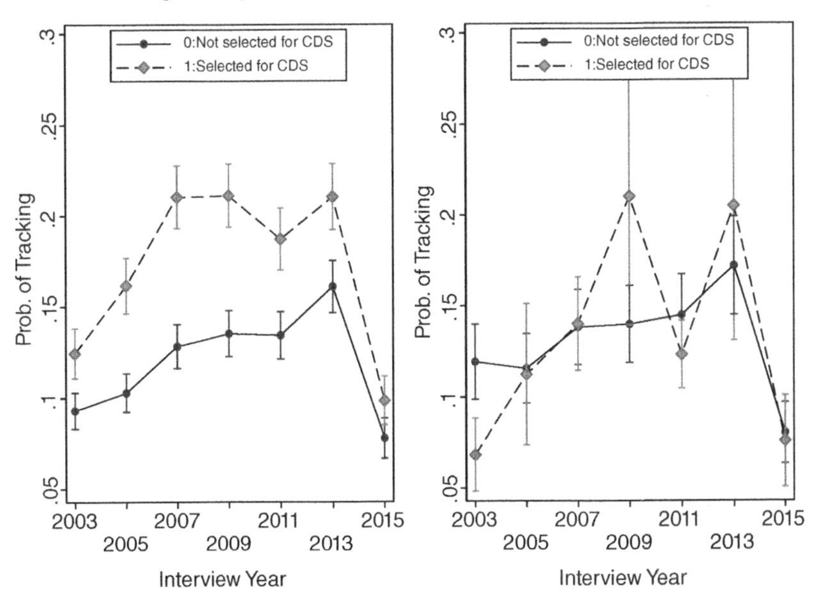

Figure 4.2 (*Continued*)

of CDS participation on subsequent attrition (Robins et al. 1995; Lunceford and Davidian 2004). These are our preferred results because of the comprehensive and flexible way in which they control for differences between observationally similar sample members who were and were not selected to participate in CDS. The results from Model 4 indicate that CDS selection is not associated with a change in attrition. However, there is a suggestion that CDS selection may be associated with higher attrition for families with the youngest children, although the higher attrition estimate is not statistically significant.

Finally, the results for Model 5 in Table 4.4 compare CDS respondents and non-respondents with those not selected for CDS in terms of propensity for subsequent attrition from PSID. These results are based on the modelling approach used for Model 4, which combines the inverse probability of treatment weights with covariate adjustment. The findings from Model 5 suggest that those who participated in CDS had a similar risk of subsequent attrition from PSID to those who were not selected for CDS; however, CDS non-respondents had a higher subsequent attrition risk, except in the comparison based on families with the youngest children. The relative risks of subsequent attrition range between 2.2 and 4.0 times higher for CDS non-respondents compared to those who not were not selected for CDS.

For each attrition model, we also examined whether there were time-varying effects of CDS selection on subsequent attrition. However, we found no evidence based on non-significant results of a joint statistical test of interactions between wave of interview and CDS selection status.

We next present results, summarised in Table 4.5, of the effects of CDS selection on the five fieldwork outcomes other than attrition (number of telephone calls, any face-to-face visits, any resistance, any interview suspension, and any tracking). We examine the four sample comparisons from the preceding set of results, but focus on two model specifications. The first model specification includes no covariates, while the second is based on our preferred specification that incorporates inverse probability of treatment weights and all covariates. More extensive results are presented in Part B of the online supplementary material.

Focusing first on the observed relationships between CDS selection and other fieldwork outcomes, based on Model 1 in Panels A–E in Table 4.5 we see that for all outcomes, and for all comparisons, there is either a statistically significant deleterious (i.e. positive) effect of CDS selection on subsequent fieldwork outcomes or no statistically significant effect. For example, the effects of CDS selection on the number of telephone calls to complete a PSID interview based on the full sample (Comparison 1) and shown in Panel A indicates that families selected for CDS required three more calls to complete the PSID interview in each year. Results (not shown) suggest that the presence of time-varying effects of CDS selection on the number of calls to complete a PSID interview, with the higher number of calls needed to complete interviews with PSID respondents who were selected for CDS increasing modestly over time. Similarly, the results for any resistance to completing a PSID interview (shown in Panel C) for the full sample (Comparison 1) indicate a 36% increase in the likelihood of resistance as a result of CDS selection. A last example in Panel E shows the effects of selection for CDS on tracking in PSID; in this case, the likelihood of tracking for the full sample (Comparison 1) is raised by 55% as a result of CDS selection. Again, there is evidence in these descriptive findings of time-varying effects of CDS selection on each of these fieldwork outcomes, although for some outcomes the effects over time of CDS selection increase while for others they decrease; in all cases, the basic conclusions are not altered in terms of the direction of the effects.

When we adjust for covariates and incorporate inverse probability of treatment weighting, in all three of these illustrative cases there is no statistically significant increase in fieldwork difficulty associated with CDS selection. This overall set of results can be seen clearly in Figure 4.2, which shows the differences in probabilities for each of these three illustrative fieldwork outcomes (number of telephone calls, resistance to completing a PSID interview, and whether tracking was necessary) between those selected and not selected for CDS in the full sample based on the observed (left panel, Model 1) and adjusted (right panel, Model 2) results.

For all three of these fieldwork outcomes, CDS selection is associated with higher fieldwork effort in the unadjusted models but in the adjusted models there is no clear difference (which is confirmed by formal statistical tests).

For the outcome of any interview suspension (Panel D), we do find a statistically significant effect at the 0.001 level of CDS selection on increasing the likelihood of an interview suspension even after incorporating inverse probability of treatment weights and covariate adjustment for the first two comparisons that are based on all cases (an odds ratio of 1.47) and on families with children (an odds ratio of 1.38). However, the results are not statistically significant for the two more focused comparisons based on younger children (Comparison 3) and older children (Comparison 4); moreover, the point estimates for these latter two comparisons are close to unity, suggest the true absence of an effect for these two subgroups.

Looking across all of the adjusted results presented in Table 4.5, we find no compelling evidence that CDS selection is associated with more difficult fieldwork for PSID. The only statistically significant result is that CDS selection is associated with greater fieldwork effort for completing a PSID interview (Panel D: any interview suspension for Comparisons 1 and 2). For all other fieldwork outcomes and all other comparisons, there is no statistically significant adverse effect of CDS selection on PSID fieldwork outcomes.

Finally, we note that in Part B of the online supplementary material we present a more complete set of results of the effects of CDS selection on PSID fieldwork outcomes that includes examining differences in effects based on cases that were CDS response compared to those that were CDS non-response. There is some scattered evidence that CDS non-response cases had worse subsequent PSID fieldwork outcomes – such as greater resistance to completing a PSID interview for Comparisons (1) and (2) and a higher likelihood of interview suspension for Comparison (2). But these results are not consistent or compelling, and are balanced by an occasional finding that CDS non-response was associated with better fieldwork outcomes (e.g. less tracking, based on Comparison 3). In addition, we examined whether there was variation over time in the effects of CDS selection on PSID fieldwork outcomes that could have affected these results but uncovered no systematic pattern of time-varying effects.

4.6 Conclusions

In this analysis, we examined whether selecting PSID respondents to participate in CDS – a major inter-wave supplemental study that collected detailed information on children's development – resulted in negative effects on response rates and fieldwork outcome measures such as the number of telephone calls to complete an interview. Overall, our results suggest that asking PSID families with children to participate in CDS resulted in a generally small, statistically insignificant increase in attrition in subsequent years but no consistent pattern of negative effects on fieldwork outcomes. Our preferred estimates are based on Model 4 in Table 4.4, which uses inverse probability of treatment weights and covariate adjustment to control for observed differences in factors associated with being selected for participation in CDS. The estimates from this model suggest a statistically significant negative effect on response rates only for families with the youngest CDS children. The reason for the higher attrition among families with the youngest children is not clear. This result represents our only statistically significant result and hence is not unexpected with a nominal Type-I error rate of 5%. Nevertheless, there are several reasons why families with very young children might view their CDS experience unfavourably and attrit at higher rates in subsequent PSID waves. For instance, the motivation for the study, which was to capture children's development in the school and neighbourhood contexts, may not be salient for children aged 0–2 years. The youngest CDS children also were asked to participate in an extra wave of data collection (CDS-III in 2007/08), which may have increased respondent burden (though the negative effect on attrition occurred before this date).

The conceptual framework we use suggests some reasons for these generally small and statistically non-significant results. We would expect to find higher attrition rates as a result of the higher costs for respondents from participating in additional study components. The fact that the attrition rates are only modestly higher, and are only statistically significant in one instance, is likely due to several of the factors identified by Groves and Couper – especially reciprocation and social exchange that is associated with respondent incentives and engagement in the on-going PSID and CDS studies. The estimated effects of CDS selection on PSID attrition were not changed much by controlling for a wide range of covariates describing sample status, prior history in the study, and demographic and socioeconomic characteristics. This finding suggests that the overall net effect of the other factors identified in the Groves and Couper framework is relatively unimportant for attrition over the subsequent eight waves of PSID. However, the detailed regression results (not shown) do reveal some statistically significant findings that support certain hypotheses from this framework.

Our results identify one particular group that is at high risk of attriting from the panel based on the request to participate in the supplement – namely, individuals from families who refused to participate in the supplement. This group was selected for CDS, but did not receive the treatment of actually participating in the supplement. Note, however, that our analysis was designed to analyse the effects of being *invited* to participate in CDS, rather than the effects of actual partition in the supplement (the former is the intent to treat, while the latter is the effects of the treatment on the treated). A likely reason for the higher attrition among the CDS non-respondents is that this group was already predisposed to ending their participation in PSID, and hence the CDS refusal is probably a marker of this intention rather than a causal factor that explains their subsequent attrition. (This latter question could be examined by adopting the modelling approach used in this paper and, in particular, by viewing the CDS non-response decision as an endogenous 'treatment' and comparing the subsequent PSID attrition among these non-respondents to observationally similar CDS respondents, although it would be difficult to account for differences in unobservable factors.) Nevertheless, this finding suggests that, for panel studies, there would be value in identifying families who are predisposed to attriting and targeting them for special retention efforts. For instance, these families might be offered higher incentives or presented with study materials that highlight key facets of the survey that are positively associated with the decision to continue panel participation; alternatively, these families could be excluded from the supplement altogether.

We have used statistical methods for causal analysis to identify the effects of CDS participation on subsequent panel attrition and fieldwork outcomes – specifically, we used inverse probability of treatment weights calculated from the propensity score for CDS selection. This method provides a convincing approach to adjusting for the effects of observed covariates. But it is not an experimental approach and, in particular, it does not account for unobserved factors. However, the CDS design did not provide a mechanism for unobserved characteristics to directly affect the selection process; rather, unobserved factors, to the extent they played a role at all, were likely to have had subtler effects, such as through the timing of fertility and thus the direct comparability of families in the groups selected and not selected for CDS. In other words, the control group may not have observationally similar families to those selected for CDS because all CDS families chose to have their children in the same window based on factors related to the effects of age, period characteristics, and other variables.

Our analysis of CDS selection on other fieldwork outcomes, such as the number of telephone calls to complete an interview, also reveal that the invitation to participation in CDS did not have any lasting deleterious effects on fieldwork processes in PSID. The analytical design and statistical methods we used were important for revealing this result, because the observational patterns suggested that selection into CDS were associated with considerably worse outcomes for many different aspects of fieldwork.

Overall, our results suggest that a major supplemental study such as CDS has had, at most, relatively minor negative effects on attrition and fieldwork difficulties in PSID over a long follow-up period during which CDS respondents were asked to complete additional waves of data collection for CDS and its successor study that tracked these children into their young adult years. At the same time, the benefits of collecting these new data are significant, at least as measured by the scientific contributions associated with over 600 publications based on these data according to the online PSID bibliography (available at www.psidonline.org). Our analysis provides support for continuing the CDS supplement to PSID, as well as an example of how the effects of supplemental studies can be evaluated for other panel studies.

Acknowledgements

Helpful comments were provided by participants at the 2016 Panel Survey Methods Workshop and the 2018 Methodology of Longitudinal Surveys II Conference, members of the PSID Board of Overseers, and Bob Schoeni. Programming assistance was provided by Meichu Chen. This project was supported by grant R01HD069609 from the *Eunice Kennedy Shriver* National Institute of Child Health and Human Development (NICHD).

References

Barber, J., Kusunoki, Y., Gatny, H. et al. (2016). Participation in an intensive longitudinal study with weekly web surveys over 2.5 years. *Journal of Medical Internet Research* 18 (6): e105. PMCID: PMC4937177. DOI: https://doi.org/10.2196/jmir.5422.

Deeg, D.J.H., van Tilburg, T., Smit, J.H. et al. (2002). Attrition in the longitudinal aging study Amsterdam: the effect of differential inclusion in side studies. *Journal of Clinical Epidemiology* 55 (4): 319–328.

Fitzgerald, J., Gottschalk, P., and Moffitt, R. (1998). An analysis of sample attrition in panel data: the Michigan panel study of income dynamics. *Journal of Human Resources* 33 (2): 251–300.

Groves, R.M. (2006). Nonresponse rates and nonresponse bias in household surveys. *Public Opinion Quarterly* 70 (5): 646–675.

Groves, R.M. and Couper, M.P. (1998). *Nonresponse in Household Interview Surveys*. New York: Wiley.

Groves, R.M., Singer, E., and Corning, A. (2000). Leverage-saliency theory of survey participation: description and an illustration. *Public Opinion Quarterly* 64 (3): 299–308.

Hart, T.C., Rennison, C.M., and Gibson, C. (2005). Revisiting respondent 'fatigue bias' in the National Crime Victimization Survey. *Journal of Quantitative Criminology* 21 (3): 345–363.

Kantorowitz, M. (1998). Is it true that nonresponse rates in a panel survey increase when supplement surveys are annexed? *ZUMA Spezial Nachrichten* 4: 121–135.

Lugtig, P. (2014). Panel attrition: separating stayers, fast attriters, gradual attriters, and lurkers. *Sociological Methods & Research* 43 (4): 699–723.

Lugtig, P., Das, M., and Scherpenzeel, A. (2014). Nonresponse and attrition in a probability-based online panel for the general population. In: *Online Panel Research: A Data Quality Perspective* (eds. M. Callegaro, R. Baker, J. Bethlehem, et al.), 135–153. New York: Wiley.

Lunceford, J.K. and Davidian, M. (2004). Stratification and weighting via the propensity score in estimation of causal treatment effects: a comparative study. *Statistics in Medicine* 23: 2937–2960.

Lynn, P. (2014). Longer interviews may not affect subsequent survey participation propensity. *Public Opinion Quarterly* 78 (2): 500–509.

McCarthy, J.S., Beckler, D.G., and Qualey, S.M. (2006). An analysis of the relationship between survey burden and nonresponse: if we bother them more, are they less cooperative? *Journal of Official Statistics* 22 (1): 97–112.

McGonagle, K.A. and Sastry, N. (2015). Cohort profile: the panel study of income dynamics' child development supplement and transition into adulthood study. *International Journal of Epidemiology* 44 (2): 415–422. PMCID: PMC4553706. DOI: https://doi.org/10.1093/ije/dyu076.

McGonagle, K.A., Schoeni, R.F., Sastry, N. et al. (2012). The panel study of income dynamics: overview, recent innovations, and potential for life course research. *Longitudinal and Life Course Studies* 3 (2): 268–284. PMCID: PMC3591471.

Ofstedal, M.B., Couper, M.P. (2008). Piling It On: The Effect of Increasing Burden on Participation in A Panel Study. Paper presented at the 2008 meeting of the Panel Survey Methods Workshop, University of Essex, U.K. (14 July 2008).

Phillips, C.B., Yates, R., Glasgow, N.J. et al. (2005). Improving response rates to primary and supplementary questionnaires by changing response and instruction burden: cluster randomised trial. *Australian and New Zealand Journal of Public Health* 29 (5): 457–460.

Porter, S.R., Whitcomb, M.E., and Weitzer, W.H. (2004). Multiple surveys of students and survey fatigue. *New Directions for Institutional Research* 2004 (121): 63–73.

Robins, J.M., Rotnitzky, A., and Zhao, L.P. (1995). Analysis of semiparametric regression-models for repeated outcomes in the presence of missing data. *Journal of the American Statistical Association* 90: 106–121.

Rolstad, S., Adler, J., and Rydén, A. (2011). Response burden and questionnaire length: is shorter better? A review and meta-analysis. *Value in Health* 14 (8): 1101–1108.

Sinibaldi, J. and Karlsson, A.Ö. (2016). The effect of rest period on response likelihood. *Journal of Survey Statistics and Methodology* 5 (1): 70–83.

5

The Effects of Biological Data Collection in Longitudinal Surveys on Subsequent Wave Cooperation

Fiona Pashazadeh[1], Alexandru Cernat[1] and Joseph W. Sakshaug[2,3,4]

[1] *Department of Social Statistics, University of Manchester, Manchester, UK*
[2] *Statistical Methods Research Department, Institute for Employment Research, Nürnberg, Germany*
[3] *Department of Statistics, Ludwig Maximilian University of Munich, Munich, Germany*
[4] *School of Social Studies, University of Mannheim, Mannheim, Germany*

5.1 Introduction

There are many considerations in the design and implementation of large-scale, nationally representative longitudinal surveys in order for researchers to maximise data quality while ensuring cost effectiveness (Buck et al. 1995; Lynn 2009a). Increasing the range of data available by adding new components to an established longitudinal survey, such as linkage to external administrative data, biological data collection or time-use diaries, can be hugely advantageous for research while also making use of existing data collection structures. However, the additional respondent burden involved may increase the likelihood of non-response and attrition, compromising the power and representativeness of the sample. Consequently, results of any analysis using the data may be biased, especially when the lack of cooperation is linked to the outcome(s) of interest (Groves 2006). This is of particular concern in longitudinal surveys where the effects of non-response can be cumulative (Watson and Wooden 2009). For these reasons, there is a need for more understanding of the mechanisms surrounding increased respondent burden and the effects on subsequent cooperation in longitudinal surveys, so that researchers can make informed decisions about whether, how, and when to implement extra data collection components. This chapter provides guidance on this matter by quantifying the effects of the addition of one such component, biological data collection, in a large-scale longitudinal survey and discussing the results in conjunction with the existing literature.

Biological data collection in longitudinal surveys has increased in recent years, and this has enhanced the potential for biosocial research by enabling the

Advances in Longitudinal Survey Methodology, First Edition. Edited by Peter Lynn.

Companion website: www.wiley.com/go/lynn/advancesinlongitudinalsurvey

investigation of links between biological mechanisms and social phenomena in the general population (Benzeval et al. 2016). For example, recent research using data from Understanding Society: the UK Household Longitudinal Study (UKHLS) includes studies on the links between job quality and biomarkers of chronic stress (Chandola and Zhang 2018), informal caregiving and markers of metabolism (Lacey et al. 2018), and unemployment and body mass index (Hughes and Kumari 2017).

However, whilst biological data are clearly of high utility, the collection involves a particularly large respondent burden, and this raises concerns about sample attrition (Marmot and Steptoe 2008; Weir 2008). Typical collection protocols for large-scale surveys include a range of anthropometric measurements, such as height, weight, and body fat, as well as blood and saliva samples (Sakshaug et al. 2015). The collection of the data may require respondents to participate in extra or longer visits, which may involve travel to a designated clinic or arranging a separate appointment with a trained nurse. There may also be requests to store the samples for future analysis. Whilst the sources of increased burden are clear, there may also be positive effects on respondents due to, for example, the value of the data for research. Understanding how, and to what extent, these competing influences may affect future cooperation with the survey will help to ensure that additional biological data collection is not detrimental to the survey as a whole.

In this chapter, we present results that quantify the effects of the additional burden of biological data collection on subsequent cooperation in an ongoing large-scale, nationally representative longitudinal survey. Our analysis involves a quasi-experimental study design using data from UKHLS that takes advantage of random subsampling for inclusion in the nurse visit in wave 2. The comparison of response propensities for waves 3–7 indicate that there is a short-term negative effect in wave 3 that can be attributed to the increased burden of the nurse visit, but this is not seen in waves 4–7. The discussion of the results includes practical considerations and suggestions for survey researchers. We also review a selection of the literature on respondent burden and survey response, focusing in particular on the issues surrounding biological data collection in large-scale longitudinal surveys.

5.2 Literature Review

It is generally hypothesised in the survey methodology literature that a higher respondent burden will lead to increased reluctance when responding to a survey request. This is likely to be especially important in the longitudinal survey context where previous experience of a survey has been found to be a significant factor in subsequent cooperation (Lepkowski and Couper 2002; Watson and Wooden 2009). If respondents feel that the burden is too high, they may not only refuse in the current wave but also become more likely to attrit in future waves (Lynn et al. 2005). The potential effect on future non-response shows the distinction between longitudinal and cross-sectional data collection and the need to consider how an increase in burden may continue to have an impact throughout the life of the survey.

The term *respondent burden* generally refers to the requirements of participating in a survey and the accumulated effect of these on the effort requested from the respondent. Bradburn (1978) discussed respondent burden in terms of the difficulty of the task presented to the respondent and how this task is perceived in each case (i.e. the same survey task may represent different levels of burden to different respondents). This is thought to be due to a combination of factors such as the length of the interview, the physical and/or emotional effort required, the stress or discomfort experienced and the number of interviews required, and how these interact with the characteristics of the respondent. There is also the idea of cognitive burden, referring to the effort required to understand and answer the questions – this will be lower if the respondent is familiar with the survey topic (Groves and Couper 1998).

Survey organisations can aim to moderate the effects of perceived burden by changing or adapting features that are under their control (e.g. by reducing the objective burden of the survey or offering incentives, which would increase motivation). However, some aspects of how burden is perceived by respondents and how this affects the decision to participate may be difficult to address. Groves and Couper (1998) consider burden as part of a cost–benefit/opportunity cost analysis that the respondent evaluates based on the information they have about the survey and this informs whether they choose to cooperate, given their characteristics and circumstances at the time. The cost–benefit analysis performed by the respondent suggests that the perceived difficulty of the task may be offset by the perceived importance of the survey, as represented by the topic, the survey sponsors and the explanations from the interviewer or advance communications, as well as any financial or other incentives offered by the survey organisation (Heberlein and Baumgartner 1978; Olsen 2005).

Leverage-salience theory seeks to formalise this decision-making process by providing a theoretical framework that incorporates a variety of survey attributes, such as financial incentives and survey sponsor or topic, and the influence of these on the decision of the respondent to participate (Groves et al. 2000). Specifically, the theory puts forward that the propensity for a respondent to cooperate, given their characteristics, is a function of the leverage of such survey attributes and whether, and to what extent, they have been made salient to the respondent. The leverage of a given survey attribute is shown as a function of relevant characteristics, as it is not usually possible to measure or observe directly. Some support for the theory comes from an illustration with an experimental study design using the 1996 Detroit Area Study, where a small financial incentive was found to have differential effects on the propensity to cooperate depending on the respondent's level of community involvement. Groves et al. (2004) also provide some support with several experiments involving the randomising of survey topics and monetary incentives to different interest groups (teachers, new parents, people aged 65+, and political campaign contributors), which yield some positive results.

Looking at empirical studies that focus on the effects of respondent burden in the previous survey literature, interview length has been by far the most frequently tested, with both observational and experimental studies (Singer and Presser 2007). Here, the mode of the survey is important, as this will affect how the potential burden is perceived. In a mail survey, the respondent can look at the full questionnaire and judge how long it will take and this can be manipulated in an experimental design. For other survey modes, the potential respondent can get an idea of the likely length from advance communications, the interviewer, and/or some indicator of progress once the survey has commenced. Importantly in longitudinal surveys, the previous interview can also give an indication of expected burden and length (Lynn et al. 2005).

Burchell and Marsh (1992), Bogen (1996), and more recently Singer and Presser (2007) provide reviews of the literature on interview length and non-response. Generally, results have shown that increasing burden, as measured by longer interview length, is associated with increasing non-response, though the relationships have often been weak. There have also been some results showing negative relationships between length and non-response, and some showing null effects (e.g. Champion and Sear 1969; Roscoe et al. 1975). Whilst much of the previous research has been on mail surveys, some more recent research has considered burden, interview length, and non-response in web surveys. Crawford et al. (2001) show evidence from an experimental web survey design that when a lower anticipated survey length is specified in the invitation, individuals are more likely to respond. Galesic (2006) found that higher levels of self-reported interest in the question topic in a web survey experiment were associated with lower dropout, whereas self-reported burden was associated with higher dropout. Higher interest level was also seen to have a positive relationship with the amount of time spent on a block of questions, though overall longer questionnaires had a higher risk of dropout.

The particular interest in this chapter is on respondent burden in longitudinal surveys where the effects may be different from cross-sectional or one-off interviews/questionnaires. Specifically, the burden of a particular survey task may affect not only completion of that task with immediate effect, but also completion of future tasks including in subsequent waves. Increases in burden in a longitudinal survey context need to be especially carefully considered, as any negative impact on participation will be detrimental to both current and future data quality, due to its cumulative effects. Several studies have been conducted to assess the risks of increased burden on longitudinal response and attrition, again focusing particularly on interview length. Branden et al. (1995) investigated the effects of survey length on attrition in the National Longitudinal Survey of Youth (NLSY) and found that the time spent on the most recent interview had a negative relationship with subsequent non-response. This was also true when using the number of questions asked as an alternative measure of interview length. There is contradictory evidence from Zabel (1998), who found that there was a negative effect on attrition in the Panel Study of Income Dynamics (PSID) when a reduction in the length of the interview was introduced in 1972. Hart et al. (2005) took a slightly different approach and used the number of crimes reported in the National Crime Victimization Survey as a measure of interview length in the first wave. They found that longer exposure to the initial interview was not associated with future non-response, when accounting for other explanatory variables.

It may be difficult to interpret results from observational studies on interview length, burden, and future cooperation due to confounding factors (e.g. it could be that interviewers spend more time with respondents who are enjoying the interview) (Lemay 2009). An early experimental study by Sharp and Frankel (1983) examined the effects of first interview length on general attitudes towards the survey and response at phase two. Respondents were randomised to either a 'short' 25-minute or 'long' 75-minute interview, and the findings suggested that, whilst increased interview length was associated with more negative feedback, it was not associated with subsequent non-response. Lynn (2014) analysed data from a more recent experimental study using the UKHLS innovation panel. Sample members were again randomised to either a short or long interview in wave 1, where the longer interview had two additional questionnaire modules. Response rates were compared for waves 2–5 but no statistically significant differences were found, implying that the longer interview had no effect on the likelihood of cooperation over time.

Furthermore, it is informative to consider not only the effects of increased burden in longitudinal surveys but also the efficacy of measures that may be taken by the survey organisation to counteract or compensate for this burden, such as the provision of financial incentives for respondents. Singer et al. (1999) conducted a meta-analysis of experimental studies, including some carried out using panel survey data, where the effectiveness of incentives at reducing non-response had been tested in various survey settings. They found that financial incentives increased response rates in surveys with higher burden, compared to no incentives, but there was no significant interaction between burden and incentives across surveys. The definition of high burden included interview length as well as survey features such as diaries, sensitive questions, and repeated interviews. The findings also held across survey modes.

For longitudinal surveys, it is also important to consider the idea of cumulative burden, or panel fatigue, which arises from asking the same group of respondents to participate in repeated waves over time in order to preserve the original sample (Apodaca et al. 1998; Lemay 2009). In addition to wave non-response, indicators of panel fatigue can include interviewer ratings of respondent enjoyment (Hill and Willis 2001) and item non-response (Loosveldt et al. 2002). There is also the possibility of a one-off occurrence or change in the survey, a shock, which can cause a sudden increase in non-response and attrition (e.g. a poor questionnaire design for a module/wave or an incident with the handling of personal data), and this could be perceived as an increase in the overall survey burden (Lugtig 2014).

The addition of biological data collection to a longitudinal survey raises a particular set of issues when evaluating the relationship with non-response. Although the data are clearly of high value, any significant impact on the representativeness of the panel through selective attrition is likely to be unacceptable due to the effects on overall data quality (Weir 2008). The increase in burden related to biological data collection clearly encompasses several of the categories identified by Bradburn (1978) compared to other potential additional components, in particular physical stress and discomfort for the respondent. There are also requirements for the completion of additional consent procedures before collecting measurements and samples, which may involve different non-response mechanisms related to respondent characteristics and interests, leading to further potential biases (e.g. Sakshaug et al. 2012).

As with surveys in general, clear differences have been found between respondents in their propensity to respond to survey requests that involve the collection of biological data, based on their personal characteristics. Results from Sakshaug et al. (2010) show that younger people and those with physical limitations were less likely to consent to the full set of measurements in the 2006 Health and Retirement Study (HRS). However, those who had diabetes were more likely to consent. They also found that indicators of resistance to the survey recorded previously by interviewers were amongst the best predictors of consent. Khare et al. (1994) found that participation in the blood sample for the National Health and Nutrition Examination Survey (NHANES) was higher for those aged 60+, but within this age group participation was lower for non-Hispanic black respondents, females, and those with higher poverty levels. Findings also from NHANES by Mcquillan et al. (2003) suggest that non-Hispanic black respondents are less likely to consent to blood sample storage for future genetic research. Results from Cernat et al. (2020) analysing nurse effects on non-response in survey-based biomeasures using UKHLS data found that there are different relationships between non-response and socio-demographic characteristics for the different stages in biological data collection (i.e. participating in the nurse visit, consenting to the blood sample, and providing the blood sample).

However, whilst there is evidence that sample members with certain characteristics are less likely to participate in biological data collection in longitudinal surveys, there do not seem to be any such reported effects on overall response rates in subsequent waves. Weir (2018) reports that response rates did not fall in the HRS with the introduction of biological data collection in 2006 but actually increased. Marmot and Steptoe (2008) state that biological data collection in the Whitehall II study did not lead to increased sample attrition, even for a sub-sample who were asked take part in more extensive tests. A high level of interest from respondents in this aspect of the study is also mentioned, a point echoed by Marmot and Brunner (2005) who speak about respondent enjoyment of the study process. Weinstein and Willis (2001) also describe how respondents who participated in the collection of biomarkers in the Taiwan Study of the Elderly did not have an increase in non-response in the following wave of the survey. Even so, they note that this may have been the result of self-selection (i.e. more motivated respondents may have been more likely to respond to both the biomarker collection and the next wave interview).

It is in this context that we conduct an empirical analysis into the effects of the additional biological data collection component on subsequent cooperation in UKHLS and investigate some of the potential mechanisms that follow from the existing literature already discussed.

5.3 Biological Data Collection and Subsequent Cooperation: Research Questions

This section quantifies the effects of an additional biological data collection component on cooperation in subsequent waves of a large-scale, nationally representative longitudinal survey in the UK. We used a quasi-experimental study design that took advantage of the randomisation of the nurse visit to a sub-sample of respondents in wave 2 of the UKHLS, allowing for a robust comparison of outcomes for each group over time.

Given the literature discussed above, the extra respondent burden from inclusion in the nurse visit sample may have arisen in various ways:

1. *Extra number of contacts*. The respondents who were included in the wave 2 nurse visit sub-sample had a minimum of two extra contact attempts from the survey organisation: an advance letter with a leaflet explaining more about the nurse visit and a short telephone interview from the contact centre. If a whole-household refusal was not received following these contacts, a nurse was

assigned to make an appointment with the household and conduct the visit. As the nurse visit was unanticipated when the respondents previously agreed to be part of the panel, the extra contact attempts would have involved giving more time to the survey than originally expected. These contacts also provided additional opportunities for respondents to consider their continued participation in the survey as a whole and may have increased cognitive burden.

2. *Time required for the nurse visit.* In addition to the time involved in the extra contact attempts, there is of course the time needed to complete the nurse visit itself – the average length for each individual was approximately one hour if all the measurements were taken (McFall et al. 2014). As discussed above, this is likely to have had different effects on respondents, depending on their personal circumstances and characteristics. Additionally, the request for extra time spent on the survey may have had a significant effect on respondents even if they chose not to participate in the nurse visit.
3. *Emotional stress and physical discomfort.* Biological data collection, in particular, may cause emotional stress, and this could occur as a result of the advance communications, even if the respondent decided not to participate. There is also the physical discomfort involved in the measurements which may have been considerable for some respondents.
4. *Extra consent procedures.* The consent procedures for the nurse visit involved the respondent reading and completing a set of leaflets and forms. This may have provoked concerns about confidentiality as well as increasing cognitive burden.

The collection of biological data may also have had some positive impact on respondents. Health surveys are generally seen as having higher response rates (Groves and Couper 1998), and the nurse visit may have worked to make the health aspect of UKHLS salient to respondents. Some respondents may also have seen the nurse visit as an opportunity for a health check, as reported by Marmot and Steptoe (2008) regarding similar data collection in Whitehall II and ELSA. Given these potentially competing influences, it was not clear a priori what should be expected in terms of the results of the analysis.

The specific research questions are:

1. Is inclusion in the UKHLS wave 2 nurse visit sub-sample associated with the likelihood of cooperation in waves 3–7?
2. Are there different relationships between inclusion in the wave 2 nurse visit sub-sample and subsequent cooperation for respondents who may be disproportionately affected by the additional burden due to their health status and/or general resistance to the survey?

The first part of the investigation involved modelling cooperation with the interviews in waves 3 to 7 and testing whether this was significantly different for those who were included in the wave 2 nurse visit sub-sample. The second part extended this analysis by including interaction terms for explanatory variables indicating health status and previous reluctance.

5.4 Data

The data used in this study (University of Essex 2014, 2017) are from UKHLS, a large and ongoing general longitudinal survey of the UK population. UKHLS started in 2009 with a sample of approximately 40 000 households and is completed annually in person by trained interviewers (McFall 2012). In waves 2 and 3, nurse visits were conducted for different subsets of the full sample – the General Population Sample, or GPS, in wave 2 and the British Household Panel Sample, or BHPS, in wave 3. This involved a separate visit by a trained nurse approximately five months after the main interview. The nurse visit involved the collection of biological data from eligible respondents including: height, weight, body fat, waist circumference, blood pressure, grip strength, lung function, and a blood sample (McFall et al. 2014).

In UKHLS wave 2, GPS respondents in England, Scotland, and Wales were eligible for inclusion in the nurse visit sample if they had completed the full main interview in English, were aged 16 and over and were not pregnant. The nurse visits were conducted over two years starting in May 2010. Due to staff constraints during the second year of data collection, it was not possible to include the full GPS and instead a random sample of 81% of the Primary Sample Units (PSUs) in England was selected (McFall et al. 2012). This random subsampling forms the quasi-experimental study design used here and enables comparisons to be made between response propensities for those who were included in the nurse visit sample and those who were otherwise eligible but not selected.

The sample members included in the analysis are those from the year 2 UKHLS sample who were eligible for the wave 2 nurse visit, had the potential to be selected into the sub-sample (i.e. they were residents in England) and also fully participated in wave 1. This is to take advantage of the random subsampling for the nurse visit in year 2 of wave 2 and also to test the effects of the nurse visit on respondents who had been actively involved in UKHLS from the beginning. The sample was further concentrated on individuals aged at least 17 at the time of the nurse visit so that everyone in the analysis had completed the adult (rather than the child) interview at waves 1 and 2. This resulted in a wave 2 sample of 12 730 individuals of which 9949 were in the nurse visit sub-sample and 2781 were not. At waves 3–7, individuals were removed if they were ineligible due to emigration or death. This led to an analytical sample size of 12 613 for wave 3, reducing to 12 137 by wave 7 (sample sizes for each wave are given in Table 5.2).

5.5 Modelling Steps

The focus of this study is on cooperation in subsequent waves so the dependent variable is defined as whether or not the respondent completed the full interview in the relevant wave, given that they were eligible to participate. Chi-squared tests were used as a first step in testing for independence between inclusion in the wave 2 nurse visit and cooperation in waves 3–7. A high level of statistical significance from the tests would suggest that there is a relationship between selection for the nurse visit sample and cooperation in the relevant survey wave.

This initial bivariate analysis was extended to multiple logistic regression models to investigate any further effects whilst accounting for other potentially significant factors. The inclusion of other explanatory variables in the regression models follows from studies that identify many factors that can contribute to the likelihood that a respondent will continue to cooperate with a longitudinal survey, including personal characteristics, household circumstances, previous survey experiences, geographical area, and the assigned interviewer (Uhrig 2008; Watson and Wooden 2009). Even though a random sub-sample was chosen for the nurse visit, this may have resulted in meaningful demographic and other differences between those who were selected and those who were not, so accounting for these other variables is desirable to separate the effect(s) of interest (Campanelli and O'Muircheartaigh 1999).[1]

As in Groves et al. (2000), it is also of interest to understand more about the potential leverage of biological data collection on certain groups. Respondents with a poorer health status may see the nurse visit as having positive leverage on their future cooperation, as it makes salient the value of the collected data for health research. Conversely, health conditions may cause the physical assessments to be unpleasant and the negative leverage of this experience (or potential experience if the respondent chose not to participate) may lead to a higher likelihood of attrition. Similarly, for those who had shown indications of resistance towards the survey, the extra burden of the nurse visit may tip the balance towards non-response. Consequently, the logistic regression models were extended by adding interaction terms between the indicator for inclusion in the nurse visit and the explanatory variables relating to health and previous reluctance.

---

1 Another potentially important consideration is the clustering by geographical area in the sampling design of UKHLS, as detailed by Lynn (2009b), and the impact this could have on the models of response. There are also several studies that show that the assigned interviewer can influence the likelihood of responding to a survey (West and Blom 2017). Both the area and interviewer can be accounted for here by using cross-classified multilevel models (Campanelli and O'Muircheartaigh 1999; Vassallo et al. 2015). However, running these more complex models did not change the substantive findings in this study so, for ease of interpretation, the models reported here are estimated using standard logistic regression techniques (estimated in R v3.3.3).

Table 5.1 Explanatory variables for the models of cooperation in UKHLS waves 3–7.

Category	Variables
Personal characteristics	Sex, age, marital status, employment status, highest qualification, ethnicity (white/non-white)
Household characteristics	Number of adults, children, housing tenure, Government Office Region (London or elsewhere), urban/rural
Health status	Longstanding illness, self-rated health
Indicators of reluctance	Number of calls for the main wave 2 interview, interviewer ratings of cooperation and suspicion following the wave 2 main interview.
Different interviewer from previous wave	–
Nurse visit indicator	–

The explanatory variables included in the models of response were guided by those used by Watson and Wooden (2009) and Uhrig (2008) in their studies of longitudinal survey response and attrition, as listed in Table 5.1.[2] As eligibility for the nurse visit included completion of the full wave 2 main interview, a range of data are available for individuals and households. The wave 2 values were also used for the regression models at each subsequent wave as it is of interest to estimate how the characteristics of the respondents and households at the time of the nurse visit may influence subsequent cooperation.[3] Full descriptive statistics are available in the online appendix. Despite the random sampling for the nurse visit, bivariate chi-squared tests do show some statistically significant differences in socio-demographic characteristics, and this may be due to the clustered sampling design.

5.6 Results

Table 5.2 shows the percentage of eligible respondents from wave 2 that completed the full interview in each subsequent wave, along with the associated chi-squared

2 It should be noted that the inclusion of the different interviewer variable does not allow estimation of the effects of interviewer continuity alone as there are often confounding reasons for the change that can affect the likelihood to respond. These include non-random events such as household relocation and difficulties in interviewer retention in certain geographical areas (Lynn et al. 2014). Therefore, this should be viewed as a composite measure representing these different aspects. There was also a significant amount of missing interviewer data so a separate category has been included.

3 74 individuals were removed from the analysis for missing data on at least one explanatory variable.

Table 5.2 Matched sample sizes at waves 3–7 with the percentage of respondents who completed the full interview at each wave and chi-squared tests of independence.

Wave	Total number of eligible individuals	Total in wave 2 nurse visit sample	Total not in wave 2 nurse visit sample	Chi-squared test of independence[a)]
3	12 613 84.4%	9851 83.8%	2762 86.7%	$\chi^2_1 = 13.991$ ($p < .001$)
4	12 443 78.8%	9717 78.3%	2726 80.8%	$\chi^2_1 = 7.712$ ($p = .005$)
5	12 343 73.8%	9638 73.5%	2705 75.1%	$\chi^2_1 = 2.857$ ($p = .091$)
6	12 262 66.1%	9573 65.7%	2689 67.4%	$\chi^2_1 = 2.473$ ($p = .116$)
7	12 137 63.1%	9480 63.0%	2657 63.9%	$\chi^2_1 = 0.907$ ($p = .341$)

a) The chi-squared tests of independence are calculated from 2 × 2 contingency tables of cooperation at each wave by inclusion in the wave 2 nurse visit sample, with Yates' continuity correction.

tests of independence. The results for wave 3 imply that inclusion in the nurse visit and cooperation at that wave were not independent ($p < .001$). This suggests that the difference between the wave 3 cooperation rate of 83.8% for the nurse visit sub-sample and the higher 86.7% for those not in the nurse visit is statistically significant. This effect is seen to a lesser extent in wave 4 but not seen in the results for the subsequent waves 5–7.

Table 5.3 displays the key results of the logistic regression models for the log-odds of cooperation at waves 3–7, which also account for the possible effects of the other explanatory variables listed in Table 5.1. The first column shows that for wave 3, the nurse visit indicator has a negative coefficient equal to −0.25 (95% CI = [−0.38, −0.12]), and this is highly statistically significant ($p < .001$). This suggests that, holding the other variables constant, inclusion in the wave 2 nurse visit reduces the probability of cooperation at wave 3, reinforcing the earlier bivariate result in Table 5.2. Using this result to calculate the odds ratio of the nurse visit indicator coefficient gives exp.(−0.25) = 0.78 (95% CI = [0.68, 0.88]). This estimates that the odds of cooperating at wave 3 for a respondent included in the nurse visit sub-sample would be reduced by a factor of 0.78 (or 1 − 0.78 = 22%) compared to a respondent with the same characteristics but not included. Full

Table 5.3 Logistic regression models of the log-odds of cooperating (with a full interview) at waves 3–7, given eligibility.[a)]

	Wave 3	Wave 4	Wave 5	Wave 6	Wave 7
Nurse visit	**−0.25***** (0.07)	−0.00 (0.07)	0.07 (0.07)	−0.06 (0.05)	0.07 (0.06)
Health					
Long-standing illness	0.12 (0.07)	**0.17*** (0.07)	**0.16*** (0.07)	**0.19***** (0.05)	**0.13*** (0.07)
Self-rated health (ref = excellent)					
Very good health	0.10 (0.08)	−0.04 (0.08)	0.07 (0.08)	0.10 (0.06)	0.02 (0.08)
Good health	−0.06 (0.08)	−0.16 (0.09)	0.01 (0.09)	−0.05 (0.06)	−0.13 (0.09)
Fair health	−0.17 (0.10)	−0.18 (0.11)	−0.08 (0.11)	−0.12 (0.08)	−0.08 (0.11)
Poor health	**−0.33*** (0.14)	**−0.35*** (0.14)	−0.14 (0.16)	**−0.42***** (0.11)	−0.27 (0.15)
Indicators of reluctance					
Number of calls at wave 2 (ref = 1–2 calls)					
3–6 calls	0.05 (0.06)	−0.07 (0.06)	0.03 (0.06)	0.03 (0.04)	**−0.13*** (0.06)
7+ calls	−0.15 (0.08)	**−0.36***** (0.09)	−0.17 (0.09)	**−0.32***** (0.07)	−0.16 (0.09)
Uncooperative	**−0.57***** (0.15)	**−0.50**** (0.17)	**−0.54**** (0.18)	**−0.54***** (0.14)	−0.24 (0.19)
Suspicious	**−0.62***** (0.13)	**−0.55***** (0.14)	−0.11 (0.17)	**−0.51***** (0.11)	**−0.40*** (0.15)
Different interviewer[b)]	**−1.16***** (0.06)	**−1.00***** (0.06)	–	**−0.26***** (0.06)	

(Continued)

Table 5.3 (Continued)

	Wave 3	Wave 4	Wave 5	Wave 6	Wave 7
Missing interviewer[b)]	**−4.96***** (0.46)	**−6.53***** (0.34)	**−9.41***** (1.00)	–	**−3.86***** (0.07)
AIC	9974.45	9414.66	8585.39	15 035.28	9589.05
Log Likelihood	−4946.23	−4666.33	−4251.69	−7478.64	−4753.53
Deviance	9892.45	9332.66	8503.39	14 957.28	9507.05
N	12 538	12 369	12 269	12 188	12 065

$^{***}p < .001$, $^{**}p < .01$, $^{*}p < .05$.

a) Variables also included in the models: intercept, sex, age, marital status, employment status, highest qualification, ethnicity (white/non-white), number of adults, children, housing tenure, Government Office Region (London or elsewhere), urban/rural. Full results available in the online supplementary material.

b) The different/missing interviewer indicators were not included in wave 6 due to a change in fieldwork agency.

results of these models appear in the online supplementary material for this chapter.

It can be helpful for interpretation to convert the results from logistic regression analysis into probabilities in order to illustrate and compare effect sizes in a more intuitive way.[4] For an individual with the reference characteristics but not selected for the nurse visit, the estimated probability of cooperating at wave 3 is relatively high at 90.2%. This decreases to 87.8% for a person with the same characteristics but selected for the nurse visit, a reduction of 2.4%. For a person estimated to be much less likely to cooperate at wave 3 (e.g. a single 20-year-old male who is unemployed with no qualifications and lives in a London furnished rental property with other adults and also showed signs of reluctance at the wave 2 interview), the estimated probability of cooperating at wave 3 is 26.8% if not included in the nurse visit and 22.2% if selected – a difference of 4.6%. Finally, for individuals with characteristics that place them in between these two hypothetical examples (e.g. a 25-year-old female with GCSE qualifications, poor self-rated health, a long-term health condition that leaves her unable to work and who lives in social housing outside London with an adult relative), the estimated probability of cooperating at wave 3 if not selected for the nurse visit is 83.6%. This probability reduces to 79.9% if included in the nurse visit sub-sample, a decrease of 3.7%.

4 A note of caution – the calculation of the predicted probabilities assumes that the model is correct, which has not been tested here explicitly.

Table 5.3 also shows the results of the logistic regressions models of cooperation at waves 4–7. The nurse visit indicator is estimated not to be significantly different from zero for these waves, largely agreeing with the results from the bivariate analysis in Table 5.2. The coefficient estimates for the nurse visit have also noticeably reduced in size.

The coefficient estimates for the other explanatory variables in Table 5.3 are generally as expected from the previous survey non-response literature. The indicators of reluctance show negative effects, as does a change in interviewer between waves. Large effects are seen for those with missing interviewer data, which is to be expected as this indicates an interviewer may not have been allocated due to previous refusal or noncontact.

Table 5.4 presents summary results for the extended logistic regression models, which include the interactions terms between the nurse visit and both the health variables and the indicators of reluctance (again, full results are in the online supplementary material for this chapter). The extended model for the log-odds of cooperation at wave 3 does not yield any new significant results and a likelihood ratio test shows that this extension is not an improvement over the original model in Table 5.3 ($p = .871$). However, the interaction models for waves 4 and 5 show a positive coefficient for the nurse visit, though this is only marginally statistically significant at ($p = .049$ at wave 4, $p = .036$ at wave 5). There is also a significant negative interaction with the different interviewer indicator. Together these estimated coefficients suggest that having a different interviewer in wave 4 or 5 changes the association between cooperation and the nurse visit from positive to negative, though the resulting effect size is small. The nurse visit is also estimated to increase the negative effect of a different interviewer for these waves.[5]

5.7 Discussion and Conclusion

This study took a quasi-experimental approach to investigate and quantify the effects of the additional burden of the nurse visit in wave 2 of UKHLS on subsequent wave cooperation. The study design took advantage of the random subsampling in year 2 of the data collection period to compare the responses of those who were included in the nurse visit with those who were not. Interactions between inclusion in the nurse visit, health status and survey reluctance, and the effects of these on subsequent cooperation, were also investigated.

Contrary to reports in the previous literature, the results of this analysis suggest that inclusion in the nurse visit sub-sample had a negative effect on the likelihood

5 It could be noted that the p-values for the change in interviewer interaction terms would not remain statistically significant at the 5% level when corrected for multiple testing ($p = .008$ for wave 4 and $p = .005$ for wave 5).

Table 5.4 Extended logistic regression models of the log-odds of cooperating (with a full interview) including interaction terms at waves 3–7, given eligibility[a)]

	Wave 3	Wave 4	Wave 5	Wave 6	Wave 7
Nurse visit	**−0.41***	**0.38***	**0.40***	0.03	0.02
	(0.21)	(0.19)	(0.19)	(0.14)	(0.21)
Interaction terms					
Nurse visit * Long-standing illness	0.02	0.17	−0.02	0.18	0.13
	(0.16)	(0.16)	(0.16)	(0.12)	(0.16)
Nurse visit * Very good health	0.24	−0.22	−0.01	0.05	0.07
	(0.21)	(0.19)	(0.20)	(0.15)	(0.20)
Nurse visit * Good health	0.27	**−0.42***	−0.19	−0.25	**−0.45***
	(0.21)	(0.20)	(0.21)	(0.15)	(0.21)
Nurse visit * Fair health	0.10	−0.04	0.16	−0.18	−0.20
	(0.26)	(0.24)	(0.25)	(0.19)	(0.26)
Nurse visit * Poor health	0.43	−0.13	−0.27	−0.35	−0.08
	(0.32)	(0.32)	(0.35)	(0.24)	(0.33)
Nurse visit * 3–6 calls	−0.08	−0.14	−0.17	−0.06	0.17
	(0.15)	(0.14)	(0.15)	(0.11)	(0.14)
Nurse visit * 7+ calls	0.04	−0.13	−0.31	−0.14	−0.10
	(0.20)	(0.20)	(0.21)	(0.15)	(0.22)
Nurse visit * Uncooperative	0.38	0.43	0.20	0.43	−0.51
	(0.39)	(0.44)	(0.47)	(0.37)	(0.53)
Nurse visit * Suspicious	−0.11	−0.51	−0.82	−0.27	−0.10
	(0.32)	(0.35)	(0.42)	(0.26)	(0.36)
Nurse visit * Different interviewer[b)]	−0.04	**−0.38****	**−0.42****	–	0.14
	(0.13)	(0.14)	(0.15)		(0.15)
Nurse visit * Missing interviewer[b)]	−0.98	**−2.55*****	9.20	–	0.17
	(0.97)	(0.72)	(138.26)		(0.17)
AIC	9990.41	9404.95	8587.19	15 042.49	9595.46
Log Likelihood	−4943.21	−4650.48	−4241.59	−7473.24	−4745.73
Deviance	9886.41	9300.95	8483.19	14 946.49	9491.46
N	12 538	12 369	12 269	12 188	12 065

$^{***}p < .001$, $^{**}p < .01$, $^{*}p < .05$.

a) Variables also included in the models: intercept, sex, age, marital status, employment status, highest qualification, ethnicity (white/non-white), number of adults, children, housing tenure, Government Office Region (London or elsewhere), urban/rural. Full results available in the online supplementary material.

b) The different/missing interviewer indicators were not included in wave 6 due to a change in fieldwork agency.

of cooperation in wave 3 of UKHLS. However, an overall effect of the nurse visit on cooperation was not seen in subsequent waves 4–7. One possible explanation is that respondents who decided the nurse visit presented too high a burden for their continued participation went on to leave the survey at the next available opportunity. Another explanation concerns the timing of the additional nurse visit. Lugtig (2014) discusses different patterns of attrition and advances that earlier waves, in particular waves 2 and 3 may be key points where those individuals who do not feel invested in the survey are more likely to drop out. In this case, the nurse visit may have given these respondents an opportunity to reconsider their participation at an earlier stage than they would have otherwise.

In the context of the survey literature, our finding of a lower response propensity in the following wave for those included in the additional nurse visit shows some agreement with a related study. Results from Lynn (2013) show a reduced response in wave 2 of the UKHLS Innovation Panel for those respondents who were subject to a more intensive mixed-mode approach by the survey organisation when making contact in wave 1. Similar to our results, the negative effect was no longer present by wave 4. This may indicate a similar mechanism of less-invested respondents dropping out of a longitudinal survey more quickly in response to a one-off increase in burden, followed by a subsequent stabilisation of attrition rates.

We extended our analysis to investigate how groups with certain characteristics may be influenced in a different way by the addition of biological data collection. The lack of significant interaction terms between the health status variables and the nurse visit is surprising given that it might be expected that those with poorer health would either perceive a higher burden for the nurse visit, leading to a negative effect, or appreciate the increased emphasis on health, leading to a positive effect. This is also the case for the interactions with the indicators of reluctance that show relatively strong negative relationships with cooperation but do not seem to be exacerbated by the extra burden of the nurse visit. However, there is some interaction between a change in interviewer and cooperation in waves 4 and 5. It could be that a selection of respondents in the nurse visit sample reluctantly remained in the survey for wave 3 but then either the assignment of a new interviewer, or reasons leading to this such as a house move, resulted in subsequent refusal.

5.8 Implications for Survey Researchers

There are some considerations for researchers planning the inclusion of biological data collection in longitudinal surveys following from the results presented here. We find a short-term negative effect on cooperation in the wave directly after the additional nurse visit in UKHLS. Whilst the size of the effect may be considered relatively small, the implications of this potential reduction in response would need to be weighed up against the benefits of the health data obtained, in particular the cumulative effect on sample representativeness. Fortunately, we did not find evidence of a longer-term impact on cooperation in the subsequent waves.

In order to mitigate the potential negative effects on response associated with the additional burden of biological data collection, measures could be put in place as part of the survey design, such as increased incentives or additional contact attempts. Our results suggest that such measures would only need to be implemented in the next wave and not throughout the life of the survey, though the potential effects of this would needed to be tested further. Additional attempts could also be made to retain interviewers in subsequent waves where possible.

Another approach would be to conduct further study into how individuals and households with certain characteristics respond to the introduction of biological data collection. Here we tested for particular effects related to health status and previous reluctance but found no evidence of differential response in subsequent survey waves for those included in the nurse visit. There may be other mechanisms by which the extra burden informs the decision to participate in future for particular sub-groups and these could be successfully addressed through a targeted approach (Lynn 2017).

A final consideration is the nature of the biological data collection itself and how objective burden could be decreased for respondents to reduce future non-response. There are currently many different designs for biological data collection implemented by survey researchers (McFall et al. 2012). Some designs are considered less burdensome than others (e.g. training interviewers to collect samples as part of the interview visit), but there is a potential trade-off with data quality as well as the range of possible measures that can be collected.

References

Apodaca, R., Lea, S., and Edwards, B. (1998). The effect of longitudinal burden on survey participation. Annual conference of the American Association for Public Opinion Research, St. Louis, MO.

Benzeval, M., Kumari, M., and Jones, A.M. (2016). How do biomarkers and genetics contribute to understanding society? *Health Economics* 25 (10): 1219–1222.

Bogen, K. (1996). The effect of questionnaire length on response rates: a review of the literature. In: *JSM Proceedings, Survey Research Methods Section*, 1020–1025. Alexandria, VA: American Statistical Association.

Bradburn, N. (1978). Respondent Burden. In: *JSM Proceedings, Survey Research Methods Section*, 35–40. Alexandria, VA: American Statistical Association.

Branden, L., Gritz, R.M., and Pergamit, M.R. (1995). *The Effect of Interview Length on Attrition in the National Longitudinal Survey of Youth.* US Department of Labor, Bureau of Labor Statistics.

Buck, N., Ermisch, J.F., and Jenkins, S.P. (1995). *Choosing a Longitudinal Survey Design: The Issues.* ESRC Research Centre on Micro-Social Change, University of Essex https://www.iser.essex.ac.uk/files/occasional_papers/pdf/op96-1.pdf (Accessed 4 November 2017).

Burchell, B. and Marsh, C. (1992). The effect of questionnaire length on survey response. *Quality and Quantity* 26 (3): 233–244.

Campanelli, P. and O'Muircheartaigh, C. (1999). Interviewers, interviewer continuity, and panel survey non-response. *Quality & Quantity* 33 (1): 59–76.

Cernat, A., Sakshaug, J. W., Chandola, T., Nazroo, J., & Shlomo, N. (2020). Nurse effects on non-response in survey-based biomeasures. *International Journal of Social Research Methodology*, 0(0), 1–14. https://doi.org/10.1080/13645579.2020.1832737.

Chandola, T. and Zhang, N. (2018). Re-employment, job quality, health and allostatic load biomarkers: Prospective evidence from the UK Household Longitudinal Study. *International Journal of Epidemiology* 47 (1): 47–57. https://doi.org/10.1093/ije/dyx150.

Champion, D.J. and Sear, A.M. (1969). Questionnaire response rate: a methodological analysis. *Social Forces* 47 (3): 335–339.

Crawford, S.D., Couper, M.P., and Lamias, M.J. (2001). Web surveys: perceptions of burden. *Social Science Computer Review* 19 (2): 146–162.

Galesic, M. (2006). Dropouts on the web: effects of interest and burden experienced during an online survey. *Journal of Official Statistics* 22 (2): 313–328.

Groves, R.M. (2006). Non-response rates and non-response bias in household surveys. *Public Opinion Quarterly* 70 (5): 646–675.

Groves, R.M. and Couper, M.P. (1998). *Nonresponse in Household Interview Surveys.* Chicester: Wiley.

Groves, R.M., Singer, E., and Corning, A. (2000). Leverage-saliency theory of survey participation: description and an illustration. *The Public Opinion Quarterly* 64 (3): 299–308.

Groves, R.M., Presser, S., and Dipko, S. (2004). The role of topic interest in survey participation decisions. *Public Opinion Quarterly* 68 (1): 2–31.

Hart, T.C., Rennison, C.M., and Gibson, C. (2005). Revisiting respondent "fatigue Bias" in the National Crime Victimization Survey. *Journal of Quantitative Criminology* 21 (3): 345–363.

Heberlein, T.A. and Baumgartner, R. (1978). Factors affecting response rates to mailed questionnaires: a quantitative analysis of the published literature. *American Sociological Review* 43 (4): 447–462.

Hill, D.H. and Willis, R.J. (2001). Reducing panel attrition: a search for effective policy instruments. *The Journal of Human Resources* 36 (3): 416–438.

Hughes, A. and Kumari, M. (2017). Unemployment, underweight, and obesity: Findings from Understanding Society (UKHLS). *Preventive Medicine* 97: 19–25. https://doi.org/10.1016/j.ypmed.2016.12.045.

Khare, M., Mohadjer, L.K., Ezzati-Rice, T.M. et al. (1994). An evaluation of non-response bias in NHANES III (1988–91). In: *JSM Proceedings, Survey Research Methods Section*, 949–954. Alexandria, VA: American Statistical Association.

Lacey, R.E., McMunn, A., and Webb, E.A. (2018). Informal caregiving and metabolic markers in the UK Household Longitudinal Study. *Maturitas* 109: 97–103. https://doi.org/10.1016/j.maturitas.2018.01.002.

Lemay, M. (2009). Understanding the mechanism of panel attrition. PhD thesis, University of Maryland, College Park.

Lepkowski, J.M. and Couper, M.P. (2002). Non-response in the second wave of longitudinal household surveys. In: *Survey Non-response* (ed. R.M. Groves), 259–272. Chichester: Wiley.

Loosveldt, G., Billiet, J., and Pickery, J. (2002). Item nonresponse as a predictor of unit nonresponse in a panel survey. *Journal of Official Statistics* 18 (4): 545–557.

Lugtig, P. (2014). Panel attrition: separating stayers, fast Attriters, gradual Attriters, and Lurkers. *Sociological Methods & Research* 43 (4): 699–723.

Lynn, P. (2009a). Methods for longitudinal surveys. In: *Methodology of Longitudinal Surveys* (ed. P. Lynn), 1–19. Chichester: Wiley.

Lynn, P. (2009b). *Sample Design for Understanding Society*. Institute for Social and Economic Research, University of Essex www.understandingsociety.ac.uk/research/publications/working-paper/understanding-society/2009-01.pdf (Accessed 17 January 2017).

Lynn, P. (2013). Alternative sequential mixed-mode designs: effects on attrition rates, attrition Bias, and costs. *Journal of Survey Statistics and Methodology* 1 (2): 183–205.

Lynn, P. (2014). Longer interviews may not affect subsequent survey participation propensity. *Public Opinion Quarterly* 78 (2): 500–509.

Lynn, P. (2017). From standardised to targeted survey procedures for tackling non-response and attrition. *Survey Research Methods* 11 (1): 93–103.

Lynn, P., Buck, N., Burton, J. et al. (2005). A review of methodological research pertinent to longitudinal survey design and data collection. ISER Working Paper Series. https://www.iser.essex.ac.uk/files/iser_working_papers/2005-29.pdf.

Lynn, P., Kaminska, O., and Goldstein, H. (2014). Panel attrition: how important is interviewer continuity? *Journal of Official Statistics* 30 (3): 443–457.

Marmot, M. and Brunner, E. (2005). Cohort profile: the Whitehall II study. *International Journal of Epidemiology* 34 (2): 251–256.

Marmot, M. and Steptoe, A. (2008). Whitehall II and ELSA: integrating epidemiological and psychobiological approaches to the assessment of biological indicators. In: *Biosocial Surveys* (eds. M. Weinstein, J.W. Vaupel and K.W. Wachter), 42–59. Washington, DC: National Academies Press (US).

McFall, S.L. (ed.) (2012). *Understanding Society – UK Household Longitudinal Study: Wave 1–2, 2009–2011, User Manual.* Colchester: University of Essex www.understandingsociety.ac.uk/system/uploads/assets/000/000/004/original/User__manual_Understanding_Society_Waves__1_2.pdf?1359115559 (Accessed 8 August 2017).

McFall, S.L., Booker, C., Burton, J. et al. (2012). *Implementing the Biosocial Component of Understanding Society – Nurse Collection of Biomeasures.* Institute for Social and Economic Research, University of Essex http://www.academia.edu/download/42153400/2012-04.pdf (Accessed 27 July 2017).

McFall, S.L., Petersen, J., Kaminska, O. et al. (2014). *Understanding Society–UK Household Longitudinal Study: Waves 2 and 3 Nurse Health Assessment, 2010–2012, Guide to Nurse Health Assessment.* Colchester: University of Essex.

Mcquillan, G.M., Porter, K.S., Ageli, M. et al. (2003). Consent for genetic research in a general population: the NHANES experience. *Genetics in Medicine* 5 (1): 35–42.

Olsen, R.J. (2005). The problem of respondent attrition: survey methodology is key 25 years of the National Longitudinal Survey – youth cohort. *Monthly Labor Review* 128: 63–70.

Roscoe, A.M., Lang, D., and Sheth, J.N. (1975). Follow-up methods, questionnaire length, and market differences in mail surveys. *Journal of Marketing* 39 (2): 20–27.

Sakshaug, J.W., Couper, M.P., and Ofstedal, M.B. (2010). Characteristics of physical measurement consent in a population-based survey of older adults. *Medical Care* 48 (1): 64–71.

Sakshaug, J.W., Couper, M.P., Ofstedal, M.B. et al. (2012). Linking survey and administrative records: mechanisms of consent. *Sociological Methods & Research* 41 (4): 535–569.

Sakshaug, J.W., Ofstedal, M.B., Guyer, H. et al. (2015). The collection of biospecimens in health surveys. In: *Health Survey Methods* (ed. T.P. Johnson), 383–419. Chichester: Wiley.

Sharp, L.M. and Frankel, J. (1983). Respondent burden: a test of some common assumptions. *Public Opinion Quarterly* 47 (1): 36–53.

Singer, E. and Presser, S. (2007). Privacy, confidentiality, and respondent burden as factors in telephone survey nonresponse. In: *Advances in Telephone Survey Methodology* (eds. J. Lepkowski, C. Tucker, M.J. Brick, et al.), 447–470. Chichester: Wiley.

Singer, E., Van Hoewyk, J., Gebler, N. et al. (1999). The effect of incentives on response rates in interviewer-mediated surveys. *Journal of Official Statistics* 15 (2): 217–230.

Uhrig, S. N. (2008). The nature and causes of attrition in the British Household Panel Survey. ISER Working Paper Series. https://www.iser.essex.ac.uk/research/publications/working-papers/iser/2008-05

University of Essex, Institute for Social and Economic Research (2014). *Understanding Society: Waves 2–3 Nurse Health Assessment, 2010–2012*. Colchester, Essex: UK Data Service https://doi.org/10.5255/UKDA-SN-7251-3.

University of Essex, Institute for Social and Economic Research, NatCen Social Research Kantar Public (2017). *Understanding Society: Waves 1–7, 2009–2016*. Colchester, Essex: UK Data Service https://doi.org/10.5255/UKDA-SN-6614-10.

Vassallo, R., Durrant, G., Smith, P.W. et al. (2015). Interviewer effects on non-response propensity in longitudinal surveys: a multilevel modelling approach. *Journal of the Royal Statistical Society: Series A (Statistics in Society)* 178 (1): 83–99.

Watson, N. and Wooden, M. (2009). Identifying factors affecting longitudinal survey response. In: *Methodology of Longitudinal Surveys* (ed. P. Lynn), 157–182. Chichester: Wiley.

Weinstein, M. and Willis, R.J. (2001). Stretching social surveys to include bioindicators: possibilities for the health and retirement study, experience from the Taiwan study of the elderly. In: *Cells and Surveys: Should Biological Measures be Included in Social Science Research* (eds. C.E. Finch, J.W. Vaupel and K. Kinsella), 250–275. Washington, DC: National Academies Press (US).

Weir, D. (2008). Elastic powers: the integration of biomarkers into the health and retirement study. In: *Biosocial Surveys* (eds. M. Weinstein, J.W. Vaupel and K.W. Wachter), 78–95. Washington, DC: National Academies Press (US).

Weir, D. (2018). Biomarkers. In: *The Palgrave Handbook of Survey Research* (eds. D.L. Vannette and J.A. Krosnick), 573–586. Cham: Springer International Publishing.

West, B.T. and Blom, A.G. (2017). Explaining interviewer effects: a research synthesis. *Journal of Survey Statistics and Methodology* 5 (2): 175–211.

Zabel, J.E. (1998). An analysis of attrition in the panel study of income dynamics and the survey of income and program participation with an application to a model of labor market behavior. *The Journal of Human Resources* 33 (2): 479–506.

6

Understanding Data Linkage Consent in Longitudinal Surveys

Annette Jäckle[1], *Kelsey Beninger*[2], *Jonathan Burton*[1] *and Mick P. Couper*[3]

[1] *Institute for Social and Economic Research, University of Essex, Colchester, UK*
[2] *IFF Research, London, UK*
[3] *Institute for Social Research, University of Michigan, Ann Arbor, MI, USA*

6.1 Introduction

Linking data from longitudinal surveys to administrative records (whether held by government or private entities) is an increasingly attractive option for several reasons. First, the rising costs of survey data collection and declining response rates require researchers to look elsewhere to supplement or replace survey data. Second, the increasing demand for more detailed and timely data raises concerns about respondent burden. Third, the increasing availability of digital data from a variety of sources, along with the development of improved tools for data linkage and protections against disclosure opens up new opportunities. The various ways in which administrative data can be used, whether to replace or enhance surveys, are detailed elsewhere (e.g. Calderwood and Lessof 2009; Gates 2011; Groves and Harris-Kojetin 2017, pp. 33–34). We take as given that the demand for administrative data linkage in surveys is likely to increase.

Administrative data linkage is not without challenges. A key one is that of obtaining informed consent for respondents to link their survey data to administrative records. While some statistical agencies (e.g. Statistics Canada; US Census Bureau; more recently, with the passage of the Digital Economy Act, the UK Office for National Statistics) are taking the view that explicit consent does not need to be solicited for such linkages, such an option is generally not available to non-government research organisations. In fact, the European General Data Protection Regulation, introduced in 2018, focuses on the need to obtain explicit informed consent. Further, regardless of the legal framework for administrative data linkage, ethical issues remain (Lessof 2009). Even if explicit consent is not required, the intended usage of the data should be transparent:

Advances in Longitudinal Survey Methodology, First Edition. Edited by Peter Lynn.

Companion website: www.wiley.com/go/lynn/advancesinlongitudinalsurvey

survey respondents should be aware that taking part in the survey implies consent for their data to be linked. This means that addressing the challenges of gaining consent, and ensuring that the consent is informed, remains a focus for methodological research.

Failure to obtain consent from respondents has several potential consequences. First, the possibility of non-consent bias may arise if those who consent are different from those who do not consent. Second, even if bias is not present, low consent rates may mean smaller effective samples for those using the administrative data. Third, if administrative data are used to replace survey questions, non-consenters may need to be asked additional questions, increasing respondent burden. Finally, the act of requesting consent to administrative data may be viewed as intrusive or threatening to respondents, potentially leading to panel attrition.

Much of the work on informed consent for administrative data linkages has focused on exploring correlates of consent at both the respondent and interviewer level and examining non-consent bias (e.g. Al Baghal 2016; Fulton 2012; Korbmacher and Schroeder 2013; Knies and Burton 2014; Mostafa 2014; Mostafa and Wiggins 2018; Sakshaug et al. 2012; Sakshaug et al. 2013; Sala et al. 2012). Most of these studies involve secondary analysis of existing data, often focusing on a single survey (Knies and Burton 2014, is an exception). These studies find considerable variation between respondents and between interviewers in consent rates, but the findings with regard to the effect of specific characteristics of respondents and interviewers on consent are inconsistent across studies. Peycheva et al. (2021, Chapter 7 in this book) find that the determinants of consent are inconsistent even within a study, and instead vary between data linkage domains.

Experimental studies examining how the consent question is worded (Das and Couper 2014; Edwards and Biddle 2021; Pascale 2011; Sakshaug et al. 2013), or whether it is asked in an earlier or later wave of a panel survey (Eisnecker and Kroh 2017; Sala et al. 2014) have generally found few consistent effects. Asking for consent after a module of questions related to the content of the data to be linked increased consent compared to asking at the end of the questionnaire (Sala et al. 2014). Wording the consent request to emphasise that not linking will reduce the value of the respondent's survey data (loss framing) increased consent compared to gain framing, which emphasises the value of linkage (Kreuter et al. 2016).

There is a relative paucity of research exploring the reasons for non-consent and identifying strategies for addressing these reasons. Similarly, there is little work focusing on how informed respondents are – whether objectively or subjectively measured – about their decision to consent or not (for exceptions, see Das and Couper 2014; Edwards and Biddle 2021; Thornby et al. 2018).

There is some evidence that consent rates are declining over time in longitudinal studies, which raises further concerns about potential bias. For example, among persons newly eligible to be asked consent to Medicare linkage in the US

Panel Study of Income Dynamics (PSID), consent rates have declined over time, from 60% in 2005, to 41% in 2007, 48% in 2009, 29% in 2011, 36% in 2013, and 32% in 2015 (Freedman et al. 2014; Katherine McGonagle, personal communication). Similarly, in the US Health and Retirement Study (HRS), 67% of those interviewed for the first time face-to-face in 2004 consented to Social Security Administration linkage, while for those interviewed for the first time in 2010 (new cohorts are introduced every six years), the corresponding number was 62%.

There is also evidence from longitudinal surveys that consent varies within respondents over time. It is common practice to re-ask consent at later waves of those who did not consent, but some surveys also repeat the request of all respondents. Mostafa and Wiggins (2018) found that 76% of respondents consented to linkage to children's health records in each of three waves of the UK Millennium Cohort Study (MCS), while only 0.5% did not consent in any of the three waves; the balance (24%) consented in one or two waves. Weir et al. (2014) report that in the HRS, 42% of respondents who had not given consent to linkage with Medicare data gave consent the second time they were asked. Similarly, the PSID, which predominantly uses the telephone mode of data collection, has sought oral consent for those eligible for Medicare (primarily persons 65 and older) since 2005 (Freedman et al. 2014). Consenting respondents were asked to provide their Medicare number. In 2005, 60% of eligible persons consented and provided a valid number. Those who did not consent or did not provide a valid number were asked again in subsequent waves. Through 2015, a further 16% consented, yielding an overall consent rate of 76% among those eligible in 2005.

As longitudinal surveys continue to respond to pressures to increase efficiency, and as new survey modes are developed, the use of mixed-mode data collection is increasing. This presents additional challenges for consent. Evidence from a pilot study for the Next Steps cohort study showed that the average consent rate across nine different requests was highest in face-to-face interviews (78%), lower in telephone (71%), and lowest in web (61%, Calderwood 2016; Thornby et al. 2018). Generally not much is known about why consent rates are lower in web, or how best to obtain informed consent from web respondents.

There is considerable variation in how consent is requested or obtained in longitudinal surveys. Knies and Burton (2014) document some of that variation in UK health surveys. Some surveys require a signature to document consent, or provision of an identification number (e.g. Social Security Number) to confirm consent and facilitate linkage, despite evidence that this may increase respondent concerns about disclosure risk (see Singer 1978, 2003) and lower consent rates (Dahlhamer and Cox 2007; Bates 2005). Sometimes the consent is for retrospective data linkage (e.g. data existing at the time of the interview); other times it is prospective (e.g. administrative data generated in the future). There is also variation in whether respondents are asked to confirm their consent and, if so, at what interval (asked just once or repeated requests); whether consent is opt-out (increasingly used by government agencies with authority to link survey and administrative data) or opt-in (often required for academic or social research); where in the survey consent is asked (at the end or in the topical section); how the consent request is framed or worded; and how to ask multiple consent questions. This variation may account for some of the inconsistent findings in the literature, but also points to a need for more research on consent to administrative record linkage, especially in the context of longitudinal surveys where multiple consent requests (both within and between waves) are common, and mixed-mode data collection is increasingly the norm.

Given this background, the focus of this chapter is on understanding the process by which respondents decide whether to give consent. We use both quantitative analyses of existing data from *Understanding Society* and qualitative work to explore the consent process. The quantitative analyses document the extent to which respondents make consistent decisions, and the extent to which the mode of data collection affects this decision, by addressing the following research questions:

- How consistent are respondents about giving consent to data linkage between topics and over time?
- Does consistency over time vary between domains?
- What is the effect of the survey mode on consent? Are differences in consent rates due to selection effects or does mode affect the consent decision?

To understand the reasons for inconsistencies in consent decisions and why the mode of data collection has an effect, the in-depth qualitative interviews address the following research questions:

- How do participants interpret consent questions?
- What do participants think are the implications of giving consent to linkage?
- What influences the participant's decision for whether or not to give consent?
- How does the survey mode influence the decision to consent?
- Why do participants change their consent decision over time?

6.2 Quantitative Research: Consistency of Consent and Effect of Mode of Data Collection

6.2.1 Data and Methods

To examine the consistency of consent over time and across topics, we use the main sample of *Understanding Society* (University of Essex 2017a). Over the course of *Understanding Society*, respondents were asked in the computer assisted personal interviews (CAPI) for consent to link administrative data in multiple domains to their survey responses. Table 6.1 shows the annual waves of data collection in which consent to link was asked of adults (aged 16+) in the individual interviews, the unweighted[1] consent rate, and the number of respondents who were asked for consent at each request. This excludes cases where consents were asked of the responsible adult on behalf of children, and consent to link to energy consumption data that was asked in the household questionnaire. For more information about data linkage on *Understanding Society*, see www.understandingsociety.ac.uk/about/data-linkage. For details on the wording of the consent questions,[2] and the participant materials used,[3] refer to the study website.

In general, at the first time of asking for consent to link to data on a specific domain, all eligible adults were asked. Eligibility for linkage to education data was based on age and country in which the person went to school; for the other linkage requests all adults were eligible. In most subsequent requests for the same domain, only those who had not given consent previously were asked. This included those who had been asked but did not consent, new entrants to the study, and those who had turned 16 and become eligible for the adult interviews since the previous request. At wave 4, however, those who had previously consented to education and health data linkages were asked to confirm their consent during the interview.

For the first seven waves of *Understanding Society*, the primary mode of data collection was face-to-face, using CAPI. At wave 7, adults in households where no one had responded at wave 6 were invited to participate online. Those who did not complete their interview online after the first few weeks were issued to CAPI interviewers. This design is referred to as a 'web-first' design (Jäckle et al. 2015). At wave 8, the group of sample members invited to participate online increased, and included a substantial proportion of those in households that had responded in the previous wave. Altogether, around 40% of households were issued 'web-first'. Consent questions were asked in both the online and CAPI interviews. For wave 8, we currently have only data from the first year of the two-year data collection cycle. Therefore, only some of the analyses include wave 8 data, as documented below.

To examine the effect of survey mode on consent, we use the ninth wave of the *Understanding Society* Innovation Panel (IP9) (University of Essex 2017b). Since wave 5 of the Innovation Panel (IP5) we have implemented a mixed-mode design in which two-thirds of the sample was issued to web first, with non-respondents

1 All analyses in this chapter are unweighted since the analysis of consent consistency aims to describe the behaviours of *Understanding Society* respondents rather than drawing inference about population level behaviours, and the analysis of mode effects relies on the randomised mode treatments for inference.

2 www.understandingsociety.ac.uk/documentation/mainstage/questionnaires

3 www.understandingsociety.ac.uk/documentation/mainstage/fieldwork-documents

Table 6.1 Consent questions in Understanding Society.

Domain	Data-holder	Wave and eligible sample members	Consent rate	Observations
State benefits	Department for Work and Pensions (DWP)	4: all	63.9%	41 988
		7: all without valid consent	68.7%[a]	19 683
Employment/ self-employment	Her Majesty's Revenue and Customs (HMRC)	5: all	62.7%	39 905
		8: all without valid consent	45.3%[b]	12 075
Vehicles	Driver and Vehicle Licensing Agency (DVLA)	5: all with driving licence, access to car/van, has at least one vehicle registered in the UK	79.7%	27 382
Education (National Pupil Database, Early Years Census)	Department for Education (DfE)	1: adults born after 1981	77.8%	6091
		4: adults born after 1983 who are new entrants or have not given consent, adults who had previously consented (confirm);	69.9%	5790
		5: 16-year-olds	52.1%	652
		6: 16-year-olds	72.5%	473
		7: all without valid consent	83.6%	7173
Education – Individualised Learner Records	Skills Funding Agency/ Department for Business, Innovations, Skills	7: 16-year-olds, OR (born after 1978 AND new entrant since Wave 4/no consent given)	84.7%	8551
Higher Education	Higher Education Statistics Agency (HESA)	5: all who completed higher education in the UK in 1995 or later	69.7%	864
NHS Data (Hospital Episode Statistics)	National Health Service (NHS)	1: all	68.1%	47 378
		4: all (previous consenters asked to confirm, previous non-consenters re-asked)	50.6%	22 454
		5: 16-year-olds	49.4%	652
		6: 16-year-olds	65.4%	480
NHS Central Register	National Health Service (NHS)	1: all	67.2%	47 356
		4: all (previous consenters asked to confirm, previous non-consenters re-asked)	50.6%	22 454
		5: 16-year-olds	49.4%	652
		6: 16-year-olds	65.4%	480

a) Due to an error with the feed-forward data, which controlled routing in the questionnaire, for a three-month period, all issued sample members were eligible to be asked for consent to the DWP linkage at wave 7, rather than just those who had not previously consented. These extra cases are included in the analyses.

b) Wave 8 data are from year 1 of wave 8 only; year 2 data are not available at the time of writing.

followed up by CAPI interviewers. The remaining one-third of sample members were issued directly to interviewers (CAPI-first). Towards the end of the fieldwork period, during the re-issue stage, non-respondents in the CAPI-first group were invited to complete their interview online. At the end of fieldwork, all non-respondents were eligible for a telephone 'mop-up', at which point interviewers would attempt to conduct a CATI interview. The initial random allocation was made at IP5, and sample members retained their mode of issue at subsequent waves.

At IP9, all adult respondents were asked for their consent to link their survey responses to administrative data that the Financial Conduct Authority (FCA) compiles from Credit Rating Agencies (see Jäckle et al. 2020). The FCA consent question was asked on the CAPI and online versions of the questionnaire, but not on the telephone.

All significance tests account for the clustering and stratification of the sample. The wording of all consent questions used in the analysis can be found in the online supplementary materials for this chapter.

6.2.2 Results

6.2.2.1 How Consistent Are Respondents about Giving Consent to Data Linkage between Topics?

This analysis uses information on all the adult-level consent requests on the main survey. Across all domains of consent and all requests for consent we have a consent rate of 66.1%. Looking at the consent rate for the first time each domain was asked, we see there is variation across domains from 84.7% for the Individualised Learner Record (education) to 62.7% for Her Majesty's Revenue and Customs (HMRC) tax records (Table 6.1).

To examine the consistency of consenting, we calculated the proportion of consent questions asked of the respondent to which they gave consent. The number of consent requests made to respondents depends on their eligibility, which varies by domain and by the number of waves a respondent has participated in. Figure 6.1 shows a cumulative distribution of this indicator: 15.0% of respondents never gave consent to linking their data, whilst almost one-half (48.0%) gave consent each time they were asked. Just over one in three respondents sometimes gave consent and sometimes withheld their consent.

The consistency of consenting documented in Figure 6.1 may be confounded by the number of consent questions asked of each respondent. It may be, for example, that those who consented to none of the requests were in fact only asked one consent question. Table 6.2 however suggests that this is not the case. The average number of consents asked of any respondent is 5.9, with the maximum being 15. However, only 4.8% of respondents were asked for consent only once, whilst

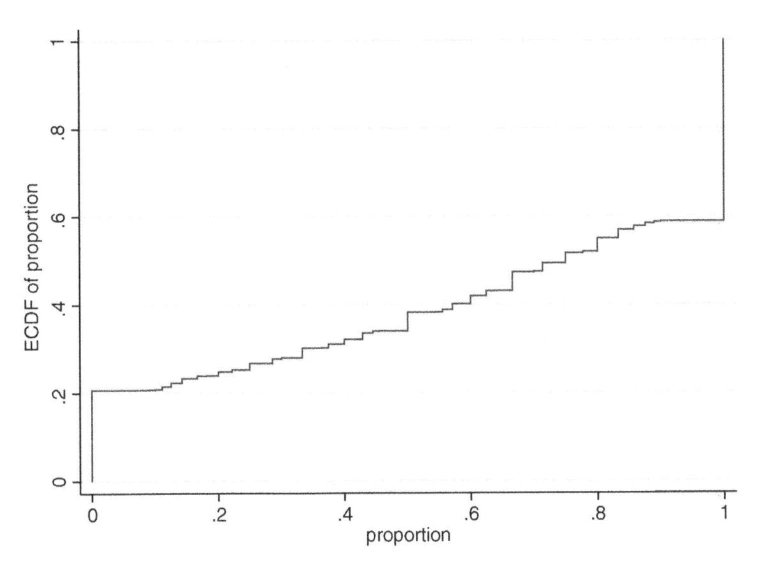

Figure 6.1 Cumulative distribution of the proportion of consent requests granted.

Table 6.2 Average consent rate by the number of consent requests.

Number of consent requests	Mean consent rate (%)	Std. Err.	95% Confidence Int.		n
1	56.7	0.008	55.1	58.3	3689
2	58.6	0.004	57.8	59.3	16 039
3	58.2	0.004	57.4	59.1	10 840
4	66.1	0.005	65.2	67.0	7953
5	79.0	0.003	78.5	79.6	11 883
6	65.8	0.003	65.2	66.5	8032
7+	53.4	0.002	52.9	53.8	18 831

one-fifth of respondents were asked twice (20.8%). Table 6.2 shows the average consent rate by the number of consent requests. The consent rate is highest for those respondents who have been asked 4–6 times. The rate is lowest for those who have had 7 or more requests; this may reflect the fact that those who do not consent to linkage to a particular domain of administrative data are re-asked the next wave this is carried.

To formally test how much of the variation in consent rates is at the respondent level, we ran a null multi-level model of the probability of giving consent, with interviewers and respondents as the two levels, and accounting for the clustered sample design. The data structure is such that there are 298 346 consent questions, nested in 77 267 respondents, nested in 1421 interviewers. The intra-class correlations derived from the null model confirm that there is large variation between respondents in the probability of giving consent (ICC = 0.595), and also some variation between interviewers (ICC = 0.156).

6.2.2.2 How Consistent Are Respondents about Giving Consent to Data Linkage over Time?

We examine consistency of consent over time using measures that examine whether people revoke consent after they have given it, whether they consent to any other request after they have revoked a specific consent, and whether they make the same consent choice at consecutive requests.

First, we examine whether people who give consent later revoke that consent. Periodically, those who have given consent are sent a letter to remind them of their consent, and to give them a chance to revoke that consent by completing a form and sending it back. In addition, sample members can withdraw their consent at any time by writing to the managers of the study. When consent is revoked, no further linkage is carried out for that sample member. If data have already been linked, and the linked data are made available to researchers, the linked data for that individual are not removed. Revoking consent is, however, rare: overall, only 3592 consents have been revoked (1.8% of consents).

Second, we examine whether people who revoke consent at one point in time, then give consent to a different data linkage request at a later time. Surprisingly, we find that over three-quarters of those who revoke consent will agree to consent the next time they are asked (78.6%, $n = 3483$). Note that those who revoke consent to a particular domain are not re-asked for consent to the same domain the next time this is asked.

Third, we look at whether respondents who do not give consent at one request will then give consent the next time they are asked *any* consent question. Less than one-third of those who do not consent at one request will then grant consent the next time they are asked any consent question (26.9%, $n = 84\,298$). This compares to 85.5% of those who do give consent to one request, who then give consent at the next request ($n = 156\,164$).

Fourth, we examine whether respondents who did not give consent at one request, then give consent the next time that *same* consent question is asked. We restrict the analysis to those consent requests that have been asked more than once (DWP, Education, National Health Service [NHS] Data, NHS Register). Usually, only respondents who did not give consent the first time are asked again. There are, however, two exceptions. First, everyone who was asked NHS consent in wave 1 was re-asked in wave 4. Second, for part of the first year of wave 7, everyone was asked for consent to education and DWP linkage again, although only previous non-consenters should have been asked. This was due to an error in the sample data that was issued for the July–September samples in the first year of wave 7. Overall, we find that just under half of those who did not consent at one request do give consent when they are asked the same request in a later interview (46.6%, $n = 31\,802$). This compares to three-quarters of those who have previously given consent consenting again (72.2%, $n = 6412$).

6.2.2.3 Does Consistency over Time Vary between Domains?

To examine whether the consistency of consent decisions over time varies between domains of data linkage, we again use those consent requests which have been asked of everyone more than once: DWP (waves 4 and 7), NHS Register (waves 1 and 4), NHS Health (waves 1 and 4), and Education (waves 1, 4, and 7).[4] Table 6.3 shows that consistency does vary between data domains. Among respondents who did not consent previously there are significant differences in the consent rates at $t + 1$ between domains, ranging from around 39% for health administrative data, to 79.4% for education data at wave 7 (Pearson design-based $F(2.45, 8147.28) = 415.25, p < .001$). Among respondents who did previously consent, there is also variation between domains, although to a lesser extent, ranging from around 67% for health-related data to 88.5% for education data at wave 7 (Pearson design-based $F(2.46, 3379.85) = 82.83, p < .001$). Note that only younger adults are eligible to be asked for consent to link to education data, whereas all responding adults are eligible to be re-asked for consent to linkage to the other three domains.

Table 6.4 shows the proportion of consents that were revoked by domain of administrative data. Generally, the revocation rate is low, suggesting that once consent has been given, respondents are unlikely to revoke it in the future.

Table 6.3 Consent given at $t + 1$ by consent status at time t.

Domain	Consent not given	*n*	Consent given	*n*
Education (W4)	54.3%	805	75.3%	733
Education (W7)	79.4%	1173	88.5%	917
DWP	64.3%	8209	82.2%	2873
NHS Register	39.5%	11 181	67.4%	5009
NHS Data	39.3%	10 962	66.8%	5246

4 There are also a small number of respondents who were first interviewed as adults at age 16 in wave 5 and so were eligible to be asked the education consent question at waves 5 and 7, and some first interviewed as adults in wave 6 who were asked at waves 6 and 7.

Table 6.4 Proportion of consents revoked by domain.

Domain	Revoke	*n* of consents revoked
NHS data	2.5%	1124
NHS register	2.5%	1107
Education	1.6%	251
DWP	1.9%	766
DVLA	0.7%	155
HESA	0.8%	5
HMRC	0.7%	183
ILR	0.0%	1

Taken together, this suggests that the decision to consent varies across domains, but also within participants. Those who consent are more likely to consent again, but are not certain to do so – and those who do not consent are less likely to consent in the future, but, again, are not certain to refuse.

6.2.2.4 What Is the Effect of Survey Mode on Consent?

To look at the effect of survey mode on consent, we use the IP9 data, which employs a mixed-mode (web-CAPI) design for two-thirds of the sample. The focus is on consent to link to information held by the FCA. In IP5, two-thirds of households were randomly allocated to a sequential web-first design, where non-respondents to the web survey were followed up by face-to-face interviewers. The remaining one-third of households was allocated to CAPI-first. In waves 6 to 9, the allocation of sample members to modes remained unchanged. Table 6.5 documents the number of interviews in each mode in IP9, by mode to which respondents were allocated. Proxy respondents (63 cases) are dropped as these are not asked for consent to data linkage. In addition, the small number of interviews that were done by telephone were excluded from the analyses (31 cases). As Table 6.5 shows, among full respondents issued to CAPI-first, 92.1% completed the survey with an interviewer. Among those issued to web-first, 73.4% completed the interview online, while 25.2% completed it with an interviewer.

Although the allocation to modes was carried forward from wave 5, by wave 9 the randomised groups were still comparable. As Table 6.6 shows there is no difference in IP9 between the CAPI-first and the web-first groups in terms of gender, educational qualifications, sample origin, labour market activity, and mean income. There are differences in terms of age, with web-first sample members being on average nearly three years younger than CAPI-first sample members. The

Table 6.5 Innovation panel wave 9, mode of interview by mode of issue (full respondents only).

	CAPI		CATI		Web		Total
Allocation	*N*	%	*N*	%	*N*	%	*N*
CAPI-first	650	92.1	11	1.6	45	6.4	706
Web-first	370	25.2	20	1.4	1078	73.4	1468
Total	1020	46.9	31	1.4	1123	51.7	2174

last columns in Table 6.6 however suggest that age is not related to the probability of giving consent. The analyses below therefore make use of the randomisation to mode of interview, to identify the effect of the mode of data collection on consent.

Examining the characteristics of respondents by mode in which they completed the wave 9 interviews shows that web and face-to-face respondents are not comparable, and some of the characteristics in which they differ are related to the probability of giving consent (Table 6.6). Web respondents had higher educational qualifications, were more likely to be in employment, younger, and had higher incomes than face-to-face respondents. In turn, those with higher qualifications and higher income were more likely to give consent. These results confirm the need to disentangle differences in consent rates that are due to different types of people answering in different modes from the causal effect of the mode on consent.

Overall, there was a 56.9% consent rate to link FCA administrative data to the IP9 survey responses. However, as shown in Table 6.7, the consent rate was 19.0 percentage points higher for those interviewed in-person (67.1%) than those who were surveyed online (48.1%).

The large difference in consent between modes may be related to the positive influence of the interviewer in being able to answer questions the participant may have about consenting, or the tendency – perhaps – for an online respondent to skim over the consent question and answer without giving much thought. That is, the difference may be due to mode differences in how consent is asked. However, the difference may also be due to selection differences into mode.

Table 6.7 documents the results of three different statistical approaches to estimating the effect of mode on consent, by accounting for differences in selection. In the analyses in Table 6.7 those individuals who were issued to a face-to-face interviewer but completed their interview online (45 cases) are excluded. For the instrumental variables method this is necessary for the assumptions underlying the method to hold, as explained below. For consistency we maintain the same sample selection criteria for the other analyses.

The first method is to estimate the effect of the 'intention to treat', that is, the consent rate by the randomised mode of issue rather than by the actual mode of

Table 6.6 Sample composition by issued mode, mode of interview, and consent outcome.

	Issued mode			Mode of interview			Consent		
	CAPI-first	Web-first	*p*-value	CAPI	Web	*p*-value	No	Yes	*p*-value
Female	54.4	53.8		54.9	55.2		46.6	53.4	
Male	45.6	46.2	0.634	45.1	44.8	0.831	38.5	61.5	0.000
Degree	24.2	25.2		21.4	27.9		36.6	63.4	
Other higher qualification	13.6	13.1		12.9	14.0		42.2	57.8	
A-level	21.6	22.8		21.5	23.3		45.3	54.8	
GCSE	24.7	25.0		25.5	24.1		44.6	55.4	
Other qualification	7.1	5.6		7.1	4.9		43.9	56.2	
No qualification	9.0	8.4	0.732	11.6	5.8	0.000	51.1	48.9	0.023
Original sample	49.1	49.2		46.6	51.5		44.2	55.9	
IP4 refreshment	25.3	22.2		22.6	23.8		40.5	59.5	
IP7 refreshment	25.7	28.5	0.394	30.9	24.8	0.063	42.9	57.1	0.515
Self-employed	7.1	8.1		7.4	7.8		46.1	53.9	
Employee	45.0	49.3		41.7	52.9		41.7	58.4	
Unemployed	3.0	3.0		3.2	2.8		48.4	51.6	
Retired	30.5	26.1		31.4	25.0		42.0	58.0	
Other	14.4	13.5	0.248	16.4	11.5	0.000	44.7	55.3	0.718
Age (mean)	51.8	49.0	0.003	52.2	48.3	0.000	49.9	50.2	0.784
Gross monthly income (mean)	1812	1836	0.699	1715	1937	0.000	1642	1968	0.000

Notes: *p*-values from chi-square tests and tests of means, adjusted for sample design.

interview. For respondents who were issued to CAPI first, the consent rate was 72.3%, for those issued to web-first 50.8%, a difference of 21.5 percentage points. The intention to treat analysis suggests that the difference in consent between modes of interview is not due to selection effects. It does however not provide an estimate of the size of the effect of mode on the consent decision.

The second approach is to control for respondent characteristics which explain the selection into mode (Vannieuwenhuyze et al. 2014). We estimate a logit model of the probability of consent, using the mode of interview and controls for differences in sample composition as identified in Table 6.6: sex, academic

Table 6.7 Effect of mode on consent to link to FCA credit rating data.

	CAPI		Web		Difference	
	% consent	*N* (base)	% consent	*N* (base)	% points	*p*-value
Mode of interview	67.1	1020	48.1	1076	19.0	<0.001
Intention to treat (issued mode)	72.3	661	50.8	1466	21.5	<0.001
Covariate adjustment	–	–	–	2073	20.6	<0.001
Instrumental variable	–	–	–	2096	30.0	<0.001

qualifications, age, total personal income, employment status, and sample origin – whether they were part of the original IP sample, the IP4 refreshment sample, or the IP7 refreshment sample. Mode of interview remains a significant predictor of consent. In fact, controlling for differences in sample composition does not explain the difference in consent rates between face-to-face and web respondents: the average marginal effect of the mode coefficient in the logistic model is 20.6 percentage points, similar to the unconditional differences in consent rate by mode of interview. This method assumes that the selection into mode is fully explained by the covariates included. This assumption is unlikely to hold, so the estimate may not fully account for mode selection. As an aside, we find that there is no statistically significant effect of sample origin, so 'time in panel' did not have an effect on consent.

The third method of controlling for selection effects accounts for differences in observed and unobserved characteristics between face-to-face and online respondents by using the random allocation to web-first versus CAPI-first as an instrument for the actual mode of interview. The design of the mixed-mode experiment corresponds to an 'encouragement design with non-compliance to treatment' in clinical trials (Greenland 2000): cases assigned to the treatment (web survey) can revert to the control treatment (face-to-face interview); however, cases in the control treatment cannot revert to the treatment. The assumptions underlying this method require an instrumental variable that is correlated with the actual mode of interview, not correlated with the variables driving the selection into mode, and not correlated with consent. The random allocation to web or face-to-face first provides such an instrument. The mode coefficient estimated from an instrumental variable regression suggests that web respondents were 30.0 percentage points less likely to give consent than respondents interviewed face-to-face.

These results show that the estimated effect of mode on consent is in fact higher when selection is accounted for (30.0 percentage points with the instrumental variable regression) than when selection and measurement are confounded (19.0 percentage points by mode of interview). This reinforces the findings in Table 6.6, suggesting that respondents who complete the survey online are people who are generally more likely to consent. However, completing the survey online rather than with a face-to-face interviewer considerably reduces their probability of giving consent.

The pattern of results reported in Table 6.7 was replicated when we estimated mode differences in consent rates to link to DWP administrative data. Respondents who had not previously given consent to link DWP administrative data to their survey responses were also asked consent to DWP linkage in IP9. This included those who had once previously withheld their consent, and those who had turned 16 since the previous wave. Despite the differences in the eligible sample for the consent request, the results for the DWP linkage consent are similar to those we find for the FCA consent and therefore are not reported.

6.3 Qualitative Research: How Do Respondents Decide Whether to Give Consent to Linkage?

In order to investigate the mechanisms underpinning respondents' decisions to consent, we carried out some qualitative research. Understanding how people make the consent decision helps to explain the reasons why respondents' consent decisions can be inconsistent between topics and over time, and why the mode of interview has such a large effect on the consent decision.

6.3.1 Methods

We conducted in-depth face-to-face interviews with 25 members of the *Understanding Society* Innovation Panel. Interviews lasted 60 minutes and took place in March/April 2017 in participants' homes. Interviews were digitally recorded and participants received a £40 thank-you voucher. To represent a range of views, participants were purposively selected from the Innovation Panel sample to include people who had given consent to data linkage, not given consent, or given consent one year but not the next (or vice versa), and to balance the mode in which they had completed their last interview (face-to-face or online). Selection was further based on location, gender, age group, educational qualifications, difficulty in meeting financial obligations each month, and whether or not participants received state benefits.

Interviewers used a semi-structured discussion guide and a range of stimulus materials and activities. Participants were shown two consent questions (in varying order), that have been used in the *Understanding Society* survey: consent to link economic data held by the DWP, and consent to link health data held by the NHS. Participants were asked for their responses and about their understanding of these questions. In addition, researchers used the projective technique of construction to access thoughts or beliefs that are less conscious or more difficult to verbalise: participants were asked to assess the benefits and risks of data linkage from the perspective of a third person. See Beninger et al. (2017) for the discussion guide, the wording of the two consent questions, and other materials used for the interviews.

To analyse the qualitative interview data, we used an iterative and inductive approach, starting with the views expressed by participants, and included elements of 'grounded theory' analysis (Charmaz 2006; Glaser and Strauss 1967). We used a thematic framework (Ritchie and Lewis 2003) approach whereby findings from interviews are coded and organised thematically within a framework developed from the aims of the research. This method of synthesising enables drawing out both the diversity of opinions and common themes emerging from the interviews. The quotations from participants referenced in the text below are documented in Table 6.8.

6.3.2 Results

6.3.2.1 How Do Participants Interpret Consent Questions?

Informed consent requires understanding four key elements of the linkage process: (i) the direction of information flows, (ii) the extent of information shared, (iii) the parties involved, and (iv) the relevance and purpose of data linkage. There was a lot of variation in how participants understood these elements, such that the initial, unprompted levels of understanding can be described as either accurate, confused, or inaccurate.

Participants whose understanding was accurate understood that the general purpose of data linkage was to obtain better data for research purposes. They also understood the direction of flow of their personal information (Q1, Q2). Only a few participants recognised that consent implied that past and future data about them could be linked (Q3).

Participants who were confused by the consent questions misunderstood key components of the request. Participants generally understood that data would be shared and that Institute for Social and Economic Research (ISER) would have access to their data held by the other organisation. They were, however, confused about whether people within those other organisations would have access to their survey data, whether they would use the data for purposes other than research, and whether third parties would have access to their linked data (Q4, Q5, Q6). A common view was that the actual survey data would be shared with government departments (Q7). Participants who were familiar with the data-holding organisation were more likely to comprehend what type of information that organisation

Table 6.8 Selected quotations from qualitative interviews.

Q1	*'They want access to your health records to help them with the answers that they [have and that] you [are] going to provide them on your health and wellbeing'.* (Female, has consented, face-to-face)
Q2	*'From what I understand, they want to get information from the DWP to cross-reference that against the answers I've given you in the survey…'* (Female, has consented, web)
Q3	*'I get it. Permission is being given for access [to] historic info. This consent would hold going forward, until I wrote a letter saying I wanted to revoke [my] consent'.* (Male, has not consented, face-to-face)
Q4	*'It's odd, because I would have thought it would be the NHS giving you info and not the other way round...It's not clear which individuals within the NHS would have it…if it was my GP I wouldn't mind because I have a good relationship (with him), but if it was others...not sure. I'd be happy to give my consent, but I would want to know...which individuals within the NHS would get it...administrative records...it's a bit vague what that means'.* (Female, has consented, web)
Q5	*'Who is actually getting the information, what is it being used for, is it being kept, is it going to be with my name?'* (Female, changed consent, web)
Q6	*'The first thing you always want to know is who would use it; the other stuff is incidental almost'.* (Male, has consented, face-to-face)
Q7	*'"Administration" sounds a bit name and address, whereas this [referring to the NHS consent question] means anything that is… [in my] file would be handed over'.* (Female, has consented, web)
Q8	*'People will say what is an employment programme, do you mean work or voluntary work'.* (Male, changed consent, web)
Q9	*'I'm not sure they would get an answer; I would love it see but the last time I went to them they (NHS) didn't have anything (his records)'.* (Male, has consented, face-to-face)
Q10	*'They want to share all this info with DWP about my records and when I've been working and not working I suppose'. (*Male, has consented, web)
Q11	*'Giving consent doesn't help me, it doesn't help us personally, you're doing it because you're happy to potentially provide a better result, and potentially a more accurate result'.* (Male, changed consent, web)
Q12	*'It would provide a wider picture. if I was the one doing the research I would want as much information as I could have'.* (Female, changed consent, web)
Q13	*'It gives the government something to think about; if you link it all together you can see, instead looking at statistics and try to number crunch this should work, you are actually getting a feel for how this is actually affecting people and how people are actually coping with it from real life people and not just a computer programme that tells you how people should be coping with it'.* (Female, has consented)
Q14	*'I can't see any benefits. How does it help planning in the community or whatever?'* (Female, changed consent, face-to-face)

(continued)

Table 6.8 (Continued)

Q15	*'The more information you share, the more chance of identity theft, whether you delete the information or not it will still be in the systems at some point'*. (Male, has consented, web)
Q16	*'You have to bear in mind each time you use data you have more chance of it being adversely used, it's a small risk but not an negligible one either because the impact can be massive'*. (Male, changed consent, web)
Q17	*'I think in terms of information getting leaked it's more annoying, I think it's annoying when people call you up because they've got your phone number and they start calling you up about a dishwasher or something or PPI [Payment Protection Insurance]...I would be more concerned about my contact details than the other details'*. (Female, changed consent, face-to-face)
Q18	*'Well, I don't know if he claimed benefits when he perhaps shouldn't have so there's always that risk of being found out…Is it an elaborate scheme to find out who's been swindling the system?'* (Female, changed consent, web)
Q19	*'I think personally if I said go ahead and then all of the sudden the dole office flicked through and my name came up, and then they would say oh will just check on all of her, what is she getting and where is she and I'll be the unlucky bugger that will happen to'*. (Female, changed consent, face-to-face)
Q20	*'Even though research is for the greater good...people aren't interested in that, they're interested in themselves...as soon as you think about risk it's just easier to close it down'*. (Female, changed consent, face-to-face)
Q21	*'My default is not to... I'd definitely, outside of the Understanding Society, say no but I trust them'*. (Male, changed consent, web)
Q22	*'It isn't in my nature; I don't tend to [worry about those kinds of things]'*. (Male, has consented, web)
Q23	*'I think I was just so in the habit, and I still am, of if a site asks you for information you don't give it, unless you trust the source'*. (Female, changed consent, web)
Q24	*'I would say yes because there is nothing I would say that could be used against me. If I was doing something dodgy, working and claiming and avoiding tax and stuff like that I would say "no, I don't want to do it."'* (Male, has consented, face-to-face)
Q25	*'I've already sold myself to the devil'*... (Female, changed consent, face-to-face)
Q26	*'If you're not prepared to give consent don't do Understanding Society in the first place, because it is a lot of personal information already'*. (Male, changed consent, web)
Q27	*'I do regard it as for the greater good of society, if people do research society. I think it is a valid form of research … I'm happy to support it'*. (Male, has consented, face-to-face)
Q28	*'I work in an environment where data protection and security is a big deal. My default position in my industry is why has that information been requested … if it doubt we are told to say no'*. (Male, changed consent, web)

(continued)

Table 6.8 (Continued)

Q29	*'Not sure because it is to with the money I'm getting and there's a lot on the telly about them wanting to cut you down. Is this one of these things when they've got ya'*. (Female, changed consent, face-to-face)
Q30	*'I don't know enough about Understanding Society, all the questions before have been multiple choice and not information as deep as that. Before and because there is so much of it (data theft) going around the world and even if Understanding Society were using it properly and in the right ways there's still so much hacking'*. (Male, has consented, web)
Q31	*'Even when it comes to getting a job, there are certain things that you don't want to share, say if it was me applying for a government job in the future, then like, I'm not saying this is the case but say I had an alcohol problem in the past and then it came up in my records'*. (Female, changed consent, web)
Q32	*'There is something that you want to keep private and if your records are open for a certain amount of time it links to the children. I know it is only for survey purposes and your confidentiality is assured but how many times in the press recently had you seen people losing records. What are they going to do with it, probably nothing but I'm a private person anyway'*. (Male, changed consent, web)
Q33	*'Mates are worried about me doing the survey so they would likely question this request'*. (Female, has not consented, face-to-face)
Q34	*'Because the impact isn't as great is it, if you read something online. It's like, you know, it's like reading an email at work – you might not take that information in. But when you're doing it face to face, and you're sitting there, you know, I'm going to read this, aren't I, because you just asked me to'*. (Female, has not consented, face-to-face)
Q35	*'I guess I am more likely to consent in person because I don't know who is asking [questions] online'*. (Male, changed consent, face-to-face)
Q36	*'They may think online it could go anywhere because we all know once it's online it is for everyone to see. I think face to face they would feel more comfortable'*. (Female, changed consent, web)
Q37	*'It would be the same thing, it doesn't matter; like I said if it was laid out like this and I had a good understanding and it was bold like that'…; 'if it was face to face or on the internet I'd just be like go ahead'*. (Female, changed consent, web)
Q38	*'Doing it online is much easier to say no because usually it is just click a box whereas if someone is talking to you about it they can either be persuasive depending on whether they are supposed to be doing that or alleviate the concerns that just can't be answered by reading on the screen. You never feel you are quite getting enough information to make an informed decision when you are doing it through a screen'*. (Male, changed consent, web)
Q39	*'If (regular interviewer) had done it face to face she usually gives extra detail and sometimes she can tell by the blank look on my face. She does tend to go into more detail sometimes'*. (Female, consented, face-to-face)
Q40	*'Subtle pressures you get in a face to face…you don't want to let that person down'*. (Female, has consented, web)

(*continued*)

Table 6.8 (Continued)

Q41	*'[Decision to consent] would depend on what was happening at the time'.* (Female, changed consent, face-to-face)
Q42	*'As time is moving on you see more breaches of personal security'.* (Male, changed consent, web)
Q43	*'Now with a lot more people hacking companies through personal details and everything developed through technology you need to be more careful of what info you are passing on to people'.* (Male, changed consent, web)
Q44	*'Perhaps I just sort of got to a point and realised you know I've been doing it for however many, 6 years, and everything's been fine thus far and they are using it for research purposes'.* (Female, changed consent, web)

has about them. Others, however, struggled to understand what information the administrative records contain (Q8). Participants were also unclear about the purpose of data linkage. Some thought ISER mistrusted the accuracy or truthfulness of the answers they had given in the survey. People who thought the organisation did not hold any data about them questioned the value of them consenting to the data linkage (Q9).

Participants whose understanding of the consent question was inaccurate tended to think that their survey responses would be shared with government departments for them to use (Q10). Participants who misinterpreted the consent request in this way, feared that data linkage was a way for government departments to monitor them.

Overall, understanding was improved if participants read the information booklet: those who read it were more likely to understand the direction and extent of data flows, and the purpose of data linkage. Understanding was, however, hampered by certain words and phrases (for details, see Beninger et al. 2017), and by the reading habits of some participants. Participants with limited literacy, dyslexia, or of older age said they tended to read small sections of the consent question that seemed important, usually just skimming the first and last sentences, rather than reading the full question text. They described this as habitual behaviour that they had adopted to manage the process of reading.

6.3.2.2 What Do Participants Think Are the Implications of Giving Consent to Linkage?

Overall, few participants understood what happens after they give consent: participants were vaguely aware that information is uploaded to a computer and used for research purposes.

Participants who understood the consent request did not identify any ways in which they would personally benefit from consenting (Q11). They did, however, identify third parties that would benefit. They generally understood that consent would lead to more accurate information about them, and it was thought that this would benefit academics, because the information would be used for research purposes (Q12). Participants also thought that more accurate information would improve government policy planning, which might benefit society more generally in the future (Q13). Some participants who misunderstood the flow of information, however, thought that consent could lead to improvement of services for them personally – for example, in the provision of health care or pensions. Other participants who did not understand the purpose of data linkage saw no potential benefits to anyone (Q14).

Most participants recognised that consenting to data linkage entailed personal risks for them, which could either result from inadequate data protection in the processes of storing, handling, and sharing information, or from misuse of personal data by government departments. The safety of personal information and the risks of identity fraud were prevalent concerns. While participants understood that websites and servers were encrypted, they thought that data are 'never truly safe', and can also be misplaced by accident (Q15, Q16). Stigma, embarrassment, or reputational damage resulting from leaked data revealing sensitive or compromising personal information was also a concern. Some participants believed that sharing their contact details put them at risk of harassment by telemarketers (Q17). The risk that government departments could misuse personal information to 'catch out' individuals for participating in illegal or inappropriate behaviour was also a concern (Q18, Q19). Participants who believed in the possibility of personal risks were less likely to consent (Q20).

6.3.2.3 What Influences the Participant's Decision Whether or Not to Give Consent?

Whether or not participants gave consent was not only determined by whether they had understood what the consent questions were asking. Instead the consent decision was based on a combination of subconscious, rational, social, and environmental factors.

Subconscious factors included personality traits and related habitual behaviours. Participants who self-identified as private, suspicious persons focused on the risks of sharing their personal data (Q21). In contrast, participants who self-identified as more open, trusting persons were dismissive of these risks (Q22). Some participants mentioned habitual behaviours, such as answering 'yes' or 'no' by default in response to consent-type questions (Q23).

Rational factors influencing the consent decision were dominated by the participants' assessment of the risks and benefits of consenting. Participants who consented were more likely to say that they 'have nothing to hide', for example health issues or financial circumstances that would make them feel vulnerable if their administrative data were accessed (Q24). Some were willing to consent because they believed that the survey organisation and the government already had so much information about them, that there was 'nothing they do not already have' (Q25). Similarly, some participants thought that having already agreed to participate in the survey, 'why stop here': their trust in the survey led them to agree to allow researchers to gather the most effective data possible (Q26).

Participants who consented also thought that the benefits of consent outweighed concerns about personal privacy or data protection (Q27). Generally, individuals who work in professions where they are aware of data exchange and data protection were more able to rationally assess the risk of consent. Some dismissed the risks, while others had heightened concerns that undermined their willingness to consent (Q28). Participants who were less likely to consent were worried about the potential personal repercussions of consenting, such as hacking and fraud (Q29, Q30). They were particularly concerned about NHS data if they had health problems that they thought could prevent them from getting jobs in the future (Q31). A few participants were also concerned about potential repercussions of data security breaches for their children or other people close to them (Q32).

Literacy levels and cognitive ability also influenced rational decision making: a few participants were not able to understand the question (as judged by the interviewer or communicated by the participant themselves) and therefore said they were unable to give consent.

Social factors influencing the consent decision included friends and family, as well as perceived norms around data sharing. Some participants said they would consult their spouse before giving consent. Others seriously considered reasons for not consenting, because they thought certain friends or family members would be hesitant (Q33). Participants were also influenced by different perceptions of established norms around data sharing, including the view that 'everyone shares everything already', and that 'in this day and age, you can never be too cautious' about sharing personal information because of the risks of identity theft, fraud, etc.

Environmental factors included trust in the organisation running this particular survey, and more broadly, stories of data security breaches reported in the media. The nature of interactions with the survey organisation, and how long participants had participated in the survey, helped establish trust: the organisation was seen as a reputable organisation that had 'too much riding on the survey to mishandle information'. Trust reduced participants' concerns about the risks of data linkage and supported the decision to consent. In contrast, news stories in the media heightened participants' awareness of phishing scams, fraud, identity theft and data leaks, which, in turn, increased concern about consenting to data linkage.

6.3.2.4 How Does the Survey Mode Influence the Decision to Consent?

We asked participants who had participated online as well as those who had not whether and how responding online would influence their decision to consent. Participants gave several reasons why they would be less willing to give consent in an online survey. The difficulty of reading a lot of content on a computer screen was one prevalent reason: participants said they were less likely to fully read information presented on screen than they would in a face-to-face setting. This tendency to skim-read online text would lead them to overlook key aspects of the consent request (Q34).

Concern about privacy and security of personal information shared online was another key reason why participants would be less likely to consent online. In contrast participants felt higher levels of trust in the security of their data in the face-to-face setting where the participant could see where their data were going as the interviewer recorded their responses (Q35, Q36). Only a minority of participants realised that the mode in which the consent question is asked does not alter the linkage procedures and would therefore not affect their willingness to consent (Q37).

The possibility of asking the interviewer clarification questions and obtaining reassurance was another reason why participants thought they would be more likely to give consent in the face-to-face setting. Although we know from previous research examining the behaviours of interviewers and respondents that respondents only rarely ask any questions about consent requests (Burton et al. 2014), the qualitative interviews indicated that the idea of support was a motivating factor (Q38, Q39). The physical presence of the interviewer also increased the willingness to consent through perceived social pressure to conform (Q40).

6.3.2.5 Why Do Participants Change their Consent Decision over Time?

Participants who have given inconsistent responses to consent questions at different points in time were asked about the reasons why they had changed their mind. These were participants who had for example not consented in past panel interviews, but did say they would give consent in the qualitative interview. The reactions of these participants highlighted the role of subconscious, environmental and social factors in driving decisions to consent. Participants were generally unable to recall whether they had consented in the past. When the interviewer reminded them of their previous responses, some were surprised about their previous decision and unable to rationalise why they had changed their mind. A few participants commented that the change was probably arbitrary and dependent on how they were feeling on the day.

For participants who, on looking back at their past consent response, were able to post-rationalise why they might have changed their consent decision, the reasons were wide-ranging. Some indicated that they might have been more or less concerned about infringements on their privacy in the past. Some thought perhaps their understanding of the purpose of linking their personal data had changed. Others mentioned life events and changes in circumstances that had changed their attitudes towards societal benefits of sharing data (Q41).

Several participants said they were more cautious about giving consent due to the perception that the increased use of technology led to a rise in hacking (Q42, Q43). Others were surprised that they had consented in the past. These participants (who misunderstood the direction of data flows) claimed they would not have consented if they had known that their information would be shared with government departments that could affect their personal finances, such as the DWP and HMRC.

Finally, a further reason expressed for changing consent was trust in the survey, built up over years of taking part, and recognition that sharing information with ISER had not led to infringements on their privacy. Participants claimed that this was reassuring and made them more likely to consent (Q44).

6.4 Discussion

The empirical results presented in this chapter show that many respondents are inconsistent in the consent decisions they make – between topics and over time. Revoking consent, although rare, is not related to whether respondents will agree to later consent requests for other domains. Similarly, a large proportion of non-consenters do give consent – for the same or other domains – when asked again in later interviews. This inconsistency in decisions over time suggests that consent is not generally based on strongly held views about data linkage. This interpretation of the empirical results is supported by findings from the qualitative interviews. These illustrate that many participants do not understand or only partially understand the linkage request and the implications of consenting. The qualitative interviews also highlight that rational consideration of the implications, risks and benefits of consenting to linkage only influence the decision in part. In addition, subconscious, social and environmental factors play an important role in determining the consent decision.

The empirical results also show that the mode of data collection has a large effect on the willingness to consent to data linkage: respondents are much less likely to give consent when completing a survey online than in a face-to-face interview. Our analyses suggest this is clearly a mode effect rather than a selection effect; in fact, respondents who complete the survey online are people who are generally more likely to consent. The qualitative interviews offer some insights into why participants are so much more reluctant to consent in a web survey. Participants seem more concerned about data security when answering survey questions online than in a CAPI interview. Only a few seem to realise that the linkage and data handling procedures are exactly the same, regardless of the mode in which they give consent. They also seem less likely to read the (often lengthy) consent questions and associated information materials and therefore more likely to overlook key pieces of information. The presence of the interviewer also supports willingness to consent, both by providing opportunity to ask clarification questions and obtain reassurance, and through social pressure to conform.

The quantitative and qualitative findings have a number of practical implications and point to needs for additional research.

In terms of practical implications, the finding that decisions about consent do not appear to be strongly held suggests it is worth re-asking consent of those who declined at some earlier wave. Doing so does not appear to be coercive; that is, the practice does not seem to persuade people to agree to something they don't want to do. A related implication is that re-asking consent of those who already consented may increase the number of non-consenters. Instead, providing an opportunity to revoke consent (as is done in *Understanding Society*) seems to give participants sufficient control over their decisions.

Another practical implication is that there is a clear need to explain the linkage process better to survey respondents (see also Thornby et al. 2018). This may be more of an issue in the web mode, where respondents are less likely to read materials and interviewers cannot compensate, by explaining the process to respondents or addressing questions or concerns they may have.

A related implication is that mode matters in the consent to administrative data linkages. This finding is unlikely to lead to a return to face-to-face surveys to maximise consent rates, but interviewer follow-up on non-consenters is one possibility to consider. The mode finding is an important consideration for those studies considering switching from an interviewer-administered mode to a self-administered one for some or all of the panel members. In multi-mode surveys it would be best to reserve consent requests for interviewer-administered instruments.

Our findings also identify a number of research needs related to administrative linkage consent. First, given the differential consent rates by mode, understanding the reasons for this difference and finding ways to overcome the lower consent rates on the web are a priority. One explanation is that the request is subject to social desirability response bias – respondents are agreeing to the consent request more in face-to-face to please the interviewer. An alternative explanation is that the interviewer is taking an active role in clarifying the process, addressing respondent concerns, and persuading the respondent to consent. It may also be that the mere presence of an interviewer conveys the legitimacy of the request in a way that is currently not done online. Understanding the role of the interviewer in the consent process, and studying the interviewer–respondent interaction during the consent process may identify effective strategies used by interviewers that could be duplicated online. Simply providing additional information to online respondents may not be sufficient if they do not read the material provided or click on links to access such information.

As an aside, there may be parallels here to the standardised versus conversational interviewing debate. As Schober and Conrad (1997) and others have found, conversational interviewing has advantages when complex questions are asked. This approach provides interviewers with an opportunity to convey uncertainty and for interviewers to clarify meaning and address concerns. The online version of consent is much like standardised interviewing, whereas (successful) interviewers may be much better at anticipating and addressing respondent concerns. Getting respondents to recognise when they need definitions for complex terms (or clarification of the consent request) online is a challenge.

As already noted, finding ways to encourage respondents to raise questions and get answers before making a consent decision may help increase how informed they are (or feel they are). Addressing the *informed* part of the informed consent process, especially in online survey administration, is a key issue for both research and practice.

Finally, we identified a number of research gaps in our review of prior literature. Our quantitative and qualitative findings suggest that these gaps may become even more important as the number and variety of consent requests increases in longitudinal studies, and as such studies increasingly make use of both interviewer- and self-administered modes of data collection. Understanding how and why people make decisions when multiple consent requests are made, and figuring out the best ways to ask consents for multiple domains is an important next step. Similarly, the observation from the qualitative research that respondents seemed more worried about the security of their data when asked online for their consent – given that the linkage process itself is identical – is intriguing, and again suggests that the context of the request matters. There is much more we need to know about the *process* of administrative data linkage consent in different modes, in addition to the *outcome* of that process.

Acknowledgements

This research was funded by the UK Economic and Social Research Council grant for *Understanding Society* [ES/N00812X/1] and by the Nuffield Foundation [OSP/43279], but the views expressed are those of the authors and not necessarily those of the Foundation.

References

Al Baghal, T. (2016). Obtaining data linkage consent for children: factors influencing outcomes and potential biases. *International Journal of Social Research Methodology* 19 (6): 623–643.

Bates, N.A. (2005). Development and testing of informed consent questions to link survey data with administrative records. In: *Proceedings of the American Statistical Association*, 3786–3793. Miami Beach, FL.

Beninger, K., Digby, A., and MacGregor, J. (2017). *Understanding Society: How People Decide Whether to Give Consent to Link Their Administrative and Survey Data. Understanding Society Working Paper 2017–13.* Colchester: University of Essex.

Burton, J., Sala, E., and Knies, G. (2014). Exploring the role of interviewers in collecting survey respondents' consent to link survey data to administrative records. Presented at the 4th Panel Survey Methods Workshop in Ann Arbor, MI (21 May 2014).

Calderwood, L. (2016). Asking for consent in a mixed mode study – issues and outcomes. Presented at the CLOSER Mixing Modes and Measurement Methods in Longitudinal Studies Workshop in London (2 November 2016).

Calderwood, L. and Lessof, C. (2009). Enhancing longitudinal surveys by linking to administrative data. In: *Methodology of Longitudinal Surveys* (ed. P. Lynn). Chichester: Wiley.

Charmaz, K. (2006). *Constructing Grounded Theory: A Practical Guide though Qualitative Analysis*. London: Sage.

Dahlhamer, J.M. and Cox, C.S. (2007). Respondent consent to link survey data with administrative records: Results from a split-ballot field test with the 2007 National Health Interview Survey. *Proceedings of the Federal Committee on Statistical Methodology Research Conference*, Arlington, VA (5–7 November 2007).

Das, M. and Couper, M.P. (2014). Optimizing opt-out consent for record linkage. *Journal of Official Statistics* 30 (3): 479–497.

Edwards, B. and Biddle, N. (2021). Consent to data linkage: experimental evidence on the impact of data linkage requests and understanding and risk perceptions. In: *Advances in Longitudinal Survey Methodology* (ed. P. Lynn). Chichester: Wiley.

Eisnecker, P.S. and Kroh, M. (2017). The informed consent to record linkage in panel studies: optimal starting wave, consent refusals, and subsequent panel attrition. *Public Opinion Quarterly* 81 (1): 131–143.

Freedman, V.A., McGonagle, K., and Andreski, P. (2014). The Panel Study of Income Dynamics linked Medicare claims data. *PSID Technical Series Paper #14–01*, Ann Arbor, MI: Survey Research Center, University of Michigan.

Fulton, J.A. (2012). Respondent consent to use administrative data. PhD dissertation. University of Maryland.

Gates, G.W. (2011). How uncertainty about privacy and confidentiality is hampering efforts to more effectively use administrative records in producing U.S. National Statistics. *The Journal of Privacy and Confidentiality* 3 (2): 3–40.

Glaser, B.G. and Strauss, A.L. (1967). *The Discovery of Grounded Theory: Strategies for Qualitative Research*. Chicago, IL: Aldine de Gruyter.

Greenland, S. (2000). An introduction to instrumental variables for epidemiologists. *International Journal of Epidemiology* 29 (4): 722–729.

Groves, R.M. and Harris-Kojetin, B. (eds.) (2017). *Innovations in Federal Statistics: Combining Data Sources while Protecting Privacy*. Washington, DC: The National Academies Press. https://doi: 10.17226/24652.

Jäckle, A., Lynn, P., and Burton, J. (2015). Going online with a face-to-face household panel: effects of a mixed mode design on item and unit non-response. *Survey Research Methods* 9 (1): 55–70.

Jäckle, A., Couper, M.P., Gaia, A. et al. (2020). Improving survey measurement of household finances: a review of new technologies and data sources. In: *Advances in Longitudinal Survey Methodology* (ed. P. Lynn). Chichester: Wiley.

Knies, G. and Burton, J. (2014). Analysis of four studies in a comparative framework reveals: health linkage consent rates on British cohort studies higher than on UK household panel surveys. *BMC Medical Research Methodology* 14 (1): 125.

Korbmacher, J.M. and Schroeder, M. (2013). Consent when linking survey data with administrative records: the role of the interviewer. *Survey Research Methods* 7 (2): 115–131.

Kreuter, F., Sakshaug, J.W., and Tourangeau, R. (2016). The framing of the record linkage consent question. *International Journal of Public Opinion Research* 28 (1): 142–152.

Lessof, C. (2009). Ethical issues in longitudinal surveys. In: *Methodology of Longitudinal Surveys* (ed. P. Lynn), 35–54. Chichester: Wiley.

Mostafa, T. (2014). Variation within households in consent to link survey data to administrative records: evidence from the UK millennium cohort study. *International Journal of Social Research Methodology* 19 (3): 355–375.

Mostafa, T. and Wiggins, R.D. (2018). What influences respondents to behave consistently when asked to consent to health record linkage on repeat occasions? *International Journal of Social Research Methodology* 21 (1): 119–134.

Pascale, J. (2011). Requesting consent to link survey data to administrative records: Results from a split-ballot experiment in the Survey of Health Insurance and Program Participation (SHIPP). In: *Study Series: Survey Methodology 2011-03*. Washington, DC: US Census Bureau.

Peycheva, D., Ploubidis, G.B., and Calderwood, L. (2021). Determinants of consent to administrative records linkage in longitudinal surveys: evidence from next steps. In: *Advances in Longitudinal Survey Methodology* (ed. P. Lynn). Chichester: Wiley.

Ritchie, J. and Lewis, J. (2003). *Qualitative Research Practice: A Guide for Social Science Students and Researchers*. London: Sage Publications.

Sakshaug, J.W., Couper, M.P., Ofstedal, M.B. et al. (2012). Linking survey and administrative records: mechanisms of consent. *Sociological Methods & Research* 41 (4): 535–569.

Sakshaug, J.W., Tutz, V., and Kreuter, F. (2013). Placement, wording, and interviewers: identifying correlates of consent to link survey and administrative data. *Survey Research Methods* 7 (2): 133–144.

Sala, E., Burton, J., and Knies, G. (2012). Correlates of obtaining informed consent to data linkage: respondent, interview and interviewer characteristics. *Sociological Methods & Research* 41 (3): 414–439.

Sala, E., Knies, G., and Burton, J. (2014). Propensity to consent to data linkage: experimental evidence on the role of three survey design features in a UK longitudinal panel. *International Journal of Social Research Methodology* 17 (5): 455–473.

Schober, M.F. and Conrad, F.G. (1997). Does conversational interviewing reduce survey measurement error? *Public Opinion Quarterly* 61 (4): 576–602.

Singer, E. (1978). Informed consent: consequences for response rate and response quality in social surveys. *American Sociological Review* 43 (2): 144–162.

Singer, E. (2003). Exploring the meaning of consent: participation in research and beliefs about risks and benefits. *Journal of Official Statistics* 19 (3): 273–286.

Thornby, M., Calderwood, L., Kotecha, M. et al. (2018). Collecting multiple data linkage consents in a mixed-mode survey: evidence from a large-scale longitudinal study in the UK. *Survey Methods: Insights from the Field*: 1–14.

University of Essex, Institute for Social and Economic Research (2017a). *Understanding Society: Waves 1–7, 2009–2016 and Harmonised BHPS: Waves 1–18, 1991–2009*, 9e. UK Data Service. SN: 6614, http://doi.org/10.5255/UKDA-SN-6614-10.

University of Essex, Institute for Social and Economic Research (2017b). *Understanding Society: Innovation Panel, Waves 1–9, 2008–2016.*, 8e. UK Data Service SN: 6849, http://doi.org/10.5255/UKDA-SN-6849-9.

Vannieuwenhuyze, J.T.A., Loosveldt, G., and Molenberghs, G. (2014). Evaluating mode effects in mixed-mode survey data using covariate adjustment models. *Journal of Official Statistics* 30 (1): 1–21.

Weir, D.R., Faul, J.D., and Ofstedal, M.B., (2014). The power of persistence: Repeated consent requests for administrative record linkage and DNA in the Health and Retirement Study. Presented at the 4th Panel Survey Methods Workshop in Ann Arbor, MI (21 May 2014).

7

Determinants of Consent to Administrative Records Linkage in Longitudinal Surveys: Evidence from Next Steps

Darina Peycheva, George Ploubidis and Lisa Calderwood

Centre for Longitudinal Studies, UCL Institute of Education, London, UK

7.1 Introduction

The value of enhancing survey data through linkage to administrative records is increasingly recognised among researchers. Linking survey and administrative data enables rich information from administrative records on a broad range of substantive areas to be combined with survey responses. Administrative data often include detailed information, such as dates of hospital admissions, school test and exam results, or financial information, which would be difficult or burdensome for survey participants to report accurately. Hence, record linkage increases the utility of survey data and the opportunities for research. In longitudinal surveys, administrative record linkage offers particular benefits for research including filling in gaps between waves or before the baseline wave, and enabling continued collection of administrative data after sample members have been lost to follow-up, and thus providing valuable information for non-respondents and non-response adjustments.

To link survey data to administrative records, participants' informed consent is usually required for ethical and legal reasons, and often this is a requirement of administrative data holders. Participants are asked to give their permission to add information held about them in administrative records to their survey data. This permission is entirely voluntary, and of great importance for the quality of the linked data. Participants' refusal to give permission for linkage (i.e. non-consent) leads to reduction in the sample size of the linked survey and administrative data, and more importantly to potential bias resulting from differential patterns of consent that may lead to non-random patterns of missing data.

Consent rates to administrative records linkage vary widely across surveys and administrative data types. Numerous studies have found differences between

Advances in Longitudinal Survey Methodology, First Edition. Edited by Peter Lynn.

Companion website: www.wiley.com/go/lynn/advancesinlongitudinalsurvey

consenting and non-consenting survey participants on key demographic and socioeconomic characteristics, suggesting that age, sex, ethnicity, education, income, are strongly related to the likelihood of consent (Kho et al. 2009; Sakshaug et al. 2012). Recent studies (Korbmacher et al. 2013; Al Baghal 2016; Mostafa 2016; Sakshaug et al. 2017) have shown that survey interviewers and interview features might also affect participants' consent decisions. Furthermore, some findings suggest that predictors of consent may also be predictive of (non-)response in general, and that both non-consent and non-response can be caused by similar factors (Jenkins et al. 2006; Dunn et al. 2004; Watson and Wooden 2009; Mostafa 2016). However, the associations found across studies on consent to data linkage are inconsistent (Kho et al. 2009; Bohensky et al. 2010; da Silva et al. 2012; Knies and Burton 2014; Al Baghal et al. 2014) and respondent's behaviour varies depending on the consent domain (Mostafa 2016). Hence, potential biases may differ according to the consent requested (Jenkins et al. 2006). Nonetheless, non-consent bias has been estimated to be small relative to other sources of bias such as non-response and measurement error (Sakshaug and Kreuter 2012).

Longitudinal surveys have an advantage over cross-sectional surveys, as they are able to use information collected over multiple survey waves to investigate determinants of consent. However, much of the literature, including those using longitudinal data, examine a relatively limited range of potential determinants, primarily guided by associations found with consent in previous studies. Most studies have examined consent in only one or two domains. Published research on consent to criminal records linkage is scarce; moreover, there is, to our knowledge, no UK study assessing the factors impacting criminal records linkage permissions. Additionally, different studies in the literature cover different populations and/or age groups, which is likely to contribute to the observed lack of consistency of findings.

This chapter extends the existing knowledge about administrative records linkage consent and addresses some of the gaps in the literature by exploring determinants of consent in a large-scale longitudinal mixed mode cohort study in the UK. Multiple consents covering health, economics, education, and criminal records were sought from the 25-year-old study members. This enables us to compare the determinants of consent over multiple domains, for a specific generation and age group. Crucially, to identify determinants of consent, we employ a data driven analytical approach, making use of all available data for the cohort, collected throughout the study.

We use data from Next Steps (previously known as Longitudinal Study of Young People in England) to look at consent across the health, education, economic, and crime domains. Unlike health, education and economic records, consent to criminal records linkage has only been asked in a limited number of studies, and no UK study findings on consent have, to the best of our knowledge, been shared with the research community. We aim to shed further light on what determines consent to link these administrative records, and investigate whether the characteristics of consenters differ between data types; as well as how consistent our findings are with the existing literature on consent patterns.

Using a data-driven – as opposed to theory driven – approach and making use of the rich life course information available for the participants, rather than an arbitrary selection of variables or being guided by the literature on consent bias, we exploit the possibility of unknown determinants of consent from various domains. Capitalising on the richness of available information in Next Steps, our results will allow researchers to consider a wider range of characteristics as auxiliary variables in principled approaches to handling missing data, such as multiple imputation, full information maximum likelihood, linear increments or inverse probability weighting to reduce bias due to selection through consent. These approaches operate under the Missing At Random (MAR) assumption, which implies that all important predictors of missingness should be included in the model, or that selection to consent is due to observables (Little and Rubin 2002; Seaman and White 2011; Carpenter and Kenward 2013). Therefore, identifying all determinants of consent will help maximise the plausibility of the MAR assumption in analysis that utilises linked consented data.

The chapter is organised as follows: Section 7.2 looks at the existing literature. Section 7.3 describes the data and methods. Section 7.4 presents the results. Our findings are summarised in Section 7.5 and their implications discussed.

7.2 Literature Review

Most of the existing research on patterns of consent and consent bias comes from medical and epidemiological studies, seeking permission to link medical records, and looking at participants' demographic and socioeconomic characteristics, as well as health conditions.

For example, a review on a number of large-scale epidemiological surveys in the UK, carried out between 1996 and 2002, on asking for permission to link to primary care medical records (Dunn et al. 2004), found very similar patterns in the different studies with respect to associations between age, gender, and symptom under investigation with consent. Males, younger people, and subjects reporting the symptom under investigation were more likely to give consent, and to be over-represented in the linked medical records.

A systematic review of 17 studies, carried out in the period 1985–2007, aimed to determine whether informed consent introduces selection bias in prospective observational studies using data from medical records (Kho et al. 2009). This also showed differences between participants and nonparticipants in terms of age, sex, race, education, income, and health status. However, consent rates varied considerably – from 36.6% to 92.9% – and there was a lack of consistency in the direction and magnitude of associations. The cause of these differences was unclear.

More recently, large-scale social surveys, primarily longitudinal, have begun to seek consent for linking administrative records providing further evidence about the determinants of administrative records consent. Findings from these surveys have also shown inconsistent associations with consent to data linkage so that an identified driver of one consent decision was not a driver for another (Jenkins et al. 2006; Al Baghal et al. 2014; Mostafa 2016).

For example, using data from the British Household Panel Survey to investigate the characteristics influencing consent to linking health and social security records, Sala et al. (2012) found that demographics were only mildly associated with the decision to consent. Men were more likely to consent to benefit data linkage, and non-white respondents were less likely to consent to health and benefits linkage than their white counterparts. Results from the UK Household Longitudinal Study, asking for administrative data linkage consent for health and education records, also pointed out that generally, but not always, ethnic minorities are less likely to consent to link their administrative records (Al Baghal et al. 2014). Ethnicity was found to have a strong impact on health and education records consent in the Millennium Cohort Study (Mostafa 2016), with non-white respondents being less likely to consent. Using data from the Australian Temperament Project to investigate the concordance between official criminal records and self-reports, Forrest et al. (2014) found that respondents who granted access to their criminal records were less likely to be male and to come from a non-English speaking background.

Household type was not found to be associated with consent by Sala et al. (2012), and the authors stated that this was in line with previous findings (Hockley et al. 2007; Tate et al. 2006; Woolf et al. 2000). However, Jenkins et al. (2006), analysing a follow-up to the British Household Panel Survey, found that for benefit records, lone parents have a lower propensity to consent than respondents in a couple, whereas, for employment records, the reverse was found.

The lack of consistent associations with consent goes beyond respondent and household characteristics and includes interviewers' and survey design features. Although respondent and household characteristics are important for the consent decision, Korbmacher and Schroeder (2013) found that in the Survey of Health, Aging and Retirement in Europe (SHARE), a large part of the variation in the data is explained by the interviewers, a finding in line with Mostafa (2016). On the contrary, interviewer demographic characteristics such as gender, age, ethnicity, or education were found to be not related to consent in the Health and Retirement Study (Sakshaug et al. 2012). Sala et al. (2012) also found no relationship between interviewers' attitudes towards persuading reluctant persons to participate in the survey and their likelihood of obtaining consent from respondents; as well as no relationship between interviewer personality traits and respondents' likelihood of consent.

The length of interview was found to be associated with both health and education records consent by Al Baghal (2016). Longer interviews lead to greater consent rate, suggesting that the respondent and the interviewer build a rapport that increases the chances of consent. However, this feature was not associated with consent in the study of Sala et al (2012), where consent to health and benefit records linkage was sought.

Despite the differences found across studies, most authors note the potential for bias in linked administrative data, resulting from differential consent using survey variables as predictors. Direct estimates of non-consent bias are limited due to the unavailability of administrative data for study members who refused to give the consents sought. However, Sakshaug and Kreuter (2012) estimated non-consent bias on administrative data from employment records used as the sampling frame for the study and found that non-consent biases were generally small relative to other sources of bias such as non-response and measurement error. Furthermore, the authors found no consistent relationship between non-consent and non-response.

It is also worth noting that non-linkage for consenting participants, resulting from missing administrative information or issues of identifying individuals' administrative records, is another potential source of bias and missing data (Calderwood and Lessof 2009), though this is not explored in this chapter.

7.3 Data and Methods

This analysis uses all available interview data from the Next Steps cohort study, sweeps 1–8, publicly available under end-user licence at the UK Data Service (NextSteps: Sweeps 1-8, 2004-2016 UK Data Service. SN: 5545).

7.3.1 About the Study

Next Steps follows the lives of 16 000 people in England born in 1989/90, sampled in 2003–2004 from state and independent schools. The sample design considered schools the primary sampling unit, with deprived schools being over-sampled by 50%. 647 state and independent secondary schools as well as pupil referral units participated in the study, from 892 selected schools. Within selected schools, pupils from minority ethnic groups (Indian, Pakistani, Bangladeshi, black African, black Caribbean, and mixed) were over-sampled to provide sufficient base sizes for analysis. The school and pupil selection approach ensured that, within a deprivation band and ethnic group, pupils had an equal probability of selection (Department for Education 2011).

Initially, cohort members were interviewed annually – at age 14 (in 2004), 15, 16, 17, 18, 19, and 20, and then when they were aged 25/26 (in 2015/16). The first seven sweeps have mainly focused on the educational experiences of the young people, attitudes to school and involvement in education; but information such as risky, criminal and anti-social behaviours, and health, has also been collected. The age 25 survey collected information about education and job training, employment and economic circumstances, housing and family life, physical and emotional health, and risky behaviours; as well as a range of administrative data linkage consents, as described below.

The interviews for the first four sweeps were conducted face-to-face, and included parents, as well as cohort members. The following sweeps, 5–8, used a sequential mixed mode approach – online, followed by telephone, and then face-to-face interviews, with the cohort member only.

Until sweep 7 in 2010, only those who participated at the previous sweep were included in the issued sample for the subsequent sweep, resulting in a reduction in the overall sample from 16 122 to 8682 at the age 20 survey. Extensive efforts were made at the age 25 survey to maximise the size and representativeness of the sample, attempting to trace and contact everyone who had ever taken part in the study. A total of 15 531 cohort members were issued to field in this most recent sweep, achieving a response rate of 51% with 7707 completed interviews, and 7495 valid consents for administrative records linkage (Centre for Longitudinal Studies, University College London 2017).

7.3.2 Consents Sought and Consent Procedure

At the end of the age 25 survey, cohort members were asked for consent to link their survey answers to nine different administrative data sources, held by a number of different government departments and non-governmental bodies. These were:

- Health records, covering visits to family doctor and other health professionals, and Hospital Episode Statistics (HES), including admissions and attendance at hospital, maintained by the National Health Service (NHS);
- Records about school participation and attainment, and pupil characteristics, and records about participation in further education and attainment, from the National Pupil Database (NPD) and the Individualised Learner Records (ILR), held by the Department for Education (DfE);
- Records covering university participation and attainment, held by the Higher Education Statistics Agency (HESA);

A single combined consent was sought for the DfE and HESA records:

- Records covering higher education applications and offers, held by the Universities and College Admissions Service (UCAS);
- Records covering payments of student support, held by Student Loans Company (SLC);
- Information on benefit and employment programs, kept by Department for Work and Pensions (DWP);
- Information on employment, earnings, tax credits, occupational pensions, and National Insurance Contributions, kept by Her Majesty's Revenue and Customs (HMRC);
- Respondents who consented to either DWP or HMRC, or SLC linkage, were also asked for their National Insurance number (NINO);
- Police National Computer (PNC) records covering arrests, cautions, and sentences, held by the Ministry of Justice (MOJ).

Administrative data linkage consents had been asked at earlier sweeps. Consent to link to NPD was sought alongside agreement to participate in the survey at sweep 1, and linked NPD data is available for research via UKDS. Consent to link to DWP and ILR records were asked to cohort members at later sweeps, but the linkages have not been enacted.

All participant materials and operational procedures involved in collecting data linkage consent were tested in exploratory qualitative work and the study pilot (Thornby et al. 2017), and approved by administrative data holders and ethical committees prior to main stage data collection.

The collection of consent involved a three stage process (pre-, during-, and post- interview). A data linkage leaflet was sent to the cohort member prior to the start of fieldwork. It gave information on the purpose, types, value, and process of data linkage, and encouraged study members to contact the study team with any questions they might have. During the interview, and following an introduction page, consents were recorded directly into the survey instrument. In the online survey, participants recorded their consent at questions within the self-completion instrument. In the telephone and face-to-face modes, consent was provided verbally by participants and recorded by the interviewer. After the interview, all participants were mailed a confirmation of their consents.

7.3.3 Analytic Sample

The dataset includes participants' characteristics and does not include interviewer or interview features, other than mode of data collection. For sweeps 1–4, alongside the cohort member's, it includes main parent's reports (second parent's report was excluded to preserve the overall analytical sample size). Mode was controlled for in the analysis to ensure that the presented estimates are independent from the effect of mode, which was strongly associated with all consent outcomes.[1] As we strictly focus on participant's characteristics, we do not report mode as part of the results of this work.

To avoid the limitation of performing this analysis on a significantly reduced analytical sample, due to the simultaneous inclusion of a large number of variables with different levels of unit and item missingness, which would have resulted in a loss of half of the age 25 achieved sample, we used Multiple Imputation (MI) with chained equations (White et al. 2011; Von Hippel 2007), and present the estimates generated from these imputed datasets, consisting of 7361 cases with valid consent data at age 25 (sweep 8).

7.3.4 Methods

We use a data-driven approach, which includes all variables available from sweeps 1–8 meeting the following selection criteria: applicable to all respondents; no category with less than 1% (categorical variables); and item non-response less than 50%. These selection criteria were adopted from the Centre for Longitudinal Studies Missing Data Strategy (Centre for Longitudinal Studies, University College London 2016). We use all variables meeting these criteria to identify those most strongly associated with consent. In total, 1152 variables, across all sweeps, met the criteria and were used in this analysis (see Table 7A.1 in the online supplementary material).

To identify the important determinants of the eight different consents asked at age 25, we employed a three step analytic strategy. The analysis at each step accounted for the clustering induced by the survey design (using SVY commands in Stata).

First, univariate logistic regression analysis was carried out, separately for each sweep and for each consent outcome, to estimate the association of each of the 1152 variables that met the predefined criteria with the response to each consent question. Variables that were significant at the 0.01 level, using a Wald chi-square test, were retained for the second step of the analysis.

The second step involved eight within-sweep (one for each sweep) multivariable log binomial regressions for each consent type. Log binomial regression was employed to avoid issues with the non-collapsibility of the odds ratio in common outcomes (Hernán et al. 2011). At this step, the variables that were retained in

1 The assumption is that the effect of each variable on consent is independent of the effect of the other variables, including mode. This assumption may, however, not hold if there is interaction between the variables (e.g. if the association of drug use and consent differs between self-administered and interviewer administered modes). Nevertheless, in this analysis we only look at main effects.

step 1 (univariate analysis), for each sweep and each consent type, were allowed to compete with each other being entered simultaneously in multivariable models. The variables that were statistically significant at the 0.01 level (Wald *p*-values <0.01), were retained for the third (and final) step of the analysis.

In the third and final step of the analysis, the variables from each sweep that were associated with consent at age 25, were entered simultaneously in models for each consent type. This analytic step allowed determinants of consent from each sweep to compete with each other and thus allowed us to identify those most strongly associated with each consent type. Acknowledging that there is not a strong theoretical justification for what constitutes a 'strong association' with consent, we defined a threshold of statistical significance at the 1% level (Wald chi-square test) as a semi conservative approach to control for family wise error rate (multiple testing) in order to balance potential type I and type II errors.

Between 31 and 40 variables were tested for each of the final models (i.e. survived from step 2), and between 13 and 16 variables were identified as determinants for each consent type.

This three-step process has been illustrated in Figure 7A.1 in the online supplementary material with an example for the NHS records linkage model.

Prior to the third and final step of analysis, we used multiple imputation (MI) with chained equations (STATA version 15) to create 70 imputed datasets. For this purpose, categorical variables were recoded (if necessary) so that each category is greater than 2%.[2] To create the imputed datasets, we fitted imputation models that included all variables statistically significant at 0.01 level in the within-sweep models (see Table 7A.2 in the online supplementary material), and also the consent outcome. Data were assumed MAR given observed values of other variables. Each of the eight models was then estimated using each of the 70 imputed datasets and the parameter estimates were combined to obtain overall estimates using Rubin's rules (1987). The results, generated from the imputed datasets were restricted to the number of cases with valid consent data (i.e. not imputed values of consent), consistent with the conservative 'impute outcome then delete' approach described by Von Hippel (2007). The results based on MI were similar to those from weighted complete records analysis. Similar results from weighted complete records analysis and MI were expected as non-response in Next Steps is mostly monotone (Centre for Longitudinal Studies, University College London 2018).

Sex and ethnicity variables (if applicable) included in the final model, were those collected at the baseline sweep – age 13/14. Repeated measures (e.g. ever tried an

---

2 An exception is the ethnicity variable, which was further collapsed into a binary variable with White and Non-white categories. This was attempted in response to a convergence problem in one of the imputation models, and the intention to present comparable parameters across the different consent types.

Table 7.1 Administrative records linkage consent rates at age 25 – pooled and by mode of participation.

Consent at Age 25 Survey	Weighted % ($n = 7495$)	Web ($n = 4630$)	Tel ($n = 660$)	F2F ($n = 2205$)
DfE/HESA records	73	61	88	88
UCAS records	70	58	87	84
NHS records	69	57	81	86
MOJ records	66	55	82	82
DWP records	63	48	81	81
SLC records	62	50	77	78
HMRC records	61	47	75	79
NINO (whole sample)	53	44	48	69
NINO (SLC, DWP or HMRC consent)	76	77	55	80

alcoholic drink or ever taken drugs) from each sweep were included, if significant at the 0.01 level in the respective within-sweep model.

7.4 Results

7.4.1 Consent Rates

Table 7.1 shows the consent rates for each of the administrative record linkages requested at the age 25 survey. The level of consent was higher for education records – DfE/HESA (73%) and UCAS (70%), and lower for economic records – DWP (63%) and HMRC (61%). The consent rate for SLC records, potentially due to the financial nature of the information requested, was similar to the economic records. NINO supply was conditional on any permission given to SLC, DWP, or HMRC records linkage.

Mode of participation was strongly associated with consent, with higher rates for all consent types obtained over the telephone and face-to-face compared to web. Due to selection into mode, we cannot determine if there is a genuine mode effect on consent propensity. However, due to the large mode differences we retain mode in the analysis as a control variable.

Table 7.2 shows the univariate associations between consent to health, criminal, education, and economic records, and a few key respondent characteristics measured at age 25. For most of the record linkages males were more likely to provide consent than females, though males were no more likely than females

Table 7.2 Consent rates by key respondent characteristics at age 25 (weighted $n = 7495$).

		Health	Crime	Education			Economic		
		NHS records	MOJ records	DfE/ HESA records	UCAS records	SLC records	HMRC records	DWP records	NINO supply
Respondent characteristics at age 25	All respondents (Col %; N = 7495)	(Row %; N = 5155)	(Row %; N = 4968)	(Row %; N = 5446)	(Row %; N = 5209)	(Row %; N = 4669)	(Row %; N = 4539)	(Row %; N = 4684)	(Row %; N = 3769)
Gender		*		*	*	*	*	*	
Male	50.8	73.0	66.9	74.6	71.4	64.4	62.5	65.0	75.9
Female	49.2	64.4	65.6	70.7	67.5	60.1	58.6	60.0	75.2
Ethnicity*									
White	84.8	71.1	69.1	74.9	71.4	64.6	63.5	65.4	76.3
Mixed	2.9	66.4	63.1	71.0	69.5	60.4	57.0	56.3	74.8
Indian	2.3	55.2	51.5	62.0	60.6	49.6	43.1	49.2	70.2
Pakistani	2.4	55.9	50.8	61.9	61.9	53.8	48.3	49.9	63.1
Bangladeshi	1.2	39.2	34.2	43.0	38.0	30.1	27.4	27.4	70.9
Black Caribbean	1.7	56.2	47.4	57.5	55.7	45.3	45.3	48.2	73.8
Black African	2.3	49.7	46.2	55.8	54.9	47.0	37.6	42.1	76.2
Other	2.5	58.8	48.8	59.2	56.9	45.9	41.0	42.1	61.8
Current activity			*	*	*	*	*	*	
In work (incl. self-employed, voluntary work, apprenticeship)	79.9	68.9	66.5	72.9	70.0	62.7	61.1	63.2	75.5

(Continued)

Table 7.2 (Continued)

		Health	Crime	Education			Economic		
		NHS records	MOJ records	DfE/ HESA records	UCAS records	SLC records	HMRC records	DWP records	NINO supply
Unemployed	6.7	66.5	62.0	67.6	63.7	55.6	56.5	58.8	73.8
Education (incl. School, college, university)	4.5	70.2	70.2	76.8	74.3	68.0	61.8	62.7	74.6
Sick or disabled	2.9	64.9	60.5	66.2	58.8	56.0	49.2	52.2	72.3
Looking after home or family	5.7	72.6	71.1	76.5	73.2	65.3	65.7	64.9	80.0
Other (incl. travelling)	0.3	37.0	42.5	64.3	59.5	55.4	24.5	24.5	81.7
Marital status		*	*	*	*	*	*	*	
Single/Never married	87.9	69.04	66.63	73.16	70.17	63.11	60.78	62.75	75.54
Married (incl. civil-partnership)	10.63	67.93	63.95	69.35	65.19	57.37	59.56	61.11	76.21
Separated/Divorced (incl. former civil partnership)	0.94	80.82	84.61	86.16	75.01	63.4	76.24	78.88	71.89
Widowed (incl. surviving civil partnership)	0.13	46.33	50.41	50.41	50.41	50.41	42.06	42.06	83.43
Refused (incl. insufficient info)	0.41	16.18	14.37	25.48	28.51	14.37	10.24	14.37	29.89

An asterisk (*) indicates $p < .01$ (Pearson χ^2 test)

to consent to criminal records linkage or to provide their NINO (among HMRC, DWP or SLC consenters). For all record types, white cohort members were more likely than all other ethnic groups to provide consent. Noting the small subgroup size, Bangladeshis were the least willing to give permissions to any records linkage. There was no evidence that participants' economic activity was related to health records consent nor NINO provision (among consenters to HMRC, DWP, or SLC records), but it yielded a statistically significant relationship with all other records consents, where anyone unemployed, sick, or disabled was significantly less likely to consent to records linkage. Separated or divorced cohort members, noting the small subgroup size, gave permission for all records linkages at a higher rate, except for NINO (among HMRC, DWP, or SLC consenters).

7.4.2 Regression Models

In the online supplementary material, Table 7A.1 indicates the number of variables involved in each of the steps of analysis and Table 7A.2 indicates the variables statistically significant at 0.01 level across the within-sweep (sweeps 1–8) multivariable regression models, and the final (all sweeps) models. The variables statistically significant at 0.01 level in the within-sweep models were included as covariates in the final models. Those retained at the final stage of analysis (i.e. statistically significant at the 1% level in the final regression models) are presented in Table 7A.3 in the online supplementary material and are summarised below.

7.4.2.1 Concepts and Variables

The variables impacting consent include study members' sociodemographic, attitudinal, and behavioural characteristics. Sociodemographic variables strongly related to consent include sex, ethnicity, educational achievement, early labour market experience, household circumstances, and time living at an address. Relevant attitudinal traits include national identity and trust, self-perceived abilities at school and school characteristics, attitudes to school and work, as well as self-perceptions of own body weight, usefulness and ability to face problems, and control over one's own life. Behavioural traits associated with consent include learning activities at home, social and family connections, social support, health behaviour, experience of bullying, willingness to answer sensitive survey questions, as well as previous consent to administrative records linkage.

Some of these characteristics were strongly related to all or most consent domains and record types, with consistency in the direction and the magnitude of their effect, while others related strongly to only some domains or records. We look first at the factors associated with all or most consent domains, then separately by domain and record type.

By 'all' consent domains we mean health and crime, and at least one permission to a record type from each of the education (e.g. school, further education and university participation, university applications and offers, or student loans) and economic (e.g. benefits or employment and tax credits) domains. By 'most' consent domains we mean three out of the four domains, with the same threshold for domains with multiple record types (i.e. at least one record per domain). We focus here on permissions to link records, but later in the section we also present the characteristics strongly associated with NINO supply.

7.4.2.2 Characteristics Related to All or Most Consent Domains

Four measures were strongly related to consent propensity in *all* domains (Table 7.3):

- Non-white study members, those resistant (at age 18/19 or 19/20) to answering questions about sexual experiences, those who feel that the government does not treat people fairly (at age 17/18), and those who have not tried Cannabis (by age 25) were less likely to give permission to any linkage request.

Previous records linkage permissions, the attitude towards having a job with regular hours in the future, and the time living at an address, were strongly related to *most* consent domains:

- Previous consent to the study (at age 16/17) to add information held by the DWP was strongly related to the likelihood of consent to all consent types, except health. Study members who have previously given this permission provided consent to all but health records linkage, at higher rate.
- Study members to whom it mattered a little (at age 19/20) to have a job with regular hours in the future were more likely to give consent to any linkage request, except for linking their economic records.
- Consent was positively impacted by the time study members lived at their address (at age 25/26). The likelihood for consent increased with the increase in the time they lived at their address. This relationship was observed for all consent domains except health.

Alongside the factors impacting consent across all or most domains, some characteristics were only strongly associated with some consent domains or record types. We summarise those characteristics below.

7.4.2.3 National Health Service (NHS) Records

The following groups were more likely to consent to health records linkage: those with higher levels of trust in other people (RR 1.02; SE 0.00), those who (at age 25/26) considered themselves underweight (RR 1.08; SE 0.03), and those meeting

Table 7.3 Risk ratios (RR) and standard errors (SE) for characteristics related to all or most consent domains.

Characteristics (statistically significant at 0.01 level)	Health	Crime	Education			Economic	
	NHS records	MOJ records	DfE/HESA records	UCAS records	SLC records	HMRC records	DWP records
Ethnicity(reference:White)							
Non-white	0.87*(0.03)	0.83*(0.03)	0.90*(0.03)	0.89*(0.02)	0.84*(0.03)	0.86*(0.04)	0.84*(0.03)
Permission to pass on details to DWP (age 16/17)(reference:No)							
Yes		1.15*(0.05)	1.16*(0.05)	1.14*(0.04)	1.14*(0.05)	1.19*(0.06)	1.14*(0.05)
How fairly British government treats people (age 17/18)							
(reference: Very fairly)							
Quite fairly	0.95 (0.02)	0.96 (0.24)	0.95*(0.02)		0.95 (0.03)	0.92* (0.03)	0.93*(0.02)
Neither fairly or unfairly	0.90*(0.03)	0.89*(0.03)	0.89*(0.02)		0.89*(0.03)	0.87*(0.03)	0.88*(0.03)
A little unfairly	0.94 (0.03)	0.92*(0.03)	0.92*(0.03)		0.92 (0.03)	0.90*(0.03)	0.89*(0.03)
Very unfairly	0.86*(0.04)	0.87*(0.04)	0.87*(0.03)		0.84*(0.05)	0.86* (0.04)	0.86*(0.04)
Willing to answer sexual experiences questions (age 18/19)(reference: Yes)							
No	0.82*(0.05)	0.85*(0.05)	0.82*(0.04)	0.84*(0.04)	0.82*(0.05)	0.77*(0.05)	0.78*(0.05)
Willing to answer sexual experiences questions (age 19/20)(reference: Yes)							
No	0.83*(0.04)	0.83*(0.04)		0.88*(0.04)	0.81*(0.05)		

Table 7.3 (Continued)

Characteristics (statistically significant at 0.01 level)	Health	Crime	Education			Economic	
	NHS records	MOJ records	DfE/HESA records	UCAS records	SLC records	HMRC records	DWP records
How much it matters to have a job with regular hours (age 19/20) (reference: Matters a lot to me)							
Matters a little to me	1.06*(0.02)	1.07*(0.02)	1.05*(0.02)	1.06*(0.02)	1.07*(0.02)		
Does not matter	1.03 (0.03)	1.05 (0.03)	1.04 (0.02)	1.06*(0.02)	1.06 (0.03)		
Ever taken Cannabis/ Marijuana (age 25/26)(reference: Yes)							
No	0.90*(0.02)	0.94*(0.02)	0.93*(0.01)	0.92*(0.01)	0.91*(0.02)	0.90*(0.02)	0.92*(0.02)
Months at current address (age 25/26)		1.00*(0.00)	1.00*(0.00)	1.00*(0.00)	1.00*(0.00)	1.00*(0.00)	1.00*(0.00)

Notes:

- The table includes variables associated with all or most consent domains at the 1% level of statistical significance ($p < .01$); $n = 7,361$.
- An asterisk (*) indicates statistical significance for the category at the 1% level ($p < .01$).
- It is a subset of Table 7A.3, available in the online supplementary material, which presents all variables associated with each consent type (at each final 'all sweeps' model) at the 1% level of statistical significance ($p < .01$).
- An empty cell indicates that the variable was either not included in this model or was not associated with consent at the 1% level ($p < .01$).

up with friends once or twice a week (at age 25/26), rather than three or more times a week (RR 1.05; SE 0.02).

On the other hand, females (RR 0.94; SE 0.02), study members who strongly disagreed (at age 17/18) with the statement 'Being British is important to me', relative to those agreeing strongly (RR 0.87; SE 0.04), and those facing their problems in the same way as usual[3] (at age 25/26), relative to those dealing with problems better than usual (RR 0.95; SE 0.02), were all less likely to consent.

7.4.2.4 Police National Computer (PNC) Criminal Records

Higher levels of consent were associated with spending free time alone in the early teenage years, having choice of school subjects less determined by the subjects they did well in, having both female and male friends at school, having higher levels of education, being under- or overweight, at the time consent was sought, and never smoking:

- Those who mainly spent their free time by themselves (at age 13/14), rather than outside with friends, gave permissions at higher rates (RR 1.10; SE 0.03).
- So did those who disagreed a little with the statement 'I only want to do subjects that I know I will do well at in exams' (at age 13/14), compared to those that agreed strongly with it (RR 1.07; SE 0.02).
- Study members whose same sex friends at school (at age 14/15) were less than half, compared to those whose all or most friends at school were the same sex, had higher likelihood of consent (RR 1.09; SE 0.04).
- University participation, by age 25, positively impacted consent (RR 1.08; SE 0.02).
- Higher consent levels were also obtained by study members who (at age 25/26) considered themselves under- or overweight, rather than being about the right weight (RR 1.10; SE 0.03 and RR 1.09; SE 0.03).
- Those who smoked occasionally, but not every day (at age 25/26), compared to those who never smoked cigarettes, were less likely to give this permission (RR 0.91; SE 0.03).

7.4.2.5 Education Records

7.4.2.5.1 Characteristics Related to Consent to All Education Records

Some of the observed variables were strongly related to consent to all types of education records (Table 7.4):

- Educational achievement impacted consent across the whole domain. Those who had ever been to university, at the time consent was asked, were more likely to consent to all three education records linkages.

---

3 An item from the General Health Questionnaire (GHQ12), used as a screening tool of probable mental ill health.

- Study members, who (at age 14/15) played computer games at home most days, compared to none, had higher propensity to consent.
- Those to whom it mattered a little or did not matter (at age 15/16) whether they will be self-employed or have own business at some point in the future, compared to those to whom it mattered a lot, were more likely to consent. (Study members' attitude towards having their own business in the future was also strongly related to consent to all economic records – see Section 7.4.2.6).

The analysis also identified characteristics that were strongly associated with consent for particular types of education records, beside those that impacted the domain as a whole.

7.4.2.5.2 Department for Education (DfE) and Higher Education Statistics Agency (HESA) records

School characteristics, family connections, and household circumstances affected consent to school, further education and university attainment records:

- Study members who had a family meal most nights (at age 13/14), compared to every night (RR 1.05; SE 0.02), were more likely to consent for their records to be linked.
- Those who lived with another adult or children, at the time consent was sought, were also more likely to give this permission (RR 1.05; SE 0.02).
- Study members who reported (at age 14/15) that some of the teachers at their school made sure homework was done, compared to most of the teachers (RR 0.95; SE 0.02), had lower likelihood to consent.

7.4.2.5.3 Universities and College Admissions Service (UCAS) records

- Study members whose same-sex friends at school (at age 14/15) were less than half, compared to those for whom all or most friends at school were the same sex, had higher likelihood to consent (RR 1.09; SE 0.03).
- Educational achievement was positively related to consent. Study members with an NVQ level 4 qualification by age 25, compared to NVQ level 1, had higher likelihood to give this permission (RR 1.11; SE 0.04).
- Those who lived with another adult or children (at age 25/26), were more likely to give permission for their records to be linked (RR 1.06; SE 0.02).
- Study members who attended a school (at age 14/15), where some of the teachers made sure homework was done, compared to most of the teachers (RR 0.94; SE 0.02), had lower likelihood to consent.
- A higher locus of control score was associated with reduced likelihood of consent (RR 0.98; SE 0.00). A higher score indicates an external locus of control – attributing success or failure to outside influences – and a lower score indicates an internal locus of control – believing that one has control over one owns life.

Table 7.4 Risk ratios (RR) and standard errors (SE) for characteristics related to consent to all education records.

Characteristics (statistically significant at 0.01 level)	Education records		
	DfE/HESA	UCAS	SLC
Days a week use a computer at home for playing computer games (Age 14/15) (reference: None)			
1–2 days	1.03 (0.02)	1.04 (0.02)	1.04 (0.02)
3–4 days	1.06 (0.02)	1.07* (0.03)	1.06 (0.03)
Most days (5 or more)	1.06* (0.02)	1.07* (0.02)	1.09* (0.03)
How much it matters ...			
To be my own boss/have my own business (Age 19/20)			
(reference: Matters a lot to me)			
Matters a little to me	1.06 (0.03)	1.07* (0.02)	1.09* (0.03)
Does not matter	1.07* (0.03)	1.09* (0.03)	1.13* (0.03)
Ever been to university (Age 25/26)			
(reference: No)			
Yes	1.08* (0.02)	1.09* (0.02)	1.10* (0.03)

Notes:

- The table includes variables associated with consent to all education records at the 1% level of statistical significance ($p < .01$); $n = 7,361$.
- An asterisk (*) indicates statistical significance for the category at the 1% level ($p < .01$).
- It is a sub-set of Table 7A.3, available in the online supplementary material, which presents all variables associated with each consent type (at each final 'all sweeps' model) at the 1% level of statistical significance ($p < .01$).

7.4.2.5.4 Student Loans Company (SLC) Records

The following characteristics were associated with the probability of granting access to student loan records:

- Considering oneself under- or overweight (at age 25/26), rather than about the right weight, was associated with higher likelihood to give this permission (RR 1.11; SE 0.04 and RR 1.11; SE 0.03).
- Having experienced a sexual assault (by age 25/26) was associated with a higher consent probability (RR 1.15; SE 0.04).
- Agreeing with a statement that having a job or career in the future is important (at age 25/26), compared to agreeing strongly, was associated with reduced likelihood to consent (RR 0.94; SE 0.02).

7.4.2.6 Economic Records

7.4.2.6.1 Characteristics Related to Consent to All Economic Records

Some of the observed factors had a strong impact on consent for both types of economic records (see Table 7.5):

- Study members to whom it mattered a little or did not matter (at age 19/20) to have their own business in the future, compared to those to whom it mattered a lot were more likely to consent to HMRC and DWP record linkage.
- Study members regularly working four or more hours in a paid job (at age 17/18) were more likely to consent to both HMRC and DWP linkage.
- Consent was more likely for those who would turn to a best friend, rather than their mother, if feeling down (at age 19/20).

Alongside the characteristics impacting the domain as a whole, the analysis identified variables that were strongly associated with propensity to consent to linkage to particular type of economic records.

7.4.2.6.2 Her Majesty's Revenue and Customs (HMRC) Records

Higher consent rates were associated with:

- Study members who (at age 18/19) agreed that 'Having a job is the best way to be an independent person', compared those who agreed strongly with the statement (RR 1.03; SE 0.02);
- Membership of a pension scheme (at age 25/26) – a scheme run by an employer or one started privately (RR 1.08; SE 0.02).

Lower consent rates were associated with:

- Study members who (at age 13/14) rated themselves as fairly or not very good in science, compared to those who rated themselves as very good (RR 0.92; SE 0.02; and RR 0.91; SE 0.02);

Table 7.5 Risk ratios (RR) and standard errors (SE) for characteristics related to consent to all economic records.

Characteristics (statistically significant at 0.01 level)	Economic records	
	HMRC	DWP
Currently doing any kind of paid job (age 18/19)		
(reference: No)		
Yes	1.07* (0.02)	1.07* (0.02)
How much it matters to be my own boss/have my own business (age 19/20)		
(reference: Matters a lot to me)		
Matters a little to me	1.10* (0.03)	1.10* (0.03)
Does not matter	1.14* (0.03)	1.13* (0.03)
People would turn to first for help if feeling a bit down (age 19/20)		
(reference: Mother)		
Father	1.09 (0.06)	1.10 (0.05)
Sibling	1.01 (0.05)	1.02 (0.04)
Partner	1.05 (0.03)	1.04 (0.03)
Best friend	1.09 * (0.05)	1.08* (0.03)
Someone else	1.07 (0.05)	1.05 (0.04)
No one	0.94 (0.07)	0.97 (0.06)

Notes:

- The table includes variables associated with consent to both economic records at the 1% level of statistical significance ($p < .01$); $n = 7{,}361$.
- An asterisk (*) indicates statistical significance for the category at the 1% level ($p < .01$).
- It is a sub-set of Table 7A.3, available in the online supplementary material, which presents all variables associated with each consent type (at each final 'all sweeps' model) at the 1% level of statistical significance ($p < .01$).

- Attending a school where some of the teachers made sure homework was done (at age 14/15), compared to one where most of the teachers did this (RR 0.92; SE 0.03);
- Having parents who (at age 16/17) received a state pension (RR0.77; SE 0.08);
- Smoking cigarettes occasionally (at age 25/26), but not every day, compared to those who never smoked cigarettes (RR 0.92, SE 0.03).

7.4.2.6.3 Department for Work and Pensions (DWP) Records

Having been to university (RR 1.08; SE 0.02) and living with other adults or children all or some of the time (RR 1.05; SE 0.02) were associated with higher propensity to consent to benefit records linkage. Those who (at age 13/14) agreed with a statement that 'School work is worth doing', compared to those who agreed strongly with that statement, consented at lower rates (RR 0.94; SE 0.02).

7.4.2.6.4 National Insurance Number (NINO)

In line with the observation that some characteristics strongly impacted consent to all or most records asked about, we found that these factors tended to have a similar impact on the participants' willingness to provide their NINO.

Greater willingness was associated with:

- Having previously (at age 16/17) given permission for the study to access benefit records (RR 1.27; SE 0.08);
- Having said that it mattered a little to them (at age 19/20) to have a job with regular hours in the future (RR 1.08; SE 0.03);
- Having lived longer at their address at the time NINO was requested (RR 1.00; SE 0.00);
- Saying it mattered a little or did not matter (at age 19/20) to have their own business in the future, compared to saying it mattered a lot (RR 1.15; SE 0.05, and RR 1.22; SE 0.05) – a characteristic already shown to affect consent to linking economic records more generally, as well as education records;
- Being a member of a pension scheme (at age 25/26) (RR 1.07; SE 0.02);
- Thinking of oneself (at age 25/26) as being underweight rather than the right weight (RR 1.14; SE 0.05).

Reduced willingness to provide a NINO was associated with:

- Feeling neither fairly nor unfairly treated by the government (at age 17/18) (RR 0.87; SE 0.04);
- Being resistant to answering questions about sexual experiences (at age 19/20) (RR 0.80; SE 0.07);
- Having never tried cannabis (by age 25/26) (RR 0.92; SE 0.02);
- Considering oneself fairly good or not very good at science (at age 13/14), rather than very good at it (RR 0.91, SE 0.02, RR 0.83; SE 0.03);

- Facing one's problems or feeling useful (at age 25/26) as usual, rather than being able to deal with problems and feelings better than usual (RR 0.90; SE 0.02 and RR 0.92; SE 0.03).

7.5 Discussion

7.5.1 Summary of Results

This work extends our understanding of determinants of consent to linkage of health, crime, education and economic records among young adults. We have used an analytical data driven approach, not previously used (to the best of our knowledge), that makes use of all available data collected during the study lifecycle to identify determinants of consent.

Our findings are consistent with the literature, according to which there are systematic differences between respondents who provide consent and those who do not, but the observed differences go beyond demographic and socioeconomic characteristics. We found a range of attitudinal and behavioural characteristics, observed throughout the study, to be strongly related to consent. This justifies the approach of looking at the widest possible range of potential predictors of consent.

Nearly half of the significant predictors for each record type were strongly related to all or most of the consents studied, consistent in the direction and the magnitude of the effect, and thus supportive of the hypothesis that consent across different record types could be determined by similar factors.

Variables strongly related to all or most of the permissions studied included ethnic background, time living at current address, attitudes towards the government and future jobs, willingness to respond to sensitive survey questions, consent decisions made in the past, and experience with drugs. This information was collected at different times of the study lifecycle, between age 16 and 26, and was provided by the cohort members themselves, rather than their parents.

The observation that factors such as ethnicity, past consent and health behaviour affect consent is not new. Their effect, however, has not always been consistent across studies and consent domains, neither has it been observed simultaneously across multiple domains. For example, ethnicity has been found to impact consent on health, education, and economic records in some studies (e.g. Mostafa and Wiggins 2015; Al Baghal 2016; Sala et al. 2014), but to have no effect on consent to health records in others (e.g. Kho et al. 2009). Previous consent has also been found to inconsistently affect future consent decisions. For instance, exploring the consistency of respondents in consenting to health records linkage over time in the Millennium Cohort Study, Mostafa and Wiggins (2015) found that consent behaviours are in general consistent – over 75% of respondents behaved consistently over time. Consent was found to not be driven by personal convictions. Rather, it depended on the circumstances of the respondent at the time of the interview. Chapter 6 of this book (Jackle et al. 2021), on the other hand, using data from the Understanding Society household panel, found that nearly 50% of respondents who did not give consent previously did consent when asked again in a later wave. The authors concluded that consent decision is not a strong or fixed decision, and can likely be influenced – thus, it can be inconsistent between topics and over time.

Previous studies have shown that participants with better health are less likely to consent to health records linkage (Dunn et al. 2004; Weissman et al. 2016). Our findings support this. Moreover, we found that participants' health affected consent propensity in other domains. Specifically, those who did not use cannabis, which is likely to indicate a better health status, were less likely to give permission to any of the consents we asked about.

Some of the characteristics that we found to be associated with consent propensity – such as ethnicity, trust in government, health behaviour and willingness to answer sensitive survey questions – are known to have an impact on the willingness to cooperate in surveys. For example, smokers, drinkers, participants' with less conservative sexual attitudes, are generally known to be more cooperative, and also systematically different from those who do not take part in sexual behaviour surveys (Dunne et al. 1997). Disengagement from government and official institutions is among the reasons for ethnic minorities to be less cooperative (Mostafa and Wiggins 2015).

Our results also highlighted that potential biases may differ according to the consent request. Notably, some predictors related more closely to the period of life and type of information the administrative record details – i.e. education records consent was associated with schooling circumstances, as well as later educational achievements, and economic records with early labour market circumstances and future work attitudes.

We found health consent decisions to be impacted by slightly different factors to most other domains. Health was the only domain not strongly affected by socioeconomic factors, and the only domain where sex, trust, and identity mattered to consent, but also a domain notably affected by factors with a clear link to (mental) health.

Our findings support the existing evidence that males are more likely to consent to health records linkage than females, and that people with poorer health are more likely to agree for their health records to be linked (e.g. Dunn et al. 2004; Al Baghal et al. 2014). We found that variables related to health, such as own body perception and ability to face problems, as well as social connections, affected health records consent. The perception of own weight is linked to self-esteem that can harm both one's physical and mental health (e.g. Mann et al. 2004). The link between social relationships and health outcomes is also well established (e.g. Umberson and Montez 2010), and the ability to face problems is itself an item from a screening tool of probable mental ill health (GHQ12).

Trust, on the other hand, has been identified in other studies as a predictor of consent to multiple domains (Al Baghal et al. 2014; Korbmacher et al. 2013; Sakshaug et al. 2012; Sala et al. 2012), including health, education and economic, but we found it to only significantly affect health, and not other permissions studied.

Crime is a domain where existing empirical evidence on patterns of consent is scarce. The only study we are aware of is an Australian longitudinal cohort study that sought written consent from its study members at age 19–20 to access their official police records. The study found that respondents who granted access were less likely to be male, to come from a non-English speaking background and to hold favourable attitudes to the criminal justice system, but were more likely to have a high degree of trust in institutions. They were less likely to have been described as dishonest, to have attention problems, to lack persistence and self-regulation, to have conduct problems or have previously reported being involved in crime (Forrest et al. 2014).

In line with these findings, we found that study members' attitude towards the fairness of treatment by the government in Britain impacted consent to criminal records, as well as all other records we asked about. We did not, however, find evidence that sex affects linkage permissions other than health. Neither did we observe behaviour problems to be related to criminal records consent. But consent to criminal records linkage was notably determined by early sweeps circumstances, namely schooling experiences, such as social connections at school and attitudes to school, as well as later educational achievement. The only other domain notably impacted by early sweep and schooling events was education.

A number of factors affected consent to all or most education records. These included socioeconomic characteristics from the most recent survey sweep – such as educational attainment and housing circumstances; as well as information collected at younger ages – like the frequency of playing computer games at home, an indicator for learning activities at home, and attitudes to work such as having own business in the future.

Our results are in line with the existing literature to the extent that consent to education records is associated with educational qualifications and housing circumstances. Previous research found that having no children is related to higher odds of consent to education records, and that household characteristics such as household size and ownership had no impact (Al Baghal et al. 2014). We found that living with someone else in the household was associated with increased likelihood to consent to education linkage.

Some characteristics affected consent to linkage to particular types of education record, typically corresponding to the period of life to which the administrative record relates. Early educational experiences impact consent to school and university participation records linkage (for example, school and family characteristics from the early study sweeps, such as strictness of school rules and family connections, impacted consent to DfE/HESA records); educational achievement and personality traits impact more notably consent decisions for university applications and offers records.

A number of factors affected consent to all economic records. These included information primarily related to work and employment such as attitude towards having one's own business in the future and early work experience, as well as social support.

Some characteristics affected consent to a particular type of economic record. Employment and tax credits records consent, for example, was determined by schooling and work characteristics, health behaviour, and sources of parental income. The characteristics related to benefits consent were similar to those determining education records consent – educational achievement, household circumstances, and attitude towards school. And the characteristics related to NINO supply were very similar to those determining health records consent – such as self-perceptions of weight, ability to face problems, feeling useful; as well as pension scheme membership and ability in science – characteristics also identified as determinants of HMRC records consent.

As stated previously, our findings cannot always be directly related to previous evidence. For example, Jenkins et al. (2006) found that lone parents had lower propensity to consent to DWP records linkage than respondents in a couple, and household type was not found to be associated with consent to benefit data by Sala et al. (2012). Our findings showed that study members who lived alone were less likely to consent to DWP record linkage. It was also observed by Jenkins et al. (2006) that respondents who had health problems were less likely to supply a NINO, while we found similarity between the factors affecting health and NINO consent, with similar direction and magnitude of effect.

Lastly, previous research found that educational qualifications were not associated with DWP or NINO supply (Jenkins et al. 2006), while we found similarity in the factors affecting education and DWP consent, and that DWP records consent is impacted by educational achievement.

7.5.2 Methodological Considerations and Limitations

The main strength of our study is the data-driven approach aimed at identifying determinants of consent using all data collected throughout the study.

There are, however, limitations that should be considered. Variables are identified based on a strictly defined threshold (i.e. *p*-value less than 0.01). An increase of the threshold (e.g. *p*-value less than 0.05), would have resulted in an increase of the number of variables associated with consent outcomes. But each consent type, at the first step of analysis, was regressed on the same variables. Considering that there is no strong theoretical justification for what constitutes a 'strong association' with consent, a threshold was needed to aid variable selection. We opted for the 1% level as a semi-conservative approach to control for family wise error rate to balance potential type I and type II errors.

Our analysis approach also highlights difficulties in comparing our findings with existing research, as the choice of variables is not based on theoretical or empirical interest, but, rather, the statistical properties of the data. But our approach has allowed us to extend existing knowledge by accounting for a much wider set of variables relating to different life stages.

7.5.3 Practical Implications

In line with previous research, our findings show the potential for over- or under-representation of population subgroups in survey data linked to administrative data. Correcting solely for sociodemographic characteristics of study members may not entirely remove bias in sample composition. We have highlighted a greater range of attributes that should perhaps be considered when handling missing data due to selective consent. Methods such as multiple imputation, full information maximum likelihood, linear increments or inverse probability weighting could be considered to reduce consent bias. Of course, findings different to ours could arise from different surveys, with a less or more rich set of predictors of consent.

References

Al Baghal, T. (2016). Obtaining data linkage consent for children: factors influencing outcomes and potential biases. *International Journal of Social Research Methodology* 19 (6): 623–643.

Al Baghal, T., Knies, G., and Burton, J. (2014). Linking Administrative Records to Surveys: Differences in the Correlates to Consent Decisions. *Understanding Society Working Paper Series No. 2014–09 December 2014.*

Bohensky, M.A., Jolley, D., Sundararajan, V. et al. (2010). Data linkage: a powerful research tool with potential problems. *BMC Health Services Research* 10: 346–353.

Calderwood, L. and Lessof, C. (2009). Enhancing longitudinal surveys by linking to administrative data. In: *Methodology of Longitudinal Surveys* (ed. P. Lynn), 55–72. New York: Wiley.

Carpenter, J.R. and Kenward, M.G. (2013). *The Multiple Imputation Procedure and its Justification*. New York: Wiley.

Centre for Longitudinal Studies, University College London. (2016). Missing Data Strategy. https://www.bing.com/search?q=cLS+missing+data+startegy&src=IE-TopResult&FORM=IETR02&conversationid= (accessed 01 June 2018)

Centre for Longitudinal Studies, University College London (2017). Next Steps Age 25 Survey Technical Report. http://doc.ukdataservice.ac.uk/doc/5545/mrdoc/pdf/5545age_25_technical_report.pdf (accessed 01 December 2017)

Centre for Longitudinal Studies, University College London. (2018). Next Steps Sweep 8 – Age 25 Survey User Guide (Second Edition). http://doc.ukdataservice.ac.uk/doc/5545/mrdoc/pdf/5545age_25_survey_user_guide.pdf (accessed 01 December 2017)

Department for Education. (2011). LSYPE User Guide to the Datasets: Wave 1 to Wave 7. http://doc.ukdataservice.ac.uk/doc/5545/mrdoc/pdf/5545lsype_user_guide_wave_1_to_wave_7.pdf (accessed 01 December 2017)

Dunn, K., Jordan, K., Lacey, R. et al. (2004). Patterns of consent in epidemiologic research: evidence from over 25,000 responders. *American Journal of Epidemiology* 159: 1087–1094.

Dunne, M.P., Martin, N.G., Bailey, J.M. et al. (1997). Participation bias in a sexuality survey: psychological and behavioural characteristics of responders and non-responders. *International Journal of Epidemiology* 26: 844–854.

Forrest, W., Edwards, B., and Vassallo, S. (2014). Individual differences in the concordance of self-reports and official records. *Criminal Behaviour and Mental Health* 24 (4): 305–315.

Hernán, M.A., Clayton, D., and Keiding, N. (2011). The Simpson's paradox unraveled. *International Journal of Epidemiology* 40 (3): 780–785.

Hockley, C., Quigley, M.A., Hughes, G. et al. (2007). Linking millennium cohort data to birth registration and hospital episode records. *Paediatric and Perinatal Epidemiology* 22: 99–109.

Jackle, A., Beninger, K., Burton, J. et al. (2021). Understanding data linkage consent in longitudinal surveys. In: *Advances in Longitudinal Survey Methodology* (ed. P. Lynn). Hoboken, NJ: Wiley.

Jenkins, S., Cappellari, L., Lynn, P. et al. (2006). Patterns of consent: evidence from a general household survey. *Journal of the Royal Statistical Society: Series A (Statistics in Society)* 169: 701–722.

Kho, M., Duffett, M., Willison, D. et al. (2009). Written informed consent and selection bias in observational studies using medical records: systematic review. *British Medical Journal* 338: b866.

Knies, G. and Burton, J. (2014). Analysis of four studies in a comparative framework reveals: health linkage consent rates on British cohort studies higher than on UK household panel surveys. *BMC Medical Research Methodology* 14: 125.

Korbmacher, J. and Schroeder, M. (2013). Consent when linking survey data with administrative records: the role of the interviewer. *Survey Research Methods* 7: 115–131.

Little, R.J. and Rubin, D.B. (2002). *Statistical Analysis with Missing Data*. New York: Wiley.

Mann, M., Hosman, C., Schaalma, H. et al. (2004). Self-esteem in a broad-spectrum approach for mental health promotion. *Health Education Research* 19 (4): 357–372.

Mostafa, T. (2016). Variation within households in consent to link survey data to administrative records: evidence from the UK millennium cohort study. *International Journal of Social Research Methodology* 19 (3): 355–375. http://dx.doi.org/10.1080/13645579.2015.1019264.

Mostafa, T. and Wiggins, R.D. (2015). How consistent is respondent behaviour to allow linkage to health administrative data over time? *CLS Working Paper 2015/03*.

Rubin, D.B. (1987). *Multiple Imputation for Nonresponse in Surveys*. New York: Wiley.

Sakshaug, J. and Kreuter, F. (2012). Assessing the magnitude of non-consent biases in linked survey and administrative data. *Survey Research Methods* 6 (2): 113–122.

Sakshaug, J., Couper, M., Ofstedal, M. et al. (2012). Linking survey and administrative records mechanisms of consent. *Sociological Methods and Research* 41 (4): 535–569.

Sakshaug, J., Hülle, S., Schmucker, A. et al. (2017). Exploring the effects of interviewer- and self-administered survey modes on record linkage consent rates and bias. *Survey Research Methods* 11 (2): 171–188. http://dx.doi.org/10.18148/srm/2017.v11i2.7158.

Sala, E., Burton, J., and Knies, G. (2012). Correlates of obtaining informed consent to data linkage: respondent, interview and interviewer characteristics. *Sociological Methods and Research* 41 (3): 414–439.

Sala, E., Burton, J., and Knies, G. (2014). Propensity to consent to data linkage: experimental evidence on the role of three survey design features in a UK longitudinal panel. *International Journal of Social Research Methodology* 17 (5): 455–473.

Seaman, S.R. and White, I.R. (2011). Review of inverse probability weighting for dealing with missing data. *Statistical Methods in Medical Research* 22 (3): 278–295.

da Silva, M.E., Coeli, C.M., Ventura, M. et al. (2012). Informed consent for record linkage: a systematic review. *Journal of Medical Ethics* 8 (10): 639–642.

Tate, A.R., Calderwood, L., Dezateux, C. et al. (2006). Mother's consent to linkage of survey data with her child's birth records in a multi-ethnic national cohort study. *International Journal of Epidemiology* 35: 294–298.

Thornby, M., Calderwood, L., Kotecha, M. et al. (2017). Collecting Multiple Data Linkage Consents in a Mixed Mode Survey: Evidence and Lessons Learnt from Next Steps. *CLS Working Paper*.

Umberson, D. and Montez, J.K. (2010). Social relationships and health: a flashpoint for health policy. *Journal of Health and Social Behavior* 51 (S): S54–S66.

Von Hippel, P. (2007). Regression with missing Ys: an improved strategy for analyzing multiply imputed data. *Sociological Methodology* 37 (1): 83–117.

Watson, N. and Wooden, M. (2009). Identifying factors affecting longitudinal survey response. In: *Methodology of Longitudinal Surveys* (ed. P. Lynn), 157–181. Chicester: Wiley.

Weissman, J., Parker, J., Miller, D. et al. (2016). The relationship between linkage refusal and selected health conditions of survey respondents. *Survey Practice* 9 (5): 1–13.

White, I.R., Royston, P., and Wood, A.M. (2011). Multiple imputation using chained equations: issues and guidance for practice. *Statistics in Medicine* 30 (4): 377–399. https://doi.org/10.1002/sim.4067.

Woolf, S.H., Rothemich, S.F., Johnson, R.E. et al. (2000). Selection Bias from requiring patients to give consent to examine data for health services research. *Archives of Family Medicine* 9: 1111–1118.

8

Consent to Data Linkage: Experimental Evidence from an Online Panel

Ben Edwards and Nicholas Biddle

Centre for Social Research and Methods, Australian National University, Canberra, Australia

8.1 Introduction

Longitudinal surveys have tended to have a strong focus on primary data collection, with most of the analytical information obtained directly from interviews. In the Australian context, this includes the Household, Income and Labour Dynamics in Australia (HILDA) survey (Watson and Wooden 2004), as well as the Longitudinal Study of Australian Children (LSAC, Edwards 2014). LSAC has made use of linked data, but this has tended to supplement rather than replace data collected through interviews. Furthermore, the research community has not made extensive use of this information.

In contrast, the next generation of longitudinal studies will likely make extensive use of linked data to augment survey responses. However, this will occur in the context of declining response rates to surveys and in consent to data linkage (Kreuter et al. 2016). There are a few unique features of longitudinal studies that make seeking consent for data linkage a different prospect than in cross-sectional surveys. Apart from the initial wave, consent requests for data linkage occur within the context of ongoing participation in the longitudinal survey, which reflects a higher level of commitment than a cross-sectional survey. Another unique feature of seeking consent for data linkage in longitudinal studies is the potential for data linkage requests to influence participation in subsequent survey waves (e.g. Das and Couper 2014). Third, more information is collected on individuals, increasing disclosure risk. To date, there have only been a limited number of studies investigating factors associated with consent to link in longitudinal studies (e.g. Chapter 7 of this book, Peycheva et al. 2021) and even fewer studies examining the effect of consent questions on subsequent attrition (e.g. Chapter 6 of this book, Jäckle et al. 2021; Sala et al. 2014).

Advances in Longitudinal Survey Methodology, First Edition. Edited by Peter Lynn.

Companion website: www.wiley.com/go/lynn/advancesinlongitudinalsurvey

Another recent development in survey methodology is probability based on online panels (Bosnjak et al. 2016; Couper 2017). Considerable investment has been made into these panels across a number of countries with examples from the United States (e.g. Pew American Trends Panel; Keeter & Weisel 2015), Germany (GESIS panel, www.gesis-panel.org), France (ELIPSS –Longitudinal Internet Studies for Social Sciences), Norway (the Norwegian Citizen Panel, http://www.uib.no/en/citizen#), Sweden (the Citizen Panel, https://lore.gu.se/surveys/citizen), Australia (Life in Australia, www.srcentre.com.au/our-research#life-in-aus), and Great Britain (NatCen Panel, http://natcen.ac.uk/our-expertise/methods-expertise/surveys/probability-panel).

Surveys on probability based online panels have several advantages in that they are cheaper and faster to implement than traditional longitudinal surveys but also have the capacity to deliver complex survey instruments. However, they also suffer from non-response – at the recruitment stage and then at subsequent waves (Couper 2017). Data linkage consent has only been the subject of limited research on these panels. To our knowledge, only one experimental study examining consent to data linkage has been conducted (Das and Couper 2014), and it concerned opt-out from data linkage, which is less commonly an option for researchers than explicit consent for data linkage.

8.2 Background

There are different types of information linked to survey data, with consent usually asked for specific sources of data. Evidence from observational studies where consent to link different types of data has been requested report consistent patterns of consent rates with consent to link highest for education records followed by health, with income or economic records having substantially lower levels of consent. For example, Sala et al. (2010) reported that in the British Household Panel Survey consent to link health and education (41% and 39%, respectively) was higher than economic records (32%), while subsequent research using Understanding Society suggested that data linkage was somewhat higher for education than for health (78% compared 67%; Baghal et al. 2014). Evidence from the Millennium Cohort Study about mother's willingness to consent to linkage of their own data also suggests that consent rates are higher for health compared to economic records (87% compared to 81%) and higher again for consent to link to their children's health and education records (93% and 94%, respectively; Mostafa 2014).

Several meta-analyses comparing response rates from different survey modes suggest response rates from online surveys are between 11 and 13 percentage points lower than other methods (Manfreda et al. 2008; Shih and Fan 2008; Wengrzik et al. 2016). However, it is unclear how differences in response rate might translate into differences in rates of consent for record linkage. Few longitudinal studies have asked for consent to data linkage solely online (e.g. LISS panel survey, Das and Couper 2014); however, there are a few examples of where this has been undertaken partly online within a mixed mode design. One of these is wave 8 of Next Steps: the Longitudinal Study of Young People in England, which involved a sequential mixed-mode design that involved online, followed by telephone and then face-to-face data collection. Wave 8 occurred when participants were aged 25. They were asked for consent to link to a very large number of administrative datasets (see Chapter 7 of this book, Peycheva et al. 2021). Data linkage consent rates for Next Steps varied between data sources. However, for income and pensions, health and education linkages consent rates on the web were around 30 percentage points lower than they were on telephone or face to face. For example, consent to link to education records was 61% online, compared to 88% via CATI or face-to-face, for health somewhat lower (online 57%; CATI 81%; and 86% face-to-face) and for government benefits even lower again (online 47%; CATI 75%; and 79% face-to-face). Another study has also examined data linkage consent in the context of a mixed-mode design (see Chapter 6 of this book, Jäckle et al. 2021). This study used an experimental design on the Understanding Society Innovation panel and will be described in Section 8.2.1.

Another sequential mixed mode study was undertaken for the 2015 cohort of the Longitudinal Surveys of Australian Youth (LSAY), where an online survey was followed by CATI (personal communication with Australian Department of Education and Training, the agency responsible for commissioning the survey). Data linkage consent was asked at age 16 years for academic achievement scores and subject choice in high school. In both instances, consent rates via online surveys were lower than CATI (academic achievement – online 81%, CATI 91%; senior secondary subjects – online 75%, CATI 88%). While sequential mixed mode designs lead to selectivity by mode, findings from these longitudinal studies are consistent with other evidence that suggests that consent rates overall are lower online than in other modes and that lower rates of consent can be expected for more sensitive records.

8.2.1 Experimental Studies of Data Linkage Consent in Longitudinal Surveys

Only a limited number of experimental studies have examined the effects of different ways of asking consent for record linkage in longitudinal studies. We were not able to identify any studies that have experimentally tested whether the type of record linkage (health, education, income and pensions, criminal records) affects consent rates. Salience of data linkage consent has been shown to increase consent rates within a questionnaire. Sala et al. (2014) in a series of randomised experiments using the Innovation Panel from Understanding Society showed that consent rates to record linkage were 7% higher when asked immediately after a series of questions on related topics, compared to asking for linkage consent at the end of the questionnaire (65% compared to 58%) with respect to record linkage of state benefits and other payments. Also, several studies have shown that rates of consent to data linkage are higher if consent is asked early in the questionnaire compared to at the end (Sakshaug and Vicari 2018; Sakshaug et al. 2013; Sala et al. 2014).

Length of the consent form for record linkage has only been the focus of limited experimental research. Das and Couper (2014) used a random subsample of 500 households from the LISS panel and conducted a two-factor experiment examining the length of consent text (short and extended text) and mode of communication (letter or email). Importantly, participants could opt out to record linkage of their pension and benefits and other data about health and well-being, rather than being asked to opt-in. The authors found that the probability of opting out was higher when asked by letter compared to email. The extended text resulted in lower opt-out rates than the shorter text. However, there was an interaction between the two conditions with the combination of a short message and email having highest opt-out rates (9.6%), but otherwise results indicated that the way in which respondents were informed did not have much effect (2.6–4.9% opt-out rate).

Das and Couper (2014) also note that there were no differences in attrition rates at the 12-month follow up, an important finding in the context of the potential concerns for implementing data linkage consent in longitudinal surveys. However, further research on length of consent form is required, as it is unclear what would occur in the context of explicit opt-in consent. Several studies note that providing explicit consent through signing a consent form and asking for explicit consent to linkage information such as social security number leads to lower rates of consent (Bates 2005; Sala et al. 2014; Singer 2003).

Das and Couper (2014) also reported that those who received the extended text had a greater understanding of what linkage was involved, and lower perceptions of risk associated with record linkage. Sala et al. (2014) found that non-consenters to data linkage were far more likely to express concerns regarding sharing of confidential data than those who consented. Their findings indicated that trust in the survey organisation was a more important reason to consent to data linkage than an understanding of the data linkage process. In a correlational study, Sakshaug et al. (2012) raised privacy and confidentiality concerns as a barrier to consent to linkage to administrative data. Using the Health and Retirement Study, they found that privacy and confidentiality concerns raised by participants was associated with lower rates of data linkage consent. In contrast, they also hypothesised that participants in the same study who received a benefit from the agency for whom data linkage was requested would be more likely to consent than those who did not have a pre-existing relationship. Findings provided mixed report for this hypothesis, with no evidence of social security recipients being more likely to consent. However, consent rates were higher for those who received other types of government income (Sakshaug et al. 2012).

Wording of the consent to data linkage in terms of losses or benefits has also been a focus of a series of experimental studies (Kreuter et al. 2016; Sakshaug et al. 2015; Sakshaug et al. 2019a; Sakshaug et al. 2019b). In the first study based on a telephone survey in the United States, respondents were told either that their survey responses would be a lot more valuable if they consented (benefit) or that they would be much less valuable if they did not consent (loss; Kreuter et al. 2016). The loss condition had a consent rate 10 percentage points higher than the benefit condition (as expected based on the literature around loss aversion). In an extension of the loss–gain framing study using a CATI survey of German workers, Sakshaug et al. (2015) tested an additional condition, referring to either the usefulness of responses already provided to the survey (referred to as sunk costs) or the responses to be provided (prospective costs). There was a significant interaction between the loss-gain and the sunk-prospective costs conditions. When the prospective-costs version of consent was used, there were significantly higher consent rates for the gain-framing than loss-framing conditions (4 percentage points). When the consent form emphasised the usefulness of responses already provided (sunk-costs), then the loss-framing condition had marginally higher consent rates (1.5 percentage points), but this was not statistically significant. Sakshaug et al. (2019a) focus on the extent to which gain and loss framing interacted with the placement of consent questions in the questionnaire (beginning or end) in a telephone and web survey. They reported that linkage consent rates in the web survey were 11 percentage points higher with loss-framing compared to gain-framing when the request was made at the end of the survey. Data linkage consent at the beginning of the survey was found to be higher regardless of survey mode and framing (at least 11 percentage points higher). Finally, Sakshaug et al. (2019b) reported that emphasising in a telephone interview the time savings from consent to data linkage yielded a small but statistically significantly higher consent rate compared to a neutral consent request. Another benefit frame that highlighted the improved study value showed no differences to a neutral request.

Research examining the effect of survey mode on consent using the ninth wave of the Understanding Society Innovation panel suggests that web surveys have substantially lower consent rates than face-to-face interviews (Chapter 6 of this book, Jäckle et al. 2021). Two-thirds of the sample were randomly allocated to web first, while the remainder was administered face-to-face interviews from wave 5 and in the ninth wave they were asked to consent to link their information about financial conduct and pensions and benefits. Jäckle et al. (2021) reported that consent rates were 30 percentage points lower (after adjustment for selection effects) for the online group compared to the face-to-face respondents, a substantial difference in consent rates.

In summary, there is a growing literature based on observational studies, but quite limited experimental research examining factors associated with data linkage consent in longitudinal surveys, particularly for online panels. Moreover, factors associated with understanding data linkage and perceptions of risk are also not well understood (Chapter 6 of this book, Jäckle et al. 2021). We do know that web surveys have lower response rates in general, that response rates are declining irrespective of survey mode, and that administrative data linkage can reduce response burden and, if collected prospectively, give an understanding of the selectivity of attrition in longitudinal surveys.

8.3 Research Questions

This chapter fills some of the gaps in our knowledge by examining several research questions:

1. Do requests for data linkage consent affect response rates in subsequent waves?
2. Do consent rates depend on (a) the type of data for which linkage is requested; (b) survey mode (online versus other); and (c) the length of the request?
3. Is understanding of the data linkage process affected by the type or length of data linkage request? Does understanding vary by other survey and demographic characteristics?
4. Are perceptions of the risk of data linkage affected by the type or length of data linkage request? Does risk vary by other survey and demographic characteristics?

We address these research questions using data from experimental studies carried out on a probability-based online panel in Australia. There are several interesting features of our contribution. First, we believe it to be the first ever study of explicit consent to data linkage using a randomised experiment on a representative online probability panel. Second, most studies have been undertaken in Europe or the United States, we are not aware of any experimental studies of data linkage consent being undertaken in Australia. This is important because Australia has a comparatively heterogeneous population with half of Australians either born overseas or having one or both parents born overseas (Australian Bureau of Statistics 2017). Finally, our study experimentally manipulates several survey design features including whether or not consent for data linkage is sought, the type of data linkage consent sought and the length of the consent question. Through two randomised experiments on an online panel in Australia, we test two sets of hypotheses. In Study 1 (at wave 11), we randomly allocated respondents either to receive no linkage consent question or to be asked for consent to link to one of three types of data (income, health or education). In Study 2 (at wave 13, and restricted to sample members who either did not participate in wave 11 or were allocated to the no-consent question treatment), we again randomly assigned sample members to consent questions regarding linkage to three different data types (income, health or education), but for two of these data types (income, health) we additionally administered either a short or a long version of the consent form. Furthermore, study 2 included additional questions about understanding of data linkage and perceptions of risk associated with data linkage. This study design allows us to address all four of the research questions set out above.

8.4 Method

8.4.1 Data

We use data from waves 11, 12, and 13 of the Life in Australia Panel (Pennay et al. 2018), Australia's first national probability-based online panel, which was established in 2016. Panel members were randomly recruited via their landline or mobile phone and provided their contact details so that they could take part in surveys on a regular basis. A dual-frame random digit dialling (RDD) sample design was employed to recruit panel members. Seventy percent were approach via mobile phone and 30% via landline. To select participants from the landline sample where two or more adults were present, an alternating next/last birthday method was used. The recruitment rate was 0.211 calculated according to the American Association of Public Opinion Research (AAPOR) standards for dual-frame surveys. LINA also includes people with and without internet access. Those with no internet access or those who are not comfortable completing surveys over the internet are able to complete surveys by telephone. Panel members receive a $20 incentive for joining the panel and another $10 to $20 for each survey they complete. Benchmarking studies of the LINA panel (Pennay et al. 2018) and other probability online panels (Chang and Krosnick 2009; Walker et al. 2009; Yeager et al. 2011) suggest that such panels can be developed to closely match key demographic and social characteristics.

8.4.2 Study 1: Attrition Following Data Linkage Consent

In wave 11 of the online sample of LiNA, 1774 completed the survey and 1078 did not (a response rate of 62%). Fieldwork for wave 11 occurred between 15 November and 3 December 2017. In this wave we randomly allocated participants at the start of the survey to the following conditions:

(1) No data linkage consent ($n = 871$).
(2) Data linkage consent:
  (a) Centrelink records (income support receipt, pensions, and benefits) ($n = 299$).
  (b) Medicare and Pharmaceutical Benefits Scheme (health records) ($n = 301$).
  (c) Education records ($n = 303$).

Consent forms were based on existing consent forms used in the Longitudinal Study of Australian Children (for income and health) and *Understanding Society* (for Education) and were provided to participants through a series of screens in the survey.[1] In wave 11 consent for data linkage was asked at the end of the survey but due to a programming error the answers were not recorded. The survey for wave 11 featured, in order, questions about important historical events in respondents' lives; alcohol use, alcohol advertising and licensing; and demographic questions.

In this study, we were able to test whether response rates at wave 12 differed between the four treatment groups. Fieldwork for wave 12 occurred between 22 January and 11 February 2018.

8.4.3 Study 2: Testing the Effect of Type and Length of Data Linkage Consent Questions

In wave 13 of LINA we repeated the record linkage consent experiment to obtain the consent to data linkage information that was not collected in wave 11 along with information about understanding and risk perceptions about the data linkage process. Those who were randomly allocated to the no-consent message treatment in wave 11 were asked data linkage consent in wave 13 ($n = 873$[2]).

For those who did not complete the survey in wave 11 ($n = 1078$), 271 were no longer active panel members by wave 13 and the remaining 807 panel members were randomly allocated to the treatment and control conditions at the sample allocation phase (data linkage: Yes, $n = 404$; No, $n = 403$).

Seventy percent of those in the data linkage condition in wave 13 participated (90.5% of those who participated in wave 11 and were not asked for data linkage consent ($n = 790$) and 29.5% of those who did not participate in wave 11 but were allocated to data linkage consent in wave 13 ($n = 119$)). Of the 909 eligible to participate, 7 were away for the duration of wave 13, and two refused to participate in the data linkage consent.

1 Details of the consent forms can be found at https://csrm.cass.anu.edu.au/sites/default/files/docs/2020/9/Consent_forms_003.pdf.
2 Two participants in wave 11 who were not randomly allocated to data linkage consent attrited from the panel by wave 13.

Respondents allocated to the data linkage consent condition were randomly allocated to one of five groups as follows[3]:

(1) Centrelink records:
  (a) Short form ($n = 151$).
  (b) Long form ($n = 149$).
(2) Health records (Medicare and Pharmaceutical Benefits Scheme):
  (a) Short form ($n = 148$).
  (b) Long form ($n = 150$).
(3) Education records ($n = 301$).

It should be noted that with approximately 150 per group our statistical power is somewhat limited to detecting large true differences.

For Centrelink and health records, the long consent version was based on real-life examples from government-funded studies already underway in Australia, and there was no equivalent long-form for education records. The education records consent form was based on the consent forms in Understanding Society.

In wave 13 consent for data linkage used the same data linkage consent forms and followed the same protocol as in wave 11. Data linkage consent was asked at the end of the survey. The survey for wave 13 focused on Australian views of foreign affairs including perceptions of other countries and leaders and finally political identification. In wave 13, participants were also asked several questions adapted from Das and Couper (2014) about their understanding of the data linkage process (question ordering was randomised and they were asked after the question on consent to data linkage). Specifically:

> *To help us understand whether the explanation we showed you about linking [GOVERNMENT AGENCY] data with study responses was clear or unclear, we would like to show you some statements about data linkage. By answering these questions you are not giving us consent to link your responses with any other data.*
> *Please specify whether you think each of the following statements is true or false based on the information you were shown.*
>
> a. *[GOVERNMENT AGENCY] will combine the information they have with your answers to this study.*
> b. *Researchers using the data will only have access to anonymous data.*
> c. *The combined data can be used by [GOVERNMENT AGENCY] to check up on you.*

---

3 One participant dropped out of the survey prior to being asked the data linkage consent question.

d. *[GOVERNMENT AGENCY] will send us the information they have about you.*
e. *Your name, address, sex, and date of birth will be saved with the linked data.*
f. *We will send your name, address, sex, and date of birth to [GOVERNMENT AGENCY].*

To test for the level and effect of understanding of data linkage, we calculated an 'understanding score' that has a minimum of −6 (all six items incorrect) to a maximum of 6. For each item, do not know was coded as 0, −1 for an incorrect answer, 1 for correct (as per Das and Couper 2014). Respondents who indicated items b, d, and f to be true were correct, while respondents who indicated that a, c, and e were false were correct.

To examine respondent perceptions of risk associated with data linkage, we also derived an index of perceived risk based on the same six questionnaire items as the understanding measure. Following Das and Couper (2014), we calculated a risk index with minimum of −6 (all six items perceived as least risky) to a maximum of 6, the greatest possible risk perception. For each item, 'don't know' was coded as 0, −1 as least risky, and 1 as risky. If respondents indicated true to the following items they were scored as risky – a, c, d, e, and f. Item b was scored as risky if the respondent indicated that this was false.

Fieldwork for wave 13 occurred between 5 March and 19 March 2018. Given the complexity of the longer consent forms, only online participants in wave 13 were eligible for this study. We used survey weights to offset the limitation of this methodological decision. Online survey completion took an average 21.2 minutes. The ANU Human Research Ethics Committee approved the conduct of this research (ANU Human Ethics Protocol 2017/910).

8.5 Results

8.5.1 Do Requests for Data Linkage Consent Affect Response Rates in Subsequent Waves? (RQ1)

Based on Study 1, there were no statistically significant differences in response rates at wave 12 for those who were not asked (93.2%) and those who were asked for data linkage consent (92.6%) at wave 11 ($\chi^2(1) = 0.26$, $p = 0.61$). Consent to data linkage was randomised in wave 11, but as an additional robustness check we also ran a logistic regression adjusting for gender, country of birth, device type, language other than English (LOTE), education, age, Aboriginal or Torres Strait Islander status, household structure, and area-level socioeconomic disadvantage

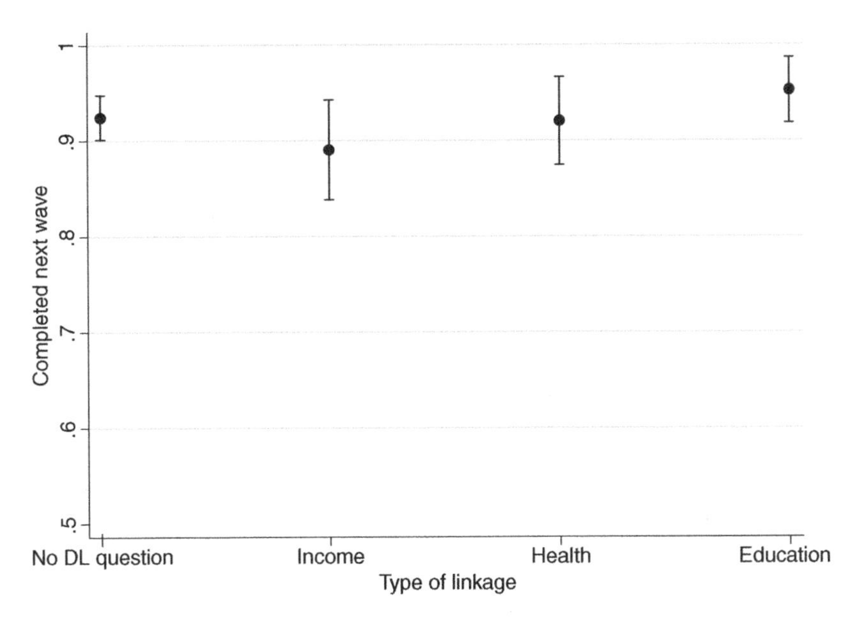

Figure 8.1 Predicted probability of completing the next survey wave by type of linkage consent requested. Source: Based on LiNA Wave 12.

(see Table A2 for details). Again, there was no statistically significant association between whether data linkage consent was requested and response rate in the subsequent wave ($t = -0.10$, $p = 0.92$) and the predicted percentages were very similar (no consent = 92.5%, 95% CI = 90.1–94.8; Consent = 92.3%, 95% CI = 89.7–94.9).

Figure 8.1 shows that while wave 12 response rates were marginally lower for respondents who were asked for consent to income record linkage at wave 11 (0.89) than for those who were asked about linkage to health (0.92) or education data (0.95), there were no statistically significant differences between the groups overall ($F(3, 1728) = 1.42$, $p = 0.24$).

In addition to the bivariate comparisons, we also undertook logistic regression on response rates that adjusted for the same variables that we used to model attrition. Results from this logistic regression also showed that the type of linkage requested did not affect response rates (Table A3).

8.5.2 Do Consent Rates Depend on Type of Data Linkage Requested? (RQ2a)

Figure 8.2 shows the proportion consenting to data linkage by type. Participants asked to consent to linkage to income support data had significantly lower consent rates (0.197, 95% CI: 0.153–0.241) than participants asked about data linkage to health (0.356, 95% CI: 0.298–0.413) or education records (0.409, 95% CI: 0.353–0.465). However, even for education records where consent rates were the highest, only two in five people consented to data linkage.

We also tested the effect of type of data linkage consent adjusting for other variables (those in Table 1) through a logistic regression model (see Table A4). We used survey weights to adjust for non-response and sample representativeness. There was a main effect of type of data linkage consent asked of participants ($F(2, 858) = 13.22$, $p = 0.0000$) consistent with findings from Figure 8.2.

Sakshaug et al. (2012) have hypothesised that having a pre-existing relationship with an administrative agency may lead to higher rates of consent to data linkage in the case of pre-existing medical or health conditions. In the instance of income support and pensions, participants who are in receipt of these payments may be concerned about the possibility that information provided about income and personal circumstances is inconsistent to survey responses and therefore be less likely to consent to data linkage. In the first wave of data collection of LINA participants were asked whether they possessed a health care card. In general, people are eligible for a health care card if they receive some form of government

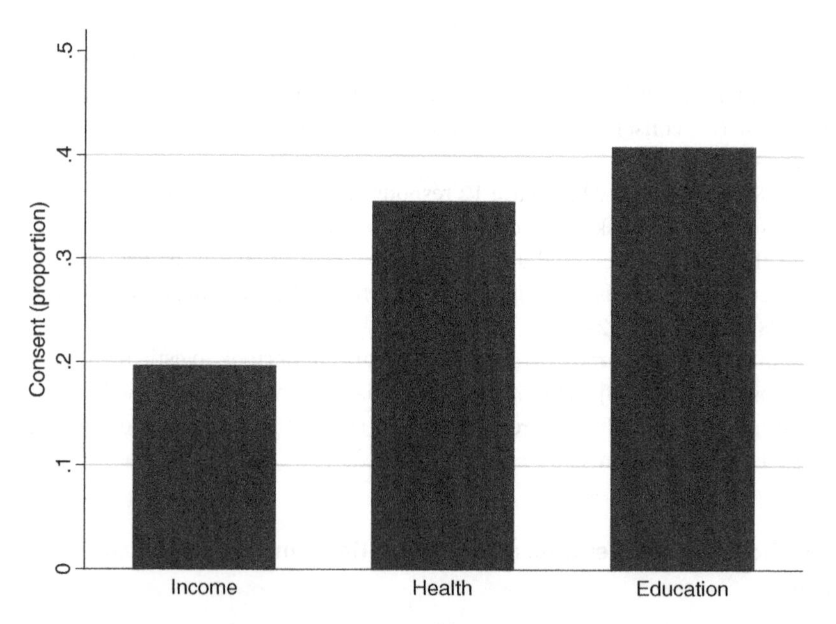

Figure 8.2 Proportion of data linkage consent by type. Source: Based on LiNA Wave 13; n = 300 (Income), 298 (health), 301 (education).

income support.[4] To further examine the pre-existing relationship hypothesis, we also tested whether higher levels of educational attainment were associated with greater propensity to consent to data linkage of education data. To test the pre-existing relationship hypothesis, interaction effects were tested between type of data linkage requested and (i) health care card and (ii) education level in a logistic regression model with the same covariates as in the previous analyses. Neither interaction effect was statistically significant (health care card: $F(2, 858) = 1.20, p = .30$; education: $F(6, 858) = 1.13, p = .34$) (see Tables A5 and A6).

8.5.3 Do Consent Rates Depend on Survey Mode? (RQ2b)

Consent rates (Figure 8.2) were much lower than have been reported in other longitudinal studies that have asked for consent to data linkage online (Next Steps: 47–61%, Chapter 7 of this book, Peycheva et al. 2021, LSAY: 75–81%, Australian Department of Education and Training 2018; Understanding Society: 48%, Chapter 6 of this book, Jäckle et al. 2021). These surveys are either cohort or panel studies that (i) had been underway for several years; and (ii) routinely have longer interviews than in an online panel such as LiNA. The only data linkage consent study that used an online panel employed an opt-out consent methodology, which had only 5.1% participants opt-out (Das and Couper 2014). Our findings suggest that asking for consent to data linkage using an opt-in format may well lead to low levels of consent; however, we only had limited encouragement to consent to link administrative data and further replication studies are needed in different countries.

8.5.4 Do Consent Rates Depend on the Length of the Request? (RQ2c)

Next we tested whether having a short or longer form for consent to data linkage to income and health data had an effect on consent rates. The consent rates for the short form (0.284, 95% CI: 0.231–0.337) were somewhat higher than for the long form (0.266, 95% CI: 0.216–0.316). We again tested for differences in rates of data linkage consent with a logistic regression model interacting type (income or health) by consent length (short or long) adjusting for the other covariates in

4 People are eligible for a Health Care Card if they live in Australia and are in receipt of one of the following – Newstart Allowance, Sickness Allowance, Youth Allowance (as a job seeker), Partner Allowance, Parenting Payment (partnered), Widow Allowance, Special Benefit, Carer Payment (for short term or irregular care less than six months), the highest rate of Family Tax Benefit Part A, Mobility Allowance (if not getting Disability Support Pension), Carer Allowance for a child under 16, Farm Household Allowance. Information on eligibility for each of these payments can be found at www.servicesaustralia.gov.au/individuals/services/centrelink.

Table 1 and using survey weights. There was no main effect of having a shorter consent form (OR = 1.46, p = .325, 95% CI: 0.69–3.09) but there was again a main effect of type, with health record linkage consent significantly more likely than income data linkage consent (OR = 3.42, p = .001, 95% CI: 1.58–6.66). The interaction between type and consent length was not statistically significant (OR = 0.74, p = .558, 95% CI: 0.26–2.05, see Table A7).

8.5.5 Effects on Understanding of the Data Linkage Process (RQ3)

Based on the understanding score developed by Das and Couper (2014), the level of understanding of data linkage in the Australian population is low. The median score on the understanding scale was 0 and the mean was 0.78. Fully 16% scored less than half correct, and only 2.5% got all six items correct.

To describe variation in understanding by demographic characteristics and survey features, we ran an OLS regression model with the 'understanding score' as the dependent variable, and a range of independent variables including type of data linkage consent, time spent completing the understanding items, and a range of socio-demographic and other indicators (see Table 8.1). Table 8.1 shows that type of data linkage consent was associated with level of understanding ($F[2, 839] = 7.05$, $p = .0009$) with those being asked income support data linkage consent having significantly lower levels of understanding than those being asked consent for linkage to education records (the difference was 0.39 standard deviation units). The consent forms for data linkage consent to income support were particularly detailed and bureaucratic in nature and included technical terminology for different payments (e.g. Newstart Allowance). In comparison, the education consent form was written in a more accessible, less technical manner.

Time spent reading consent to data linkage information was also associated with greater levels of understanding (Effect Size, ES = 0.23) while greater time spent completing the six items was associated with lower levels of understanding (ES = 0.09). In both instances, the size of these associations was small. University educated participants had significantly higher levels of understanding when compared to those with a vocational qualification. There were no other differences in understanding by levels of education. There were significant differences in level of understanding by age ($F(7, 839) = 3.18$, $p = .0025$) with those over 65 years having the lowest levels of understanding. These age groups had significantly lower levels of understanding of data linkage than those 34 years and under, who had the highest levels of understanding (Figure 8.3). There were also differences in levels of understanding for individuals living in different types of households ($F(5, 839) = 2.56$, $p = .026$). Single parents had significantly lower levels of understanding than those in couple households (couple only or with children). Level of socioeconomic disadvantage by area was also associated with

Table 8.1 Factors associated with understanding and perceived risks of data linkage.

	Understanding		Risks	
Variables	**B**	**P**	**B**	**P**
Type of data linkage (ref: Income)				
Health	0.28	0.161	0.15	0.537
Education	0.77	0.000	−0.59	0.022
Gender (ref: Male)				
Female	0.29	0.081	−0.78	0.000
Other/refused			−0.85	0.484
Age group (ref: 18–24 yr)				
25–34	0.36	0.370	0.06	0.899
35–44	−0.54	0.181	0.50	0.288
45–54	−0.14	0.718	−0.51	0.278
55–64	−0.27	0.495	−0.18	0.697
65–74	−1.00	0.018	−0.31	0.529
75+	−1.27	0.006	−0.69	0.261
Refused	0.10	0.866	−2.86	0.106
Education (ref: University)				
Trade/certificate/diploma qualification	−0.48	0.022	0.65	0.003
Year 12 or equivalent	−0.33	0.318	−0.16	0.637
Year 11 or less	−0.32	0.194	0.67	0.068
Aboriginal or Torres Strait Islander	−0.34	0.473	−1.99	0.002
Speaks a language other than English at home	−0.27	0.291	0.36	0.300
Country of birth (ref: Australian born)				
Mainly non-English speaking background	−0.39	0.155	−0.05	0.884
Mainly English speaking background	−0.17	0.537	−0.19	0.533
Refused	−1.23	0.013	2.83	0.025

(Continued)

Table 8.1 (Continued)

	Understanding		Risks	
Variables	**B**	**P**	**B**	**P**
Household structure (ref: Living alone)				
Couple only	0.18	0.606	−0.20	0.538
Couple with children	0.04	0.925	−0.13	0.717
Single parent	−0.73	0.090	0.67	0.142
Non-related adults sharing	−0.48	0.371	0.14	0.800
Other household type	−0.32	0.499	0.78	0.172
Healthcare card	0.41	0.064	0.08	0.758
SEIFA (Ref: Quintile 1 –Most disadvantaged)				
Quintile 2	0.90	0.005	−0.20	0.608
Quintile 3	0.16	0.614	0.26	0.506
Quintile 4	0.65	0.017	−0.05	0.892
Quintile 5 –Least disadvantage	0.38	0.168	−0.04	0.905
Missing	0.16	0.751	0.84	0.377
Device type (ref: Desktop or undefined)				
Mobile	0.04	0.825	0.10	0.66
Tablet	−0.24	0.356	0.33	0.347
Time spent reading (minutes)	0.02	0.011	−0.0002	0.030
Time completing questions (minutes)	−0.01	0.003	0.00	0.18
Intercept	0.39	0.505	−0.70	0.245

*$p < .05$, **$p < .01$, ****$p < .001$; Understanding, $R^2 = .15$; $N = 840$; Risk, $R^2 = .09$; N = 840.

understanding of data linkage ($F(5, 839) = 2.40$, $p = .0357$). Those living in the most disadvantaged quintile had significantly lower levels of understanding than those in the second and fourth most disadvantaged quintile.

To test whether length of form made a difference, we tested the same regression model restricting the sample to the Centrelink and Medicare groups (because the form length experiment was restricted to those groups). We included indicators for type of data linkage request, length of consent protocol, and the interaction

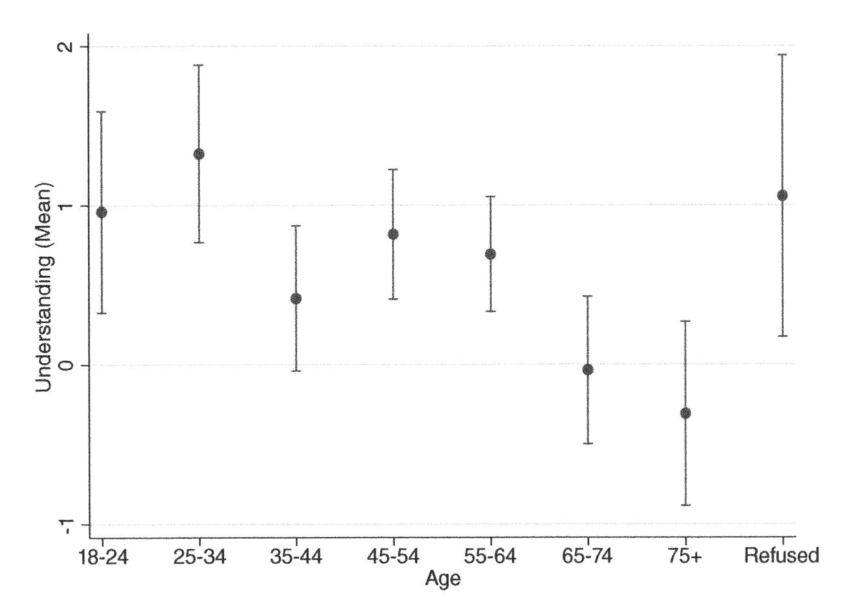

Figure 8.3 Predicted mean levels of understanding by age. Source: Based on LiNA Wave 13.

between these two variables. There was no statistically significant difference in level of understanding for these factors.

8.5.6 Effects on Perceptions of the Risk of Data Linkage (RQ4)

Based on the risk perception index, the Australian population perceives data linkage to have only a limited risk. The median score for the risk index was −1, with a mean of −1.14, and only 18% of respondents scoring 1 or above. Only 1.8% of the population perceived data linkage to have the greatest risk (score of 6).

To model the variation in the risk index by demographic characteristics and survey features, we ran the same regression model as with understanding. Type of data linkage consent was associated with level of risk ($F(2, 558) = 4.81$, $p = .008$) with those being asked about education data linkage having a lower perception of the level of risk (ES = 0.20) compared to income support. Perceptions of risk when linking to health records were comparable to income. Women perceived less risk associated with data linkage than men (a 0.27 standard deviation difference). Aboriginal or Torres Strait Islander people had much lower perceptions of risk of data linkage (ES = 0.68). Those who spent more time reading information about consent also perceived lower levels of risk due to data linkage.

The final regression model suggested that there was no evidence of a difference in level of perceived risk by length of data linkage request ($p = .93$) and no interaction between length and type of data linkage request (Centrelink or Medicare).

8.6 Discussion

In this study there is little evidence to suggest that asking for opt-in consent to data linkage increases attrition rates in online panel surveys. This accords with earlier findings from Das and Couper (2014) from an opt-out data linkage consent experiment in the LISS panel that showed the same response rates for those who were asked for consent and for those that were not. We extend these findings by examining whether the type of record linkage consent requested leads to different response rates and find that there is again no statistically significant difference in response rates by type of data linkage consent when compared to no request.

Consent rates from the online panel in our study (27.8%) were less than half the opt-in consent rates in household panel and child cohort studies based on face-to-face interviewing, such as Understanding Society and the Millennium Cohort Study (67–94%, Baghal et al. 2014; Mostafa 2014). This is consistent with other research that suggests the online surveys have lower response rates than other modes of data collection (Manfreda et al. 2008; Shih and Fan 2008; Wengrzik et al. 2016) and underscores the importance of the interviewer to explain and engender trust.

Our findings also suggest that the more sensitive the data linkage requested, the lower the consent rate, with consent rates substantially lower for linkage to income support and pensions (20%) than for health or education records (36% and 41%, respectively). The pattern of differences between types of record linkage accords with findings from observational studies that employ face-to-face interviewers (Baghal et al. 2014; Peycheva et al. 2021; Sala et al. 2010). However, the consent rates were much lower in LINA than in other studies that have asked for record linkage consent online (Next Steps: 47–61%, Chapter 7 of this book, Peycheva et al. 2021, LSAY: 75–81%, Australian Department of Education and Training 2018; Understanding Society: 48%, Chapter 6 of this book, Jäckle et al. 2021).

In our experiment, the only information provided to participants was a consent form designed to be administered during a face-to-face interview. In LSAC on which two of three consent forms were based, participants were provided with a primary approach letter indicating that record linkage consent would be requested, and the consent forms were provided ahead of the interview so that participants could have time to read the consent. In other studies, such as Next Steps, a 16-page booklet was sent in advance along with a link to an animated video explaining data linkage and the importance of the process. A future study might replicate our randomised approach with a range of approach letters and information booklets, and test experimentally for differences in consent rate. Nonetheless, the existing evidence suggests best-survey practice should lead to higher rates of consent, and certainly for education record linkage the consent rates appear high enough to warrant the effort (around 80%). Although replication of these findings is needed, there is a question about the value of asking online for consent for linkage to more sensitive records such as income support and pensions. These lower consent rates, coupled with the fact that a record linkage match rate of 95% of those who have consented is considered best practice, further limits the utility of income record linkage for online panels.

Contrary to some earlier studies (Das and Couper 2014) we did not find that a longer consent form was significantly associated with higher rates of data linkage consent. Indeed, it may be noted that consent rates for the shorter form were about 2 percentage points higher. We also found no effect of longer consent forms on understanding or perceptions of risk. Importantly, we did find that the time spent actually reading the form was associated with higher levels of understanding and lower perceptions of risk. Further research might usefully test ways of encouraging more thorough reading of consent forms, either through prompting or checking of understanding. We also identified that there were significant differences in understanding between those being asked for their consent to record linkage to income support and benefits compared to education records. With higher levels of understanding reported for the education record linkage than the income support linkage. In part, this may be due to the complexity of the Australian income support and benefit system when compared to the more intuitive education records.

One specific aspect of our study is that the level of informational support provided was limited when compared to contemporary practice. In our study we provided a consent form as the sole piece of information about the record linkage at the time of the survey. In practice, contemporary studies will provide information in pre-approach letters, provide detailed brochures and animated videos so that participants have a variety of information channels and time to make an informed decision (e.g. Bailey et al. 2017). It is highly likely that this will increase overall consent rates; however, it is unlikely to affect the validity of our findings as these studies also report that consent rates depend on the sensitivity of particular record linkages (Australian Department of Education and Training 2018; Chapter 7 of this book, Peycheva et al. 2021).

There are numerous areas of further research including but not limited to replication of our findings in other contexts. It may be worthwhile re-asking those who have declined to consent, as other research suggests that people who decline will often consent at a subsequent wave (Jäckle et al. 2021; Sala et al. 2014). Another area that should be explored is the acceptability and ethics of opt-out record linkage consent. There is some evidence from other studies that people are much more willing to consent to record linkage if they are passively consenting (Hunt et al. 2013; Sala et al. 2014). This fits within the broader behavioural economics literature on the power of defaults (Johnson and Goldstein 2003). However, we do not know whether this is due to apathy and disinterest and a lack of understanding of the implications of their decision, or whether it is due to the default being seen as an explicit endorsement by a trusted organisation. With opt-in consent to record linkage, both our study and Das and Couper (2014) reported low levels of understanding of the record linkage process.

The newly introduced European General Data Protection Regulation, with its emphasis on explicit consent, may well mean that opt-out consent is no longer an option for many surveys. Therefore, considerable effort may be required to boost consent rates. For instance, in Next Steps, multiple channels of information about the data linkage process were provided. However, further research on the extent that this approach leads to higher consent rates is recommended in the sequential mixed mode context, given that the most cooperative respondents are likely to complete their survey online. Further research should systematically measure the effects of the complexity of the data linkage consent forms.

In summary, asking for consent to record linkage appears to make no difference to subsequent attrition but linkage to income support and pensions data is particularly sensitive and has lower consent rates than linkage to education or health data. Our findings suggest that the length of consent forms does not seem to aid participant understanding, although this may be due to our study being somewhat underpowered to detect smaller effects. There is also a suggestion that if participants read the form, higher levels of understanding and reduced perceptions of risk may result.

References

Australian Bureau of Statistics (2017). *Cultural Diversity in Australia*. Australian Bureau of Statistics: Canberra, Australia.

Australian Department of Education and Training. (2018). Personal communication.

Baghal, T., Knies, G., and Burton, J. (2014). Linking administrative records to surveys: Differences in the correlates to consent decisions. *Understanding Society Working Paper Series,* No. 2014–09.

Bailey, J, Breeden, J, Jessop, C. et al. (2017), Next Steps Age 25 Survey – Technical Report, http://www.cls.ioe.ac.uk/shared/get-file.ashx?id=3282&itemtype=document

Bates, N.A. (2005). Development and testing of informed consent questions to link survey data with administrative records. In: *Proceedings of the American Statistical Association*, 3786–3793. Miami Beach, FL: (12–15 May, 2005). Available at: https://

pdfs.semanticscholar.org/b48c/5befb8bb8fd4e19db9955955f7176932a1c9.pdf (accessed April 10, 2018).

Bosnjak, M., Das, M., and Lynn, P. (2016). Methods for probability-based online and mixed-mode panels: selected recent trends and future perspectives. *Social Science Computer Review* 34 (1): 3–7.

Chang, L. and Krosnick, J. (2009). National Surveys via RDD telephone interviewing versus the internet: comparing sample representativeness and response quality. *Public Opinion Quarterly* 73 (4): 641–678.

Couper, M.P. (2017). New developments in survey data collection. *Annual Review of Sociology* 43: 121–145.

Das, M. and Couper, M.P. (2014). Optimizing opt-out consent for record linkage. *Journal of Official Statistics* 30 (3): 479–497.

Edwards, B. (2014). Growing up in Australia: the longitudinal study of Australian children: entering adolescence and becoming a young adult. *Family Matters* 95: 5–14.

Hunt, K.J., Schlomo, N., and Addington-Hall, J. (2013). Participant recruitment in sensitive surveys: a comparative trial of 'opt in' versus 'opt out' approaches. *BMC Medical Research Methodology* 13 https://doi.org/10.1186/1471-2288-13-3.

Jäckle, A., Beninger, K., Burton, J. et al. (2021). Understanding data linkage consent in longitudinal surveys. In: *Advances in Longitudinal Survey Methodology* (ed. P. Lynn). Chichester: Wiley.

Johnson, E.J. and Goldstein, D. (2003). Do defaults save lives? *Science* 302 (5649): 1338–1339.

Keeter, S. and Weisel, R. (2015). *Building Pew Research Center's American Trends Panel*. Pew Research Centre http://www.pewresearch.org/files/2015/04/2015-04-08_building-the-ATP_FINAL.pdf.

Kreuter, F., Sakshaug, J.W., and Tourangeau, R. (2016). The framing of the record linkage consent question. *International Journal of Public Opinion Research* 28 (1): 142–152.

Manfreda, L., Bosnjak, M., Berzelak, J. et al. (2008). Web surveys versus other survey modes: a meta-analysis comparing response rates. *International Journal of Market Research* 50 (1): 79–104.

Mostafa, T. (2014). Variation within households in consent to link survey data to administrative records – Evidence from the UK Millennium Cohort Study. *Centre for Longitudinal Studies Working Paper,* 2014- No. 8.

Pennay, D.W., Neiger, D., Lavrakas, P.J., et al. (2018). The online panels benchmarking study: A total survey error comparison of findings from probability-based surveys and non-probability online panel surveys in Australia. ANU Centre for Social Research and Methods Working Paper No. 3/2018.

Peycheva, D., Ploubidis, G.B., and Calderwood, L. (2021). Determinants of consent to administrative records linkage in longitudinal surveys: evidence from next steps. In: *Advances in Longitudinal Survey Methodology* (ed. P. Lynn). Chichester: Wiley.

Sakshaug, J.W. and Vicari, B.J. (2018). Obtaining record linkage consent from establishments: the impact of question placement on consent rates and bias. *Journal of Survey Statistics and Methodology* 6: 46–71.

Sakshaug, J.W., Couper, M.P., Ofstedal, M.B. et al. (2012). Linking survey and administrative records: mechanisms of consent. *Sociological Methods and Research* 41 (4): 535–569.

Sakshaug, J.W., Tutz, V., and Kreuter, F. (2013). Placement, wording, and interviewers: identifying correlates of consent to link survey and administrative data. *Survey Research Methods* 7: 133–144.

Sakshaug, J.W., Wolter, S. and Kreuter, F. (2015). Obtaining record linkage consent: Results from a wording experiment in Germany. Survey Insights: Methods from the Field. Retrieved from http://surveyinsights.org/?p=7288

Sakshaug, J.W., Schmucker, A., Kreuter, F. et al. (2019a). The effect of framing and placement on linkage consent. *Public Opinion Quarterly* 83 (S1): 289–308. https://doi.org/10.1093/poq/nfz018.

Sakshaug, J.W., Stegmaier, J., Trappmann, M. et al. (2019b). Does benefit framing improve record linkage consent rates? A survey experiment. *Survey Research Methods* 13 (3): 289–304. https://doi.org/10.18148/srm/2019.v13i3.7391.

Sala, E., Burton, J., and Knies, G. (2010). Correlates of obtaining informed consent to data linkage: Respondent, interview and interviewer characteristics. *ISER Working Paper Series,* No. 2010–28.

Sala, E., Knies, G., and Burton, J. (2014). Propensity to consent to data linkage: experimental evidence on the role of three survey design features in a UK longitudinal panel. *International Journal of Social Research Methodology* 17: 455–473.

Shih, T.-H. and Fan, X. (2008). Comparing response rates from web and mail surveys: a meta-analysis. *Field Methods* 20 (3): 249–271.

Singer, E. (2003). Exploring the meaning of consent: participation in research and beliefs about risks and benefits. *Journal of Official Statistics* 19: 273–285.

Walker, R., Pettit, R., and Rubinson, J. (2009). The foundations of quality initiative. *Journal of Advertising Research* 49 (4): 464–485.

Watson, N. and Wooden, M. (2004). *Assessing the quality of the HILDA Survey wave 2 data HILDA Project Technical Paper Series 5/04*. Melbourne: Melbourne Institute of Applied Economic and Social Policy, University of Melbourne http://www.melbourneinstitute.com/hilda/hdps/htec504.pdf.

Wengrzik, J., Bosnjak, M., and Manfreda, L. (2016). Web surveys versus other survey modes—a meta-analysis comparing response rates, General Online Research Conference, Dresden, Germany (March 2–4, 2016).

Yeager, D.S., Krosnick, J.A., Chang, L. et al. (2011). Comparing the accuracy of RDD telephone surveys and internet surveys conducted with probability and non-probability samples. *Public Opinion Quarterly* 75 (4): 709–747.

9

Mixing Modes in Household Panel Surveys: Recent Developments and New Findings

Marieke Voorpostel[1], Oliver Lipps[1] and Caroline Roberts[2]

[1] *Surveys Unit, FORS, Lausanne, Switzerland*
[2] *Institute of Social Sciences, University of Lausanne, Lausanne, Switzerland*

9.1 Introduction

Survey researchers have always adapted their methods in response to the latest societal and technological developments (Groves 2011), but the growth of alternative, lower-cost modes of data collection provided by the internet has resulted in a sea change in how surveys are carried out. Web-based surveys allow data to be gathered from (more) respondents more quickly and cheaply than is feasible with interviewer-administered surveys, but they need to be used in combination with other methods to ensure adequate representation. While mixing web with other modes of data collection may be effective at reducing fieldwork costs and selection errors, it can create a number of complications for researchers conducting the survey (in terms of questionnaire development and fieldwork management) and analysing the survey data (because data gathered in different modes may not be comparable) (De Leeuw 2005). These complications are particularly challenging in the context of longitudinal surveys of households, and many large-scale academic panel studies have been reluctant to move to mixed mode designs. Yet recent years have seen a growing interest in the potential benefits to be gained from mixed-mode panels, and a growing empirical base from which to draw conclusions about their actual impact on various outcomes of interest.

This chapter addresses the complexities of mixing modes in household panel surveys with a focus on how collecting data using multiple modes affects response rates and sample composition in any given panel wave, as well as longitudinally. We first outline the major challenges of mixing modes in longitudinal household panels. We then provide an overview of the combinations of modes that are used in the household panels included in the Cross-National Equivalent File (CNEF)

Advances in Longitudinal Survey Methodology, First Edition. Edited by Peter Lynn.

Companion website: www.wiley.com/go/lynn/advancesinlongitudinalsurvey

(Frick et al. 2007)[1] and a number of other ongoing household panels that have mixed modes for data collection purposes. Next, we present the design and findings of the first wave of a two-wave pilot study for the Swiss Household Panel, testing the feasibility of using the web both as the primary mod, as well as sequentially, in combination with telephone and face-to-face.

9.2 The Challenges of Mixing Modes in Household Panel Surveys

Combining multiple modes of data collection (e.g. telephone, face-to-face, mail, and web) in surveys need not be problematic per se. What matters for ensuring the effectiveness of mixed-mode surveys and for minimising the negative consequences is *how* different modes are combined. If the goals are to reduce costs and selection errors, then certain mode combinations are likely to be more effective than others. For example, so-called 'push-to-web' or 'web first' designs are intended to maximise the proportion of the sample responding in the lowest-cost mode, and hence, the cost-saving benefits of mixing modes (Dillman 2017). Similarly, certain follow-up modes may be unsuitable for certain subgroups in the population (e.g. due to noncoverage or low mode penetration), so the choice and sequencing of mode combinations is key to effectively reducing selection errors. Irrespective of these considerations, once individual sample units complete a given set of survey questions via a different mode of administration, there is a risk of differential measurement error between modes, which may affect the accuracy of aggregate estimates derived from the data and hinder comparisons between groups of interest more or less likely to respond in one mode over another (De Leeuw 2005).

The peculiarities of mixed-mode survey designs present both opportunities and challenges for longitudinal surveys. Yet, whereas there is now an extensive body of literature on mixed-mode survey designs in cross-sectional studies (see e.g. Hox et al. 2017; Tourangeau 2017), empirical findings on the effects of mixing modes in longitudinal studies are more limited (although, see Jäckle et al. 2017, for a recent review). In particular, the ways in which mixed-mode designs affect participation over time and the sample composition of panels has not been widely or systematically studied in different contexts. Watson and Wooden (2009, p. 164) remarked that while 'a large number of large panel studies … have changed mode, rarely has the impact of this change on response rates been reported' and that they were 'unaware of any experimental evidence gathered from a longitudinal

1 We include only the CNEF household panels that are still active. Hence, for example the discontinued Canadian Survey of Labour and Income Dynamics is excluded from this overview.

design'. Dex and Gumy (2011) similarly pointed out the lack of analyses on the consequences of the mode of interview for subsequent wave attrition in longitudinal surveys. However, many longitudinal studies have now started to implement mixed-mode designs in an effort to reduce non-response and fieldwork costs, and insight into the consequences for response rates and sample composition is growing.

There are several reasons why longitudinal studies have been slow to adopt mixed-mode survey designs. First, the focus of longitudinal surveys is to produce data that is comparable over time, to permit the measurement of change. Attrition must be kept to a minimum to ensure a sufficient number of cases for longitudinal analyses. This makes it less attractive or even risky to change modes between waves or to experiment with alternative modes, as the mode switch could affect respondents' willingness to participate and disrupt the continuity of measurement, leading to bias in measures of change (in different ways and for different respondents, depending on how modes are mixed) (Dillman 2009). For household panels, a further complexity is that members of the same household may complete the questionnaire in different modes. Besides affecting the comparability of measurements within households and over time, this could potentially complicate the operation of the survey, and compound the already complex data structures of household panel studies. Moreover, mixed-mode measurement could complicate the estimation of survey weights for adjusting longitudinal analyses if these are based on data from questionnaire variables.

Despite these drawbacks, however, the longitudinal context of panel studies can facilitate the implementation of mixed-mode designs. A high-response-rate mode can be used for recruitment and lower-cost modes for subsequent data collection. Contact details such as email addresses can be collected in one wave for use in a subsequent wave, and it may be possible to offer people modes based on their expressed preferences at an earlier wave (Olson et al. 2012), thereby reducing respondent burden. Furthermore, questionnaire data collected in earlier waves can be used for the purposes of isolating differential mode effects on selection and measurement (Tourangeau 2017).

Given these considerations, more and more panel studies have been exploring how to incorporate alternative modes, but only cautiously, trying to change as little as possible to maximise comparability over time. Often, this has involved using alternative modes for add-on questionnaires or specific sub-samples (samples incorporated from other studies, samples of extremely loyal respondents, rarely using the core samples). In addition, a number of longer-running panels have made the switch from face-to-face interviewing using paper questionnaires to face-to-face interviewing using computers (Laurie 2003; Schräpler et al. 2006; Watson and Wilkins 2015). More recently, more substantial changes to data collection methods have been incorporated into a number of longitudinal studies. We provide an overview of how modes have been mixed in the following household panels[2]:

- the German Socio-Economic Panel (SOEP)
- the Household, Income, and Labour Dynamics Australia (HILDA) Survey
- the Panel Study of Income Dynamics (PSID)
- the UK Household Longitudinal Study (UKHLS)
- the Korean Labor and Income Panel Study (KLIPS)
- the Swiss Household Panel (SHP)

9.3 Current Experiences with Mixing Modes in Longitudinal Household Panels

9.3.1 The German Socio-Economic Panel (SOEP)

The SOEP is a long-running household panel that started in 1984. The main mode of interview is face-to-face interviewing. Initially, the interviewers used paper questionnaires. computer-assisted personal interviewing (CAPI) was introduced in 1998 with a controlled experiment. After confirming that mode-effects were negligible (Schräpler et al. 2006), the SOEP progressively replaced the paper questionnaire with CAPI interviewing (Wagner et al. 2007).

The most recent core samples of the SOEP (since 2011) only use face-to-face interviewing. The main reason for opting for face-to-face interviewing is to ensure good data quality and allow questionnaire modules that are not viable in other modes, such as cognitive tests and behavioural experiments (Huber 2017). For the older samples, alternative modes are still available. For the core samples that started before 2011, respondents can complete the questionnaire in paper format, and return it by mail (a solution first used as a refusal conversion strategy for sample members who did not agree to further interviewer visits). For these older samples, the SOEP also allows other household members to complete paper questionnaires while the interviewer is interviewing another household member, to decrease the total length of the interviewer visit. How successful these alternative

2 Other household panels conduct interviews exclusively face-to-face and hence are not discussed in this chapter: the Russian Longitudinal Monitoring Survey of HSE (RLMS-HSE), the South-African National Income Dynamics Study (NIDS), the Morocco Household Survey Panel (MHSP), the China Family Panel Studies (CFPS), the Singapore Panel Study on Social Dynamics (SPSSD), and the Ukrainian Longitudinal Monitoring Survey (ULMS). The Japan Household Panel Survey (JHPS/KHPS) originally used different modes in two separate samples but has a unified approach since 2014 (drop-off and pick-up self-completion questionnaire). To our knowledge, no studies have been published on the impact the different modes had on participation rates and the sample composition of the JHPS.

modes are in terms of limiting (selective) attrition, or whether there are any mode effects on measurement has not been documented.

Alternative modes are also used in other samples included in the study. In 2014, two samples of the study 'Families in Germany', which started in 2010, were integrated in the SOEP. These samples consist of single parents, households with three or more children, and low-income households, who are approached by a telephone/web hybrid mode, followed by face-to-face interviewing. In the hybrid group, telephone interviewers contact each household to encourage online participation and collect information on the composition of the household. If the respondent cannot or does not wish to participate online, the telephone interviewer arranges a face-to-face interview.

In a quasi-experimental setting, the SOEP recently tested the effects on response rates of switching from CAPI to web-based data collection, for samples that had participated in the panel for at least four years (Lüdtke and Schupp 2016). A web-plus CAPI sequential design was used in order not to lose panel members who were not able or willing to respond online. Non-response rates were higher in the mixed-mode group compared with the standard CAPI mode, but there were no systematic patterns of non-response based on panel members' characteristics. However, the study revealed an interesting effect of interviewer continuity. While having the same interviewer return to the same households is in general positively related to continued participation (Rendtel 1990), the reverse effect was observed in the wave of the mode switch among certain sub-samples: households with high interviewer continuity were less willing to participate under the new mode conditions.

Moreover, there were effects of internet literacy on the mode choice in the sequential design. The study therefore recommends that longitudinal surveys planning to switch modes in the next wave use a sequential mixed-mode design with the principal mode from previous waves offered as alternative option. Moreover, there may be cost advantages of implementing the online alternative as early as possible in the lifetime of a face-to-face panel since the likelihood of participating in web mode appears to decrease as rapport builds between panel members and interviewers (Lüdtke and Schupp 2016).

9.3.2 The Household, Income, and Labour Dynamics in Australia (HILDA) Survey

The HILDA survey is a nationwide household panel survey in Australia that started in 2001. The HILDA interviews all household members aged 15 or older face-to-face in their homes. The respondents also fill out an additional self-completion questionnaire. Initially, interviewers used paper questionnaires. Since 2009, interviewers use laptops or tablets. This change went well: moving from PAPI to CAPI, a change that went together with a change of fieldwork agency and an increase in the value of incentives, resulted in a higher re-interview rate in wave 9 (2009). No effect was found on item non-response overall, except for an increase in the amount of missing data on a monetary question (Watson 2010). Effects on interview length depended on the extent to which dependent data could be used, and whether the interviewer used a laptop or a tablet (Watson and Wilkins 2015).

The HILDA Survey also uses computer-assisted telephone interviewing (CATI), but this mode is used as a last resort when the respondent prefers it, and in cases where a household moves outside of the reach of the interviewer network. Hence, CATI is used to keep as many respondents in the survey as possible (Watson and Wooden 2004). For waves 1–14, the percentage of individual questionnaires completed by telephone ranged from 0.5 (wave 1) to 10.1 (wave 8) (Watson and Wooden 2015). A study conducted on the first four waves showed that a CATI interview did not decrease the likelihood of next wave non-response. Instead, in fact, there was a very small positive effect (Watson and Wooden 2009).

Whether interviews were conducted by telephone instead of face-to-face affected the response rates of the self-completion questionnaires, which are used for questions that require more time, such as on expenditure, and questions concerning attitudes. Interviewers who visit respondents' homes either take the completed questionnaires with them when they leave or visit the address at least once more to pick them up. When the interview is conducted by telephone, the self-completion questionnaire is mailed and respondents are asked to return the completed questionnaire by mail, which generally results in lower response rates compared to when interviewers collect them in person (Watson and Wooden 2015).

The HILDA Survey has also experimented with conducting the self-completion questionnaire component of the survey by web, using their 'Dress Rehearsal Sample', which is about one-tenth the size of the main sample and is used to test survey content and methods. The reason for the web trial was mostly that interviewers reported that respondents had asked for it, rather than driven by cost saving considerations. Respondents in the trial were offered a choice between face-to-face and web. Not many chose web, and the overall response rate was lower in the group offered a choice of mode. Currently, there are no plans to introduce web as a mode in the main sample of the HILDA Survey.

9.3.3 The Panel Study of Income Dynamics (PSID)

The American PSID started in 1968 as a face-to-face survey, and became primarily a telephone survey in 1973. Computer-assisted telephone interviewing was introduced in 1993, although every wave about 2–3% of households are contacted or interviewed in person in their homes. The PSID was run as an annual survey until

1997 after which it switched to a biennial frequency. The study differs in design from the other household panels in the CNEF consortium in that only one respondent per household is interviewed, who provides information on all household members (McGonagle et al. 2012).

The PSID implemented its first mixed-mode study in 2014 (McGonagle and Freedman 2017), aimed at testing the advantages of offering panel members a web-based questionnaire. They invited all household heads who completed the 2013 wave of the PSID, and their spouses/partners, to complete a supplemental study recalling childhood experiences. Respondents who reported in 2013 that they had an internet connection at home were assigned to the web protocol (73%), and the remainder was assigned to the 'choice' protocol. Both groups were initially invited to participate by web, but the choice group was informed that a paper questionnaire would be sent after two weeks, while the web group was informed that they could request a paper questionnaire. All nonrespondents in the web group received a paper questionnaire in a final mailing after 10 weeks. The majority of respondents from the web group who responded only after receiving the final mailing used the paper questionnaire (76%). Overall, response rates were higher and item non-response rates lower in the web group compared with the choice group.

In 2016, the PSID implemented a second mixed-mode supplemental survey, focused on well-being and daily life (Freedman et al. 2017). These data were collected among household heads and spouses/partners aged 30 years or older in the PSID in 2015. There were again two protocols: web and 'choice'. Respondents assigned to the web protocol were invited to complete the study by web and sent a paper copy after six weeks. The 'choice' group was also invited to participate by web but were informed that a paper copy would be sent in two weeks. Panel members were assigned the treatments on the basis of their predicted probability of responding in a given mode, obtained by modeling the mode of response to the first mixed-mode study as a function of internet and device use and a number of demographic characteristics. Respondents were either assigned to a 'likely web' group ($p > .70$, assigned to the web treatment), 'likely paper' group ($p < .30$, assigned to 'choice' treatment), and a 'no likely mode' group for the remaining respondents, who were assigned at random to either of the two protocols. Compared to the likely web and likely paper groups, those with no likely mode had lower response rates and required more fieldwork effort. Overall response rates from the 'no likely mode' group were similar for both protocols, but there was a higher share of completions by web in the web-first protocol.

With the aim of maintaining high response rates and data quality, the PSID has also taken steps to develop a web version of the instrument as a data collection tool for the main part of the study (see McGonagle et al. 2017). They tested a web version of the PSID in 2015–2016 on two samples: a sample from an external internet panel, to test the translation of the telephone interview to self-administered web, and a sample of highly cooperative respondents from the PSID, representing those most likely to complete the questionnaire by web without extensive follow-up, who, additionally, were from households with fewer than five members that had not experienced any family composition changes since the 2013 PSID interview. The initial sample consisted of 200 individuals, 160 of whom tried to log in. Of these 160 respondents, 61 were not eligible due to changes in their household composition and, hence, were not invited to complete the web questionnaire. All the remaining 99 respondents completed the questionnaire.

As the PSID telephone interview takes between 60 and 90 minutes, a main concern with the implementation of the web questionnaire was survey length. It took respondents longer on average to complete the web questionnaire compared to the telephone interview, mainly because they took breaks. However, the subjective experience of the respondents was the opposite: interview length was perceived as shorter by web. In sum, the results were encouraging, but were based on a small sample of highly cooperative respondents who had previously participated in a web questionnaire and who would have little difficulty completing the household roster, as they did not need to report any changes in household composition.[3]

9.3.4 The UK Household Longitudinal Study (UKHLS)

The UKHLS is a large household panel study that started in 2009, incorporating the former British Household Panel Study that started in 1991. The UKHLS always uses CAPI in the first wave in which a household participates, and up until its most recent waves, it interviewed all household members aged 16 and older face-to-face, while household members aged 10–15 filled out a self-completion questionnaire. CATI was used for a very small group of households (between 0% and 2.4% in waves 1–6), where members did not agree to a house visit by an interviewer.

The Innovation Panel of the UK Household Longitudinal Study (UKHLS-IP) has been experimenting with mixed-mode data collection since 2009 (Jäckle et al. 2017). In wave 2 of the UKHLS-IP (2009), CAPI was compared with two telephone protocols, one aiming to maximise the number of household members interviewed by phone, the second switching to face-to-face as soon as it became clear a house visit would be needed for at least one household member (Lynn 2013). This experiment showed that response rates were lower in the two sequential mixed-mode designs that used CATI followed by CAPI than in the single-mode CAPI protocol.

3 The Child Development Supplement (CDS) of the PSID also uses a mixed-mode approach. The CDS interviews every five years children 0–17 years of age (face-to-face) and their caregivers (face-to-face and telephone) in the PSID families (McGonagle and Sastry 2015). Older children complete their own interviews using audio-computer-assisted interviewing or interactive voice response technology.

In wave 5 (2012) of the UKHLS-IP, an experiment included web as an alternative mode (Jäckle et al. 2015). There were two treatment groups: the standard CAPI protocol and a protocol inviting respondents to participate online with a CAPI follow-up. It was found that the mixed-mode design initially resulted in a lower proportion of households fully responding (to the household and individual questionnaires), lower individual response rates and higher item non-response rates. Yet, after three waves, these initial differences disappeared, leading to comparable cumulative response rates between the CAPI only and mixed-mode groups (Bianchi et al. 2017).

Following these encouraging results, the UKHLS has recently started to implement a sequential mixed-mode design in the main study (Carpenter and Burton 2017; Jäckle et al. 2017). This web-first protocol was first introduced in wave 7 (2015/2016), when it was offered to persons in households that did not respond in wave 6. In wave 8, 40% of the main sample was invited to participate online.[4] This included all households in which any person had completed the survey by web in wave 7 and non-responding households from wave 7, with the allocation of remaining households to mode protocol based on probabilities to complete the web survey, estimated from UKHLS-IP data. A face-to-face follow-up was used for anyone who did not complete the web survey. Also, 20% of the sample was only invited to complete a face-to-face interview. Within households, the same protocol was applied for all household members.

Initial findings from the main sample show no evidence of damage to response rates over time. Rather, performance of web as a mode of interviewing improves each wave. The web survey is particularly successful for previous non-respondents. Respondents are offered an additional bonus incentive to complete the survey online before a certain date, and receive additional mailings. These measures have been judged to save more costs than they add, by averting the need to send face-to-face interviewers to the field (Benzeval et al. 2017). Preliminary findings suggest there are substantial effects on selection, but there seem to be very few mode effects on measurement (personal communication).

9.3.5 The Korean Labour and Income Panel Study (KLIPS)

Like most household panel studies, the KLIPS is an annual face-to-face study that interviews all household members. Although there is no intention to change the mode of interviewing in the KLIPS in the near future, the KLIPS team has explored the use of web as a potential parallel mode.

4 This percentage was increased to 60% at wave 9 and 70% at waves 10 and 11.

In this experiment, a sample of 20931 individuals (not KLIPS respondents) representative for South Korea in terms of gender, age, occupation, education, and income level were sent a preliminary survey (by email), collecting information on a number of demographic characteristics, intention to participate in the experiment, and mode preferences (face-to-face, web, or 'Either will do'). Of this sample, 4224 individuals replied and 816 households were selected so that the ratio by region and household size matched that of the KLIPS 19th wave. Of these 816 households, 280 were assigned to CAPI and 536 households to web. Target responding sample sizes for the two groups were 150 households for the CAPI control group and 300 in the web group. The fieldwork continued until the target sample sizes were reached, resulting in 322 individuals from 151 households in the control group (CAPI) and 509 individuals from 307 households in the experimental (web) group. The two groups did not differ significantly in terms of household size, education, gender, or occupation, but the web group was on average younger than the CAPI group.

Households in the web group were more likely to start the interview (92%) compared with the CAPI group (64%), but the web group had a significantly lower completion rate (defined as a completed household questionnaire and a completed individual questionnaire from at least one household member). Ninety-nine percent of eligible household members in responding households completed the individual questionnaire in the CAPI group, versus 90% in the web group.

9.3.6 The Swiss Household Panel (SHP)

The SHP started in 1999 and conducted interviews exclusively by telephone. Since 2010, however, it has offered alternative modes to respondents who were reluctant to participate in prior waves. Households that are unwilling to respond by telephone are allowed to complete the household and individual questionnaires with a face-to-face interviewer, while a web-based version of the individual questionnaire is offered after an initial refusal or stated reluctance to participate. In practice, these alternative modes are rarely used (for example, 0.5% of the household questionnaires were completed face-to-face and 1.3% of the individual questionnaires were completed by web in 2015). At the start of the second refreshment sample in 2013, the SHP also offered households without telephone numbers face-to-face (8.9% of the recruited households completed the household questionnaire face-to-face in 2013). The majority of the face-to-face respondents from wave 1 in 2013 have subsequently participated by telephone.

The small sample of household members who completed the individual questionnaire online had different characteristics compared with respondents who completed the telephone interview, which is not surprising, given that it was only offered to initially refusing panel members. Respondents who completed the survey by web were more likely to be male, younger, in couples with children, and employed full-time (Dangubic and Voorpostel 2017).

9.4 The Mixed-Mode Pilot of the Swiss Household Panel Study

9.4.1 Design of the SHP Pilot

Unlike most of the other household panels (with the exception of the PSID), the SHP is a telephone survey. Declining response rates in telephone surveys and the increasing costs associated with boosting response have led the SHP to consider alternative modes of interviewing for its next refreshment sample (the SHP_IV, scheduled to start in 2020). Furthermore, advances in information and communication technologies and the increasing use thereof, suggest that web-based data collection offers a particularly promising alternative to telephone in the Swiss context. However, differences between CATI and web are large in terms of the features most likely to result in measurement differences – e.g. interviewer presence, interview pace, and the channels of communication used (oral or visual) (Couper 2011), so comparability between the modes should be evaluated.

Also, while most studies have assessed the effect of a switch in modes after a panel has run for a number of years, the SHP is considering *recruiting* a refreshment sample in a different mode than the one used until now, which for large-scale household panel surveys has not yet been tested (though see Lynn 2020). It is often assumed that web mode would be inappropriate for panel recruitment purposes (Tourangeau 2017) because of its typically lower response rates, but it is unclear whether this argument still holds in the context of a panel recruited by telephone. Like all large-scale household panels, the SHP has a complex design, including two levels of interview (household and individual), for which there are three different questionnaires (including the household 'grid', for which a suitable household reference person must be selected), and the longitudinal aspect of the study. In addition to helping to convince households to participate, an interviewer may be useful in guiding the responding households through the different steps. Online self-completion of the grid questionnaire may put the respondent off participating in the rest of the study, and may provoke motivated misreporting (Tourangeau et al. 2012, 2015). Thus, a comparison of the effects of CATI and web on panel cooperation was needed before proceeding.

To test the ways in which offering web as an alternative mode affects response rates, sample composition and measurement, the SHP mounted a two-wave pilot study, incorporating a mixed-mode experiment. Here, we present the design and the preliminary results of the first (recruitment) wave of data collection with respect to response rates and sample composition. The second wave will take place in 2019. The results of this pilot after two waves will inform the design for the SHP_IV.

For the purposes of the study, the Swiss Federal Statistical Office (SFSO) provided a random probability sample of 4195 adult individuals from their frame based on population registers.[5] The sampling frame contains a small number of socio-demographic variables on all sampled individuals as well as on their household members. The sampled person was the first contact in the household and was asked to complete the grid. If the sampled person was a young adult presumably living with his/her parents, he or she was replaced by one of the presumed parents as the first point of contact in the household, on the assumption that the latter would have better knowledge of the financial situation of the household, and would be better able to complete the household questionnaire, once the grid was completed.

To test how different ways of including web would compare to the current design of the SHP, sample members were randomly assigned to one of three treatment groups. In the control group (group 1), the design was as similar as possible to the current SHP; thus, the dominant mode of data collection was telephone interviewing. Telephone numbers are matched to the sampling frame and supplied with the sample by the SFSO, but for legal reasons, they are only able to provide listed landline numbers. The SFSO delivered a telephone number for 60% of the households. Households without a known landline number were sent an invitation letter asking them to communicate a telephone number. If the fieldwork agency could not obtain a telephone number, the household was assigned to a face-to-face interviewer. At the start of the fieldwork, 29% of households in group 1 were allocated to face-to-face interviewers (some of these cases were later referred to telephone interviewers, where telephone numbers were subsequently obtained). Sampled individuals received a mailing inviting them to participate in the study with a cash incentive of Sfr.10 and were contacted by telephone (or face-to-face) to complete the grid and household questionnaire. Refusals on the household level were re-contacted by telephone if a number was available. In a second stage, all eligible household members received a personal invitation to participate in the study with an incentive of Sfr.10. If a household reference person completed the interview by telephone, household members could respond to the individual questionnaire by phone or by web. Household members could only participate in a face-to-face interview if the reference person was also interviewed in this mode. Reluctant and

5 As the SFSO drew an individual sample rather than a sample of households, the sample is not representative for household size. Individuals living in single-person households are underrepresented (6.3% in the sample versus 16% in the population) and in larger households overrepresented (22% lives in households of at least five persons, versus 15% in the population).

refusing household members were given the option to complete the questionnaire online.

The two treatment groups tested alternative ways of including web-based data collection. Group 2 followed the current design of the SHP for the household level, and included web only on the individual level. The main idea behind this approach was to recruit the households offline, and then move the second part of the data collection online. Hence, in group 2 the recruitment was done by telephone or face-to-face. At the start of the fieldwork, 31% of the households were allocated to a face-to-face interviewer. Households in group 2 received a request for one person to participate in a short, 15-minute interview to complete the grid and household questionnaire, and in a second step, each adult in the household was invited to complete an online individual questionnaire. If a household agreed to participate in the study, the interviewer continued with the grid and, once the household reference person was established, the household questionnaire. Although we could have pressed to continue with the individual questionnaire for the household reference person at that point and only asked other household members to participate online, we chose to push all individual questionnaires to web. The main reason for this was to avoid having the reference person responding in a different mode to other household members, which would likely introduce systematic bias in the data. All household members, including the reference person, received a letter with login details inviting them to complete the individual questionnaire by web. If an individual was not willing or able to participate online, the interviewer conducted the interviews by telephone or face-to-face (depending on the mode used for the household questionnaire). Hence, a combination of modes within households was possible, but also rather exceptional (see below).

In group 3, we tested a fully online design in which household members were asked to complete the grid, household questionnaire and individual questionnaires online. In a first step, the sampled individual received a letter inviting them to participate in the study with login details and a cash incentive of Sfr.10 (the letter also mentioned they could contact the fieldwork agency to request a telephone interview if so desired). Non-respondents received two reminders two weeks apart, and those with a known landline number were re-approached by telephone for a CATI interview. After the grid and household questionnaire was completed, all eligible household members, including the reference person, received a letter inviting them to complete the individual questionnaire by web with login details and an incentive of Sfr.10. Household members without a known telephone number received two reminders by mail. If a telephone number was available, the household member was approached by phone for a telephone interview after one reminder.

The groups were not of equal size. Group 1 contained 790 households, 2192 households were assigned to group 2, and 1213 households to group 3. As in wave 2, group 2 will be divided into two experimental groups; this group is the largest. As response rates, in web surveys are generally lower compared to telephone surveys, group 3 included more households than group 1 to obtain a large enough sample size to compare responding households on various characteristics.

9.4.2 Results of the First Wave

The pilot was in the field for two months at the beginning of 2018. Table 9.1 shows the obtained response rates in the three groups.

9.4.2.1 Overall Response Rates in the Three Groups

We calculated response rates on the household level (for the grid and the household questionnaire) as a percentage of the total sample (all sampled households were assumed to be eligible at the time of sampling). Response rates on the household level were comparable in group 1 and 2 (53% and 52%), but lower in group 3 (47%). When we compare households with at least one individual questionnaire completed (of households completing the grid questionnaire), the respective response rates were 90%, 81%, and 78%. The lower percentage in group 2 compared with group 1 is most likely due to the mode switch after the household questionnaire. In group 3, as in group 2, logins and incentives for the individual questionnaires were sent in separate mailings after the household grid had been completed, because first the presence and eligibility of all household members had to be assessed. This, not surprisingly, led to a larger share of households for which only a grid was completed but no other questionnaires (10% of the households in group 1, 19% in group 2, and 22% in group 3). With respect to response rates at the individual level, we assessed the number of individual questionnaires completed as a share of all eligible household members in participating households (based on information on age in the household grid). The share of individual questionnaires completed amongst participating households was 69% in group 1, 67% in group 2, and 62% in group 3.

9.4.2.2 Use of Different Modes in the Three Groups

Of the responding households, 14% completed the household grid face-to-face in group 1 and 15% in group 2. For the household questionnaire, these percentages were almost identical (13% and 15%, respectively). In group 1, the majority of individual questionnaires were conducted by telephone (85%), and 9% of the interviews were conducted face-to-face. In line with the current design of the SHP, web was only offered if no telephone number was available or as a way to convince reluctant sample members, and was used by 6% of the respondents. In group 2, all

Table 9.1 Completed questionnaires in different modes for three treatment groups.

	Group 1		Group 2		Group 3	
	N	%	N	%	N	%
Total household sample	**790**		**2192**		**1213**	
	Household Grid					
Telephone	371	86	982	85	92	15
Face-to-face	58	14	171	15	–	
Web	–		–		509	85
Total	**429**	100	**1153**	100	**601**	100
Response rate grid		**54**		**53**		**50**,***
	Household Questionnaire					
Telephone	363	87	970	85	88	15
Face-to-face	54	13	169	15	–	
Web	–		–		486	85
Total	**417**	100	**1139**	100	**574**	100
Response rate household		**53**		**52**		**47**,*****
	Individual Questionnaire					
Total eligible household members	**1027**		**2699**		**1425**	
Telephone	603	85	195	11	87	10
Face-to-face	65	9	24	1	–	
Web	39	6	1579	88	792	90
Total	707	100	1798	100	879	100
Response rate individual		**69**		**67**		**62***,*****
Grid completed but no individual questionnaire[a]	44	10	218	19***	133	22***
Complete households (at least one individual interview)[a]	**385**	**90**	**935**	**81*** **	**468**	**78*** **
Complete households (all eligible household members)[a]	**226**	**53**	**669**	**58***	**289**	**48,*** **

Notes: All household-level response rates are calculated as a percentage of the total sample of households. All individual-level response rates are calculated as a percentage of the total sample of eligible household members in participating households (according to information on age in the grid). ***, **, * significant on 1%, 5%, 10% level (first set of asterisks denote significant difference to first group, second set of asterisks denote significant difference to second group).

a) Base: households with completed grid.

eligible household members were initially directed to web, and of all respondents to the individual questionnaire, 88% completed this questionnaire by web. Still 11% chose to complete the interview by telephone, and only 1% face-to-face. In group 3, which had web as the main mode, 15% of households and 10% of individuals still chose to participate by telephone, either after contacting the fieldwork agency to request this or in the follow-up phase when non-respondents with telephone numbers were approached by phone.

Within a household, different persons could complete the individual questionnaire in different modes. However, this rarely happened. For the vast majority of households (97%), all household members responded in the same mode. In 47 households, individual questionnaires were completed by phone and by web; in 4 households face-to-face and by web. Households with a combination of modes among household members were more likely to be part of groups 1 and 2 than group 3. Of all households in which two persons participated, 95% used a single mode for all individual questionnaires; this was 94% of all three-person households and 96% for all households of four persons or more.

9.4.2.3 Household Nonresponse in the Three Groups

We used information from the grid and the register-based sampling frame to compare respondents to the total sample in each group on a number of characteristics.[6] We assessed differences in age, civil status, language, gender, nationality, household size, and presence of a registered landline (Table 9.2).

Table 9.2 shows that overall, the sample composition of the responding households in all three groups reflects the gross sample composition relatively well. Where differences exist, they tend to go in the same direction in all groups. Households with a registered telephone number, with reference persons who are married, hold Swiss nationality, and were born in Switzerland are overrepresented, and single-person households are underrepresented in all three groups. Yet, in group 3 the differences between the gross sample and the respondents with regard to the presence of a listed telephone number were smaller. With regard to the age distribution, group 2 was the only group that did not show any significant differences with the gross sample. Whereas unmarried individuals were less likely to participate in groups 1 and 2, group 3 included a lower share of widowed individuals compared to the gross sample. No significant differences with respect to gender emerged.

6 We used information from the grid questionnaire for the respondents and from the sampling frame for the non-respondents. These two sources of information are almost identical (between 98.4% and 100% identical, depending on the variable considered).

Table 9.2 Socio-demographic characteristics of selected and responding households within each group.

Auxiliary Variables	(1) Gross sample Group 1 %	(2) Respon-dents Group 1 %	(3) Gross sample Group 2 %	(4) Respon-dents Group 2 %	(5) Gross sample Group 3 %	(6) Respon-dents Group 3 %
N	790	429	2192	1153	1213	601
Age group						
<=30	9.2	5.6 ***	8.5	7.4	8.0	6.2 *
31–44	24.1	22.1	24.8	23.7	23.7	25.1
45–58	35.3	35.2	34.3	34.7	35.0	37.1
59+	31.4	37.1 **	32.4	34.3	33.2	31.6
Marital Status						
Single	19.2	14.5 **	20.6	18.6 *	20.4	18.5
Married[a)]	65.2	70.4 **	61.2	64.7 **	62.7	68.4 ***
Widowed	5.4	5.1	5.9	4.9	5.8	4.2 *
Divorced	10.1	10.0	12.3	11.8	11.0	9.0
Language						
(Swiss-) German	70.9	70.2	72.4	71.0	68.6	64.9 *
French	23.8	24.2	22.9	23.9	26.1	28.5
Italian	5.3	5.6	4.7	5.1	5.3	6.7
Gender						
Women	49.5	49.0	53.1	51.7	51.3	50.1
Telephone number						
No listed fixed-line number	43.3	35.9 ***	44.8	37.7 ***	45.8	41.3 **
(First) Nationality						
Swiss	75.8	79.7 *	76.5	80.1 ***	75.8	79.9 **
Bordering country	10.4	9.1	9.3	8.9	9.0	9.2
Other	13.8	11.2	14.3	10.9 ***	15.2	11.0 ***
Household size						
1	17.6	14.9	18.1	16.8	18.4	13.0 ***
2	32.2	35.4	35.6	35.6	35.2	38.9
3	17.6	19.8	17.1	17.8	18.1	18.5
4+	32.7	29.8	29.2	29.7	27.4	29.6

Notes: Auxiliary variables are characteristics of the assigned target person in the household. For responding households this is the household reference person; for nonresponding households the person approached after replacing young adults by their presumed parents.

a) Includes legal partnerships. two-sided Z test: *** $p < .01$, ** $p < .05$, * $p < .1$.

To get an overall indication of the representativeness of the three groups, we calculated r-indicators (Schouten et al. 2009). The r-indicator is based on the individual predicted probabilities ρ to participate. To calculate ρ we use a multivariate logit model using the socio-demographic frame variables in the table above as covariates. The r-indicator is defined as

$$r(\rho) = 1 - 2S(\rho)$$

with S the standard deviation of the predicted response probabilities. The r-indicator has a range between 0 and 1, with the value 1 indicating perfect representativeness and the value 0 being the maximum deviation from perfect representativeness. The r-indicators on the household level for the three groups were 0.746, 0.807, and 0.725, respectively. The McFadden pseudo r^2 for the three groups were in line with the r-indicators: 0.048, 0.027, and 0.056, respectively. This shows that, taking the subcategory sample sizes into account, the r-indicator suggests slightly better representativeness in the second group and worse in the third, compared with group 1. A limitation of the r-indicator, however, is that it is difficult to draw conclusions about the extent to which small differences in magnitude translate into important biases on key variables of interest in the survey. Moreover, the r-indicator can be sensitive to the choice and categorisation of covariates. Nevertheless, where auxiliary data are available from the frame, they provide a useful indicator of the relative risk of non-response bias across different survey designs.

9.4.2.4 Individual Nonresponse in the Three Groups

Table 9.3 shows how responding individuals compare to all eligible household members in the gross sample for the respective groups.

Table 9.3 shows that also on the individual level, a number of characteristics show some bias, but generally in the same direction in the three groups. The three designs seem to perform equally well at first sight. All groups show a relatively low share of younger (and unmarried) respondents, although this difference is less pronounced in group 3. Whereas on the household level group 1 included a larger share of older reference persons, on the individual level group 2 overrepresents the oldest age group among household members. Group 2 also underrepresented household members living in households without a landline. In groups 1 and 3, households with and without landline numbers were equally represented. Nationality also mattered: Group 2 had a significantly lower share of non-Swiss individuals, group 3 included fewer foreigners with a nationality from countries not sharing a border with Switzerland. Individuals from single-person households were overrepresented in group 2, and this group underrepresented individuals from large households of more than three people.

Table 9.3 Socio-demographic characteristics of the household members in the gross sample and responding individuals to the gross sample within each group.

Auxiliary Variables	(1) Gross sample Group 1 %	(2) Respon-dents Group 1 %	(3) Gross sample Group 2 %	(4) Respon-dents Group 2 %	(5) Gross sample Group 3 %	(6) Respon-dents Group 3 %
n	1027	707	2699	1798	1425	879
Age group						
<=30	27.8	24.6 *	27.8	25.7 **	26.3	24.9
31–44	17.6	18.0	19.5	18.9	20.6	20.0
45–58	27.8	29.4	26.8	26.9	28.1	28.0
59+	26.7	28.0	25.8	28.5 ***	25.1	27.1
Marital Status						
Single	33.3	30.0 *	35.4	32.9 **	34.4	32.4
Married[a)]	58.3	61.1	55.7	57.8 *	57.9	58.9
Widowed	2.6	2.4	2.6	2.8	2.5	2.2
Divorced	5.7	6.5	6.4	6.5	5.3	6.5
Language						
(Swiss-) German	69.9	67.8	71.2	71.2	63.4	63.0
French	24.4	26.6	23.7	23.6	29.2	30.1
Italian	5.6	5.7	5.1	5.1	7.4	6.8
Gender						
Women	52.0	53.9	50.5	52.1	50.0	51.9
Telephone number						
No listed fixed-line number	33.0	31.9	33.8	30.1 ***	37.6	35.6
(First) Nationality						
Swiss	81.3	82.9	80.7	86.2 ***	80.9	85.1 ***
Bordering country	9.1	9.6	7.6	5.9 ***	9.4	8.4
Other	9.6	7.5 *	11.7	7.9 ***	9.7	6.5 ***
Household size (persons)						
1	6.3	7.4	7.4	9.1 ***	5.6	6.5
2	29.8	30.8	30.2	32.0 *	32.5	33.8
3	20.8	20.7	20.0	19.6	20.8	21.3
4+	43.0	41.2	42.3	39.3 ***	41.1	38.5

Notes: Auxiliary variables are defined at the individual level and taken from the sampling frame for nonrespondents and from the individual questionnaire for respondents. Correspondence between the two sources was 95.5% for age (140 cases had an age difference of one year) and between 98.9% and 100% for the other variables.

a) Includes legal partnerships. two-sided Z test: *** $p < .01$, ** $p < .05$, * $p < .1$.

9.5 Conclusion

The results of the first wave of the two-wave pilot of the SHP shed new light on questions relating to using web-based questionnaires in household panels: Can households be recruited completely online, is it better to collect only part of the information online, or should the SHP continue to predominantly rely on its traditional use of telephone interviewing? The response rates obtained in the first wave suggest that the original design with telephone as dominant mode yields the highest response rates. Nonetheless, a design using telephone on the household level followed by online questionnaires on the individual level produced results that came close to those obtained in the current design of the SHP. The 'break' between the household and the individual level in this design was likely detrimental to obtained response rates. Analyses of measurement differences between modes based on Wave 1 will inform whether such a break for the reference person can be avoided in the next wave of data collection. None of the groups stood out with regard to the composition of the sample. Some differences between the groups emerged, but they were relatively modest. Even in a fully online approach, while the response rates were slightly lower, the achieved household and individual level samples resembled the respective gross samples more or less as well as in the other groups.

Lower per-interview costs may compensate for lower response rates in a mixed-mode design. Taking into account costs and response rates, the same budget could yield more interviews in group 3, by using larger sample sizes but obtaining lower response rates. But only after wave 2 can we can assess how successful the three designs are in re-interviewing the respondents. If re-interview rates are lower in group 3, then the panel may not be able to survive for a prolonged period of time.

This chapter provided an overview of the use of different modes in large-scale household panel surveys. In order to observe real change over time, household panel studies tend to be rather conservative about adapting their survey design. But, whereas all panels have a dominant mode of interviewing, mostly face-to-face, only a few use a single mode. The motivation to keep panel members in the study seems the most important consideration for combining the dominant mode with others, with cost considerations playing a smaller role. Other considerations vary from interview costs and length, restrictions to following households in the primary mode, to the suitability of measurement instruments for different modes.

Whereas consequences for participation rates have received some research attention, we know little about how the use of different modes affects measurement in longitudinal surveys (but see Chapter 10 in this book, Cernat and Sakshaug (2021)). The need to obtain more measurement points of the same household over time generally prevails over concerns about whether data are strictly comparable between modes. However, with increasing use of multiple modes in household panels, more studies comparing response rates, representativeness, and mode effects on measurement are underway, and knowledge in this domain is advancing.

References

Benzeval, M., Bianchi, A., Brewer, M. et al. (2017). *Understanding Society Innovation Panel Wave 9: Results from Methodological Experiments. Understanding Society Working Paper 2017-07.* Colchester, UK: Institute for Social and Economic Research.

Bianchi, A., Biffignandi, S., and Lynn, P. (2017). Web-face-to-face mixed-mode design in a longitudinal survey: effects on participation rates, sample composition, and costs. *Journal of Official Statistics* 33 (2): 385–408. https://doi.org/10.1515/jos-2017-0019.

Carpenter, H. and Burton, J. (2017). Moving Understanding Society to Mixed Mode: Effects on Response and Attrition. Paper presented at the Understanding Society Conference, July 2017, Colchester.

Cernat, A. and Sakshaug, J. (2021). Estimating the measurement effects of mixed modes in longitudinal studies: Current practice and issues. In: *Advances in Longitudinal Survey Methodology* (ed. P. Lynn). Chichester: Wiley.

Couper, M.P. (2011). The future of modes of data collection. *Public Opinion Quarterly* 75 (5): 889–908. https://doi.org/10.1093/poq/nfr046.

Dangubic, M. and Voorpostel, M. (2017). *Computer-Assisted Web Interviewing (CAWI) in the Swiss Household Panel: Demographics, Participation and Data Quality. Swiss Household Panel Working Paper 1_17.* Lausanne: FORS.

De Leeuw, E.D. (2005). To mix or not to mix? Data collection modes in surveys. *Journal of Official Statistics* 21 (2): 1–23.

Dex, S., and Gumy, J. (2011). On the Experience and Evidence about Mixing Modes of Data Collection in Large-Scale Surveys where the Web is used as one of the Modes in Data Collection. National Centre for Research Methods Review Paper.

Dillman, D.A. (2009). Some consequences of survey mode changes in longitudinal surveys. In: *Methodology of Longitudinal Surveys* (ed. P. Lynn), 127–140. Chichester, UK: Wiley.

Dillman, D.A. (2017). The promise and challenge of pushing respondents to the web in mixed-mode surveys. *Survey Methodology* 43 (1): 3–30.

Freedman, V.A., McGonagle, K.A., and Couper, M.P. (2017). Use of a targeted sequential mixed mode protocol in a nationally representative panel study. *Journal of Survey Statistics and Methodology* 6 (1): 98–121. https://doi.org/10.1093/jssam/smx012.

Frick, J.R., Jenkins, S.P., Lillard, D.R. et al. (2007). The Cross-National Equivalent File (CNEF) and its member country household panel studies. *Schmollers Jahrbuch* 127 (4): 627–654.

Groves, R.M. (2011). Three eras of survey research. *Public Opinion Quarterly* 75 (5): 861–871. https://doi.org/10.1093/poq/nfr057.

Hox, J., De Leeuw, E., and Klausch, T. (2017). Mixed mode research: issues in design and analysis. In: *Total Survey Error in Practice* (eds. P. Biemer, E. De Leeuw, S. Eckman, et al.), 511–530. New York: Wiley.

Huber, S. (2017). An overview of the SOEP samples: fieldwork report 2016 from kantar public. In: *SOEP Wave Report 2016* (eds. J. Britzke and J. Schupp), 28–36. Berlin: DIW.

Jäckle, A., Lynn, P., and Burton, J. (2015). Going online with a face-to-face household panel: effects of a mixed mode design on item and unit non-response. *Survey Research Methods* 9 (1): 57–70. http://dx.doi.org/10.18148/srm/2015.v9i1.5475.

Jäckle, A., Gaia, A., and Benzeval, M. (2017). *Mixing Modes and Measurement Methods in Longitudinal Studies. CLOSER Resource Report*. London: UCL, Institute of Education.

Laurie, H. (2003). *From PAPI to CAPI: Consequences for data quality on the British Household Panel Survey, Working Papers of the Institute for Social and Economic Research, paper 2003-14*. Colchester: University of Essex.

Lüdtke, D. and Schupp, J. (2016). Wechsel von persönlichen Interviews zu webbasierten Interviews in einem laufenden Haushaltspanel Befunde vom SOEP. In: *Methodische Probleme von Mixed-Mode-Ansätzen in der Umfrageforschung* (eds. S. Eifler and F. Faulbaum), 141–160. Wiesbaden: Springer.

Lynn, P. (2013). Alternative sequential mixed-mode designs: effects on attrition rates, attrition bias, and costs. *Journal of Survey Statistics and Methodology* 1 (2): 183–205. https://doi.org/10.1093/jssam/smt015.

Lynn, P. (2020). Evaluating push-to-web methodology for mixed-mode surveys using address-based samples. *Survey Research Methods* 14 (1): 19–30. https://doi.org/10.18148/srm/2020.v14i1.7591.

McGonagle, K.A. and Freedman, V.A. (2017). The effects of a delayed incentive on response rates, response mode, data quality, and sample bias in a nationally representative mixed mode study. *Field Methods* 29 (3): 221–237. https://doi.org/10.1177/1525822X16671701.

McGonagle, K.A. and Sastry, N. (2015). Cohort profile: the panel study of income dynamics' child development supplement and transition into adulthood study. *International Journal of Epidemiology* 44 (2): 415–422. https://doi.org/10.1093/ije/dyu076.

McGonagle, K.A., Schoeni, R.F., Sastry, N., and Freedman, V.A. (2012). The panel study of income dynamics: overview, recent innovations, and potential for life course research. *Longitudinal and Life Course Studies* 3 (2): 268–284.

McGonagle, K. A., Freedman, V. A., Griffin, J., and Dascola, M. (2017). Web development in the PSID: Translation and testing of a web version of the 2015 PSID telephone instrument. *PSID Technical Series Paper 17-02*.

Olson, K., Smyth, J.D., and Wood, H.M. (2012). Does giving people their preferred survey mode actually increase survey participation rates? An experimental examination. *Public Opinion Quarterly* 76 (4): 611–635. https://doi.org/10.1093/poq/nfs024.

Rendtel, U. (1990). Teilnahmebereitschaft in Panelstudien: Zwischen Beeinflussung, Vertrauen und sozialer Selektion. *Kölner Zeitschrift für Soziologie und Sozialpsychologie* 42 (2): 280–299.

Schouten, B., Cobben, F., and Bethlehem, J. (2009). Indicators for the representativity of survey response. *Survey Methodology* 35 (1): 101–113.

Schräpler, J.-P., Schupp, J., and Wagner, G.G. (2006). *Changing from PAPI to CAPI: A Longitudinal Study of Mode-Effects Based on an Experimental Design DIW Discussion Papers, No. 593*. Berlin: DIW.

Tourangeau, R. (2017). Mixing modes: Tradeoffs among coverage, non-response, and measurement error. In: *Total Survey Error in Practice: Improving Quality in the Era of Big Data* (eds. P.P. Biemer, E.D. De Leeuw, S. Eckman, et al.), 115–132. Hoboken, NJ: Wiley.

Tourangeau, R., Kreuter, F., and Eckman, S. (2012). Motivated underreporting in screening interviews. *Public Opinion Quarterly* 76 (3): 453–469.

Tourangeau, R., Kreuter, F., and Eckman, S. (2015). Motivated misreporting: Shaping answers to reduce survey burden. In: *Survey Measurements: Techniques, Data Quality and Sources of Error* (ed. U. Engel), 24–41. Frankfurt: Campus Verlag.

Wagner, G.G., Frick, J.R., and Schupp, J. (2007). *The German Socio-Economic Panel Study (SOEP) - Evolution, Scope and Enhancements (July 1, 2007). SOEPpaper No.1*. Berlin: DIW.

Watson, N. (2010). The Impact of the Transition to CAPI and a New Fieldwork Provider on the HILDA Survey. HILDA Project Discussion Paper Series No. 2/10 December 2010. Melbourne: University of Melbourne.

Watson, N. and Wilkins, R. (2015). Design matters:the impact of CAPI on interview length. *Field Methods* 27 (3): 244–264. https://doi.org/10.1177/1525822X15584538.

Watson, N. and Wooden, M. (2004). Wave 2 Survey Methodology. In: *HILDA Project Technical Paper Series No. 1/04*. Melbourne: University of Melbourne.

Watson, N. and Wooden, M. (2009). Identifying factors affecting longitudinal survey response. In: *Methodology of Longitudinal Surveys* (ed. P. Lynn), 157–181. Chichester, UK: Wiley.

Watson, N. and Wooden, M. (2015). Factors Affecting Response to the HILDA Survey Self-Completion Questionnaire. In: *HILDA Project Discussion Paper Series No.1/15*. Melbourne: University of Melbourne.

10

Estimating the Measurement Effects of Mixed Modes in Longitudinal Studies: Current Practice and Issues

Alexandru Cernat[1] and Joseph W. Sakshaug[2,3,4]

[1] *Department of Social Statistics, University of Manchester, Manchester, UK*
[2] *Statistical Methods Research Department, Institute for Employment Research, Nürnberg, Germany*
[3] *Department of Statistics, Ludwig Maximilian University of Munich, Munich, Germany*
[4] *School of Social Studies, University of Mannheim, Mannheim, Germany*

10.1 Introduction

A mixed-mode survey is a survey in which two or more modes of data collection are used to collect the same data from different respondents. This is in contrast to a multi-mode survey, which uses multiple modes to collect different data from the same respondents. Currently, the most popular data collection modes are computer-assisted personal interviewing (CAPI), computer-assisted telephone interviewing (CATI), paper self-administered questionnaires (SAQ), and web surveys (web), and combinations of these modes are frequently implemented within mixed- and multi-mode surveys (De Leeuw 2005).

There are several advantages of mixing modes. First, it may reduce coverage error because not all population members have a telephone or access to the internet and thus offering an alternative mode may increase the likelihood of establishing contact, especially with hard-to-reach subgroups. Second, mixing modes may reduce non-response error because some sample members may have a strong preference for responding in an alternative mode and therefore implementing multiple modes may increase response compared to implementing only a single mode. A third advantage is that mixed-mode surveys have the potential to save costs relative to a single-mode survey if a large enough subset of respondents is interviewed using a less-expensive alternative mode (e.g. SAQ or web) (Wagner et al. 2014). Implementing a multi-mode survey also has advantages, particularly in surveys that collect sensitive items. For example, switching respondents from an interviewer-administered mode to a self-administered mode (e.g. SAQ, interactive

Advances in Longitudinal Survey Methodology, First Edition. Edited by Peter Lynn.

Companion website: www.wiley.com/go/lynn/advancesinlongitudinalsurvey

voice response) is commonly done to provide respondents a more private setting to answer personal questions and reduce the risk of social desirability bias.[1]

However, despite their advantages, introducing a mixed-mode design can potentially lead to lower-quality data compared to a single-mode design. This is due to the fact that survey modes can differentially influence the answers that respondents provide. In other words, the same respondent might answer the same question differently in one mode compared to another mode. Different modes can produce different measurement errors, and the difference between measurement errors can lead to so-called measurement effects in mixed-mode surveys (for an overview of possible mechanisms, see Tourangeau et al. 2000). To illustrate, consider the scenario in which different modes are used to interview different sets of respondents. A common approach is to deploy a self-administered mode (e.g. SAQ, web) in the first phase of data collection and deploy an interviewer-administered mode (e.g. CATI, CAPI) in the subsequent non-response follow-up phase. These mode types tend to elicit different measurement errors. For instance, interviewer-administered modes have been shown to elicit responses that are considered to be more socially desirable compared to responses elicited in self-administered modes (Kreuter et al. 2008; Cernat et al. 2016; Heerwegh 2009). Deploying both mode types in a sequential (or concurrent) mixed-mode design will therefore produce survey data with different measurement error properties reflecting mode-specific influences on reports of socially undesirable behaviours/attitudes.

Another cause of measurement effects in mixed-mode surveys is due to the auditory vs. visual presentation of the survey questions. In auditory modes (e.g. CATI), interviewers usually read the questions and response options out loud to respondents, who must memorize the information before providing a response. One consequence is that respondents have a tendency to choose the last read categories as these are easier to remember than the earlier read categories (Dillman and Christian 2005). In contrast, for questions presented to respondents in visual modes (e.g. SAQ, web) there is a tendency for respondents to select the first response categories listed without giving equal consideration to the subsequent categories listed (de Leeuw 2005). Thus, visual and auditory modes can produce different types of measurement errors, and the difference between these measurement errors has the potential to introduce measurement effects in a mixed-mode design.

Such measurement effects threaten the validity of inferences drawn from mixed-mode surveys, particularly inferences based on comparisons of respondents interviewed under different data collection modes, comparisons of mixed-mode survey data with single-mode survey data, or unit-level change

1 Nevertheless, in this chapter our focus is on mixed-mode surveys, as they offer the most opportunities as well as the most complexity to data collection.

inferred from measurements taken in different modes. The last type of inference is particularly important in the case of longitudinal surveys, which aim to collect data from the same respondents over multiple time points in order to measure within- and between-individual change over time. While longitudinal surveys strive to minimize measurement effects by maintaining the same question wording in every wave, it is not always possible to maintain the same mode design in every wave. For example, budget pressures can force longitudinal studies to deviate from expensive, single-mode interviewer-administered designs (e.g. CAPI) in favour of mixed-mode designs that introduce a less expensive alternative mode (e.g. web). Longitudinal data can be used either to investigate a particular point in time or to investigate change by using measurements collected at different points in time. If researchers are interested in analyzing just one wave of data then the concerns regarding measurement error are the same as in a cross-sectional study. On the other hand, when estimating change, standard analysis methods assume that measurement error is constant in time; for an overview of measurement error and its underlying assumptions in estimates of change, see Cernat and Sakshaug (2021). Implementing a new mode of data collection or a mixed-mode design can change the stability of the measurement error.[2]

Although selection effects and costs may be reduced by implementing a mixed-mode design compared to a single-mode design, what remains largely unclear is the impact of mixing modes on the validity of change estimates derived from longitudinal surveys. The primary challenge in studying measurement effects in longitudinal mixed-mode surveys is separating the measurement effects from true measures of change. For example, consider a scenario where a longitudinal survey has been implemented over several waves under a single-mode, CAPI design. Due to budget constraints, the survey may decide to employ a mixed-mode design (similar to the one described above) where web is the first mode offered to respondents with CAPI follow-up for the remaining nonrespondents. Now, suppose an increase is observed in the estimated prevalence of individuals who reported losing their job in the last past months in the mixed-mode wave, compared to the previous single-mode wave. The apparent increased prevalence of job loss could be attributed to the introduction of the self-administered web mode, which may have facilitated respondents to report more candidly about their employment situation than in the previous wave, or the difference could reflect an actual increase in job losses in the population between waves, perhaps due to worsening economic conditions. Both explanations are plausible, but the effects

2 Standard longitudinal analysis methods always make the assumption that measurement error is constant in time. Even with single-mode data this might not be true. For example, panel conditioning might change the reliability of measurement in time, leading to incorrect estimates of change.

of mixing modes and actual change in the population are completely confounded, which poses challenges in assessing the validity of longitudinal estimates.

In light of this challenge, this chapter addresses the issue of estimating measurement effects in longitudinal mixed-mode designs. First, we provide a brief overview of different types of mixed-mode designs used in longitudinal studies and examine how these designs can potentially lead to measurement effects on estimates of change. We then summarize the literature on current practices for identifying measurement effects in longitudinal surveys. Lastly, we review different methods of measuring and adjusting for mixed-mode measurement effects in longitudinal estimates of change. We conclude this chapter with a set of recommendations and best practices for survey practitioners and data analysts on minimizing measurement effects in the design and analysis of longitudinal data mixed-mode designs.

10.2 Types of Mixed-Mode Designs

There are various types of mixed-mode designs used in longitudinal studies. In this section, our aim is not to provide an exhaustive review of every possible design, but rather, focus on a selection of typical features that distinguish between different mixed-mode designs. Specifically, we consider three important distinctions between different designs: (i) whether sample units can switch modes between waves (multiple times, just once, or not at all); (ii) whether mode switches that occur between waves are completely controlled by design, or are partly determined by respondents; and (iii) which waves are mixed mode (all, most, some, or none). These distinguishing characteristics are noted for a selection of longitudinal studies presented in Table 10.1. For example, the Health and Retirement Study (HRS) implements a crossover mixed-mode design between waves, whereby a random-half of units is assigned to a face-to-face interview and the other half is assigned to a telephone interview. In each subsequent wave, the mode assignment is reversed. Thus, respondents switch between face-to-face and telephone modes multiple times over the course of the panel, and the mode switches are completely controlled by the design. The MRC National Survey of Health and Development (NSHD) is another longitudinal study in which mode switches have occurred between waves by design. Although the NSHD has implemented multiple modes of data collection, including face-to-face and mail, over time, respondents are always interviewed in the same mode; that is, the NSHD has not implemented a mixed-mode design within a single wave.

Other mixed-mode designs allow for sample units to switch modes between waves, but the switching is partly determined by respondents. For example, the German Panel Study Labour Market and Social Security (PASS) implements a sequential mixed-mode design involving CATI and CAPI in each wave of the

Table 10.1 Examples of surveys implementing mixed-mode designs.

	Sample unit can switch modes between waves	Mode switches are completely controlled by design, or partly determined by respondents	Mixed-mode waves
1970 British Cohort Study (BCS70)	Multiple times	Design	None
English Longitudinal Study of Ageing (ELSA)	Not at all	Design	None
GESIS Panel	Not at all	Respondents	All
The Household, Income and Labour Dynamics in Australia (HILDA) Survey	Multiple times	Respondents	Most
Health and Retirement Study (HRS)	Multiple times	Design	Most
Millennium Cohort Study (MCS)	Not at all	Design	None
1958 National Child Development Study (NCDS)	Multiple times	Respondents; Design (Age 46 Survey)	Some
MRC National Survey of Health and Development (NSHD)	Multiple times	Design	None
Next Steps (previously known as the Longitudinal Study of Young people in England; LSYPE)	Multiple times	Respondents	Some
Panel Study Labour Market and Social Security (PASS)	Multiple times	Respondents	All
The Survey of Health, Ageing and Retirement in Europe (SHARE)	Not at all	Design	None
German Socio-Economic Panel (SOEP)	Multiple times	Respondents	Some
Understanding Society – The UK Household Longitudinal Study (UKHLS)	Multiple times	Design (experimentally); Respondents	Some
Understanding Society Innovation Panel (UKHLS Innovation Panel)	Multiple times	Design (experimentally); Respondents	Most

Study links: BCS70 (https://cls.ucl.ac.uk/cls-studies/1970-british-cohort-study); ELSA (https://www.elsa-project.ac.uk); GESIS Panel (https://www.gesis.org/en/gesis-panel/gesis-panel-home); HILDA (https://melbourneinstitute.unimelb.edu.au/hilda); HRS (http://hrsonline.isr.umich.edu); MCS (https://cls.ucl.ac.uk/cls-studies/millennium-cohort-study); NCDS (https://cls.ucl.ac.uk/cls-studies/1958-national-child-development-study); NSHD (https://www.nshd.mrc.ac.uk); Next Steps (https://nextstepsstudy.org.uk); PASS (https://fdz.iab.de/en/FDZ_Individual_Data/PASS.aspx); SHARE (http://www.share-project.org/home0.html); SOEP (https://www.diw.de/en/soep); UKHLS (www.understandingsociety.ac.uk); UKHLS Innovation Panel (www.understandingsociety.ac.uk/documentation/innovation-panel).

study. Sample units are first approached in one mode and if they do not respond in this mode, then they are approached in the other mode. Thus, it is entirely possible for the respondent to dictate which mode they are interviewed in based on whether or not they respond in the first mode. Next Steps is another survey that, in some waves, has implemented a sequential mixed-mode design, first approaching sample units by web, followed by telephone and then face-to-face. Some surveys implement sequential mixed-mode designs as part of an experiment in comparison to a single-mode design. For example, *Understanding Society* – The UK Household Longitudinal Study (UKHLS) Innovation Panel – has experimented with sequential mixed-mode designs on separate occasions. For instance, in wave 2 sample units were randomized to a sequential telephone followed by face-to-face interview, compared to face-to-face only. Since wave 5, sample units were randomized to a sequential web followed by face-to-face design, while the remaining subset was assigned to face-to-face only. Thus, mode switches between the waves are controlled both by the design (randomized mode assignment) as well as the respondent (sequential mode design).

In contrast to the above types of mixed-mode designs, there are also studies that implement multiple modes of data collection within a wave, but each sample unit is interviewed in the same mode over the course of the panel. Thus, sample units do not switch modes between waves. This is the design of the GESIS Panel, whereby sample units are assigned to either a mail or web mode based on their preference in their first wave of panel membership, and are interviewed in the same mode in all subsequent waves.

Each of the distinguishing features of the above mixed-mode designs can have implications for measurement effects generally, and for estimates of change in particular. If respondents do not switch modes between waves, either by design or determined by the respondent, then estimates of individual-level change should not be affected by measurement mode effects. However, if switching between modes influences respondents to give different answers to the same question over time, then estimates of change will be confounded by measurement mode effects. If the mode switching is controlled by the design (as in the randomized mixed-mode design of the HRS), then respondents themselves will have less of an impact on measurement mode effects. However, if mode switching is partly determined by the respondent and occurs over multiple waves, then controlling for measurement mode effects becomes more complicated as we discuss in the following sections.

10.3 Mode Effects and Longitudinal Data

Mixed-mode designs are becoming more popular in survey research and increasingly common in longitudinal studies. In this section, we discuss how switching modes can impact estimates of change, how to investigate mode effects, and research findings that are relevant to longitudinal data.

10.3.1 Estimating Change from Mixed-Mode Longitudinal Survey Data

As mentioned in Section 10.1, a change of modes in longitudinal surveys can lead to a confounding of true change and change in measurement error. This is extremely important because a main reason for carrying out longitudinal surveys is to estimate individual-level change.

Two main types of statistics are estimated in this context. The first one is the level of change. A simple example could be the proportion of people that transition from one state to another in two waves of data collection; for example, the proportion of people who were employed and become unemployed in the next wave. A second type of statistic that is often estimated in this context is the random effect of change. This can be interpreted as the amount of variation between individuals in their change trajectories. If most people follow a similar trajectory, then this effect will be low. The more people differ in their trajectories, the higher this coefficient will be. Analyses such as random effects, multilevel models of change, and latent growth models estimate both the individual trajectory and variation between individuals.

These two coefficients are also important because they influence the effect size of correlates of change (e.g. fixed effects coefficients). For example, the effect of gender on a person's change in satisfaction depends on both the average change and the between variation of this. If average change and the between variation in change are influenced due to the mode design, then the observed effects of other variables on estimated change could also be influenced.

Changing the data collection mode for some respondents in a longitudinal study can impact both of these coefficients. For example, if a portion of respondents are switched to a web mode (for example, by shifting from a CAPI-only survey to a web-CAPI one), then we may artificially increase the number of people that report being unemployed (because it is easier to admit this in a self-administered mode than face-to-face). This, in turn, would artificially increase the amount of individual change we see in our data, changing both the trend and the between-person variation.

10.3.2 General Concepts in the Investigation of Mode Effects

There is a growing literature regarding mixed-mode designs and the trade-offs they entail. Nevertheless, there is a pervasive issue that makes such investigations difficult. That is, the confounding of selection and measurement effects when using mixed modes.

Typically, mixed-mode designs are introduced in order to decrease costs or ameliorate non-response. While some survey modes, such as web, mail, and telephone are cheaper than face-to-face, they also impact the likelihood of participation either due to coverage or non-response (De Leeuw 2005; Roberts 2007). Finally, for those that participate in the survey, different modes can lead to different responses. For example, people tend to be more honest (contribute less social desirability bias) in self-administered modes such as web or mail (Cernat et al. 2016; Heerwegh 2009; Tourangeau et al. 2000). The issue that much of the literature on mixed mode effects faces is how to separate the positive effects, typically due to different mode selection propensities, and those that are less desirable, such as some effects on measurement.

In order to understand this issue, the literature discussing causal effects can be very useful (Morgan and Winship 2007; Pearl 2009; Vannieuwenhuyze et al. 2014). In this framework, we are typically interested in the counterfactual – that is, what people would have done in a different context. In the mixed-mode literature, that would entail answering the question: What would someone have answered if that person were participating using a different mode? The difference between the observed answer and the counterfactual is the mode effect.

There are typically three ways of identifying the counterfactual effect. The first one is called the *back-door method*. In the mixed-mode context, this would entail controlling for the different selection probabilities of the modes (e.g. controlling for the tendency of older respondents to answer by face-to-face or mail and of younger respondents by web). All the differences between modes after the selection correction (e.g. controlling for age) would represent the effect of the modes on measurement. This can be done in multiple ways. One approach is the use of regression analysis and controlling for all the selection confounders (e.g. Jäckle et al. 2010). Alternative ways are the use of weights or matching (Lugtig et al. 2011). There are two essential assumptions of this approach: (i) that the model completely explains any selection differences; and (ii) the variables used to make the corrections are mode insensitive.

The second approach is called the *front-door method* and is relatively underused in mixed-mode research (Cernat 2015b; Vannieuwenhuyze et al. 2014). This entails explaining the measurement differences between modes first before any selection correction. If, for example, it is possible to control for the higher social desirability effects in face-to-face surveys (compared to web or mail), then any difference remaining between modes is due to different selection propensities of the modes. This approach also makes two assumptions: (i) the model explains all the differences in measurement due to the modes; and (ii) there is no relationship between the measurement and selection effects. This method has been applied using regression calibration by Vannieuwenhuyze et al. (2014) and structural equation modelling (Cernat 2015b).

The last approach is through the use of an instrumental variable (Angrist et al. 1996; Morgan and Winship 2007; Pearl 2009; Vannieuwenhuyze and Révilla 2013). In this case, a variable that explains the selection process but is not related with the variable of interest is needed. This can be included in a more complex regression model in order to control for the selection effects. Vannieuwenhuyze and Révilla (2013) present an example in the context of mixed-mode designs.

In addition to the statistical estimation of counterfactuals, experimental designs have also been used to investigate mode effects. If the intention is to compare the measurement of two modes, an ideal design would first ensure the participation of the respondents and then it would allocate respondents to one mode or another. This is an experimental design that has not been used often in the mixed-mode literature. A more common approach is to give different modes or mode designs to random subgroups of respondents. While this design has the advantage of being easier to implement and in line with real-world practice, it does give rise to the confounding of selection and measurement of the modes. Another experimental design proposed in this context is the use of a re-interview in a different mode administered a short time after the initial survey (Schouten et al. 2013; Klausch et al. 2015). While this design is attractive, it comes with considerably higher costs and burden for respondents.

The use of experimental designs and methods from the causal literature is equally useful both in cross-sectional studies and longitudinal ones. Nevertheless, the difficulties in pinpointing measurement mode effects are magnified when using longitudinal data. This is because while quasi-experimental designs can be implemented, we again have confounding of selection, which can also include differential attrition patterns (Lynn 2013) and measurement effects (Cernat 2014). Additionally, the statistics that are of main interest, how people change over time, are based on data that can come from multiple waves that can have multiple mode(s) (designs).

10.3.3 Mode Effects on Measurement in Longitudinal Data: Literature Review

We have presented the typical mixed-mode designs used in longitudinal studies as well as the most common approaches to estimating the effect of modes on data quality. Next, we present the evidence available, so far, regarding the effects of modes on measurement in longitudinal studies. As we shall see, there is currently a very limited amount of research in this area.

Two of the papers that investigated the effects of modes on estimates of change have used the aforementioned experimental design from the UK Household Longitudinal Study Innovation Panel (UKHLS-IP); a more detailed discussion of this is provided in the next section. This is a yearly household panel study that included a number of mixed-mode experiments (Al-Baghal et al. 2016). The first four waves of the data were collected using CAPI, with the exception of an experimental group that received a sequential mixed-mode CATI-CAPI in wave 2. Cernat (2014) investigated how the quality of 46 variables that were measured in all four waves was influenced by the mixed-mode design. Data quality was estimated using the longitudinal reliability of the questions. The results showed that for most of the variables, the reliability (data quality) and the stability (estimates of change) of the groups were equal, indicating that the introduction of the mixed-mode design in one of the waves did not have a significant impact on measurement.

Using the same data, Cernat (2015a) investigated data quality in 12 variables measuring health (the SF12 scale) using multigroup Confirmatory Factor Analysis (CFA). Only small differences were found in the measurement of health between the groups that answered only using CAPI in four waves and those that had a combination of CATI and CAPI. It is interesting to note that some differences in measurement between groups were found in subsequent waves even when the data was collected using CAPI for everyone. This might indicate that the introduction of a mixed-mode design in a longitudinal study might have lasting effects on data quality. Nevertheless, it is not clear if the differences found are due to selection (Lynn 2013) or to other measurement effects such as panel conditioning.

Finally, Cernat et al. (2016) used a quasi-experimental design in the HRS to compare CAPI, CATI, and web data collection in three waves. They found significant differences between CAPI/CATI on the one hand and web on the other for two multi-item scales measuring depression and physical activity, which can be explained by higher levels of social desirability bias in the former.

These studies mainly investigated data-quality differences. Nevertheless, it is important to know how the difference in data quality affects substantive statistics, such as estimates of change. There is surprisingly little research on this topic. Cernat (2015a) using the same health variables and data from UKHLS-IP investigated how the estimates of change in the four waves were influenced by implementing the mixed mode design in wave 2. He found that 4 out of the 12 questions showed more change over time in the group that had the mixed-mode design in wave 2 compared to the single-mode group. While there is a need for more evidence in this area, there seems to be an indication that estimates of change can be influenced by the change of the mode design.

A study carried out by Buelens and Van den Brakel (2017) has investigated how the change of modes influences estimates of the unemployment rate in the Labour Force Survey in the Netherlands (see also the chapter by Van den Brakel et al. in this book). They show that correcting for these changes in the relative sizes of the modes can lead to markedly different statistics.

10.4 Methods for Estimating Mode Effects on Measurement in Longitudinal Studies

Mixed-mode designs in longitudinal studies can be complex with different modes or mixed-mode designs used at different points in time. This renders the task of investigating mode effects on measurement error even more complex than with cross-sectional data. Thus, in order to understand the impact of mode of answering on measurement error in longitudinal studies, we can compare mode designs either within each wave, as in a cross-sectional study, or between waves. The latter is essential for estimates of change.

The simplest way to investigate the effect of answering in a particular mode on responses would be by separately comparing modes within each wave. In this context, we would be able to use the same approaches as in the cross-sectional context where combinations of experimental designs and statistical models can be used to infer how the mode of the survey influences data quality.

We use the UKHLS Innovation Panel (Al-Baghal et al. 2016) as an example dataset to facilitate the different comparisons that are possible. As previously mentioned, this is a nationally representative longitudinal study collected every year in the UK. Table 10.2 shows a quasi-experimental design used in the first four waves of the study. The notation is that proposed by Campbell and Stanley (1963). We have two groups that are randomly allocated: the single-mode CAPI and the sequential mixed-mode CATI-CAPI. For the first group, CAPI was used to collect the information in all four waves (indicated by O), while for the second group this was used in waves 1, 3, and 4. In wave 2 of the study, the sequential mixed-mode design was implemented (noted by XO_2).

The survey started out as single-mode CAPI survey in wave 1. In the second wave, the designers experimented with a mixed-mode design. In order to facilitate comparisons a random portion of the sample received face-to-face single mode while another portion received a sequential CATI-CAPI design. After the second wave, everyone received the same CAPI survey. To investigate change over time in the first four waves of the UKHLS-IP, we typically assume that measurement error is constant over time. If introducing the mixed-mode design in wave 2 changes

Table 10.2 Quasi-experimental design in a longitudinal study.

Group	Wave 1	Wave 2	Wave 3	Wave 4
R_{CAPI}	O_1	O_2	O_3	O_4
$R_{CAPI\text{-}CATI}$	O_1	XO_2	O_3	O_4

Source: Adapted from Cernat (2014).

the way people answer questions (e.g. we observe more recency effects for those participating by telephone), then our estimates of change will be biased due to the mode design.

Of central importance here is whether the wave 2 mixed-mode design, through the use of CATI on a portion of the respondents, changed measurement error. We can investigate this in multiple ways. Firstly, it is possible to compare the single-mode group and the mixed-mode group in wave 2 using the methods discussed above. This would answer the question: *Does using the CATI-CAPI mixed-mode design lead to the same measurements as the single-mode design?* This can be done by using either the quasi-experimental design, in which case we can think of it as an intention-to-treat effect (CATI was offered to everyone but some did not accept it), or by combining it with one of the approaches proposed in the causal literature (as explained in a previous section).

An alternative comparison that can be made in wave 2 is between those who answered using CATI and those who did so using CAPI. If we are interested in estimating measurement differences due to mode, this would be the most direct way to estimate it. The issue with this is once again the confounding of selection and measurement. It is possible to use the back-door method, for example, by using a regression model and controlling for variables from wave 1 that account for selection. Thus, any differences left between the two groups would represent mode measurement effects. In this way, we answer the question: *Do CAPI and CATI respondents answer in the same way in wave 2?*

Another useful comparison that can be made is between wave 1 and wave 2 within the mixed-mode group. While this comparison would normally confound mode effects with change over time, it is possible to bypass this by using a method such as multi-group CFA (presented below). This would enable the comparison of measurement quality between mode designs, while controlling for the change over time. This would help answer the question: *Is measurement quality/error the same over time for the CAPI and CATI-CAPI groups?* This is an important question, as it is one of our assumptions when estimating change over time using this data collection design.

Fourth, it is possible to compare the single- and mixed-mode groups in waves 3 and 4. Such comparisons would be done if there are concerns about long-term effects of mixing modes, for example, due to panel conditioning.

Fifth, it is possible to compare change over time from wave 1 to wave 2 between the single-mode and the mixed-mode designs. This would answer the question: *Are estimates of change over time different when introducing the mixed-mode design?*

This section highlighted how the investigation of mode effects on measurement error in longitudinal studies can be complex. At the moment, there is relatively little research in this area, especially regarding the last type of comparison, which is essential for longitudinal studies. The next section will cover how to make such comparisons and then how to control for any differences found.

10.5 Using Structural Equation Modelling to Investigate Mode Differences in Measurement

The previous section outlined the multiple types of comparisons that are possible and necessary in order to ensure that estimates of change over time are not biased due to a mixed-mode data collection. While such comparisons are possible using the popular regression approach an alternative approach is Structural Equation Modelling (SEM). This can be better suited for multiple types of comparisons (Bollen 1989; Kline 2010). This approach has already been extensively used in mixed-mode research, both cross-sectionally and longitudinally (Cernat 2014, 2015a, 2015b; Cernat et al. 2016; Heerwegh and Loosveldt 2011; Hox et al. 2015; Klausch et al. 2013; Révilla 2010; Révilla et al. 2014; Révilla and Saris 2012; Vannieuwenhuyze and Révilla 2013).

In addition to incorporating the typical regression models, SEM has a number of advantages. Firstly, it can incorporate latent/unobserved/factor variables. The advantage of this is that measurement error is explicitly modelled by separating the observed scores and the latent ones. Thus, coefficients such as factor loadings and residuals can indicate the relative sizes of measurement error (Bollen 1989). Secondly, SEM enables the comparison of coefficients, such as random or systematic error. Thus, it is possible to statistically test hypotheses regarding the effects of modes on estimates. Thirdly, it offers a number of estimates of goodness-of-fit indices that enable to better understand the statistical models used.

An example of factor analysis can be seen in Figure 10.1. Here the circle represents the latent factor (which can also be a latent class – i.e. categorical) while the squares represent the observed variables. The one way arrows represent regression coefficients. The loadings (lambda, λ) coefficients are indicators of the relationship between the unobserved variable of interest and the observed variables in the data. Ideally, our indicators measure the latent variable perfectly and this relationship is 1. The residuals (epsilon, ε) are unexplained variance and are indicators of random error. Finally, the intercepts (nu, ν) are associated with the mean and systematic error.

Using the SEM framework enables the comparison of any of these coefficients across different groups. While this on its own does not solve the confounding of measurement and selection mode effects (there is still a need for experimental designs or controls for selection), it does make explicit the testing of differences in measurement quality.

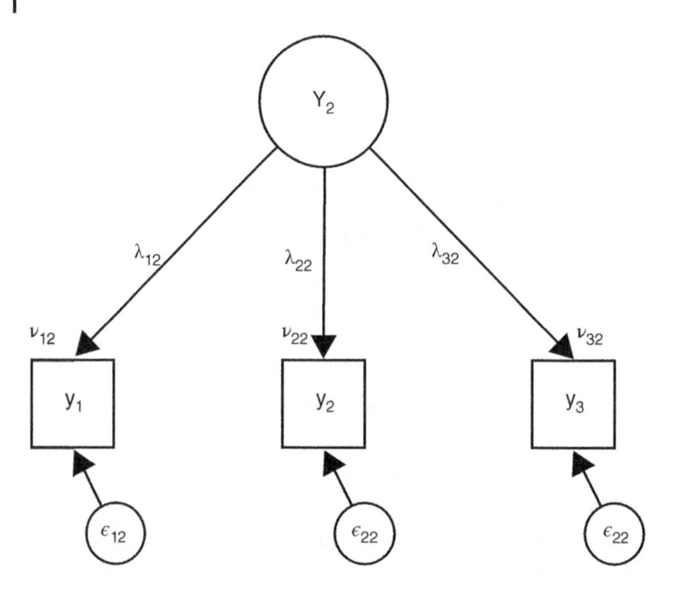

Figure 10.1 Confirmatory factor analysis (CFA) model.

All four research questions mentioned above can be answered in this framework. Typically, this would be done using equivalence testing (Cernat 2015a; Davidov et al., 2014; Hox et al. 2015; Klausch et al. 2013; Steenkamp and Baumgartner 1998; Vannieuwenhuyze and Révilla 2013). This is done using a series of models that constrain different coefficients of the model presented in Figure 10.1 to be equal between groups. Thus, one could test, for example, if random error is equal between a mixed-mode and a single-mode design.

Nevertheless, there is one particular type of equivalence testing that is absent in the survey methods literature. That is longitudinal equivalence (Little 2013; Newsom 2015). This is unfortunate, as such a test can explicitly estimate and control for exactly the types of measurement effects of interest with longitudinal mixed-mode designs.

Figure 10.2 shows how the testing of equivalence of the measurement over time would be set up. The same concept (Y) is measured at two points in time. At each point it is estimated using three observed variables. It is expected that the concept is correlated with itself over time as are the errors (these are the double arrows in the figure). The aim of this model is to investigate if the concept of interest is measured in the same way at multiple points in time. As such, this model investigates *if measurement error is constant over time.* While this might not be the main research question for applied researchers it is nevertheless a condition for making unbiased comparisons over time and estimating correct change coefficients.

As an example, it is possible to apply this to the experimental data in the UKHLS-IP to investigate if a concept of interest, such as health measured using

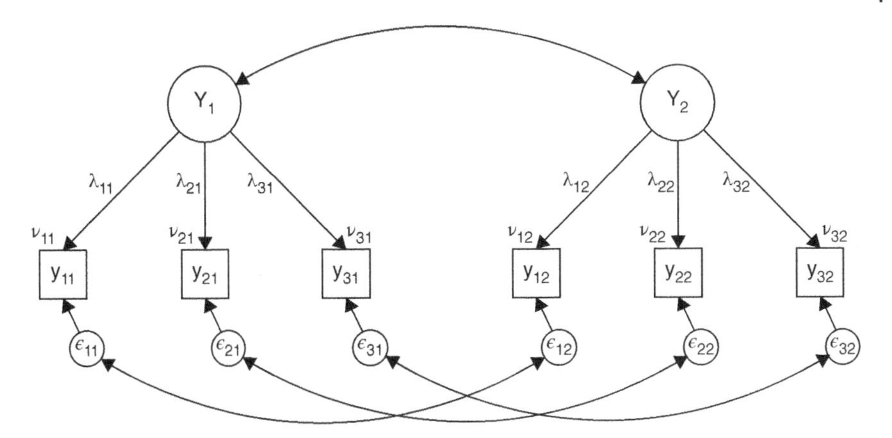

Figure 10.2 Example of longitudinal factor analysis that can be used for equivalence testing.[4]

the SF12, is measured in the same way in waves 1–4. If the measurement is the same (the loadings and the conditional means are the same), then the mixed-mode design in wave 2 did not impact the measurement quality of our concept. This means that estimates of change can be calculated for our concept (Little 2013; Newsom 2015).

A valid criticism of the above approach could be that often there are no multiple variables that measure the same concept, or that there are no clearly defined and accepted concepts. In such situations, it is possible to use an alternative model: the quasi-simplex model (Alwin 2007; Cernat et al. 2014; Heise 1969; Wiley and Wiley 1974).

As can be seen in Figure 10.3, this model[5] has only one variable measured at different points in time. At each point, it estimates the relationship between an observed variable (A) and a true score (T). Again, the model calculates loadings (lambda, λ) and residuals (epsilon, ε) for this relationship.[6] Additionally, the model explicitly estimates change over time of the true score assuming a lag-1 effect (e.g. wave 3 is only influenced by wave 2 and not wave 1) through the beta coefficients, β. This model can be estimated under certain restrictions, for example, if the

4 All the models in this section (except the cross-sectional CFA) need longitudinal data. While the figures do not include individual subscripts (like those used in the multilevel model notation) the same information measured for the same individuals is needed in order to estimate the correlation and regression coefficients between the different points in time.
5 This model has a number of different related names, such as auto-regressive models, hidden Markov chain, lag-1 model, etc. It uses longitudinal data to investigate the stability of a characteristic in time (the autoregressive part represented by the beta coefficient) while controlling for measurement error (the "hidden" or "latent" part that estimates measurement error using lambda).
6 We do not show the conditional mean (nu) in order to keep the figure easy to read.

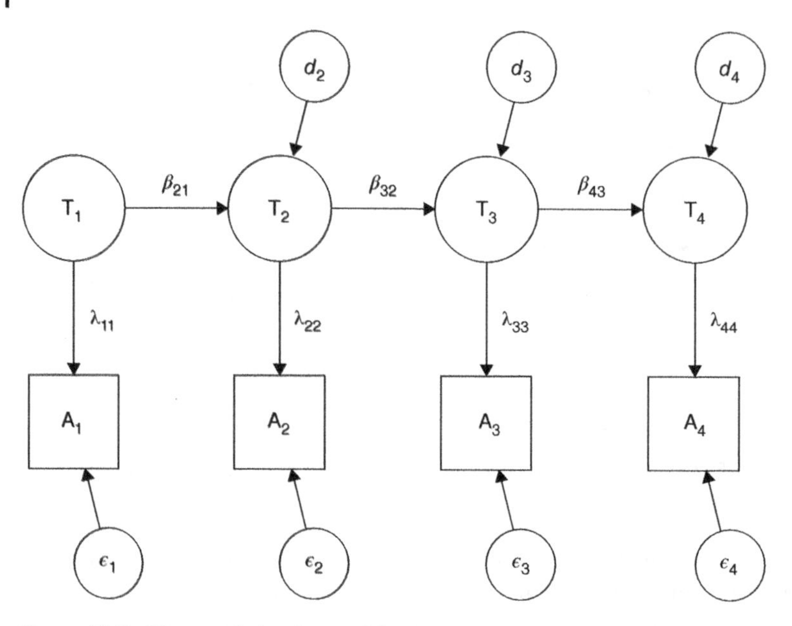

Figure 10.3 The quasi-simplex model.

variance of the residual is equal in the first and last wave (Alwin 2007; Cernat et al. 2014; Heise 1969; Wiley and Wiley 1974). If the assumptions hold, this model is extremely attractive. Firstly, it does not need multiple items that measure the same concept but only one variable measured at a minimum of three points over time. Secondly, it explicitly estimates the change in time, using the beta coefficients, and random error. Also, it is part of the general SEM framework that makes it possible to compare coefficients across groups. An example of using multi-group quasi-simplex models to estimate mode effects can be found in Cernat (2014).

Finally, SEM can also be used together with more complex quasi-experimental designs. An example is given by Cernat et al. (2016). Figure 10.4 shows the experimental design from the HRS, waves 2010 and 2012, and an off-year wave (2011) where a web survey was conducted. There are two randomized groups. The first group received a sequence of CAPI, web, and CATI over the three waves while the second group received a sequence of CATI, web, and CAPI. The researchers created three mode groups: telephone (TEL), face-to-face (F2F), and web, which are a combination of information from all three waves. Thus, as seen in Figure 10.4, the telephone group includes responses in year 2010 from group 1 and from group 2 in 2012 while the web group includes all the responses from 2011. It is then possible to compare the three groups, using equivalence testing (mentioned earlier), to investigate mode effects. This is an example of

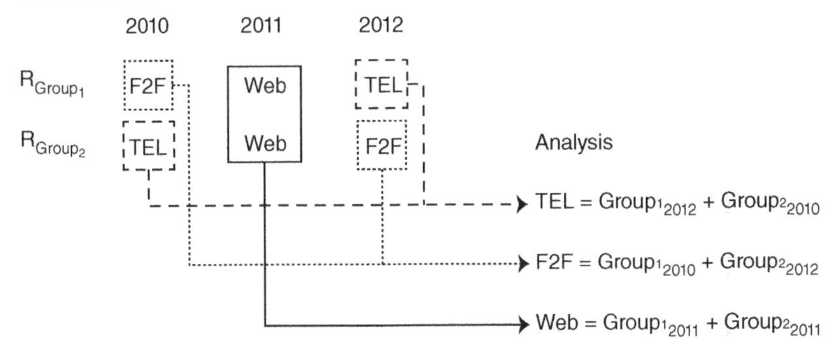

Figure 10.4 Quasi-experimental design in the Health and Retirement Study. Source: Adapted from Cernat et al. (2016).

using longitudinal data, quasi-experimental designs, and SEM to investigate measurement error.[7]

We have seen that SEM enables estimation of measurement error directly as well as the comparison of measurement errors across groups. This can be fruitfully used to understand mode effects on measurement. Additionally, it is also possible to use this framework to investigate the effects of modes directly on estimates of change. Figure 10.5 shows how a multilevel model of change would be modelled using SEM. This is the latent growth model (Little 2013; Newsom 2015).

To highlight how the model in Figure 10.5 is the same as the multilevel model of change, which is often used in the social sciences, we present the formulas for both models below. The first one represents the notation in multilevel model form (Singer and Willett 2003), where γ_{00} represents the average of Y, which varies by i individuals and j points in time, at the start of the study, and γ_{10}represents average change over time. These are also known as the fixed effects. This can be easily extended by including other predictors to explain change in time. The random part of the model is represented by ξ_{0i},which shows the between-people variation at the start of the study, ξ_{1i},which shows between-variation in estimates of change, and ε_i, the within-individual variation.

$$Y_{ij} = \gamma_{00} + \gamma_{10}Time_{ij} + \xi_{0i} + \xi_{1i}Time_{ij} + \varepsilon_i$$

Using the latent growth model in Figure 10.5 leads to the same equation (Newsom 2015), although now the values at the start of the study and the change in time are estimated as latent variables:

$$y_{ti} = \alpha_0 + \alpha_1\lambda_{ti} + \zeta_{0i} + \zeta_{1i}\lambda_{t1} + \varepsilon_{ti}$$

7 Note that this approach assumes that any change in measurement error that is not due to the mode is linear. Please refer to the paper for a more elaborate discussion.

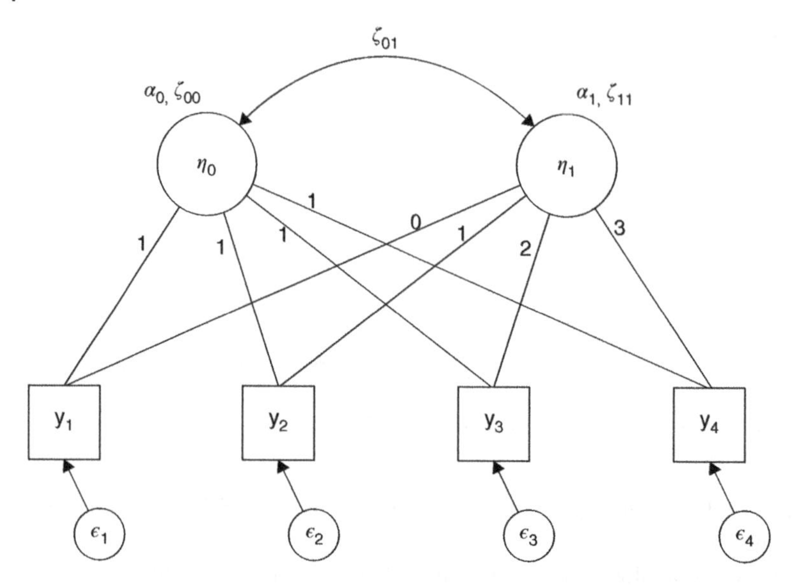

Figure 10.5 Estimation of change over time using longitudinal data in SEM.

The α (which is the mean of the latent variables) has the same interpretation as the γ, namely, the average of the fixed effects, while ξ (the variance of the latent variables) is the same as ζ, referring to the between-people variation. The two models have two minor differences. Firstly, by default, latent growth models estimate within variation at each point in time, ε_{ti}, while, by default, multilevel models estimate only one such coefficient ε_i (although this can be changed). Secondly, change over time is defined using the λ coefficients for latent growth models, while this is done using the time variable in a multilevel model (this is because one uses wide-form data while the other long-form data).

There are two reasons why latent growth models might prove useful in the investigation and correction for mode effects in longitudinal data. Firstly, it is relatively easy to compare estimates of change across different groups. For example, in the UKHLS Innovation Panel data presented above, it would be possible to create a multi-group latent growth model where one group consists of respondents that received single mode for all four waves while the other is made by those that received the mixed mode in wave 2. Structural equation modelling facilitates model comparisons and the introduction of restriction to see if indeed the two groups exhibit the same degree of change in time. An example of such an investigation can be found in Cernat (2015a).

The second reason why using latent growth models might prove useful in the investigation and correction of mode effect is because they can be combined with other models, such as equivalence testing shown in Figure 10.2. Researchers could investigate if the measurement is the same over time using longitudinal equivalence testing and, if they find at least two questions that are measured in the same way over time, they could use partial equivalence to correct for differences (Byrne et al. 1989). This model can be extended by estimating a latent growth model on the factor, also known as a second-order latent growth model (Little 2013; Newsom 2015). In this way, differences in the way the concept of interest is measured over time can be controlled for while still estimating change over time.

10.6 Conclusion

Mixed-mode surveys are becoming increasingly popular due to their potential cost savings and improved coverage. This is also true for longitudinal studies. We have shown that a large number of large-scale longitudinal studies are using mixed-mode designs. While this trend will probably continue, there are also some concerns about how the use of mixed modes in longitudinal studies can impact the measures collected. This is especially important when estimating change over time, one of the main reasons why longitudinal data are collected in the first place.

One key assumption when estimating change over time is measurement equivalence. That is, questions measure the same concept and have the same measurement error at each point in time. Even when using the same mode, this assumption might not hold (e.g. due to panel conditioning or the maturation of the sample), but this is especially problematic when we are using a mixed-mode design. This is because true change might become confounded with mode change.

We have reviewed the existing research on this topic, which is currently very limited. We have seen some evidence of the long-term impact of introducing a mixed-mode design in an otherwise single-mode longitudinal study (Cernat 2015a; Lynn 2013). Looking at estimates of change, there is some limited evidence based on quasi-simplex models indicating that there are no differences in measurement quality or estimates of change between a single-mode CAPI longitudinal survey and a sequential mixed-mode CATI-CAPI one (Cernat 2014). Using the same data, it was also shown using latent growth models that the mixed-mode design can lead to overestimation of chance (Cernat 2014, 2015a). Nevertheless, the current evidence is very limited and more research is needed in this area.

Implementing quasi-experimental designs during data collection helps us disentangle selection and measurement effects and makes it easier to understand possible causes for these differences. We have also discussed how SEM can be a useful tool to investigate differences across modes and mode designs, both cross-sectionally and longitudinally.

These tools can also be useful for correcting for mode differences in measurement. If lack of measurement equivalence is found, then this can be included in the models using a 'partial invariance' approach where some of the coefficients are restricted to be the same and some are different. Alternative approaches can be used, such as that proposed by Buelens and Van den Brakel (2017), which effectively keeps constant the mode-specific bias. Other approaches, such as calibration, can be considered as well (Buelens and van den Brakel 2015; Kolenikov and Kennedy 2014; Vannieuwenhuyze and Loosveldt 2013).

The literature review has highlighted that there is a need for more research regarding estimates of change in mixed-mode surveys and how these could be biased due to the different modes or combination of modes used. Future research should focus in this area. We also believe that SEM is underused in this literature, even though it has a number of tools that could help investigate mode effects.

Acknowledgement

This work is supported by the Russian Science Foundation under grant No 17-78-20172.

References

Al-Baghal, T., Jackle, A., Burton, J. et al. (2016). *Understanding Society the UK Household Longitudinal Study Innovation Panel, Waves 1–8, User Manual.* Colchester Essex: Institute for Social and Economic Research University of Essex Available at: www.understandingsociety.ac.uk/d/291/6849_ip_waves1-8_user_manual_June_2016.v3.1.pdf?1468270407.

Alwin, D. (2007). *The Margins of Error: A Study of Reliability in Survey Measurement.* New Jersey: Wiley-Blackwell.

Angrist, J.D., Imbens, G.W., and Rubin, D.B. (1996). Identification of causal effects using instrumental variables. *Journal of the American Statistical Association* 91 (434): 444–455. https://doi.org/10.1080/01621459.1996.10476902.

Bollen, K. (1989). *Structural Equations with Latent Variables.* New York: Wiley-Interscience.

Buelens, B. and van den Brakel, J.A. (2015). Measurement error calibration in mixed-mode sample surveys. *Sociological Methods & Research* 44 (3): 391–426. https://doi.org/10.1177/0049124114532444.

Buelens, B. and Van den Brakel, J.A. (2017). Comparing two inferential approaches to handling measurement error in mixed-mode surveys. *Journal of Official Statistics* 33 (2): 513–531. https://doi.org/10.1515/jos-2017-0024.

Byrne, B.M., Shavelson, R.J., and Muthén, B. (1989). Testing for the equivalence of factor covariance and mean structures: the issue of partial measurement invariance. *Psychological Bulletin* 105 (3): 456–466.

Campbell, D.T. and Stanley, J.C. (1963). *Experimental and Quasi-Experimental Research*.Boston, MA: Houghton Mifflin Company.

Cernat, A. (2014). The impact of mixing modes on reliability in longitudinal studies. *Sociological Methods & Research* 44 (3): 427–457. https://doi.org/10.1177/0049124114553802.

Cernat, A. (2015a). Impact of mixed modes on measurement errors and estimates of change in panel data. *Survey Research Methods* 9 (2): 83–99. https://doi.org/10.18148/srm/2015.v9i2.5851.

Cernat, A. (2015b). Using equivalence testing to disentangle selection and measurement in mixed modes surveys. *Understanding Society* Working Paper Series (2015–01): 1–13.

Cernat, A., Lugtig, P., Uhrig, S.N., et al. (2014). Assessing and relaxing assumptions in quasi-simplex models. ISER Working Paper (2014–09): 1–22.

Cernat, A., Couper, M.P., and Ofstedal, M.B. (2016). Estimation of mode effects in the health and retirement study using measurement models. *Journal of Survey Statistics and Methodology* 4 (4): 501–524. https://doi.org/10.1093/jssam/smw021.

Cernat, A. and Sakshaug, J.W. (2021). *Measurement Error in Longitudinal Data*. Oxford: Oxford University Press.

Davidov, E., Meuleman, B., Cieciuch, J. et al. (2014). Measurement equivalence in cross-National Research. *Annual Review of Sociology* 40 (1): 55–75. https://doi.org/10.1146/annurev-soc-071913-043137.

De Leeuw, E.D. (2005). To mix or not to mix data collection modes in surveys. *Journal of Official Statistics* 21 (5): 233–255.

Dillman, D.A. and Christian, L.M. (2005). Survey mode as a source of instability in responses across surveys. *Field Methods* 17 (1): 30–52.

Enders, C.K. (2010). *Applied Missing Data Analysis*. New York: Guilford Press.

Heerwegh, D. (2009). Mode differences between face-to-face and web surveys: an experimental investigation of data quality and social desirability effects. *International Journal of Public Opinion Research* 21 (1): 111–121. https://doi.org/10.1093/ijpor/edn054.

Heerwegh, D. and Loosveldt, G. (2011). Assessing mode effects in a National Crime Victimization Survey using structural equation models: social desirability bias and acquiescence. *Journal of Official Statistics* 27 (1): 49–63.

Heise, D.R. (1969). Separating reliability and stability in test-retest correlation. *American Sociological Review* 34 (1): 93–101.

Hox, J.J., De Leeuw, E.D., and Zijlmans, E.A.O. (2015). Measurement equivalence in mixed mode surveys. *Frontiers in Psychology* 6 (87): 1–11. https://doi.org/10.3389/fpsyg.2015.00087.

Jäckle, A., Roberts, C., and Lynn, P. (2010). Assessing the effect of data collection mode on measurement. *International Statistical Review* 78 (1): 3–20. https://doi.org/10.1111/j.1751-5823.2010.00102.x.

Klausch, T., Hox, J.J., and Schouten, B. (2013). Measurement effects of survey mode on the equivalence of attitudinal rating scale questions. *Sociological Methods & Research* 42 (3): 227–263. https://doi.org/10.1177/0049124113500480.

Klausch, T., Schouten, B., and Hox, J.J. (2015). Evaluating bias of sequential mixed-mode designs against benchmark surveys. *Sociological Methods & Research* 46 (3): 456–489. https://doi.org/10.1177/0049124115585362.

Kline, R. (2010). *Principles and Practice of Structural Equation Modeling*, 3rd edition. The Guilford Press.

Kolenikov, S. and Kennedy, C. (2014). Evaluating three approaches to statistically adjust for mode effects. *Journal of Survey Statistics and Methodology* 2 (2): 126–158. https://doi.org/10.1093/jssam/smu004.

Kreuter, F., Presser, S., and Tourangeau, R. (2008). Social desirability Bias in CATI, IVR, and web surveys: the effects of mode and question sensitivity. *Public Opinion Quarterly* 72 (5): 847–865.

Little, T.D. (2013). *Longitudinal Structural Equation Modeling*. New York: Guilford Press.

Lugtig, P.J., Lensvelt-Mulders, G.J.L.M., Frerichs, R. et al. (2011). Estimating nonresponse bias and mode effects in a mixed mode survey. *International Journal of Market Research* 53 (5): 669–686.

Lynn, P. (2013). Alternative sequential mixed-mode designs: effects on attrition rates, attrition bias, and costs. *Journal of Survey Statistics and Methodology* 1 (2): 183–205. https://doi.org/10.1093/jssam/smt015.

Morgan, S.L. and Winship, C. (2007). *Counterfactuals and Causal Inference: Methods and Principles for Social Research*. New York: Cambridge University Press.

Newsom, J.T. (2015). *Longitudinal Structural Equation Modeling: A Comprehensive Introduction*. Routledge.

Pearl, J. (2009). *Causality: Models, Reasoning and Inference*, 2nd edition. New York: Cambridge University Press.

Révilla, M. (2010). Quality in Unimode and mixed-mode designs: a multitrait-multimethod approach. *Survey Research Methods* 4 (3): 151–164.

Révilla, M.A. and Saris, W.E. (2012). A comparison of the quality of questions in a face-to-face and a web survey. *International Journal of Public Opinion Research* 25 (2): 242–253.

Révilla, M.A., Saris, W.E., and Krosnick, J.A. (2014). Choosing the number of categories in agree-disagree scales. *Sociological Methods & Research* 43 (1): 73–97. https://doi.org/10.1177/0049124113509605.

Roberts, C. (2007). Mixing Modes of Data Collection in Surveys: A Methodological Review. *NCRM Methods Review Papers*. Southampton: ESRC National Centre for Research Methods.

Schouten, B., van den Brakel, J., Buelens, B. et al. (2013). Disentangling mode-specific selection and measurement bias in social surveys. *Social Science Research* 42 (6): 1555–1570.

Singer, J. and Willett, J. (2003). *Applied Longitudinal Data Analysis: Modeling Change and Event Occurrence*. Oxford University Press.

Steenkamp, J.E.M. and Baumgartner, H. (1998). Assessing measurement invariance in Cross-National Consumer Research. *Journal of Consumer Research* 25 (1): 78–107. https://doi.org/10.1086/209528.

Tourangeau, R., Rips, L.J., and Rasinski, K. (2000). *The Psychology of Survey Response*. Cambridge University Press.

Vannieuwenhuyze, J.T.A. and Loosveldt, G. (2013). Evaluating relative mode effects in mixed-mode surveys: three methods to disentangle selection and measurement effects. *Sociological Methods & Research* 42 (1): 82–104. https://doi.org/10.1177/0049124112464868.

Vannieuwenhuyze, J.T.A. and Révilla, M. (2013). Relative mode effects on data quality in mixed-mode surveys by an instrumental variable. *Survey Research Methods* 7 (3): 157–168.

Vannieuwenhuyze, J.T.A., Loosveldt, G., and Molenberghs, G. (2014). Evaluating mode effects in mixed-mode survey data using covariate adjustment models. *Journal of Official Statistics* 30 (1): 1–21. https://doi.org/10.2478/jos-2014-0001.

Wagner, J., Arrieta, J., Guyer, H., and Ofstedal, M.B. (2014). Does sequence matter in multimode surveys: results from an experiment. *Field Methods* 26 (2): 141–155.

Wiley, J. and Wiley, M. (1974). A note on correlated errors in repeated measurements. *Sociological Methods & Research* 3 (2): 172–188. https://doi.org/10.1177/004912417400300202.

11

Measuring Cognition in a Multi-Mode Context

Mary Beth Ofstedal, Colleen A. McClain and Mick P. Couper

Survey Research Center, University of Michigan, Ann Arbor, MI, USA

11.1 Introduction

Challenges arise as large-scale surveys that have traditionally used interviewer-administered modes increasingly move towards including a web option. Striking a balance between taking advantage of the opportunities of the self-administered, computer-based mode and maximizing comparability with interviewer-administered modes presents operational and substantive challenges for researchers. This is a particular concern in the context of longitudinal surveys when the introduction of a new mode may disrupt time series estimates of trends and trajectories that are of primary value in such surveys.

Although multi-mode studies may present complications for a variety of survey measures, tests of cognitive ability are particularly problematic. Measures of cognitive ability have been incorporated in many surveys, and they are especially common in longitudinal surveys of health and ageing. These measures can be methodologically challenging to administer even without the complication of mixing modes (Herzog et al. 1999) and the introduction of a new mode adds further complexity. In some cases, the tests that have formed the core of interviewer-administered research designs are difficult or impossible to administer in an online setting, raising questions about how to design measures that minimize measurement error, and respondent difficulty while maximizing comparability and response quality across modes. Furthermore, these issues may be exacerbated for older respondents who may be unfamiliar with technology or have cognitive impairments that could affect the quality and completeness of the data differentially across modes. Despite these challenges, there are few mode comparisons that focus on cognitive measures, and the implications of mixed-mode design decisions for the measurement of cognition in longitudinal surveys remain largely unclear.

Advances in Longitudinal Survey Methodology, First Edition. Edited by Peter Lynn.

Companion website: www.wiley.com/go/lynn/advancesinlongitudinalsurvey

Using data from the Health and Retirement Study (HRS), collected from the same respondents over the course of three years via web, telephone (computer-assisted telephone interviewing [CATI]), and face-to-face (FTF) (computer-assisted personal interviews [CAPI]) administration, we address the following questions: What are the implications of mixing modes for measurement of cognitive performance in a longitudinal setting? Do the same tests administered to the same individuals in different modes produce different response distributions and response behaviour? Finally, are any of the observed mode differences consequential for the substantive conclusions that would be drawn?

Comparisons of measurement error in interviewer-administered and web modes have become more common in recent years (Dillman and Christian 2005; Duffy et al. 2005; Fricker et al. 2005; Chang and Krosnick 2009; Heerwegh 2009; Cernat et al. 2016); however, few studies have included tests of cognitive ability in their comparisons (for a recent exception, see Al Baghal 2019). We conduct an initial examination of several cognitive tests, with the goals of outlining considerations for test selection in future mixed-mode studies and highlighting particular areas of concern when mixing modes in studies that measure cognitive ability or decline as a key outcome or predictor.

11.2 Motivation and Previous Literature

11.2.1 Measurement of Cognition in Surveys

Cognitive assessments are an integral part of a number of large-scale longitudinal surveys. (For a list of cognitive tests included in selected studies, see Table 1 in McClain et al. 2018.) In general, cognitive assessments in surveys share several qualities that complicate implementation across modes. First, a complete assessment is multidimensional, often including tests of memory, reasoning, orientation, calculation, language, knowledge, and fluid intelligence (see Perlmutter 1988; Salthouse 1999). Second, many of the tests employed are relatively time-consuming, and may require special materials or conditions for full administration; for example, showcards, sound capabilities (in web or computer-assisted self-interviewing (CASI) surveys), programmed adaptive tests that display different questions based on previous answers, or the ability to measure response time. Finally, detailed instructions or interviewer assistance may be necessary to successfully administer certain tests.

Any attempt to use multiple modes will introduce potential measurement comparability issues. Several recent papers have assessed the role of computerisation in cognitive testing for the field as a whole (Wild et al. 2008; Zygouris and Tsolaki 2015) or for the development of specific tests (Runge et al. 2015; Ruano et al. 2016).

However, a systematic investigation of the factors that influence the measurement of cognition in large-scale surveys is lacking in the field.

11.2.2 Mode Effects and Survey Response

Survey mode is recognised as multidimensional, with effects on numerous aspects of total survey error. In particular, the integration of web surveys raises many considerations that are specific to the issues and approaches of this mode as it evolves (Couper 2000, 2011). The pressure to move to the web often results in mixed-mode formats that may trade off error sources and costs (Dillman 2000; de Leeuw 2005).

Mode effects studies often serve one (and sometimes both) of two purposes (see Jäckle et al. 2010). First, a number of studies test for *comparability* of estimates across modes, either considered in reference to a standard or existing mode or against a 'gold standard' or benchmark. These studies may be in the form of testing for data 'completeness', differences in marginal effects, effects on overall psychometric properties, comparisons of relationships between variables and trajectories by mode, or formal equivalence tests (e.g. Klausch et al. 2013; Cernat 2015a, 2015b; Mariano and Elliott 2017).

Second, other studies examine *how* and *why* these effects arise, often using experimental designs to draw conclusions that are narrower in scope. These include specific tests of mode features, such as visual versus auditory processing by controlling the use of showcards (e.g. Jäckle et al. 2010), or of hypotheses for mode differences such as the role of social distance and social desirability in studies comparing interviewer- and self-administration (e.g. Aquilino 1994; Holbrook et al. 2003; Kreuter et al. 2008; Heerwegh and Loosveldt 2011). While most of these studies involve experiments, others take advantage of studies that reinterview respondents in different modes, either by design or given their longitudinal nature (see, e.g. Cernat et al. 2016; Al Baghal 2019). We take the latter approach and consider mode effects on measurement, holding the sample source constant and taking advantage of multiple modes of administration within and across waves in a single study.

11.2.3 Cognition in a Multi-Mode Context

A number of features of survey modes have been identified as potentially affecting measurement (e.g. Couper 2011; Jäckle et al. 2010; Tourangeau et al. 2000, Chapter 5). Several of these are particularly relevant for the measurement of cognition.

First, modes differ in the presence of an interviewer. Interviewers are associated with higher levels of socially desirable responding, but interviewers also reduce missing data and other errors in surveys. While cognitive ability may not be particularly vulnerable to social desirability influences, respondents might be sufficiently hesitant to perform poorly on a cognitive test that they are unwilling to attempt the test in the presence of an interviewer. Similarly, respondents might be more likely to guess on the web or to admit that they don't know (if provided the option to do so). On the other hand, the presence of an interviewer may introduce time pressures and increase performance anxiety. Interviewers clearly play an important role in explaining the structure of a cognitive test, providing instructions, and ensuring that a test is completed once started. Still, interviewers may also (whether subtly or overtly) assist respondents in the completion of a cognitive test. Little is known about the behaviour of interviewers during cognitive testing and the effect they may have on outcomes.

Furthermore, when completing a cognitive assessment on the web, respondents may feel more comfortable using distributed cognition (see Clark and Chalmers 1998; Hutchins 1995), for example, by using aids such as a pencil and paper, a calculator, looking up answers online or consulting others. Estimates of seeking outside help, often considered to be 'cheating', range from minimal to pervasive in experimental studies of knowledge questions (e.g. Clifford and Jerit 2016; Munzert and Selb 2017).

Another key dimension on which modes vary is the medium of communication. This includes both the presentation of stimulus material and the delivery of the response. Telephone surveys (for example) use oral communication for both. FTF surveys could include visual materials (e.g. show cards). A self-administered (e.g. CASI) component could include both visual and oral/aural (sound) stimulus material. Online administration can include both visual and oral stimuli, although (as we discuss later) the mode is predominantly visual in nature. The entry of responses can also differ across modes, from oral responses in telephone and FTF to clicking or typing in CASI and web. But on a finer level, even the direct selection of an object using a touchscreen (e.g. on a tablet) may change performance on a task relative to using a mouse and pointer, depending on respondent dexterity and familiarity with technology. This is especially true if speed is an element of the test. In other words, the mode of administration may change the nature of the test.

Time pressures present competing potential influences on data quality. Time pressures that are likely present in the interviewer-administered modes, and especially on the telephone, may lead to worse performance or higher missing data compared to a self-administered survey if respondents take the opportunity to think further about the questions on the web. In contrast, respondent focus might be lessened without the presence of the interviewer; and if satisficing (Krosnick 1991) is more prevalent in self-administered than in interviewer-administered cognitive tests (Heerwegh 2009), more missing data, speeding, or straightlining could result on the web.

Visual presentation and processing may not only contribute to primacy response order effects (Schwarz et al. 1992), but could also fundamentally change the cognitive task at hand – for example, from a recall task to a recognition task, with the latter generally considered to be an easier cognitive task. The effects of cognitive burden may be fairly complex. For example, Chang and Krosnick (2010) found that respondents with lower cognition levels responded with higher levels of concurrent validity to a set of questions on the web than when the survey was administered over an intercom by an interviewer, suggesting that the demands posed by working memory might yield lower response quality among those at the lower end of a range of cognitive abilities in auditory modes.

Mode effects may vary across subgroups of the population, and may be especially problematic given the importance of measuring cognitive decline at the low end of the spectrum. Respondents who are less computer literate, or who have physical or cognitive limitations that may make it difficult to use a computer, may be particularly susceptible to mode effects when the web is utilised. Cognitive tests must thus be designed in a manner accessible to those with disabilities, health problems, visual or aural impairments, or low levels of literacy, cognition, and computer skills, if they are to be considered equivalent across modes.

11.2.4 Existing Mode Comparisons of Cognitive Ability

Despite the fact that cognitive measures are widely used in surveys and have the potential to be subject to a variety of mode effects, few mode comparisons of cognitive measures have been carried out that compare interviewer and web administration. Those that have been conducted are limited to individual measures and specific contexts.

While comparisons between different interviewer-administered modes exist, they are also limited. Rodgers et al. (2003) found higher scores (indicative of greater cognitive ability) on the telephone than via FTF interviewing, but acknowledge that mode may be confounded with cognitive ability due to lack of clean random assignment. An earlier randomised mode experiment in HRS/AHEAD found no differences in average cognitive scores between telephone and FTF interviewing (Herzog et al. 1999).

More recent comparisons, including web administration, are rare. In one such study, Runge et al. (2015) compared the performance of female respondents from the HRS to an independent sample of women who participated in the 2013 Women's Health Valuation (WHV) study, recruited from an existing online panel of US adults using quotas for age and race/ethnicity. The WHV replicated the HRS immediate and delayed recall tasks on the web with one key difference: WHV respondents were required to type in the words they recalled rather than saying them aloud to an interviewer. The authors found that WHV respondents had higher immediate and delayed recall scores than HRS respondents from equivalent years. Key predictors of recall were largely similar across studies, with some exceptions. For example, in the WHV, those with poor self-reported memory recalled fewer words than similar HRS counterparts

A more controlled mode comparison of one cognitive measure was carried out by Gooch (2015). Visitors to a research facility were randomised after agreement to participate to either an FTF interview or a web survey that took place on the facility's computers. Among other measures, respondents were asked to respond to the Gallup–Thorndike Verbal Intelligence Test, also called the 'Wordsum' test. Gooch found that the modest-to-difficult questions in the battery were answered correctly more often on the web than in the FTF survey, though the pattern reversed for easier questions (they were answered correctly more often in FTF administration). However, the ordering of points in an item response theory model was identical across modes and the mode effects did not yield significant differences when used in multivariate substantive models.

More recently, Al Baghal (2019) conducted an analysis of cognitive measures in the *Understanding Society* Innovation Panel (IP7, conducted in 2014), a mixed-mode design, with cases randomly assigned to a sequential mixed-mode protocol (web then FTF) or FTF only. In the FTF mode, the cognitive measures were self-administered as part of the CASI module. The measures he examined included the number series, verbal analogies, numeracy, and four forward digit-span tasks, testing working memory capacity. Among other differences, those responding via the web in the mixed-mode design were significantly younger, more educated, and more likely to be daily (as opposed to less frequent) internet users. To control for self-selection, Al Baghal examined responses to the cognitive measures in IP1 (where all were interviewed in-person) and used inverse probability of treatment weighting (IPTW) methods to account for differential selection into modes. He found that web respondents perform significantly better than FTF respondents in the measures of inductive reasoning, numeracy, and recall (measured by forward digit span). The study maintained a consistent visual presentation across all modes, so differences are not attributable to aural versus visual modes of delivery.

Given the widespread use of cognitive measures in both methodological and substantive models, there is need for more systematic mode comparisons to evaluate the ways in which survey mode may affect the usefulness of the measures and their validity with respect to an individual's true cognitive state. We do this in the current study by examining measurement differences for four cognitive tests. We restrict our analysis to a set of respondents who completed the same measures in both interviewer- and self-administered modes in order to distinguish measurement effects from any potential selection effects, especially those related to cognitive ability.

11.3 Data and Methods

11.3.1 Data Source

The HRS is a longitudinal study of older adults in the United States conducted by the University of Michigan. It surveys respondents over the age of 50 and their spouses, with successive cohorts of respondents added over time to maintain a representative sample of the study population. Respondents participate in core surveys every other year. Starting in 2006, individuals under the age of 80 were randomised so that roughly half received an FTF interview and half a telephone interview. The mode flipped in 2008 and has followed this rotation every wave thereafter, so that half receive an FTF interview in 2006, 2010, 2014, etc. and a telephone interview in 2008, 2012, 2016, etc. and the other half is on the opposite schedule. The content of the telephone and in-person interviews are essentially identical, with the exception of added physical measures, biomarker collection, a psychosocial leave-behind questionnaire and data linkage consent requests in the FTF administration. In years that do not contain a core interview, the study team fields a variety of off-year efforts, including web and mail surveys. The web surveys, which were fielded in alternate years between 2003 and 2013, contain some questions from the core interview, as well as a range of new topics. (See the HRS website for more information: http://hrsonline.isr.umich.edu.)

11.3.2 Analytic Sample

Our analysis utilises data from 4223 respondents between the ages of 50 and 80 who self-completed (i.e. did not have a proxy respondent) each of the 2012 core, 2013 web, and 2014 core survey requests. The 2013 web survey was administered to a sub-sample of HRS participants who reported in their most recent core interview that they had access to the internet. A random 80% of those with internet access were selected for the 2013 web sample ($n = 7744$) and 75% of those sampled completed the survey ($n = 5813$). We remove those younger than 50 (age-ineligible spouses) as well as those who completed their core interview by proxy (for whom the standard cognitive tests were not administered). In order to cleanly perform between-sample comparisons, and to leverage the benefits of randomisation, we remove respondents older than 80 (who are nearly always assigned to FTF data collection) as well as those who did not complete the 2012 and/or 2014 core interview in the mode to which they were assigned.[1]

Because our focus is on within-subject comparisons, a key strength of our analysis is that there is no confounding with selection effects and sampling variance is

1 Most respondents in the analytic sample completed the core survey in the assigned mode: 96% in 2012, 95% in 2014.

reduced. In addition, the cross-over design of the mode assignment in HRS means that we can control for change over time at the sample level.

11.3.3 Administration of Cognitive Tests

Since the primary core modes of HRS are telephone and FTF, the main considerations for test selection have been to include items that can be administered over the phone (i.e. not reliant on visual aids) and to keep the cognitive battery sufficiently short to minimise respondent burden. Originally, tests were primarily drawn from the Telephone Interview for Cognitive Status (TICS) screen, based on the Mini-Mental State Exam (Brandt et al. 1988). HRS cognitive measures were expanded starting in 2010 to provide more differentiation at the higher end of functioning (Fisher et al. 2013). The following four tests are utilised in the present analysis.

The ***quantitative number series*** measures quantitative reasoning and fluid intelligence, and is a six-item, block-adaptive test based on answers to 6 out of 15 possible items. All respondents start with the same three items, but the difficulty of the items shown in the second set of three items depends on the answers to the first set. Respondents are asked to fill in the blank in a series – 'for example, if I said the numbers "1 2 BLANK 4," then what number would go in the blank?' We utilise the 'W-scores' in the HRS dataset. These are standardised scores designed to be comparable to the Woodcock-Johnson III (WJ-III) test battery on which this task was based (Fisher et al. 2013). The score ranges from 409 to 569. A 10-point decrease in the score represents halving the probability that a respondent answers a given item (or one of equal difficulty) correctly. Unlike other tests, the respondent is asked to write down the series of numbers in the interviewer administered modes before communicating the answer to the interviewer,[2] introducing visual processing into the telephone and FTF contexts. The number series test is administered in alternate waves in the core interview (2012, 2016, etc.). It was administered to a random sub-sample of participants in the 2013 web survey.

Numeracy is measured via respondents' answers to three questions developed by Lipkus et al. (2001; see also Huppert et al. 2004): how many individuals will be expected to get a disease out of 1000 given a 10% chance ('chance of disease'); how much money will be received by each of five winners of a $2 million dollar lottery prize ('lottery split'); and how much money will result after two years if a sum of $200 yields 10% interest per year ('compound interest'). In the core interview, this last item is only asked if either of the first two is answered correctly. In the web survey, a more restrictive rule was applied that asked the third question only if *both*

2 We have no indicator of the extent to which respondents complied with this instruction.

of the first two questions were answered correctly. To eliminate this confound of mode and test administration, we restrict our analysis to respondents who would have received the same treatment in both modes (i.e. we exclude respondents who answered only one of the first two questions correctly).[3] We follow the 0–4 scoring of Levy et al. (2014) in which partial credit is given for a 'nearly correct' answer to the compound interest question (i.e. $240 garners one point; a correct answer of $242 garners two points). The numeracy items are administered in alternate waves in the core interview (2010, 2014, etc.). They were administered to the full 2013 web sample.

The ***Serial 7s*** test measures working memory (Ofstedal et al. 2005) and requires respondents to subtract 7 from 100 five consecutive times, with the composite score ranging from 0 to 5 (representing the number of correct answers). The Serial 7s test is administered in every wave of the core interview and was administered to the full 2013 web sample.

The ***verbal analogies*** test is administered in similar fashion to the quantitative number series and measures verbal reasoning. The six-item test is block-adaptive and administered from a set of 15 possible items, with the difficulty of the second set of items dependent on the respondent's answers to the first set. For this test, respondents are asked to fill in the missing word in an analogy such as 'Mother is to Daughter as Father is to…' The standardised W-score ranges from 435 to 560. The verbal analogies test was administered to a small random sub-sample in the 2012 core interview and to a random half sample in the 2013 web survey. Starting in 2014, verbal analogies is administered to the full core sample in alternate waves (2014, 2018, etc.).

Due to the alternate wave and subsampling design for the cognitive measures in the core waves and 2013 web survey, the resulting analytic sample size permitting within-respondent analysis across waves varies substantially across tests. Sample sizes for each test are provided in each of the tables.

11.3.4 Methods

All of the analyses presented are unweighted and do not take the complex survey design into consideration, as we are purposefully working with a subset of cases that are not representative of the entire HRS sample. We describe each set of analyses in turn below.

---

3 We obtain similar results when applying an alternative assumption, that respondents who answered only one of the first questions correctly in the core interview would have answered the third wrong.

11.3.4.1 Item Missing Data

Given the varied nature of the cognitive tests used both in number of items and question design, an overall assessment of item-missing data across modes requires definitions of 'missing' values that are specific to each test. Additionally, since no explicit 'don't know' response was provided to web survey respondents, we are unable to determine whether missing data was due to the respondent not knowing the answer, refusing to answer the question, or for some other reason. For the number series, Serial 7s and verbal analogies, we define missing based on the first item for each test. If the respondent skipped the first item (in web) or responded either 'Don't know' or 'Refused' (in interviewer mode), that respondent is coded as missing for that test. (If the respondent answered the first item but had missing data on a subsequent item, they are coded as *not* missing for the test for purposes of the item missing data comparisons.) For numeracy, we consider answers to the individual questions separately using the same guidelines.

We conduct *paired t-tests* when examining differences in levels of item missing data between interviewer-administered waves (telephone and FTF) and web administration (which are based on the same individuals) and *between-sample t-tests* when examining differences between the telephone and FTF modes within a wave.

11.3.4.2 Completion Time

We use between- and within-sample *t*-tests to compare completion times for each of the cognitive tests. Times were summed from the raw paradata at the page- or field-visit level to the test level (see McClain et al. 2018). Outliers beyond the 95th percentile at the test level in a given wave are top-coded to the 95th percentile and comparisons are restricted to cases with a positive time spent on the test in all three waves. For all tests, time to read (or be read) introductions and practice questions was counted as part of the test administration time.

11.3.4.3 Overall Differences in Scores

We again use between and within-sample t tests when comparing mean scores for each of the cognitive tests across modes. To keep the sample size and composition constant across comparisons, only cases with a substantive answer for the test (including 0, where relevant) in all three waves are included; that is, any case with an overall missing score on any wave is excluded.

11.3.4.4 Correlations Between Measures

Because the subsampling design described previously precludes most confirmatory factor analysis that could assess the consistency of the factor structure of the cognitive tests across modes, we focus on correlations between *pairs* of tests within a wave, as well as correlations for a given test across waves. Again, we restrict the sample to those who substantively answered the relevant tests in all waves examined in order to keep sample size and composition constant.

11.3.4.5 Trajectories over Time

As Rabbitt et al. (2004) summarise, many longitudinal studies of cognition have one of three aims – to assess trajectories of change, and particularly whether cognitive decline accelerates in old age (e.g. Hertzog and Schaie 1986); to assess how rates of change differ between baseline mental abilities; and to assess whether these trajectories are affected by a wide range of demographic, social, and environmental factors. We provide an initial assessment of change between paired waves to assess whether a given respondent performed better, worse, or about the same between the waves, where change is defined as approximately a one-standard deviation difference from the previous wave.

Two types of longitudinal models are subsequently used to test the hypothesis that the trend over time varies with mode. First, a random intercept repeated measures model with an autoregressive error structure is fit to observations clustered by respondent, controlling for whether or not the individual began in 2012 with FTF or telephone administration. This approach is commonly applied to repeated measures data and, when it incorporates random coefficients or 'slopes' as well with larger numbers of time points, parallels the use of latent growth curve models to assess individual-level trajectories (for one discussion of their similarities, see Hox and Stoel 2005). We fit a simpler model here given only three time points, though further analysis could estimate individual trajectories. Second, a latent class growth analysis (Nagin et al. 2016), similar to the growth mixture modelling approach, is fit to the same data to assess whether respondents appear to fall into different classes of trajectories over time. This model tests the hypothesis that latent groups of respondents can be defined by assessing differences in their trajectories in cognitive scores (e.g. examining the shape of trajectories and one's probability of following a given trajectory), specifying various trajectory forms and class solutions in model selection.

11.3.4.6 Models Predicting Cognition as an Outcome

Finally, we compare results from models that utilise a set of covariates to predict cognitive ability across tests and mode of administration. Predictors of cognition were drawn from relevant substantive literature (e.g. Rodgers et al. 2003; Langa et al. 2017) and include age, gender, race/ethnicity, education, work status, chronic conditions, depressive symptoms, household income, and a count of types of internet activities the respondent reported (e.g. financial, other commerce, social network, contact [e.g. email], news/entertainment, and/or work-related tasks). We use OLS regression models for this analysis.

Due to space constraints, we are not able to present results from all of the analyses in this write-up. However, the full set of results is available in McClain et al. (2018) (https://hrs.isr.umich.edu/publications/biblio/9606).

11.4 Results

11.4.1 Item-Missing Data

Table 11.1 presents the percentage of respondents with item-missing data on each test, using the criteria described earlier. The trend in item-missing data varies by the type of test. Verbal analogies and number series, the two longer, adaptive tests, have higher rates of missing data on the web than in the interviewer-administered modes. However, numeracy and Serial 7s yield *less* item-missing data on the web.

Table 11.1 Percent of cases with item-missing data, and tests of differences.

	2012		2013	2014	
	Tel % Missing (N)	FTF % Missing (N)	Web % Missing (N)	Tel % Missing (N)	FTF % Missing (N)
Number Series (First item) (n = 994)	3.6% (19)	2.2% (10) ^{W}	5.3% (53) ^{F12}		
Numeracy Item 1 (Chance of disease) (n = 2030)			0.6% (12) ^{T14,F14}	2.1% (21)^{W}	1.8% (18) ^{W}
Numeracy Item 2 (Lottery split) (n = 2030)			4.5% (91) ^{T14,F14}	7.2% (72)^{W}	7.4% (76) ^{W}
Serial 7s (First item) (n = 2147)	1.7% (19) ^{W}	0.8% (9)	0.6% (14) ^{T12}	1.1% (12)	1.4% (15)
Verbal analogies (First item) (n = 429)	0.0% (0) ^{W}	0.0% (0)	2.1% (9) ^{T12}	0.0% (0)	1.0% (2)

Superscripts denote significant differences from within-subject difference-of-means tests ($p < 0.001$). Within-subject tests are conducted between web scores (W) and a given respondent's scores in telephone in 2012 (T12), FTF in 2012 (F12), telephone in 2014 (T14), and FTF in 2014 (F14).

11.4.2 Completion Time

An analysis of completion time for the cognitive tests across modes suggests that all of the tests except the number series take *longer* to administer on the web – going against conventional wisdom that web administration might be more efficient than interviewer administration (see Table 6 in McClain et al. 2018). As the sole exception, the number series is administered more quickly on the web than via telephone, though it takes about the same amount of time as in FTF administration. No significant differences are observed between telephone and FTF administration.

11.4.3 Differences in Mean Scores

Table 11.2 illustrates the differences in mean cognitive scores across modes and waves, utilizing only cases for which a substantive response was given. In all cases, within-subject tests suggest that mean scores are significantly higher for web than for telephone and FTF ($p < .001$). While most between-subject comparisons for telephone versus FTF in a given wave yield no significant differences, verbal analogies exhibit significantly higher scores via FTF than telephone administration ($p < .05$), but only in 2014. Overall, the pattern suggests that respondents do better on cognitive tests on the web, though in some cases differences are small and may not be substantively important.

Table 11.2 Means, standard errors, and tests of differences for cognitive scores.

	2012		2013	2014	
	Tel mean (SE) (N)	**FTF mean (SE) (N)**	**Web mean (SE) (N)**	**Tel mean (SE) (N)**	**FTF mean (SE) (N)**
Number series	535.0 (1.1) *(520)* ^{W}	532.5 (1.2) *(453)* ^{W}	541.4 (0.7) *(973)* ^{T12,F12}		
Numeracy			2.95 (0.03) *(1069)* ^{T14,F14}	2.56 (0.05) *(526)* ^{W}	2.67 (0.05) *(543)* ^{W}
Serial 7s	4.12 (0.04) *(1072)* ^{W}	4.05 (0.04) *(1041)* ^{W}	4.43 (0.02) *(2113)* ^{T12,F12,T14,F14}	4.20 (0.04) *(1041)* ^{W}	4.07 (0.04) *(1072)* ^{W}
Verbal analogies	512.0 (1.7) *(201)* ^{W}	515.2 (1.7) *(212)* ^{W}	520.5 (1.2) *(413)* ^{T12,F12,T14,F14}	513.9 (1.9) *(212)* ^{W}	519.4 (1.9) *(201)* ^{W}

Superscripts are defined in previous table.

Table 11.3 Within-test correlations across waves of data collection.

	2012		2014		2012/2014
Cognitive test	**Tel*2013 Web**	**FTF*2013 Web**	**Tel *2013 Web**	**FTF *2013 Web**	**Iwer*Iwer**
Serial 7s	0.22 (1072)	0.27 (1041)	0.31 (1041)	0.22 (1072)	0.52 (2113)
Verbal analogies	0.43 (201)	0.41 (212)	0.44 (212)	0.33 (201)	0.63 (413)

Sample size restricted to respondents answering a given test in all three waves of administration.

We also examined the percentage of respondents achieving the *maximum score* for each test and found that this percentage is substantially higher on the web than in either telephone or FTF interviews (see Table 8 in McClain et al. 2018).

11.4.4 Correlations Between Measures

Results shown in Table 11.3 suggest that cognitive test scores are more weakly correlated with each other when web administration is involved, suggesting a potential decline in reliability when modes are mixed. Where tests are available across all three waves of data collection, correlations between administrations of a single test in interviewer-administered waves are higher than the correlations between web and interviewer administration, despite the fact that the latter administrations are closer in time than the former. We also examined between-test correlations and correlations for each test with self-rated memory *within* a wave (see Table 10 in McClain et al. 2018). For all tests except the number series, correlations are uniformly lower in the 2013 web administration than in other waves and modes.

11.4.5 Trajectories over Time

Figure 11.1 displays the distribution of respondents performing worse (left portion of bar), about the same (middle), and better (right) from one wave to the next, where change is defined as a one standard deviation difference. Changes in either direction are relatively evenly distributed when examining the two interviewer-administered waves at the top of the figure. *More* change is observed overall for the one-year intervals involving switches between interviewer- and self-administration, and the direction of this change appears to diverge by mode sequence. Respondents tend to do better in a subsequent wave when moving from interviewer- to self-administration (middle panel), and worse when moving from self- to interviewer-administration (bottom panel).

To further address these trajectories over time, we estimate several longitudinal models for the two cognitive tests for which we had measurements at three time

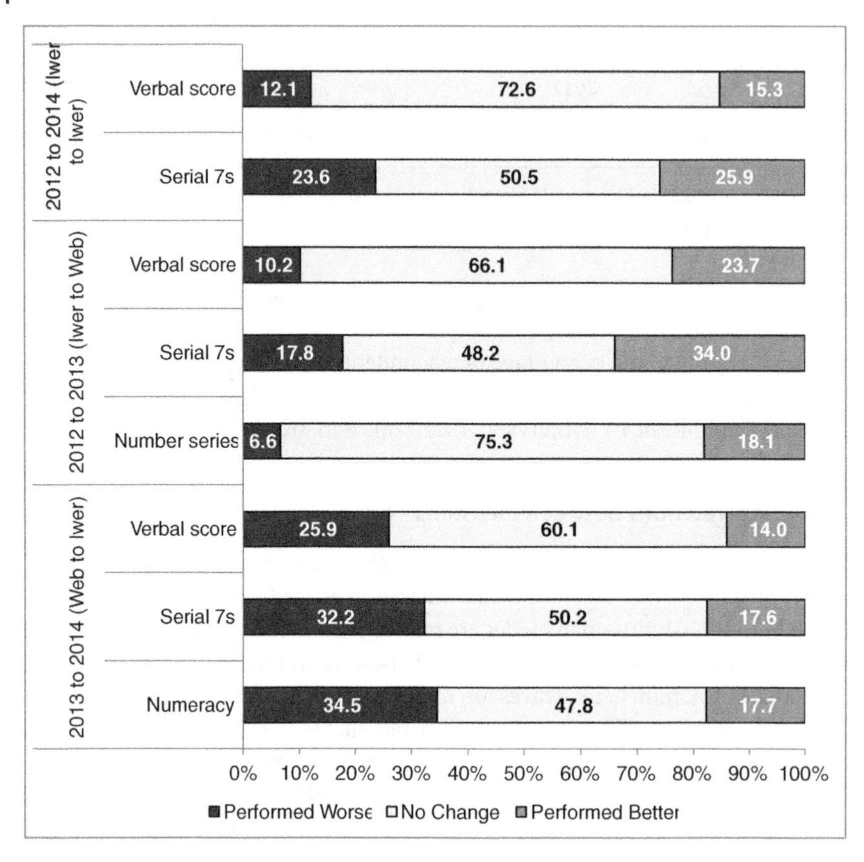

Figure 11.1 Percentage of respondents performing worse, better, and with no change between paired waves.

points (Serial 7s and verbal analogies). First, we fit a random effects model with random intercepts for each respondent, a non-linear fixed effect of time, and an autoregressive error structure. Results from this model suggest a potential curvilinear pattern over time for each cognitive outcome (Table 11.4). While there is a significant, positive fixed effect for the 2013 time point as compared to 2012, there is no such effect for the 2014 time point as compared to 2012 for Serial 7s. In other words, while scores for the interviewer-administered waves do not differ with time, changing from interviewer to self-administration is associated with an increase in cognitive score. While the verbal analogies score does display significant increases in both years, the increases are substantively quite small (especially between 2012 and 2014) and may not be practically significant. We estimated the same random effects model with a set of socio-demographic and health covariates (age, gender,

Table 11.4 Longitudinal models for cognitive outcomes.

	Serial 7s	Verbal score
Fixed Effects		
Intercept	4.081 (0.032)***	513.73 (1.624)***
Time: 2013 (vs. 2012)	0.345 (0.032)***	6.889 (1.335)***
Time: 2014 (vs. 2012)	0.043 (0.028)	2.947 (1.255)*
2012 FTF (vs. telephone)	0.017 (0.038)	−0.183 (2.018)
Variance Components		
Random intercept	0.252	272.64
Autoregressive errors	0.264	0.1302
Residual variance	1.137	374.08
Model Fit and Sample Size		
BIC	19 367.4	11 279.6
N	2113	413

***$p < .001$; **$p < .01$; *$p < .05$

race, education, chronic conditions, count of internet activities; results not shown) and observed the same pattern of fixed effects of time across waves.

Second, we ran a set of latent class growth models, specifying quadratic trajectories for each class, and testing solutions with one to four classes. The Bayesian Information Criterion indicated a three-class solution as the best fit for each cognitive outcome. For Serial 7s, only 15% of respondents fall into a class that suggests a practically significant curvilinear trajectory; the other two classes showed essentially no change across the three waves (see Figure 3 in McClain et al. 2018). The three-class solution for the verbal analogies score shows a small increase in scores between 2012 and 2013, followed by a decrease in 2014 for one class, predicted to represent 44% of respondent trajectories (see Figure 4 in McClain et al. 2018).

For both the random effects and latent class growth models, whether a respondent completes the survey via telephone or FTF in 2012 does not predict which class of trajectories he or she has the highest probability of falling into.

11.4.6 Substantive Models

Finally, we estimated a set of OLS regression models to assess whether predictors of cognitive scores were consistent across modes. Results from these models yield somewhat inconsistent results, with many of the predictors displaying different associations across modes (see Table 12 in McClain et al. 2018). For example, education is positively related to Serial 7s in the interviewer-administered waves, but it is unrelated to Serial 7s on the web. Women had higher verbal scores than men in the web administration, but not in the interviewer-administered waves. Hispanics performed significantly worse than non-Hispanic whites on Serial 7s in the interviewer-administered waves, but there was no difference on the web. Of note, scores on all tests are positively related to internet usage.

11.5 Discussion

Our results suggest that survey mode *does* affect estimates of cognitive ability. The main differences are between web vs. interviewer-administered modes, although we also observe some differences between telephone and FTF administration. For all of the cognitive measures, respondents performed better on the web than in either interviewer administered mode. This was true regardless of whether the web administration occurred before or after interviewer administration. As a result, measures of trajectories over the three waves we study are adversely affected. A sizeable proportion of respondents are characterised by an inverse U-shaped trajectory, whereby cognitive performance increases between T1 (interviewer) and T2 (web) and declines between T2 and T3 (interviewer). This pattern, which is contrary to the expected pattern of age-related cognitive decline, is especially apparent for verbal analogies.

Whereas performance on the cognitive measures is consistently higher on the web than interviewer modes, correlations between pairs of tests or within-test correlations across waves are consistently lower for measures administered on the web. Likewise, the strength of key predictors (e.g. age, sex, education) of cognition differs across modes. Thus, mode appears to affect not only levels of cognitive performance but also other properties of the cognitive measures.

It is unclear which mode yields more valid results. On one hand, respondents may feel more anxiety or pressure with an interviewer present and perform worse than their true ability. On the other hand, without an interviewer to observe, web respondents may use aids (e.g. calculator, online searches, write down words, etc.) and/or take more time to think about their answers when completing the tests and perform better than their true ability. Al Baghal's (2019) findings of differences between CASI and web administration for identical tests and our finding that HRS respondents take longer to complete most of the tests on the web than in interviewer modes, suggest that either or both of these circumstances may be at play. Satisficing (or sub-optimal responding) is another potential confounder, and on this our evidence is mixed. Our findings of higher performance and longer completion times on the web seem to suggest less satisficing; however, higher missing data rates on the web suggest more satisficing. The lower between-test correlations and lower associations with predictors of cognition also suggest lower reliability for the web administration.

Regardless of which mode yields more valid and reliable results, however, our findings suggest that a switch from interviewer to web-based administration (or vice versa) will affect measures of cognitive performance. If the mode differences were limited to levels and occurred uniformly across respondents, then a simple calibration or control for mode may suffice. However, because we find mode differences in associations – both between pairs of cognitive tests and between cognitive performance and known predictors of cognition – this suggests that a more nuanced approach is needed. Given the nature of our analytic sample – who are disproportionately white, educated, employed, healthy, young, with higher average income, and higher in cognitive functioning than the full HRS sample – it is possible that our results are conservative and that those with lower cognitive abilities would, if pushed to the web, display even larger mode effects.

The study has some limitations. A major one is that, in our analysis, mode is confounded with time. For two of the tests (number series and numeracy), the web measurement occurs either before or after the interviewer-administered measurement, but not both. This makes the assessment and interpretation of mode effects difficult. However, the availability of three measurement points for both Serial 7s and verbal analogies (interviewer, web, interviewer) allows us to disentangle mode effects from true change in cognitive function. A second limitation is that the analytic sample is somewhat selective of younger, better educated and higher functioning individuals compared to the full HRS sample (see Table 3 in McClain et al. 2018), so results may not be generalizable to the general population. A third limitation is the small sample sizes for some of the comparisons.

Despite these limitations, our results provide lessons for longitudinal surveys that plan to add web as a new mode, as well as for new multi-mode studies that are getting underway. The first is to be clear about priorities. The need to maximise comparability across modes calls for a different design approach than the desire to take maximum advantage of the capabilities of individual modes. For example, self-administration allows the use of visual stimuli, which is not possible in telephone interviews. On the other hand, tests that require verbal communication (reading or repeating words, counting backwards, naming animals) may be difficult or impossible to replicate in a web survey, where it is not practical (or ethical) to control settings on respondents' computers. If it is not critical for the study's purposes to administer the same measures in all modes, then using multiple modes could enhance the measurement of cognition by allowing for a broader set of measures than would be possible in a single mode. However, for most studies, this is likely not an option, or at least not the top priority.

If comparability of tests is a priority, it is important to identify tests that are suitable for administration across different modes. Regardless of how comparable the tests are in terms of administration protocol, however, there are still likely to be some measurement differences by mode. At a minimum, this means that careful attention to mode differences and possible calibration of the data will be needed. To the extent possible, experiments that are designed to test for mode effects and evaluate the effectiveness of calibration would be valuable to incorporate in longitudinal studies administering cognitive measures.

Acknowledgements

Data used in the study come from the Health and Retirement Study (HRS), which is funded by a grant from the National Institute on Ageing (U01 AG009740) with supplemental support from the Social Security Administration. HRS is conducted by the University of Michigan.

References

Al Baghal, T. (2019). The effect of online and mixed-mode measurement of cognitive ability. *Social Science Computer Review* 37 (1): 89–103.

Aquilino, W.S. (1994). Interview mode effects in surveys of drug and alcohol use: a field experiment. *Public Opinion Quarterly* 58 (2): 210–240.

Brandt, J., Spencer, M., and Folstein, M. (1988). The telephone interview for cognitive status. *Neuropsychiatry, Neuropsychology, and Behavioral Neurology* 1 (2): 111–117.

Cernat, A. (2015a). The impact of mixing modes on reliability in longitudinal studies. *Sociological Methods & Research* 44 (3): 427–457.

Cernat, A. (2015b). Impact of mixed modes on measurement errors and estimates of change in panel data. *Survey Research Methods* 9 (2): 83–99.

Cernat, A., Couper, M.P., and Ofstedal, M.B. (2016). Estimation of mode effects in the Health and Retirement Study using measurement models. *Journal of Survey Statistics and Methodology* 4 (4): 501–524.

Chang, L. and Krosnick, J.A. (2009). National surveys via RDD telephone interviewing versus the internet: comparing sample representativeness and response quality. *Public Opinion Quarterly* 73 (4): 641–678.

Chang, L. and Krosnick, J.A. (2010). Comparing oral interviewing with self-administered computerized questionnaires: an experiment. *Public Opinion Quarterly* 74 (1): 154–167.

Clark, A. and Chalmers, D. (1998). The extended mind. *Analysis* 58 (1): 7–19.

Clifford, S. and Jerit, J. (2016). Cheating on political knowledge questions in online surveys: an assessment of the problem and solutions. *Public Opinion Quarterly* 80 (4): 858–887.

Couper, M.P. (2000). Web surveys: a review of issues and approaches. *The Public Opinion Quarterly* 64 (4): 464–494.

Couper, M.P. (2011). The future of modes of data collection. *Public Opinion Quarterly* 75 (5): 889–908.

Dillman, D.A. (2000). *Mail and Internet Surveys: The Total Design Method*. New York: Wiley.

Dillman, D.A. and Christian, L.M. (2005). Survey mode as a source of instability in responses across surveys. *Field Methods* 17 (1): 30–52.

Duffy, B., Smith, K., Terhanian, G. et al. (2005). Comparing data from online and face-to-face surveys. *International Journal of Market Research* 47 (6): 615–639.

Fisher, G.G., McArdle, J.J., McCammon, R.J. et al. (2013). *New Measures of Fluid Intelligence in the HRS*. Ann Arbor: Institute for Social Research, University of Michigan https://hrs.isr.umich.edu/sites/default/files/biblio/dr-027b.pdf.

Fricker, S., Galesic, M., Tourangeau, R. et al. (2005). An experimental comparison of web and telephone surveys. *Public Opinion Quarterly* 69 (3): 370–392.

Gooch, A. (2015). Measurements of cognitive skill by survey mode: marginal differences and scaling similarities. *Research & Politics* 2 (3): 1–11.

Heerwegh, D. (2009). Mode differences between face-to-face and web surveys: an experimental investigation of data quality and social desirability effects. *International Journal of Public Opinion Research* 21 (1): 111–121.

Heerwegh, D. and Loosveldt, G. (2011). Assessing mode effects in a national crime victimization survey using structural equation models: social desirability bias and acquiescence. *Journal of Official Statistics* 27 (1): 49–63.

Hertzog, C. and Schaie, K.W. (1986). Stability and change in adult intelligence: I. Analysis of longitudinal covariance structures. *Psychology and Aging* 1 (2): 159–171.

Herzog, A.R., Rodgers, W.L., Schwarz, N. et al. (1999). Cognitive performance measures in survey research on older adults. In: *Cognition, Aging, and Self-Reports* (eds. N. Schwarz, D. Park, et al.), 327–340. Philadelphia: Psychology Press.

Holbrook, A.L., Green, M.C., and Krosnick, J.A. (2003). Telephone versus face-to-face interviewing of national probability samples with long questionnaires: comparisons of respondent satisficing and social desirability response bias. *Public Opinion Quarterly* 67 (1): 79–125.

Hox, J. and Stoel, R.D. (2005). Multilevel and SEM approaches to growth curve modeling. In: *Encyclopedia of Statistics in Behavioral Science* (eds. B.S. Everitt and D.C. Howell), 1296–1305. Chichester: Wiley.

Huppert, F.A., Gardener, E., and McWilliams, B. (2004). Cognitive function. In: *Retirement, Health and Relationships of the Older Population in England: The 2004*

English Longitudinal Study of Ageing (eds. J. Banks, E. Breeze, et al.), 217–242. London: Institute for Fiscal Studies.

Hutchins, E. (1995). *Cognition in the Wild*. Cambridge, MA: MIT Press.

Jäckle, A., Roberts, C., and Lynn, P. (2010). Assessing the effect of data collection mode on measurement. *International Statistical Review* 78 (1): 3–20.

Klausch, T., Hox, J.J., and Schouten, B. (2013). Measurement effects of survey mode on the equivalence of attitudinal rating scale questions. *Sociological Methods & Research* 42 (3): 227–263.

Kreuter, F., Presser, S., and Tourangeau, R. (2008). Social desirability bias in CATI, IVR, and web surveys the effects of mode and question sensitivity. *Public Opinion Quarterly* 72 (5): 847–865.

Krosnick, J.A. (1991). Response strategies for coping with the cognitive demands of attitude measures in surveys. *Applied Cognitive Psychology* 5 (3): 213–236.

Langa, K.M., Larson, E.B., Crimmins, E.M. et al. (2017). A comparison of the prevalence of dementia in the United States in 2000 and 2012. *JAMA Internal Medicine* 177 (1): 51–58.

de Leeuw, E.D. (2005). To mix or not to mix data collection modes in surveys. *Journal of Official Statistics* 21 (2): 233–255.

Levy, H., Ubel, P.A., Dillard, A.J. et al. (2014). Health numeracy: the importance of domain in assessing numeracy. *Medical Decision Making* 34 (1): 107–115.

Lipkus, I.M., Samsa, G., and Rimer, B.K. (2001). General performance on a numeracy scale among highly educated samples. *Medical Decision Making* 21 (1): 37–44.

Mariano, L.T. and Elliott, M.N. (2017). An item response theory approach to estimating survey mode effects: analysis of data from a randomized mode experiment. *Journal of Survey Statistics and Methodology* 5 (2): 233–253.

McClain, C.A., Couper, M.P., Hupp, A. et al. (2018). A typology of web survey paradata for assessing total survey error. *Social Science Computer Review*, online first, doi: https://doi.org/10.1177/0894439318759670.

McClain, C.A., Ofstedal, M.B., and Couper, M.P. (2018). *Measuring Cognition in a Multi-Mode Context*. Ann Arbor: Institute for Social Research, University of Michigan https://hrs.isr.umich.edu/publications/biblio/9606.

Munzert, S. and Selb, P. (2017). Measuring political knowledge in web-based surveys: an experimental validation of visual versus verbal instruments. *Social Science Computer Review* 35 (2): 167–183.

Nagin, D.S., Jones, B.L., Lima Passos, V. et al. (2016). Group-based multi-trajectory modeling. *Statistical Methods in Medical Research*. Advance online publication. doi: https://doi.org/10.1177/0962280216673085.

Ofstedal, M.B., Fisher, G.G., and Herzog, A.R. (2005). *Documentation of Cognitive Functioning Measures in the Health and Retirement Study*. Ann Arbor: Institute for Social Research, University of Michigan http://hrsonline.isr.umich.edu/sitedocs/userg/dr-006.pdf.

Perlmutter, M. (1988). Cognitive potential throughout life. In: *Emergent Theories of Aging* (eds. J.E. Birren and V.L. Bengtson), 247–268. New York: Springer.

Rabbitt, P., Diggle, P., Holland, F. et al. (2004). Practice and drop-out effects during a 17-year longitudinal study of cognitive aging. *The Journals of Gerontology Series B: Psychological Sciences and Social Sciences* 59 (2): P84–P97.

Rodgers, W.L., Ofstedal, M.B., and Herzog, A.R. (2003). Trends in scores on tests of cognitive ability in the elderly US population, 1993–2000. *The Journals of Gerontology Series B: Psychological Sciences and Social Sciences* 58 (6): S338–S346.

Ruano, L., Sousa, A., Severo, M. et al. (2016). Development of a self-administered web-based test for longitudinal cognitive assessment. *Scientific Reports* 6, Article number 19114. Retrieved from https://www.nature.com/articles/srep19114.

Runge, S.K., Craig, B.M., and Jim, H.S. (2015). Word recall: cognitive performance within internet surveys. *JMIR Mental Health* 2 (2) https://doi.org/10.2196/mental.3969.

Salthouse, T.A. (1999). Theories of cognition. In: *Handbook of Theories of Aging* (eds. V.L. Bengtson and K.W. Schaie), 196–208. New York: Springer.

Schwarz, N., Hippler, H.J., and Noelle-Neumann, E. (1992). A cognitive model of response-order effects in survey measurement. In: *Context Effects in Social and Psychological Research* (eds. N. Schwarz and S. Sudman), 187–201. New York: Springer.

Tourangeau, R., Rips, L.J., and Rasinski, K. (2000). *The Psychology of Survey Response*. Cambridge: Cambridge University Press.

Wild, K., Howieson, D., Webbe, F. et al. (2008). Status of computerized cognitive testing in aging: a systematic review. *Alzheimer's & Dementia* 4 (6): 428–437.

Zygouris, S. and Tsolaki, M. (2015). Computerized cognitive testing for older adults: a review. *American Journal of Alzheimer's Disease & Other Dementias* 30 (1): 13–28.

12

Panel Conditioning: Types, Causes, and Empirical Evidence of What We Know So Far

Bella Struminskaya[1] *and Michael Bosnjak*[2]

[1]*Department of Methodology and Statistics, Utrecht University, Utrecht, The Netherlands*
[2]*Leibniz-Institute for Psychology and Department of Psychology, University of Trier, Trier, Germany*

12.1 Introduction

Panel surveys offer several important advantages over cross-sectional surveys, such as measuring gross-level and individual-level change as well as providing measures of stability and instability (Lynn 2009, p. 5; Lynn and Lugtig 2017, p. 280). These advantages can be endangered by measurement error due to panel effects: systematic panel attrition and specific types of panel conditioning. This chapter focuses on the latter.

Panel conditioning is an umbrella term summarising learning effects that can occur due to the participation in surveys, resulting in different answers given by respondents who have already taken part in a survey from the answers that these respondents would have given if they were participating for the first time (Kalton et al. 1989; Lynn 2009; Struminskaya 2020; Waterton and Lievesley 1989).

Panel conditioning has been a concern since the recognition of the advantages offered by panel surveys over cross-sectional surveys. For instance, Lazarsfeld (1940) noted that repeated interviewing can affect respondents' opinions and behaviour distinguishing between the issues with which a respondent is already concerned and those on which he or she does not have an opinion. The effect he hypothesised would be either an opinion change or commitment to a particular position once an opinion had been acquired.

Panel conditioning can be advantageous or disadvantageous for data quality (Struminskaya 2016). Advantageous panel conditioning refers to an increased quality of reporting, for example, due to improved understanding of the questionnaire and the surveying procedure. Disadvantageous panel conditioning refers to decreases in data quality such as learning to answer filter questions in a way

Advances in Longitudinal Survey Methodology, First Edition. Edited by Peter Lynn.

Companion website: www.wiley.com/go/lynn/advancesinlongitudinalsurvey

that would allow avoiding follow-up questions. Misreporting to filter questions in order to avoid the burden of follow-up questions is an example of conditioning that can occur across repeated interviews as well as over the duration of a single interview.

There are three types of changes that can result from panel conditioning: (i) changes in reporting behaviour; (ii) changes in actual behaviour, attitudes, or knowledge; and (iii) a combination of actual changes and changes in reporting. Even if changes in attitudes, behaviour, or knowledge are beneficial for an individual, for example, learning about pensions or dangerous bacteria following survey participation (Toepoel et al. 2009), they are disadvantageous from a data quality perspective, because systematic bias is introduced into the sample compared to a case with no prior surveying. Hence, it is important to study the causes of panel conditioning and its possible detrimental effects on longitudinal data.

The literature is rather inconsistent and vague about the existence, direction, and magnitude of panel conditioning. The reasons for the apparently equivocal evidence-base on panel conditioning include a lack of a coherent theoretical framework specifying the antecedents, mechanisms, and consequences of panel conditioning and a multitude of study designs to explore the phenomenon (Bergmann and Barth 2018; Halpern-Manners and Warren 2012; Holt 1989; Struminskaya 2016; Uhrig 2012).

12.2 Methods for Studying Panel Conditioning

Several research designs have been used for studying panel conditioning (Struminskaya 2014; Uhrig 2012), that vary in their ability to tackle confounding factors:

1. *Cross-sectional comparison sample.* Responses are compared to a temporally corresponding cross-section with the same or similar questions (e.g., Coombs 1973; Corder and Horvitz 1989; Fisher 2019; Kruse et al. 2009; Pevalin 2000; Wilson and Howell 2005) or to fresh respondents in another panel (e.g., Toepoel et al. 2009).
2. *Rotating panel designs.* Some respondents are surveyed for the first time while others are surveyed for the second or subsequent time, where sample design and procedure as well as the question wordings are identical (e.g., Bailar 1975; Halpern-Manners et al. 2014; Waterton and Lievesley 1989).
3. *Multiple cohorts design.* Groups of various tenure are compared within a single panel so that one group with minimal tenure has been surveyed/received the questionnaire content one time and other groups with higher tenure have received it multiple times due to, for example, the recruitment of refreshment samples (e.g., Clinton 2001; Dennis 2001; Kruse et al. 2009).

4. *Predictive modelling*: Specific effects are predicted from theory using a single sample with no rotation groups, such as increased stability of attitudes due to increased opinionation (e.g., Sturgis et al. 2009; Kroh et al. 2016) or naming fewer friends in the social network modules of a questionnaire (Van der Zouwen and Van Tilburg 2001).
5. *Randomised experiments*. The experimental and the control group(s) receive different questionnaire content in the first or several waves of the panel and the same content in the wave that serves as a basis for comparison; randomised experiments have been implemented in small-scale studies (e.g., Bridge et al. 1977; Kraut and McConahay 1973; Traugott and Katosh 1979; Yalch 1976) and large-scale panels (e.g., Axinn et al. 2015; Crossley et al. 2017; Silber et al. 2019; Torche et al. 2012; Uhrig 2012; Yan and Eckman 2012).

Design approaches 1 through 3 have a similar underlying idea of comparing first-time respondents to respondents who have taken several surveys within the panel. However, designs 2 (rotating panel) and 3 (multiple cohorts) have some advantages over the first one (cross-sectional comparison). In design 1, question wordings between panel surveys and a cross-section used for comparison usually differ. Even if question wordings are the same, differential non-response to the two surveys can still be problematic due to differences in required commitment (higher in a panel survey) or in factors such as topic interest, incentives, and other study characteristics. Strategies 2 and 3 help mitigate the problem of differences in non-response between the two surveys being compared: differences in sponsor, sampling, and recruitment protocols, and question wordings are no longer an issue, though differences due to differential burden (number of waves) may remain. Design 4 (predictive modelling) is distinct from the first three because it involves comparisons of empirical findings with theoretical expectations rather than with a separate sample. However, with this design it is not possible to distinguish between true changes in the population and panel conditioning. Furthermore, the analytic strategy when using this design usually involves assuming that increased reliability and stability of attitudes is a result of panel conditioning, but this finding might be a result of sample maturation: as people get older, their attitudes become more stable (Sturgis et al. 2009). Maturation of the sample could only rarely be excluded as an alternative explanation (e.g., Kroh et al. 2016). Design 5 (randomised experiments) is the most favourable in terms of providing the possibility of excluding, or controlling for, alternative explanations of the differences that can be found because all other study design features are held constant.

Within these five general designs, there is still considerable room for variation in methods. For example, studies may be able to use validation data, that is, data from a registry, administrative data, or data from other sources such as employment records. And studies may vary in whether and how they handle sample maturation, and in methods to control for differential attrition.

Designs 1 through 3 are especially affected by the difficulty of disentangling attrition bias from conditioning bias (Kalton et al. 1989): differences in responses of panel respondents and respondents from a cross-section or a different panel wave or rotation group can be caused by the fact that the composition of the sample has changed as a result of panel attrition. Researchers have used several strategies in studies of panel conditioning to correct for attrition (Struminskaya 2014, p. 111).

The first strategy to control for attrition is to exclude equivalent sets of non-respondents from each sample, such as focusing on the groups of respondents who participate in all waves (e.g. Sturgis et al. 2009) or in case of a rotating panel excluding respondents from the 'fresh' group who would attrite in the later waves (e.g., Waterton and Lievesley 1989).

The second strategy is to decompose the total bias into panel conditioning and attrition by imposing additional assumptions on the attrition process, such as that subsequent-wave non-response is missing at random conditional on observable characteristics or that the relation between attrition and the variable of interest for which panel conditioning is studied is stable over time (Das et al. 2011). These assumptions, however, are not testable (Das et al. 2011).

The third strategy also relies on assumptions about the attrition process, with the goal of correcting for attrition bias through weighting. For example, observations are weighted from both conditioned and unconditioned groups so that the distributions of relevant demographic variables match for 'old' and 'new' respondents (Sikkel and Hoogendoorn 2008), applying post-stratification weighting (Clinton 2001) or propensity-score weighting that includes covariates that can be predictive of attrition beyond demographics (e.g., Bergmann and Barth 2018; Struminskaya 2016).

In terms of controlling for design factors and attrition that can confound panel conditioning, randomised experiments are superior to the other designs. However, there is still a possibility of differential attrition due to the experimental manipulation, that is, if attrition is related to the questionnaire content for at least one group. Researchers can try to minimise the risk of differential attrition by asking questions that are similar in their characteristics, such as length, level of difficulty, or sensitivity. Moreover, the attrition process can be modelled, for example, using a regression predicting attrition as a function of treatment group membership, including covariates that are thought to be related to attrition (Torche et al. 2012) and if the groups differ, the weighting procedures recommended by, for example Sikkel and Hoogendoorn (2008), can be applied. Earlier studies on panel conditioning largely used non-experimental approaches (e.g., Bailar 1975; Coombs 1973; Pevalin 2000; Silberstein and Jacobs 1989; Waterton and Lievesley 1989) and those that were experimental were mostly based on small samples and focused on election behaviour (e.g., Bridge et al. 1977; Clausen 1968; Kraut and McConahay 1973). However, in recent years there has been a growth in experimental studies and studies specifically designed to study panel conditioning (e.g., Axinn et al. 2015; Bach and Eckman 2018; Crossley et al. 2017; Kroh et al. 2016; Silber et al. 2019; Torche et al. 2012; Yan and Eckman 2012).

The objective of this chapter is to provide an overview of findings about the mechanisms, presence, and magnitude of the effects of panel conditioning and to offer practical guidelines regarding survey design that would allow the effects of disadvantageous panel conditioning to be minimised. We review the literature, with purposeful selection of studies, including studies whose design allows us to draw inferences about the phenomenon and to infer recommendations for survey practice.

This chapter will be of interest to survey managers and survey commissioners of longitudinal studies as well as survey methodologists. We first provide a review of the mechanisms from a perspective of the framework of survey response; the sections whose headings start with 'empirical evidence' illustrate the mechanisms of panel conditioning providing empirical evidence regarding the conditions under which questions are susceptible to panel conditioning and survey features have been found to influence the undesirable effects of panel conditioning. We conclude by providing some tips on how to minimise the negative effects of panel conditioning on data quality and by giving advice on optimal designs for studying panel conditioning. Also, throughout the chapter, we point out the areas for future research that have received little attention to date.

12.3 Mechanisms of Panel Conditioning[1]

12.3.1 Survey Response Process and the Effects of Repeated Interviewing

According to the survey response process model (Tourangeau 1984; Tourangeau et al. 2000), respondents need to go through at least four stages in order to answer a survey question: comprehension of the question and the survey task, retrieval of relevant information from memory, generating judgment based on the

1 We exclude the so-called 'mere measurement' and 'self-fulfilling prophecy' effects (e.g., Greenwald et al. 1987; Sherman 1980) from our review. These effects refer to changes in actual behavior that occur as a result of measuring behavioural intentions prior to performing the behaviour (e.g., a question 'Are you planning to donate blood?' is asked in the first wave and the behavior – having donated blood – is measured in the second wave), a situation rather unusual to panel surveys, with the except of voting intention studies, findings of which related to panel conditioning are included in Section 12.3.

information that was retrieved, and reporting an answer.[2] Not every respondent necessarily follows all of these steps sequentially; some respondents can backtrack to the earlier steps in the process or skip some stages, but this is a typical order that respondents usually follow if they are answering the questions diligently. Given that answering a survey question is a cognitively demanding task, at each of these stages there is room for error. Respondents may misunderstand or misinterpret the question, forget relevant information, make erroneous judgements based on the retrieved information, or map their answer on an erroneous category during the reporting stage. Repeated interviewing affects the error potential at each of these stages.

During the *comprehension stage*, respondents attend to questions and instructions, represent logical form of the question, identify the focus of the question or the information being sought, and link the key terms in the question to relevant concepts (Tourangeau et al. 2000). Administering the same questions multiple times can increase comprehension of the question and comprehension of the answer options, while the mere fact of being interviewed can increase comprehension of the respondent task. Waterton and Lievesley (1989) argue that respondents become more familiar with the survey task and are less likely to answer 'don't know' because they learn that their opinion, even if not well-formed, is important for the researchers. Binswanger et al. (2013) find that the number of 'don't know' answers decreases in complex attitudinal open questions for panel respondents compared to inexperienced respondents.

During the *retrieval stage*, respondents generate the retrieval strategy and retrieval cues, draw specific and generic memories from their short-term and long-term memory, and fill in missing details (Tourangeau et al. 2000). Repeated interviewing influences responses as it increases the accessibility of the relevant information (Bergmann and Barth 2018; Kroh et al. 2016). Respondents' memories can be viewed as an associative network, where ideas can be seen as nodes or intersection points of a network, and these ideas are connected by links that correspond to the relations between them (Tourangeau 1984, p. 80). Bergmann and Barth (2018) argue that repeated interviewing changes the information processing within respondents' associative networks leading to changes in attitudes and behaviours. The inferences respondents make during the retrieval and judgement stages depend on their prior knowledge activated by the question and its context (Tourangeau 1984). Respondents might – as a result of being exposed to the survey questions – acquire new information by, for example, paying more

2 The survey response process model has been extended to include two additional stages prior to comprehension: encoding and storage (Tourangeau 2017). During the encoding stage, a respondent notices and interprets some aspects of an experience; the storage stage refers to retention of information in the long-term memory. While acknowledging that these stages exist, we discuss panel conditioning relating it to the original model.

attention to the topics raised during the interview in the media or discussing them with friends and family, which can lead to changes in opinions they initially held or stimulate them to acquire new opinions in case they did not have any prior to the survey (Sturgis et al. 2009). Moreover, respondents can prepare themselves for the following interviews (Kalton and Citro 1993) by looking up relevant information, thereby aiding recall during the next interview. Studies that use external data sources for validation of the information provided during the interview show the increases of accuracy in reporting of respondents' income (Frick et al. 2004; Rendtel et al. 2004) and receiving unemployment benefits (Yan and Eckman 2012) as panel experience increases.

During the *judgement* stage, respondents assess the completeness and relevance of the retrieved information, draw inferences based on the accessibility of this information, combine the retrieved material or supplement the incomplete information they recalled (Tourangeau et al. 2000). Examples of how repeated interviewing influences survey responses at this stage are the desire of the respondent to appear consistent over waves of the panel, a phenomenon referred to as 'freezing' of attitudes (Sturgis et al. 2009; Waterton and Lievesley 1989) and anchoring effects, that is, relying on a certain starting value which is adjusted to provide the final answer (Tversky and Kahneman 1974, p. 1128), where the response in the previous interview acts as the anchor.

During the final *reporting stage*, respondents map the judgement they made onto one of the offered response categories and select an answer (Tourangeau et al. 2000). Respondents can intentionally misreport an answer as a result of social desirability, or a desire to appear in a positive light (DeMaio 1984) or be more or less willing to provide an answer at all. Repeated interviewing can increase the familiarity with the survey and trust in the survey organisation, prompting respondents to give more truthful answers (Waterton and Lievesley 1989).

All the aforementioned examples of potential influences of repeated interviewing on response strategies during the response process assume that respondents are motivated to proceed through these stages. However, not all respondents may be highly motivated. According to Tourangeau et al. (2000, p. 17), there are two general ways to produce an answer to a question: a 'high road' and a 'low road'. The high road of arriving at an answer is a conscientious process of carefully going through the aforementioned stages. The low road, in contrast, is providing an answer without giving it much thought and being guided by motives other than accuracy. Krosnick and Alwin (1987) refer to such opposing strategies used by respondents as optimising and satisficing. Optimising respondents seek the best possible answers they can produce, whereas satisfiers select acceptable answers to minimise the psychological costs of seeking and providing accurate answers. The choice of strategy depends on the difficulty of the task, the respondents' ability, and their motivation (Krosnick 1991). If respondents are highly motivated, they will take the high road and use their prior experience, for example, to prepare for future interviews (Cantor 2008). In contrast, if respondents have low levels of motivation, they can use the knowledge gained in prior panel waves to try to reduce burden (Bradburn 1978) and, for example, misreport to filter questions in order to shorten the interview.

In the following, we review the most prominent mechanisms of panel conditioning that have received attention in the literature in more detail, namely: changes in attitudes, behaviour, and knowledge due to *reflection* about the interview; *reduction of social desirability bias* in reporting as a result of learning survey procedures and questionnaire content, and changes in reporting due to *satisficing*.

12.3.2 Reflection/Cognitive Stimulus

The mechanism of reflection or alternatively a cognitive stimulus model (Sturgis et al. 2009) has received the largest attention in the literature. The mechanism of reflection is based on a notion that repeated interviewing changes knowledge and attitudes due to raising consciousness, whereby attitude changes can also result in behavioural changes. We first discuss reflection in the context of survey measurement of attitudes, since the cognitive stimulus model was developed in this context, and then discuss behavioural changes and changes in knowledge.

The mechanism of reflection/cognitive stimulus model states that respondents are prompted by interviewing to gather information to form opinions where none existed before or change pre-existing opinions (Jagodzinski et al. 1987; Waterton and Lievesley 1989; Sturgis et al. 2009). As a result, attitudes are expected to become more consistent over the waves of the panel (crystallised) and attitude strength is predicted to increase. After an initial interview, respondents may discuss the issues raised in an interview with friends and family, deliberate privately, or acquire information by paying closer attention to the media. The result of performing these actions is that respondents arrive at rational preference-based judgement as opposed to attitudes constructed 'on the spot' (Sturgis et al. 2009). The cognitive stimulus model has three empirical implications: (i) attitudes become more reliable; (ii) attitudes become more stable; and (iii) opinionation on social and political issues increases over time (Sturgis et al. 2009). Moreover, respondents can commit to a position expressed in an initial interview, a phenomenon referred to as *freezing of attitudes* (Bridge et al. 1977; Waterton and Lievesley 1989).

Alternatively, attitudes can change as a result of increased knowledge. This possibility is demonstrated by deliberative polls, a methodology developed by James Fishkin in 1988, who questioned the adequacy of measuring preferences through polls (Fishkin 1995, p. 46) and proposed a method that combines traditional polling with small-group discussions. First, respondents answer survey questions in a traditional way, then they are asked to deliberate on the issues asked in the poll in person (while the newer studies also allow online deliberation) after being provided briefing materials, and in the third step, after participants become more engaged with the issues and pay more attention to the information, hence arriving at more thoughtful conclusions, participants' opinions are measured again through a traditional survey (see Merkle 1996). The methodology of deliberative polling is not meant to describe public opinion; rather, it shows what conclusions people would reach if they were better informed and examined issues raised by the survey more thoroughly (Fishkin 1995, p. 162).

Several authors have argued that changes in attitudes as a result of panel conditioning are more likely to occur for attitudes that are not strong to begin with (Bridge et al. 1977; Lazarsfeld 1940; Warren and Halpern-Manners 2012; Waterton and Lievesley 1989) since an interview can increase the saliency of the topic for the respondents, while if the topic is perceived as not important or is already very salient with a respondent having well-formed attitudes on it, changes in attitudes are less likely.

Although the cognitive stimulus model was developed specifically for attitudinal questions, raising consciousness by previous interviewing can, as we have mentioned earlier, affect behaviour and knowledge as well. Cantor (2008) hypothesises that respondents with a high awareness of the topic and those respondents who associate the behaviour with a higher cost (i.e., expensive or difficult) are the ones least likely to change their actual behaviour but might change their reporting behaviour to align with the norms of importance of such behaviour (e.g., vaccination) conveyed by the interview. Also, respondents might be more likely to change those behaviours to which they attach positive utility (Warren and Halpern-Manners 2012).

12.3.3 Empirical Evidence of Reflection/Cognitive Stimulus

12.3.3.1 Changes in Attitudes Due to Reflection

Waterton and Lievesley (1989) compared members of the UK Social Attitudes Panel over the course of four waves to fresh cross-sections and found that respondents became 'politicised' as a result of an interview: conditioned respondents showed increased proportions of supporters of a political party. In line with the expectation that changes are more likely to occur if opinions are not well-formed, the conditioning effect was more pronounced among respondents with lower initial political knowledge scores. Waterton and Lievesley (1989) conducted in-depth follow-up interviews with panel members and found out that the survey generated a considerable amount of discussion between panel respondents and their friends and relatives.

Sturgis et al. (2009) used the data from the British Household Panel Survey and found – in line with the cognitive stimulus model – that attitudes on gender roles, work-family balance, and interest in politics become more reliable and stronger over time. Freezing of attitudes, when respondents feel pressure to provide consistent answers over the waves, did not find empirical support (Sturgis et al. 2009; Waterton and Lievesley 1989).

Sturgis et al. (2009) further found – in line with their prediction – that opinionation increased with repeated interviewing, that is, respondents provided fewer 'don't know' responses on the left–right attitude scale. The decrease of 'don't know' responses over the waves has been reported elsewhere as well (Bailar 1989; Binswanger et al. 2013; Cantor 1989; Porst and Zeifang 1987; Traugott and Katosh 1979; Waterton and Lievesley 1989). However, while Waterton and Lievesley (1989) explain the decreases in 'don't know' responses through learning the rules that govern an interview (respondents learn that their opinions are of importance to the researchers even if they are not well-formed), Sturgis et al. (2009) attribute these findings on increased opinionation to the subject matter rather than learning the interview procedure. However, neither of the proposed explanations of 'don't know' decrease has been tested empirically.

Contrary to the explanation of paying more attention to the media after being interviewed, Clinton (2001) found that respondents with longer tenure in an online panel reported lower levels of news consumption than respondents who had been with the panel for a shorter time period. Multiple other studies also failed to find the increased stability of attitudes over time (for a review, see Warren and Halpern-Manners 2012). Kruse et al. (2009) found that inter-wave stabilities increased in one out of the four items they analysed.

The studies reviewed above are not randomised experiments. Waterton and Lievesley (1989) and Kruse et al. (2009) used a rotating panel design, a design in which some respondents are surveyed for the first time while other respondents are interviewed for the second or subsequent time, both groups are then compared based on the assumption that 'fresh' respondents reflect possible true opinions in the population in the given time periods while the differences between those interviewed for the first time and those interviewed for the n-th time reflect the effect of panel conditioning. Such a comparison is better than comparing panel respondents to an unrelated cross-section both because questions, timing of the interviews, and recruitment procedures are identical and because differential attrition can be better controlled in the analysis. The researcher can compare respondents with the same attrition patterns: for example, in a three-wave panel survey, for a given time-point, wave 2 respondents can be compared not with all respondents of wave 1 but with the subset of those wave 1 respondents who also responded at the subsequent wave 2. Though Kruse et al. (2009) studied differential attrition patterns between the groups and used weighting to control for attrition, they did not rule out the possibility of the influence of the political climate or other external factors on responses. Sturgis et al. (2009) tested a set of hypotheses derived from the cognitive stimulus model (increased reliability and attitude strength, increased opinionation) on a single dataset. The results of their study can be confounded by sample maturation – that is, as people get older they provide more consistent attitudes – or changes in attitudes can relate to trends occurring in the population so that panel conditioning is indistinguishable from real change.

Kroh et al. (2016) applied the method of testing hypotheses on a single dataset using 17 instruments that have been administered for over 30 years in the German Socio-Economic Panel (GSOEP) studying the internal consistency of multi-item constructs, which they referred to as person-fit statistics that evaluate the consistency of individual response patterns. Due to repeated refreshments, Kroh et al. (2016) were able to control for possible sample maturation and used sophisticated modelling approaches to control for differential attrition: by using fixed-effects models to capture within-person change in person reliability; by basing the analysis on a balanced sample of respondents; and, in order to obtain an effect of survey experience conditional on future panel attrition, by including a measure that reflects whether person reliability is lower in respondents who will refuse to participate in the following wave of the survey compared to those who will continue participation. They found that repeated interviewing increased the reliability of individual survey responses over time, suggesting that reflection might have contributed to an increased inter-item homogeneity.

Bridge et al. (1977) randomly assigned respondents to two groups that differed in the topics of the questions: half of the respondents received questions about cancer, while the other half received questions about burglary prevention. At the second wave, the group interviewed about cancer reported increased ratings they attached to the personal importance of good health, while the group that received questions about burglary prevention did not show different ratings on the personal importance of safety from crime compared to the control group. Bridge et al. (1977) state that respondents had a greater concerns about crime in the initial interview than about cancer, hence the change in attitudes occurred only for the topic whose salience was lower at the first measurement. As with lower saliency topics having greater potential for panel conditioning, so do attitudes that are less crystallised: Bergmann and Barth (2018) showed, using the data from German Longitudinal Election Study, that vote intentions were more likely to form over the course of the panel for those respondents who had weak attitude strength at the beginning of the panel survey.

12.3.3.2 Changes in (Self-Reported) Behaviour Due to Reflection

Changes in behaviour can result from increased awareness or changes in attitudes about the behaviour in question. Several election studies find that (reported) voter turnout increases in subsequent interviews after pre-election interviews

(Bartels 1999; Clausen 1968; Kraut and McConahay 1973; Traugott and Katosh 1979; Yalch 1976). It seems that being asked about voting behaviour motivates respondents to vote. Clausen (1968) compared the turnout in a University of Michigan Survey Research Center (SRC) sample interviewed before and after the election with a sample of the Census Bureau that was interviewed only after the election and proposed a stimulus hypothesis that pre-election interview stimulated interest and participation based on the findings of increased turnout reported in the second interview. In a more recent study, Bartels (1999) used the data from two waves of the National Election Study (NES) and compared the outcomes to a parallel cross-section controlling for attrition. He found that out of six variables (ratings of Bill Clinton's 'morality'; correct relative placements of the presidential candidates on the NES's seven-point ideology scale; perceptions of the state of the nation's economy; interest in the campaign; reported turnout; and reported presidential vote), two variables, campaign interest and voter turnout, showed significant increases in the panel compared to the cross-section. Bartels (1999) demonstrated that panel effects taken together, that is, attrition and conditioning, could reduce the inferential value of the data by 20 percentage points. However, neither of the studies could use records to ascribe the increased turnout to an actual behaviour rather than changes in reporting. At least three studies on voter turnout had used administrative records, which provided the validation measure for the actual behaviour: Kraut and McConahay (1973) conducted a randomised experiment in which they randomly assigned one group to a pre-election and post-election interviews while the other group only took part in a post-election interview. Reported voter turnout was then compared to the actual turnout from the individual records. Their findings of increased turnout of the interviewed group confirmed the stimulus hypothesis (the predecessor of the cognitive stimulus hypothesis of Sturgis et al. 2009), according to which the interview raised the salience of politics for a respondent, leading those respondents who usually were not interested in politics to be more aware of the upcoming elections, and this awareness along with the 'citizen conscience' activated by the interview stimulated them to vote (Clausen 1968; Kraut and McConahay 1973). Similar conclusions were reached by Yalch (1976), who compared survey data to the records maintained by the political organisations involved in the campaign during local elections, and Traugott and Katosh (1979), who also used individual administrative records of voters and in addition to finding support for the stimulus hypothesis; moreover, they found that the highest participation rates were observed among those who had been interviewed the most number of times, which only partly could be attributed to panel attrition of those less interested and less involved in politics.

Behavioural changes due to panel conditioning have been observed in other domains as well. Crossley et al. (2017) randomly assigned the respondents of the Dutch LISS Panel to two groups: receiving a module with questions on retirement and not receiving such a module and then linked the survey data to the administrative data on wealth to compare savings behaviour of those respondents who received the module (conditioned) and those who did not (nonconditioned). Crossley et al. (2017) found that conditioned respondents started spending more, especially older and highly educated respondents, while the younger and poorly educated respondents marginally increased their savings. This effect is attributed to the 'salience shock' produced by the retirement questionnaire, so that respondents started to reflect on expenditure needs in retirement and retirement planning after taking the survey and realising that they were saving 'too much' (older and higher educated) or 'too little' (younger and lower educated). Battaglia et al. (1996, pp. 1010–1013) compared immunisation records for young children whose parents were interviewed about their children's vaccination status in the National Immunization Survey and those of the general population and found increased children's vaccination levels in the group of interviewed persons. Differences in vaccination levels between the two quasi-experimental groups were strongest for children with mothers who had lower education, children in families with low annual income, and children with an unmarried mother. Yan and Eckman (2012) used the data from four waves of a large-scale German panel survey on labour market outcomes (PASS), in which a refreshment sample was drawn at each wave. They compared the answers of the respondents from a refreshment sample and from the panel waves the to the administrative data on employment and unemployment benefit receipt status. Persons who received unemployment benefits were more likely to move off the benefits after being interviewed.

However, some studies found no apparent behavioural changes or only changes in certain aspects of a behaviour. Axinn et al. (2015) conducted a randomised experiment in a journal-keeping study of young women with weekly surveys on romantic, sexual, and contraceptive behaviour. The study lasted for two and a half years, and a smaller group sampled from the same sampling frame was assigned to keep a journal for 12 months while the other group was assigned to no journal keeping condition, and both were interviewed after those 12 months. Axinn et al. (2015) found that out of 36 behavioural measures, only three showed significant differences, and two of these were not in the expected direction, thus, concluding that repeated interviewing does not alter behaviour.

12.3.3.3 Changes in Knowledge Due to Reflection

Changes in knowledge as a result of panel conditioning due to reflection have received far less attention in the literature than attitudinal and behavioural questions. One reason might be that knowledge is a relatively rare subject of interest in social surveys; another reason might be that increases in knowledge in knowledge questions are somewhat tautological in nature. Nevertheless, it has been shown that repeated interviewing leads to increased knowledge levels of respondents, as the respondents' awareness about the issues and topics addressed increases during the survey. Coombs (1973) found increased knowledge of contraception methods when she compared panel respondents with a fresh cross-section; Kruse et al. (2009) found some evidence of panel conditioning for a question on political knowledge. Panel respondents were more likely to correctly name Obama's religion than cross-sectional respondents; Yan et al. (2011) found increases in awareness about the US Census after interviewing on this topic. However, not all knowledge questions are equally susceptible to panel conditioning effects. Toepoel et al. (2009) found that increases in reported knowledge did not occur for all items: for example, on the topic of knowledge of certain bacteria, increases were found for campylobacter but not for salmonella. Coombs (1973) reports that the greatest increase in knowledge about contraceptive methods was observed in women who knew of no method at all. Struminskaya (2016) compared two groups with lower experience and higher experience within an online panel, with both groups receiving the knowledge items for the first time hypothesising that more experienced group would more willingly admit the deficiency in knowledge due to social desirability reduction (see next section). Conditioning effects were observed for two out of the four items, with experienced respondents choosing the 'don't know' option at a higher rate, which could not be explained by satisficing. The two items for which conditioning effects could not be observed were relatively widely known, while the other two less so. These findings, taken together, suggest that baseline knowledge should be relatively low for panel conditioning to occur.

12.3.4 Social Desirability Reduction

As has been noted in the section on the response process earlier, respondents may be embarrassed by certain survey questions and in order to avoid negative feelings they misreport – an effect known as social desirability bias (DeMaio 1984; Sudman and Bradburn 1982). Waterton and Lievesley (1989) argue that repeated interviewing can reduce social desirability bias by way of respondents learning the rules of the interview procedure: With each successive wave of interviewing, respondents develop trust in the survey organisation and the survey procedure, which, in turn, increases respondents' trust and willingness to disclose unflattering information.

12.3.5 Empirical Evidence of Social Desirability Effects

In line with their hypothesis, Waterton and Lievesley (1989) found that respondents were more willing to reveal racial prejudice and disclose their income in subsequent panel waves. Further empirical support for social desirability bias reduction is provided by Brannen (1993), who found in qualitative interviews with mothers returning to work, that over the course of a three-year multi-wave study, respondents reported to have searched information about their children in order to compare them with other children, became more politicised and more aware of the state of the country's childcare system. Pevalin (2000), however, compared the estimates from the British Household Panel Study (BHPS, now UK Understanding Society) to the Health Surveys for England (HSE) and found no effects of social desirability reduction. Reporting of certain behaviours such as smoking or drug use has been found to exhibit more social desirability bias in later waves with respondents denying having ever performed such behaviour (a phenomenon also referred as *recanting,* Fendrich and Vaughn 1994; Mensch and Kandel 1988; Wagstaff et al. 2009). However, recanting might also indicate a decrease in social desirability. In a recent study, Mavletova and Lynn (2019) found that the reported lifetime alcohol use of children age 10–15 decreased by about 15 percentage points between the third and the fourth wave of the UK Household Longitudinal Study. This effect might be due to increased trust and more honest reporting: for some children, due to certain group norms, it might be socially desirable to claim having drunk alcohol even if they have not, but after participating in the study for several years, children might have developed trust and commitment to the survey and thus answer more honestly. More honest answers to this question could, however, alternatively be a result of children's maturation: Mavletova and Lynn's (2019) study was not based on a randomised experiment.

Torche et al. (2012) conducted a randomised experiment in the Longitudinal Survey of Drug Use in Chile: Half of the respondents received questions on exposure to, perceptions of, and experiences with alcohol, cigarette, marijuana, and cocaine use, while the other half received a comparable in length module on behaviour and personality. In the second wave one year later, everyone received drug-use questions. Torche et al. (2012) found that adolescents who answered questions about drug use at wave 1 were considerably less likely than members of a control group to report lifetime substance use ('Have you ever …?') when re-interviewed.

An explanation to such contradicting findings might also be provided by Uhrig (2012), who argues that panel conditioning in sensitive questions may be a function of a question itself rather than a process of developing trust in the survey organisation or survey instrument as Waterton and Lievesley (1989) had proposed. According to Uhrig (2012), the initial shock of being asked a potentially embarrassing question can subside in the following waves as respondents learn that the answer that was slightly corrected for social desirability can be given without negative consequences; respondents then are no longer motivated to misreport and experience negative consequences associated with lying, thus, providing an accurate answer. Furthermore, panel conditioning may operate differently for behaviours that are at the lighter end of a social desirability spectrum, such as height and bodyweight, compared to more threatening questions such as drug use, abortion, or sexual behaviour (Uhrig 2012). The respondents of the Understanding Society Innovation Panel were randomly assigned to two groups across two waves, in which the conditioned group received questions on height and bodyweight asked in a personal interview twice and the unconditioned (control) group received these questions once. Uhrig (2012) found that conditioned women reported being heavier and taller than unconditioned women, rounding and item non-response of bodyweight were also less likely. That provided partial support for Uhrig's hypotheses, since for men only lightweight conditioned men reported lower weights, while for height the effect was in the unexpected opposite direction: men reported greater heights (although this effect was only significant for taller men).

Further evidence of the influence of question sensitivity being related to panel conditioning comes from Halpern-Manners et al. (2014). They hypothesise that surveys on non-normative stigmatising behaviour cause respondents to reflect on that behaviour (mechanism described in detail in Section 12.3.2) and subsequently use less or report inaccurately lower levels to appear more norm-conforming. Halpern-Manners et al. (2014) randomly assigned respondents of the probability-based online GfK KnowledgePanel (now Ipsos KnowledgePanel) to receiving part of the substance-use questions in the first wave of the survey and receiving all questions in the second wave. The spacing of the waves was experimentally varied: one month and one year after the initial interview. In the group that was followed up after one year, Halpern-Manners et al. (2014) found no changes. In the group that was re-interviewed after one month, respondents were more likely to report lifetime stealing or lifetime drunk driving, while the other two items (arrest and marijuana) showed no effect. As with Uhrig (2012), the effect sizes Halpern-Manners et al. (2014) found were modest.

12.3.6 Satisficing

According to Krosnick's framework of survey satisficing (1991), whether respondents provide accurate answers or look for the first acceptable option to answer a survey question depends on the following factors: difficulty of the task, the respondents' ability, and respondents' motivation. Respondents might not be motivated from the beginning of the interview, or they might become fatigued during the interview and shift their strategy from optimising to satisficing. For example, Galesic and Bosnjak (2009) found in a randomised experiment that in longer questionnaires, response quality decreased towards the end of the survey.

Respondents with lower levels of motivation can use their prior experience within a panel survey and knowledge of the questionnaire content to reduce response burden. Response burden refers to respondents' effort and stress, interview length, and frequency of being interviewed (Bradburn 1978); respondent effort can be required to answer monotonous (boring) questions, and stress may be caused by the need to answer sensitive questions. To lower response burden, panel members might employ strategies that would allow them to shorten the interview, for example, in the way that they answer filter questions – that is, questions that determine the following sequence of questions – in such a way that allows to skip the follow-up questions.

12.3.7 Empirical Evidence of Satisficing

12.3.7.1 Misreporting to Filter Questions as a Conditioning Effect Due to Satisficing

Multiple studies showed that filter questions that were followed by the follow-up questions were triggered in fewer cases by panel respondents than by inexperienced (first-time) respondents. For example, Mathiowetz and Liar (1994) found that panel respondents reported fewer functional limitations compared to first-time respondents; Warren and Halpern-Manners (2012) showed that filter questions that were followed by the follow-up questions elicited for panel respondents to hold fewer jobs, have fewer household members, and be members of political parties at lower rates; Wilson and Howell (2005) compared the prevalence of arthritis reported by the panel members of the Health and Retirement Survey to the cross-sectional National Health Interview Survey in the USA and reported a decline among the panel members. Their study gave rise to the debate about methodological aspects of comparing an unrelated cross-section to a panel survey (Weir and Smith 2007); nonetheless, evidence from experimental and quasi-experimental studies reviewed below provides some support for the satisficing explanation.

Halpern-Manners and Warren (2012) used rotating panel design of the Current Population Survey (CPS) and linked CPS records controlling for attrition and mode effects and found that unemployment rates for respondents interviewed for the first time are higher than for respondents who have been in the panel longer. After participating in an initial interview, some panel members switched their labour force status from 'unemployed' to 'out of the labour force'. This can be the result of respondents' satisficing in order to avoid emotional burden in responding questions associated with what might be thought of as undesirable status of being unemployed or an attempt to shorten the interview if respondents had thought they would receive follow-up questions if answered 'unemployed' rather than 'out of labour force'. Halpern-Manners and Warren (2012) showed that the unemployment rate estimated based on the CPS would be 0.75 percentage points higher if calculated for individuals who have not previously participated in the CPS than if estimated among otherwise similar individuals who have previously participated and were susceptible to panel conditioning.

Nancarrow and Cartwright (2007) compared groups with different exposure to questions on personal care and brand recall in an online panel. Groups that were interviewed more frequently than the control group (two to three times vs. one-time) had somewhat higher level of brand recall for both known and dummy brands; however, the highest frequency group (interviewed monthly during the five months) demonstrated lower recall and lower frequency of such a regular behaviour as using toothpaste. However, the reported use of other five personal care products did not change for the high-frequency group in the course of the repeated interviewing. The authors attributed the effect to greater boredom or reporting such a mundane behaviour.

12.3.7.2 Misreporting to More Complex Filter (Looping) Questions

Another particularly tedious (boring) type of question in which respondents may attempt to manipulate survey instruments in order to reduce the interview time is a network/names generator question. A network generator question asks respondents to provide (a certain number of) names of members of respondents' networks (e.g., friends, individuals with whom they discuss important issues, etc.), upon which follow-up questions are asked about those members and/or about the relationships between named persons. Thus, it is a filter question with a looping component, where series of follow-up questions are repeated a number of times. Van der Zouwen and Van Tilburg (2001) used the data from the longitudinal Dutch survey 'Living Arrangements and Social Networks of Older Adults' and verbatim protocols of interviews to study changes of the reported network size. They found that the network size decreased over the four waves of the panel and this effect could not be attributed to the respondents' panel conditioning, but was caused by the interviewers. Similar results about interviewer influence on the network size were found by Brüderl et al. (2013) using the German longitudinal pairfam data and by Marsden (2003), who used cross-sectional data from the US General Social Survey (GSS).

To study the possible influence of interviewers on panel conditioning, Eagle and Proeschold-Bell (2015) replicated the names generator from the GSS in a longitudinal three-wave Clergy Health Initiative survey administered by telephone and online (or mail if requested) with a random assignment to survey mode. They found, firstly, that respondents named significantly fewer network members in the interviewer-administered condition than in the self-administered condition, and that the network size among online respondents declined over the waves of the panel. In this study, respondents were asked the follow-up questions only on the first five names they provided, and Eagle and Proeschold-Bell (2015) argued that if respondents switched to naming only five names in subsequent waves, it would provide evidence of panel conditioning. Indeed, the authors found that respondents who changed their answers in the later wave of the panel survey most commonly gave exactly five names, indicating that respondents remembered that they had received follow-up questions about five network members and thus reported the maximum number of five members in the following waves. However, interviewer learning does not always imply disadvantageous panel conditioning: Valente et al. (2017) in a study on the use of skilled birth attendance in Ghana found that for some interviewers, the number of elicited names decreased, while some interviewers were able to increase the number of recorded names as they became more experienced.

Struminskaya (2016) replicated the names generator from the German General Social Survey (ALLBUS) with slight adjustments in a probability-based panel of Internet users and compared respondents with more experience within the panel who had been exposed to filters to those who had less experience. She found no significant differences in the number of reported names between the groups and high data quality for both groups, suggesting they were equally highly motivated. Using paradata, that is, process data collected automatically during the interview (Callegaro 2013; Kreuter 2013), she furthermore found no evidence of satisficing (e.g., speeding or going back and changing the number of names provided). However, neither Eagle and Proeschold-Bell (2015) nor Struminskaya (2016) implemented a control group not receiving the names generator questions. Silber et al. (2019) conducted a randomised experiment in a German online access panel varying the exposure to the network module: one third of the respondents received the names generator three times, one third received it twice, and the control group received it once, receiving questions on an unrelated topic in order to make the interview length comparable. Silber et al. (2019) found slight decreases in network size in the conditioned groups only for respondents with high network sizes. Data quality indicators that could be indicative of satisficing such as network density, non-differentiation, item non-response, and response time did not differ between the experimental groups; neither did the direct test of moderation by respondent's ability and motivation based on Krosnick's framework (1991) show any effects.

12.3.7.3 Within-Interview and Between-Waves Conditioning in Filter Questions

The learning effect of shortening an interview through providing certain answers to filter questions has been observed in cross-sectional one-time interviews as well.

Duan et al. (2007) randomly assigned respondents of a mental health service use survey to two conditions: an interleafed format that placed follow-up questions immediately after each individual filter (screening) question and a grouped format that first asked multiple filter (screening) questions in the beginning of the survey and was followed by questionnaire blocks of follow-up questions relevant for each filter. The authors hypothesised that the interleafed format would be more prone to the within-survey conditioning than the grouped format, since respondents would have a chance to learn that each screener is followed up by series of questions and adapt their answering behaviour accordingly, if they wished to reduce burden. Duan et al. (2007) find, as hypothesised, that respondents who received questions in the interleafed format reported lower service use than in the grouped format. Kreuter et al. (2011) tested the effects of grouped and interleafed formats in a telephone multi-topic survey with validation data available for some of the items. They found that respondents were more likely to affirmatively answer filter questions in a grouped format than in interleafed format with increasing effect as the number of filters increased. However, this negative effect only took place within each of five topical sections, the new section on another topic 'reset' respondents' reactions to filters. Similarly, Eckman et al. (2014) linked answers of the respondents to a telephone survey in Germany, which focused on clothing purchases, employment history and income sources, to administrative data containing information about employment. They found evidence of within-survey conditioning with respondents underreporting in an interleafed format; however, the effect was not universal across the three topics. The recommendation the authors derived from their study was, despite their findings, to use the interleafed format that generally is more familiar to respondents while searching for the ways to minimise its negative effects. Eckman and Kreuter (2018) conducted an experiment in a web survey, where they asked looping questions about employers and places the respondent had lived in a grouped format (first all filter questions, then all follow-ups) and interleafed format (each filter question with immediate follow-up). Eckman and Kreuter (2018) found that the grouped format elicited more employers and living locations reported by respondents than the interleafed format. Having linked the survey data to administrative records, Eckman and Kreuter (2018) were able to examine the accuracy of reports and found that the interleafed format produced more underreports (i.e., more undesirable conditioning) than the grouped format. However, the grouped format elicited higher number of 'don't know's and higher item non-response than the interleafed format, indicating a trade-off between the number of events reported (higher in the grouped format) and the quality of answers to the follow-up questions (better in the interleafed format).

There is only one published study that we know of that simultaneously tested within-survey and panel conditioning: Bach and Eckman (2018) conducted a cross-over experiment in two consecutive waves of the Dutch LISS Panel randomly assigning half of the respondents to the grouped and half to the interleafed format in the first wave and again randomly assigning half to either format in the second wave. As expected, the number of filters that triggered follow-up questions was significantly higher in the grouped format than in the interleafed format; however, contrary to the expectation, they did not find empirical evidence that respondents interviewed in the same format in both waves misreported more to filter questions when they were re-interviewed in the second wave. It is difficult to say why the results of earlier studies on panel (and not survey) conditioning in filter questions could not be replicated by Bach and Eckman (2018); however, given the true experimental design and the elimination of alternative explanations (e.g., controlling for attrition bias), these results are promising for the survey practitioners and would be useful to replicate increasing the exposure to several waves. Overall, satisficing as a reason for panel conditioning seems to be dependent on survey features such as mode of administration, question format (e.g., forming respondent's expectations about an appropriate number of names in the names generator), time intervals between waves and number of repeated administrations of the questions rather than being an inherent respondent characteristic.

12.4 Conclusion and Implications for Survey Practice

Overall, the results of our narrative literature review of studies on panel conditioning suggest that the effects appear rather small and context-dependent; however, they cannot be ignored entirely. For certain items and certain subgroups the potential consequences can be drastic. Based on the detailed accounts of the studies reviewed and the diversity of findings, it can be summarised that panel conditioning effects appear to be:

- *Relevant for certain subgroups more than for others*. A consistent finding across studies that looked at subgroups seems to be that lower educated respondents are more prone to panel conditioning.
- *Subject-dependent*. For highly salient topics, conditioning effects are less likely to occur than for topics that are of low salience to the respondent.
- *Dependent on the level of question sensitivity*. Reduced social desirability bias is more likely for questions at the lower end of the social desirability spectrum, while for more stigmatising behaviours a reduction in social desirability bias is less likely.
- *For attitudes, dependent on initial attitude strength*. Panel conditioning is generally less likely for attitudes that are strong and when respondents have higher knowledge levels about the issue.
- *For knowledge, dependent on the levels of previous knowledge*. Panel conditioning is less likely when the knowledge levels are high.

- *Dependent on mode.* In interviewer-administered surveys, panel conditioning can be caused by interviewers, for example, providing fewer names in network generator questions to reduce the interviewer burden and interview length but also eliciting more names as interviewers become more experienced.
- *Dependent on the interval between the waves.* Panel conditioning is more likely when the waves are administered closer together in time.
- *Possible also within-interview.* Repetitive looping questions are especially prone to within-interview conditioning. Questions in a grouped format (first all filter questions, then all follow-ups) seem to elicit more namings than in an interleafed format (follow-up questions asked directly after a filter question); however, data quality is better in the interleafed format.
- *Subject to a non-linear increase as the burden increases.* For particularly monotone and tedious questions, respondents can 'cap' advantageous conditioning (such as reporting more names in name generators or more frequent use of products) after a certain number of question administrations has been reached.

Since panel conditioning seems to depend heavily on the design features of the study and content of the questions, and several mechanisms could be at play that could cause it, it would be unwise to propose a one-size-fits-all set of recommendations. The safest bet would be to implement a small-scale experiment alongside the main data collection phase such, for example, was done by Axinn et al. (2015), who sampled a small group of persons from the same sampling frame as was used for the main study and assigned a small group of respondents of a two-year panel study to only one-time follow-up after one year. This strategy will be successful for assessing the extent of possible panel conditioning for the specific questionnaire content and a particular survey design. Once the existence and the magnitude of panel conditioning are known, there is a possibility to correct for it. Having the data from such small-scale randomised experiments would give rise to studies that focus on statistical methods that help adjust for the effects of panel conditioning and are few to date.

One prospective solution on a larger scale would be to implement a survey design that allows decreasing respondent burden in order to avoid negative consequences, such as providing incorrect answers to filter questions. This would take more than changing or eliminating individual items. The burden of self-report, reliance on memory, and the influence of social desirability – all relevant to panel conditioning effects, as has been shown in this chapter – can be eliminated by using administrative data where possible (e.g., Bethlehem 2008) or passive measurement of behavioural data using smartphones (e.g., measurement of financial behaviour by Jäckle et al. 2019, see also Jäckle et al. 2021, Chapter 14 in this book; tracking of online behaviour, Révilla et al. 2017; measuring physical activity, Scherpenzeel 2017, and physical movement, Geurs et al. 2015; measuring labour market behaviour, Kreuter et al. 2020). With rising number of surveys administered online and on mobile devices (e.g., Link et al. 2014) and lower willingness to take lengthy surveys (Kelly et al. 2013), the possibilities of split questionnaire design when the questionnaire is divided into smaller topical modules and administered in multiple sessions (e.g., Eberl 2016; Toepoel and Lugtig 2016) also has a promise of reducing burden; whether panel conditioning effects of learning the procedure in such setting would outweigh the benefits of reducing burden is something that needs to be investigated.

For a long time, research on panel conditioning was rather atheoretical with panel conditioning studies being a by-product of substantial research rather than studies being specifically designed for the purposes of prospectively testing panel effects. However, recently the research field has changed towards more experimental and quasi-experimental designs focusing on the causes of panel conditioning, studies that test the proposed mechanisms and assess the magnitude of panel conditioning.

In this chapter, we reviewed various types of studies in an attempt to provide a broad picture while guiding the reader through empirical evidence and paying attention to description of the data and study designs in as much detail as the scope of this chapter would allow.

While this review did not specifically aim to compare the methodological rigour of prior studies, we hope we were able to highlight the designs that we regard as best-practice examples of study designs available to date. These are:

- Experimental studies with one or more of the following features:
  - random assignment to questionnaire content,
  - random assignment to interviewer-administered or self-administered modes to exclude interviewer influence,
  - varying intervals between survey administrations,

  while other study design features (e.g., questionnaire length, burden, incentives) are held constant;
- Studies with adequate controls for attrition;
- Studies that focus on questions that imply more or less formed attitudes and more or less accessible knowledge, as well as differential level of social desirability.

Though we recognise the importance of studying panel conditioning across diverse topics and various question types, we encourage survey methodologists to replicate earlier studies using optimal study designs to further develop the theoretical framework (as has recently been done by e.g., Bach and Eckman 2018; Bergmann and Barth 2018; Halpern-Manners and Warren 2012; Halpern-Manners et al. 2014; Kroh et al. 2016; Sturgis et al. 2009; Uhrig 2012; Warren and Halpern-Manners 2012). We further encourage more experimental carefully designed studies with methods to control for panel attrition (e.g., as suggested by Das et al. 2011), influence of interviewers (e.g., Eagle and Proeschold-Bell 2015), separate learning the content effects from learning the survey procedure through survey design (e.g., Struminskaya 2016), and separate real change effects from reporting using validation (administrative) data (as e.g., Crossley et al. (2017), Yan and Eckman (2012), and others have done) to provide evidence-based and actionable recommendations for the survey practice. Furthermore, nearly all studies reported here (exceptions being e.g., Binswanger et al. 2013; Uhrig 2012) focus on analysing main effects using full samples. Studying whether panel conditioning is homogenous across respondent subgroups may be a fruitful area for future research.

References

Axinn, W.G., Jennings, E.A., and Couper, M.P. (2015). Response of sensitive behaviors to frequent measurement. *Social Science Research* 49: 1–15.

Bach, R. and Eckman, S. (2018). Motivated misreporting in web panels. *Journal of Survey Statistics and Methodology* 6 (3): 418–430. https://doi.org/10.1093/jssam/smx030.

Bailar, B.A. (1975). Effects of rotation group bias on estimates from panel surveys. *Journal of the American Statistical Association* 70: 23–30.

Bailar, B.A. (1989). Information needs, surveys, and measurement errors. In: *Panel surveys* (eds. D. Kasprzyk, G. Duncan, G. Kalton and M.P. Singh), 1–24. New York, NY: Wiley.

Bartels, L. (1999). Panel effects in the American national election studies. *Political Analysis* 8: 1–20.

Battaglia, M.P., Zell, E.R., and Ching, P.L.Y.H. (1996). Can participating in a panel sample introduce bias into trend estimates? In: *Proceedings of the Survey Research Methods Section,* Chicago (4–8 August 1996). Alexandria, VA: American Statistical Association.

Bergmann, M. and Barth, A. (2018). What was I thinking? A theoretical framework for analysing panel conditioning in attitudes and (response) behaviour. *International Journal of Social Research Methodology* 21 (3): 333–345. https://doi.org/10.1080/13645579.2017.1399622.

Bethlehem, J. (2008). Surveys without questions. In: *International Handbook of Survey Methodology* (eds. E.D. De Leeuw, J.J. Hox and D.A. Dillman), 500–511. New York, NY: Taylor & Francis.

Binswanger, J., Schunk, D., and Toepoel, V. (2013). Panel conditioning in difficult attitudinal questions. *Public Opinion Quarterly* 77: 783–797. https://doi.org/10.1093/poq/nft030.

Bradburn, N.M. (1978). Respondent burden. In: *Paper presented at the 138th Annual Meeting of the American Statistical Association*, , San Francisco (17–20 August 1978)*Proceedings of the Survey Research Methods Section*. Alexandria, VA: American Statistical Association.

Brannen, J. (1993). The effects of research on participants: findings from a study of mothers and employment. *The Sociological Review* 41: 328–346.

Bridge, G.R., Reeder, L.G., Kanouse, D. et al. (1977). Interviewing changes attitudes – sometimes. *Public Opinion Quarterly* 41: 56–64.

Brüderl, J., Huyer-May, B., and Schmiedeberg, C. (2013). Interviewer behavior and the quality of social network data. In: *Interviewers' Deviations in Surveys. Impact, Reasons, Detection and Prevention* (eds. P. Winkler, R. Porst and N. Menold), 147–160. Frankfurt: Peter Lang.

Callegaro, M. (2013). Paradata in web surveys. In: *Improving Surveys with Paradata: Analytic Uses of Process Information* (ed. F. Kreuter), 261–280. New York: Wiley.

Cantor, D. (1989). Substantive implications of selected operational longitudinal design features: the National Crime Survey as a case study. In: *Panel surveys* (eds. D. Kasprzyk, G. Duncan, G. Kalton and M.P. Singh), 25–51. New York, NY: Wiley.

Cantor, D. (2008). A review and summary of studies on panel conditioning. In: *Handbook of Longitudinal Research: Design, Measurement, and Analysis* (ed. S. Menard), 123–138. Burlington, MA: Elsevier.

Clausen, R.A. (1968). Response validity – vote report. *Public Opinion Quarterly* 32 (4): 588–606.

Clinton, J. D. (2001). Panel bias from attrition and conditioning. https://my.vanderbilt.edu/joshclinton/files/2011/10/C_WP2001.pdf

Coombs, L.C. (1973). Problems of contamination in panel surveys: a brief report on an independent sample. *Studies in Family Planning* 4 (10): 257–261.

Corder, L.S. and Horvitz, D.G. (1989). Panel effects in the National Medical Care Utilization and expenditure survey. In: *Panel surveys* (eds. D. Kasprzyk, G. Duncan, G. Kalton and M.P. Singh), 303–318. New York, NY: Wiley.

Crossley, T.F., de Bresser, J., Delaney, L. et al. (2017). Can survey participation alter household saving behavior? *The Economic Journal* 127: 2332–2357. https://doi.org/10.1111/ecoj.12398.

Das, M., Toepoel, V., and Van Soest, A. (2011). Nonprametric tests of panel conditioning and attrition bias in panel surveys. *Sociological Methods and Research* 40 (1): 32–56.

DeMaio, T.J. (1984). Social desirability and survey measurement: a review. In: *Surveying Subjective Phenomena*, vol. 2 (eds. C.F. Turner and E. Martin), 257–281. New York: Russell Sage Foundation.

Dennis, M. (2001). Are internet panels creating professional respondents? *Marketing Research* 13: 34–38.

Duan, N., Algeria, M., Canino, G. et al. (2007). Survey conditioning in self-reported mental health service use: randomized comparison of alternative instrument formats. *Health Services Research* 42 (2): 890–907.

Eagle, D.E. and Proeschold-Bell, R.J. (2015). Methodological considerations in the use of name generators and interpreters. *Social Networks* 40: 75–83.

Eberl, M. (2016). Shorter smarter surveys: Fragebögen durch Modularisierung und Stiching für mobile Endgeräte fit machen [Making the questionnaires fit for mobile devices through modularization and stiching]. In: *Marktforschung der Zukunft - Mensch oder Maschine? [Market research of the future - Humans or machines?]* (eds. B. Keller, H. Klein and S. Tuschl), 217–230. Wiesbaden: Springer https://doi.org/10.1007/978-3-658-14539-2.

Eckman, S. and Kreuter, F. (2018). Misreporting to looping questions in surveys: recall, motivation and burden. *Survey Research Methods* 12 (1): 59–74.

Eckman, S., Kreuter, F., Kirchner, A. et al. (2014). Assessing the mechanisms of misreporting to filter questions in surveys. *Public Opinion Quarterly* 78 (3): 721–733.

Fendrich, M. and Vaughn, C.M. (1994). Diminished lifetime substance use over time: an inquiry into differential underreporting. *Public Opinion Quarterly* 58: 96–123.

Fisher, P. (2019). Does repeated measurement improve income data quality? *Oxford Bulletin of Economics and Statistics* 81: 989–1011.

Fishkin, J. (1995). *The Voice of the People: Public Opinion and Democracy*. New Haven, CT: Yale University Press.

Frick, J.R., Goebel, J., Schechtman, E. et al. (2004). *Using Analysis of Gini (ANoGi) for Detecting Whether Two Sub-Samples Represent the Same Universe: The SOEP Experience. IZA Discussion Paper Series*, vol. 1049. Bonn: Institute for the Study of Labor (IZA).

Galesic, M. and Bosnjak, M. (2009). Effects of questionnaire length on participation and indicators of response quality in a web survey. *Public Opinion Quarterly* 73: 349–360.

Geurs, K.T., Thomas, T., Bijlsma, M.I., and Douhou, S. (2015). Automatic trip and mode detection with MoveSmarter: first results from the Dutch mobile mobility panel. *Transportation Research Procedia* 11: 247–262. https://doi.org/10.1016/j.trpro.2015.12.022.

Greenwald, A.G., Carnot, C.G., Beach, R., and Young, B. (1987). Increasing voting behavior by asking people if they expect to vote. *Journal of Applied Psychology* 72 (2): 315–318.

Halpern-Manners, A. and Warren, J.R. (2012). Panel conditioning in longitudinal studies: evidence from labor force items in the current population survey. *Demography* 49: 1499–1519. https://doi.org/10.1007/s13524-012-0124-x.

Halpern-Manners, A., Warren, J.R., and Torche, F. (2014). Panel conditioning in a longitudinal study of illicit behaviors. *Public Opinion Quarterly* 78: 565–590.

Holt, D. (1989). Panel conditioning: discussion. In: *Panel Surveys* (eds. D. Kasprzyk, G. Duncan, G. Kalton and M.P. Singh), 340–347. New York: Wiley.

Jäckle, A., Burton, J., Couper, M. P., Lessof, C. (2019). Participation in a mobile app survey to collect expenditure data as part of a large-scale probability household panel: Response rates and response biases. *Survey Research Methods* 13: 23–44.

Jagodzinski, W., Kühnel, S.M., and Schmidt, P. (1987). Is there a 'Socratic Effect' in nonexperimental panel studies? *Sociological Methods & Research* 15: 259–302.

Jäckle, A., Couper, M.P., Gaia, A. et al. (2021). Improving survey measurement of household finances. In: *Advances in Longitudinal Survey Methodology* (ed. P. Lynn). Chichester: Wiley.

Kalton, G. and Citro, C.F. (1993). Panel surveys: adding the fourth dimension. *Survey Methodology* 19: 205–260.

Kalton, G., Kasprzyk, D., and McMillen, D.B. (1989). Nonsampling errors in panel surveys. In: *Panel Surveys* (eds. D. Kasprzyk, G. Duncan, G. Kalton and P. Singh), 249–270. New York: Wiley.

Kelly, F., Johnson, A., and Stevens, S. (2013). Modular survey design: Bite sized chunks 2. Paper presented at the CASRO Online Research Conference, San Francisco, March 7–8.

Kraut, R.E. and McConahay, J.B. (1973). How being interviewed affects voting: an experiment. *Public Opinion Quarterly* 37 (3): 398–406.

Kreuter, F. (2013). Improving surveys with paradata: introduction. In: *Improving Surveys with Paradata: Analytic Uses of Process Information* (ed. F. Kreuter), 1–9. New York: Wiley.

Kreuter, F., McCulloch, S., Presser, S., and Tourangeau, R. (2011). The effects of asking filter questions in interleafed versus grouped format. *Sociological Methods and Research* 40: 88–104. https://doi.org/10.1177/0049124110392342.

Kreuter, F., Haas, G.-C., Keusch, F., Bähr, S., Trappmann, M. (2020). Collecting Survey and Smartphone Sensor Data With an App: Opportunities and Challenges Around Privacy and Informed Consent. *Social Science Computer Review* 38(5): 533–549.

Kroh, M., Winter, F., and Schupp, J. (2016). Using person-fit measures to assess the impact of panel conditioning on reliability. *Public Opinion Quarterly* 80 (4): 914–942.

Krosnick, J.A. (1991). Response strategies for coping with the cognitive demands of attitude measures in surveys. *Applied Cognitive Psychology* 5: 213–236.

Krosnick, J.A. and Alwin, D.F. (1987). An evaluation of a cognitive theory of response order effects in survey measurement. *Public Opinion Quarterly* 51: 201–219.

Kruse, Y., Callegaro, M., Dennis, M. J. et al. (2009). Panel conditioning and attrition in the AP-Yahoo! News election panel study. https://www.researchgate.net/publication/253431812_Panel_Conditioning_and_Attrition_in_the_AP-Yahoo_News_Election_Panel_Study

Lazarsfeld, P.F. (1940). "Panel" studies. *Public Opinion Quarterly* 4: 122–128.

Link, M., Murphy, J., Schober, M. F. et al. (2014). Mobile technologies for conducting, augmenting and potentially replacing surveys. AAPOR Task Force Report. American Association for Public Opinion Research.

Lynn, P. (2009). Methods for longitudinal surveys. In: *Methodology of Longitudinal Surveys* (ed. P. Lynn), 1–19. Chichester, West Sussex: Wiley.

Lynn, P. and Lugtig, P. (2017). Total survey error for longitudinal surveys. In: *Total Survey Error in Practice* (eds. P. Biemer, E.D. de Leeuw, S. Eckman, et al.), 279–298. Hoboken, NJ: Wiley.

Marsden, P.V. (2003). Interviewer effects in measuring network size using a single name generator. *Social Networks* 25: 1–16.

Mathiowetz, N.A. and Liar, T.J. (1994). Getting better? Changes or errors in the measurement of functional limitations. *Journal of Economic and Social Measurement* 20: 237–262.

Mavletova, A. and Lynn, P. (2019). Item nonresponse rates and panel conditioning in a longitudinal survey among youth. *Field Methods* 31 (2): 95–115.

Mensch, B.S. and Kandel, D.B. (1988). Underreporting of substance use in a National Longitudinal Youth Cohort: individual and interviewer effects. *Public Opinion Quarterly* 52: 100–124.

Merkle, D.M. (1996). The polls – review: the National Issues Convention deliberative poll. *Public Opinion Quarterly* 60 (4): 588–619.

Nancarrow, C. and Cartwright, T. (2007). Online access panels and tracking research: the conditioning issue. *International Journal of Market Research* 49: 573–594.

Pevalin, D.J. (2000). Multiple applications of the GHQ-12 in a general population sample: an investigation of long-term retest effects. *Social Psychiatry and Psychiatric Epidemiology* 35 (11): 508–512.

Porst, R. and Zeifang, K. (1987). A description of the German general social survey test-retest study and a report on the stabilities of the sociodemographic variables. *Sociological Methods and Research* 15: 177–218.

Rendtel, U., Nordberg, L., Jäntti, M. et al. (2004). Report on quality of income data. Chintex Working Paper 21.

Revilla, M., Ochoa, C., and Loewe, G. (2017). Using passive data from a meter to complement survey data in order to study online behavior. *Social Science Computer Review* 35 (4): 521–536. https://doi.org/10.1177/0894439316638457.

Scherpenzeel, A. (2017). Mixing online panel data collection with innovative methods. In: *Methodische Probleme von Mixed-Mode-Ansätzen in der Umfrageforschung [Methodological Problems of Mixed-Mode Approaches in Survey Research]* (eds. S. Eifler and F. Faulbaum), 27–49. Wiesbaden: Springer.

Sherman, S.J. (1980). On the self-erasing nature of errors of prediction. *Journal of Personality and Social Psychology* 39 (2): 211–221.

Sikkel, D. and Hoogendoorn, A. (2008). Panel surveys. In: *International Handbook of Survey Methodology* (eds. E.D. De Leeuw, J.J. Hox and D.A. Dillman), 479–499. New York: Taylor & Francis Group.

Silber, H., Schröder, J., Struminskaya, B. et al. (2019). Does panel conditioning affect data quality in ego-centered social network questions? *Social Networks* 56: 45–54. https://doi.org/10.1016/j.socnet.2018.08.003.

Silberstein, A.R. and Jacobs, C.A. (1989). Symptoms of Repeated Interview Effects in the Consumer Expenditure Interview Survey. In: *Panel Surveys* (eds. D. Kasprzyk, G. Duncan, G. Kalton and M.P. Singh), 289–303. New York: John Wiley & Sons.

Struminskaya, B. (2014). Data quality in probability-based online panels: Nonresponse, attrition, and panel conditioning. Doctoral dissertation. Utrecht University.

Struminskaya, B. (2016). Respondent conditioning in online panel surveys: results of two field experiments. *Social Science Computer Review* 34 (1): 95–115. https://doi.org/10.1177/0894439315574022.

Struminskaya, B. (2020). Panel Conditioning. In: The SAGE Encyclopedia of Research Methods, ed. by P. A. Atkinson, S. Delamont, R. A. Williams, A. Cernat. SAGE. DOI: 10.4135/9781526421036915183

Sturgis, P., Allum, N., and Brunton-Smith, I. (2009). Attitudes over time: the psychology of panel conditioning. In: *Methodology of Longitudinal Surveys* (ed. P. Lynn), 113–126. New York, NY: Wiley.

Sudman, S. and Bradburn, N.M. (eds.) (1982). *Asking questions: A practical guide to questionnaire design*. San Francisco, CA: Jossey-Bass.

Toepoel, V. and Lugtig, P. (2016). Adapting Surveys to the Mobile World: Data Chunking in the Dutch Probability-based LISS Panel. Paper presented at the American Association for Public Opinion Research (AAPOR) 71st Annual Conference. Austin, TX.

Toepoel, V., Das, M., and van Soest, A. (2009). Relating question type to panel conditioning: comparing trained and fresh respondents. *Survey Research Methods* 3: 73–80.

Torche, F., Warren, J.R., Halpern-Manners, A., and Valenzuela, E. (2012). Panel conditioning in a longitudinal study of adolescents' substance use: evidence from an experiment. *Social Forces* 90: 891–918. https://doi.org/10.1093/sf/sor006.

Tourangeau, R. (1984). Cognitive science and survey methods. In: *Cognitive Aspects of Survey Design: Building a Bridge Between Disciplines* (eds. T. Jabine, M. Straf, J. Tanur and R. Tourangeau), 73–100. Washington, DC: National Academy Press.

Tourangeau, R. (2017). The survey response process from a cognitive viewpoint. Presentation at OECD seminar on "Improving the Quality of Data Collection in Large Scale Assessments", May 11, 2017. Paris, France. Retrieved from: https://www.oecd.org/skills/piaac/The%20Survey%20Responses%20Process%20from%20a%20Cognitive%20Viewpoint_Roger%20Tourangeau.pdf

Tourangeau, R., Rips, L.J., and Rasinski, K. (2000). *The Psychology of Survey Response.* Cambridge: Cambridge University Press.

Traugott, M.W. and Katosh, J.P. (1979). Response validity in surveys of voting behavior. *Public Opinion Quarterly* 43: 359–377.

Tversky, A. and Kahneman, D. (1974). Judgement under uncertainty: heuristics and biases. *Science* 185 (4157): 1124–1131.

Uhrig, S.N. (2012). Understanding panel conditioning: an examination of social desirability bias in self-reported height and weight in panel surveys using experimental data. *Longitudinal and Life Course Studies* 3 (1): 120–136.

Valente, T.W., Dougherty, L., and Stammer, E. (2017). Response bias over time: interviewer learning and missing data in egocentric network surveys. *Field Methods* 29 (4): 303–316.

Van der Zouwen, J. and Van Tilburg, T. (2001). Reactivity in panel studies and its consequences for testing hypotheses. *Sociological Methods and Research* 30: 35–56.

Wagstaff, D.A., Kulis, S., and Elek, E. (2009). A six-wave study of the consistency of Mexican/Mexican-American preadolescents' lifetime substance use reports. *Journal of Drug Education* 39: 361–384.

Warren, J.R. and Halpern-Manners, A. (2012). Panel conditioning effects in longitudinal social science surveys. *Sociological Methods and Research* 41: 491–534.

Waterton, J. and Lievesley, D. (1989). Evidence of conditioning effects in the British social attitudes panel. In: *Panel Surveys* (eds. D. Kasprzyk, G. Duncan, G. Kalton and M.P. Singh), 319–339. New York, NY: Wiley.

Weir, D.R. and Smith, J.P. (2007). Do panel surveys really make people sick? A commentary on Wilson and Howell (60: 11, 2005, 2623–2627). *Social Science & Medicine* 65 (6): 1071–1077.

Wilson, S.E. and Howell, B.L. (2005). Do panel surveys make people sick? US arthritis trends in the health and retirement survey. *Social Science and Medicine* 60: 2623–2627.

Yalch, R. (1976). Pre-election interview effects on voter turnout. *Public Opinion Quarterly* 40 (3): 331–336.

Yan, T. and Eckman, S. (2012). Panel conditioning: Change in true value versus change in self-report. In: *Proceedings of the Survey Research Methods Section*, San Diego (28 July–2 August 2012). Alexandria, VA: American Statistical Association http://www.iab.de/897/section.aspx/Publikation/k140206302.

Yan, T., Datta, R., and Hepburn, P. (2011). Conditioning effects in panel participation. Paper presented at the 66th Annual Conference of the American Association for Public Opinion Research, Phoenix, AZ, USA.

13

Interviewer Effects in Panel Surveys

Simon Kühne[1] *and Martin Kroh*[1,2]

[1]*Faculty of Sociology, University of Bielefeld, Bielefeld, Germany*
[2]*Research Fellow of SOEP at DIW Berlin*

13.1 Introduction

Many popular cross-sectional surveys in the social sciences, such as the European Social Survey (ESS, Stoop et al. 2010), and particularly large-scale panel surveys, such as the Socio-Economic Panel Survey (SOEP, Goebel et al. 2018), UK Understanding Society (Buck and McFall 2012), the Household, Income and Labour Dynamics in Australia (HILDA, Wooden et al. 2002), and the Survey of Health, Ageing and Retirement in Europe (SHARE, Börsch-Supan et al. 2013) rely on face-to-face interviews.[1] Interviewers contribute to (panel) data quality by maintaining high participation rates, handling complex survey instruments and questionnaires, and providing assistance and clarifications to respondents (Fowler and Mangione 1990). This is why many panel studies rely on interviewers despite their high costs compared to other modes of data collection, such as mail and web surveys (for an overview of the dis/advantages of different modes of data collection, see Groves 2004; Groves et al. 2009). However, survey research also points to a number of disadvantages of interviewer-administered surveying, showing, among others, that interviewers can trigger undesired respondent behaviour and measurement error through their presence and their characteristics. Measurement error occurs, for instance, in the form of responses collected by the same interviewer being more similar compared to responses collected by different interviewers. There are numerous studies on these interviewer effects (see West and Blom 2017 for a research synthesis in the topic), and social scientists have investigated

1 Other large-scale panel surveys such as the Panel Study of Income Dynamics (PSID, McGonagle et al. 2012) and the Swiss Household Panel (SHP, Voorpostel et al. 2017) rely mainly on telephone interviewing.

Advances in Longitudinal Survey Methodology, First Edition. Edited by Peter Lynn.

Companion website: www.wiley.com/go/lynn/advancesinlongitudinalsurvey

interviewer effects since at least the 1940s (e.g. Katz 1942). Nonetheless, very little is known about potential peculiarities of interviewer effects in panel surveys.

This chapter contributes to closing these gaps in the research on interviewer effects. We focus on two main objectives. First, we compare different strategies in order to identify interviewer effects in panel data. So far, the advantages and disadvantages of identification strategies are neither sufficiently discussed nor empirically compared. In this regard, we propose a novel method that relies on the panel data structure itself. Second, we analyse how interviewer effects develop in panel surveys: Are they comparatively stable or do they increase or decrease over time? Specifically, do they change the more times a respondent participates or the more times the same interviewer interviews a respondent? How do they compare to the magnitude of established estimates of interviewer effects in cross-sectional surveys? Finally, we make use of rich information about interviewers and investigate the influence of characteristics beyond socio-demographics, including their opinions on a variety of topics.

The chapter is structured as follows. In Section 13.2, we illustrate the background, define concepts and summarise the state of research on interviewer effects (in panel surveys). Section 13.3 describes the data and variables. The methods and results are presented in Sections 13.4 and 13.5, followed by a summary and discussion in Section 13.6.

13.2 Motivation and State of Research

In this section, we provide an overview of the current state of research on interviewer effects focusing on theories of response behaviour that account for interviewer effects, on the potential peculiarities of interviewer effects in panel surveys, and on identification strategies. For an extensive summary of the literature we refer to West and Blom (2017) as well as Schaeffer et al. (2010).

13.2.1 Sources of Interviewer-Related Measurement Error

The Total Survey Error approach (see e.g. Groves and Lyberg 2010) aims to conceptualise the different possible sources of error in a survey, including specification, frame, non-response, measurement, and processing error. Typically, interviewer effects are located at the level of measurement, i.e. interviewers' presence, behaviour, and characteristics elicit error in the responses of interviewees.[2] In this regard, interviewer-related error includes both the variance and bias

2 There is a large body of literature on interviewer effects on unit non-response. See O'Muircheartaigh and Campanelli (1998) and Campanelli et al. (1997) for an overview. More recent contributions include Vassallo et al. (2015) and Jäckle et al. (2013). Interviewers may

that an interviewer contributes to the measurement of a survey variable. While random measurement error introduced by interviewers lowers the precision of survey estimates, thus increasing their standard errors, bias associated to interviewers refers to a systematic deviation from the true and unknown score of a survey measure. Thus, interviewer effects generally threaten both the reliability as well as the validity of survey measures.

Reviewing the existing literature, scholars point to three main mechanisms through which interviewers affect responses and introduce measurement error into survey data: interviewer deviations, social desirability, and priming.

13.2.1.1 Interviewer Deviations

Interviewers may introduce measurement error through deviations from the standardised interviewing script (see Schaeffer 2018 for an overview on the topic). For instance, some interviewers may probe heavily while others do not; some interviewers read out all answer options whereas others do not. Interviewer training can contribute to minimise such deviations by making interviewers aware of the importance of following the standardised interview protocol. Nevertheless, it is practically impossible to eliminate all differences in how interviewers behave while interviewing.

Several interviewer characteristics are related to the occurrence of interviewer deviations. First, there is evidence that the interviewers' motivation and their attitudes towards their job affect the quality of the data collected (Blom and Korbmacher 2013). Second, the interviewers' job experience is associated with data quality; however, evidence is inconclusive (Bilgen 2011; Ackermann-Piek and Massing 2014; Essig and Winter 2009; Lipps 2007; Olson and Bilgen 2011). Third, the interviewers' expectations about respondent answers may lead them to deviate, for instance by probing suggestively (see Kelley 2017; Uhrig and Sala 2011; Biemer and Lyberg 2003). These 'expectancy effects' are well known to (social) psychologists (e.g. Rosenthal and Rubin 1978) and are also addressed by survey researchers (Singer and Kohnke-Aguirre 1979; Sudman et al. 1977; Esbensen and Menard 1991; Smith and Hyman 1950), but mainly focusing on interviewers' general (prior) expectations; for instance, regarding expected difficulties in asking specific questions. Finally, the interviewers' own attitudes, opinions, and traits may lead them to deviate from a standardised behaviour, which then affects responses. Interviewers may (sub)consciously persuade or manipulate respondents to answer in a way that reflects their own attitudes on

introduce sampling or non-response error, as they often recruit respondents and need to convince them to take part in the survey. Interviewer-related sampling/non-response error is not part of this chapter. See West et al. (2013), and West and Olson (2010), for a discussion and disentangling of different error sources.

a topic or issue, as well as, potentially, their preconceptions of the respondent's likely answer, for instance, by probing in a suggestive manner (Smit et al. 1997).

13.2.1.2 Social Desirability

Interviewer-related error may furthermore occur as some respondents choose to adjust their 'true' answers and provide answers that are seen as desired or favourable by others, i.e. the interviewer. Thus, some respondents apply impression management strategies; for instance, adjusting an answer or lying, in order to appear in a good light in front of the interviewer. Such 'socially desirable responding' (see Krumpal 2013 for an overview) is a classic type of interviewer effect. Thus, the presence of an interviewer – both in face-to-face and telephone interviews – may provoke unwanted answering behaviour, especially in the context of sensitive questions (Tourangeau and Yan 2007). While survey research literature usually refers to social norms as the main source of desirability (Krumpal 2013, e.g. stealing is socially undesirable), the respondents' own expectations and judgements, as well as anticipated interviewer expectations, may also function as a benchmark for desirability (Cannell et al. 1981; DeMaio 1984, p. 258). Regarding the latter, Fendrich et al. (1999) propose the 'Social Attribution Model' in relation to interviewer effects and social desirability bias. It is based on the assumption that respondents 'make inferences about interviewers based on interviewers' observable characteristics' and that they furthermore 'use these inferences, in conjunction with general cultural stereotypes, to tailor (or edit) their answers to elicit interviewer approval' (Fendrich et al. 1999, p. 1014).

Many of the studies on interviewer effects implicitly assume this mechanism underlies the occurrence of interviewer effects. Consequently, scholars investigate how specific interviewer characteristics affect response behaviour and answers. In this regard, the majority of empirical insights relates to the effects of interviewer socio-demographics, i.e. personal and observable characteristics. Many study the effects of interviewers' race/ethnicity (e.g. Athey et al. 1960; Williams Jr. 1964, 1968; Schuman and Converse 1971; Weeks and Moore 1981) and gender (e.g. Kane and Macaulay 1993; Catania and Binson 1996; Webster 1996) on responses to thematically related questions and surveys (see West and Blom (2017) for an extensive overview).[3] For example, Schuman and Converse (1971) show that reported levels of militancy and hostility towards whites are higher among black respondents when interviewed by black interviewers compared to white interviewers.

Few studies focus on effects of not directly observable (not visible) interviewer characteristics, such as attitudes, opinions, and traits on responses to related questions. This is due, in part, to the fact that information about interviewers beyond typically visible socio-demographics is often not available. Katz (1942, p. 267), using face-to-face interview data, reveals that, for instance, middle-class

3 Other contributions focus on the religious clothing of interviewers (such as hijabs) on responses to related questions (see Blaydes and Gillum 2013).

and white-collar interviewers obtain a greater incidence of conservative attitudes compared to working class interviewers. Nybo Andersen and Olsen (2002) find negligible interviewer effects when investigating the effects of computer-assisted telephone interviewing (CATI) interviewers' health beliefs and personal habits on response data on smoking and alcohol consumption in a study of pregnant women. And Healy and Malhotra (2014) find no evidence of an effect of interviewers' partisan leanings on responses to related questions in CATI interviews. In contrast, based on a CATI panel, Lipps and Lutz (2010) show that responses to four political questions are associated with the respective interviewers' opinions on the topics. Moreover, based on face-to-face interview data, Himelein (2016) finds that interviewers' opinions about political/social issues, such as corruption and women's rights, strongly affect the responses of interviewees to questions on these topics. Using large-scale panel survey data, Hilgert et al. (2016) analyse whether the interviewers' personality traits (Big Five inventory) affect respondent answers to the same Big Five inventory items, revealing rather small, but significant, positive effects of the interviewers' scores on openness, conscientiousness and agreeableness on the respective respondents' scores. Turner et al. (2015) investigated the impact of interviewer personality traits on response variability and found little or no effect. Finally, Kühne (2020) shows that interviewer effects in questions on opinions toward political and social issues can be largely explained by how an interviewer is seen by respondents, i.e., what respondents think the interviewer thinks about a given topic or issue.

The question arises of how respondents are able to form expectations and make judgements about interviewer characteristics, such as their opinions, in cases where a question topic is not (mainly) related to an observable socio-demographic interviewer characteristic. Research in social psychology suggests that humans continuously detect and employ cues made available by others in order to categorise them (Brunswik 1956; Nestler et al. 2012). Humans use verbal and nonverbal cues and hints, such as overall appearance, gestures, and facial expressions to make judgements about others; for instance, regarding their personality. Judgements (or estimates) of such characteristics that are not visible are proven to be astonishingly accurate. Even at 'zero acquaintance' – which is the near complete lack of information on another person, except hints and information obtained in the first encounter – interacting partners are able to produce quite accurate estimates of each other (Levesque and Kenny 1993; Ambady et al. 1995). This is verified, for instance, in the case of personality traits (Nestler and Back 2013), intelligence (Borkenau and Liebler 1993), and political orientation (Samochowiec et al. 2010). Applying these findings to face-to-face interviews, it appears plausible that respondents and interviewers can use verbal and nonverbal cues available in the social interaction to infer opinions and attitudes about each other. Consequently, some respondents may adjust their answers in accordance to these inferences.

13.2.1.3 Priming

Whereas socially desirable responding is usually framed as an active and conscious act to adjust answers towards norms or anticipated interviewer expectations, interviewers may also affect response behaviour at a subconscious level. So-called 'priming' effects (Tulving and Schacter 1990) – the pre-activation due to a stimulus that itself affects the processing of further stimuli – may explain the interviewers' influence on responses that are formed spontaneously; for instance, in attitudinal questions. Interviewer characteristics or their behaviour may activate certain memory systems, thereby influencing respondent answers (as hypothesised by Schuman and Converse (1971) in their investigation of race-of-interviewer effects). With respect to attitudinal questions, visible characteristics of interviewers may additionally affect respondent's 'belief-sampling' (see Tourangeau et al. 2000) – a psychological process in which answers on attitude questions are initially formed on the basis of a sample of relevant and (most) accessible beliefs, feelings, impressions, general values, and prior judgements (referred to as considerations about an issue).

13.2.2 Moderating Factors of Interviewer Effects

In which interview settings are interviewer effects most pronounced? This question clearly is of interest to both survey researchers and practitioners, as it opens the possibility to minimise measurement error, thus increasing overall data quality. In addition to interviewer characteristics, such as the interviewers' motivation or their own opinions on topics, the magnitude of interviewer effects may also depend on the interview mode, respondent characteristics, as well as interactions between interviewer and interviewee characteristics.

With respect to survey mode, it is known that interviewer effects are less likely in cases where the respondent perceives the interview situation as more private. This is why in some studies relying on face-to-face interviewing, highly sensitive questions are integrated into a computer-assisted self-interviewing module (see De Leeuw et al. 2003). In line with this, interviewer effects are usually larger in face-to-face surveys than telephone-administered surveys (Groves 2004).

Few studies examine whether there are general respondent characteristics that are associated with the occurrence of interviewer bias. Are certain respondents especially likely to adjust answers towards anticipated interviewer traits, behaviours, or opinions? From a rational choice perspective, respondents are expected to provide an unadjusted (true) answer if its expected utility is higher compared to providing an adjusted answer. Perceived costs and benefits of both options are likely dependent on the respondent's need for social approval (Groves and Magilavy 1986; Tourangeau et al. 2000; Stocké 2004). The need for social approval is associated with the characteristic of conformity (Kelman 1958;

Strickland and Crowne 1962; Hogg 2010). Conformity can be related to specific personality traits. DeYoung and Peterson (2002) find that conformity is positively related to the traits of agreeableness, conscientiousness, and emotional stability and negatively related to extraversion and openness to experience.

With respect to the interaction of respondent and interviewer characteristics, the social distance model (see e.g. Fendrich et al. 1999), is probably the most prominent explanation of heterogeneous interviewer effects and makes assumptions about the conditions under which respondents are most likely to adjust an answer towards the interviewer. In the linear social distance model, respondents who perceive a greater social distance between themselves and the interviewer are thought to be more likely to edit their responses for the purpose of impression management (see also Williams Jr. 1968). Taking an alternative perspective, the nonlinear social distance model, based on the work of Dohrenwend et al. (1968), expects interviewer effects to be most likely under conditions of very high and very low social distance between respondent and interviewer. For both models, conclusive empirical evidence is almost non-existent. Moreover, social distance may not only be framed as differences in social status but more generally in terms of similarity or dissimilarity between respondent and interviewer. For example, the interviewer's gender may generally affect response behaviour in gender-equality-related questions but may furthermore be mediated by the respondent's gender as well: Male respondents interviewed by a female interviewer may answer systematically differently compared to male respondents interviewed by a male interviewer, whereas female respondents may answer the same regardless of the gender of the interviewer. Thus, interviewer effects may also result from an interaction of respondent and interviewer characteristics (see West and Blom 2017).

13.2.3 Interviewer Effects in Panel Surveys

Surprisingly, little is known about potential peculiarities of interviewer effects in panel surveys; as Durrant et al. (2010, p. 26) note, it is a 'largely unexplored area.' Thus far, we do not know whether interviewer effects are stable over time in an ongoing panel survey or whether they increase or decrease. There are a number of aspects inherent to panel surveys that potentially moderate the interviewers' impact on response behaviour.

First, respondents gain survey experience through their ongoing panel participation. For instance, respondents who have already participated for many years in a panel may be less likely to be affected by an interviewer's presence and their specific characteristics. Related to this, respondents may provide more reliable answers to survey questions with growing panel experience (Kroh et al. 2016; Sturgis et al. 2009), possibly because attitudes and opinions become more crystallised and stronger or they gain trust in the interviewer/survey (e.g. Fisher

2019; Chadi 2013). Lipps and Lutz (2010) hypothesise that interviewer effects may decrease over time; i.e. that respondents are increasingly less likely to express an opinion in the direction of the interviewer. However, they do not find a significant interaction effect of interviewer opinions and time in the panel on respondent reported opinions.

Second, the interviewer staff in a panel survey gains (work) experience as well, which in turn may affect their interviewing behaviour. Interviewers may become increasingly trained to apply a standardised interviewing protocol, which would then lead to decreasing interviewer effects over time. However, it could be that their motivation to do so may also decrease over time and they might tend to apply individual (i.e. deviating) interviewing behaviours, which, in turn, would increase interviewer effects over time. Existing studies offer rather inconclusive results (West and Blom 2017). Regular interviewer training on the survey most likely affects this potential relationship between interviewer experience and data quality.

Finally, in many panel surveys, the same interviewers are allocated to the same respondents across many panel waves.[4] The reasons for this are diverse. First, organisational or financial limitations can result in interviewer continuity. It is usually more cost-effective to allocate interviewers who live near the respondents (primary sample clusters), thereby reducing travel costs. Thus, if neither the interviewer nor the respondent moves, allocating the interviewer living geographically closest results in interviewer continuity. Second, it is argued that interviewer continuity contributes to the quality of the data collected (see e.g. Lynn and Lugtig 2017). Indeed, there is evidence that interviewer continuity reduces panel attrition rates (e.g. Rendtel 1995; Lynn et al. 2014; Campanelli and O'Muricheartaigh 1999; Vassallo et al. 2015). In this regard, scholars hypothesise that interviewer continuity promotes the development of trust (e.g. Fisher 2019; Halpern-Manners et al. 2014), emotional closeness, and loyalty (DeMaio 1984), as well as interview rapport between respondents and interviewers (e.g. Holbrook et al. 2003), and that this, in turn, increases the respondents' motivation to answer thoughtfully and truthfully. Hajek and Schumann (2014) and Schräpler (2004) show that interviewer continuity can decrease item non-response rates. Only a few studies examine possible effects on measurement error. Kühne (2018), Chadi (2013), and Lipps (2007) provide evidence for a decrease in socially desirable responding towards social norms with ongoing interviewer continuity. Thus, respondents seem to be less likely to adjust their answers towards social norms when becoming familiar with their interviewers. In contrast, Uhrig and Lynn (2008) reveal an increase in social desirability bias for six out of eight items tested. In line with this, Mensch and Kandel (1988) find lower reports of drug

4 Including the SOEP (Goebel et al. 2018), Understanding Society - the UK Household Longitudinal Study (Mitchell et al. 2015, p. 9), and the Household, Income and Labour Dynamics in Australia (HILDA, Watson and Wooden 2009).

usage behaviour with growing interviewer continuity. In this regard, the authors speculate that, 'interviewer familiarity increases salience of normative standards and that participants respond not only in terms of their past familiarity but also in terms of their subjective expectations regarding the probability of a future encounter with the interviewer' (Mensch and Kandel 1988, p. 100).

13.2.4 Identifying Interviewer Effects

Survey researchers apply different strategies in order to identify interviewer effects on responses, focusing either on random measurement error (unreliability) or systematic bias (invalidity).

13.2.4.1 Interviewer Variance

Many studies on interviewer effects apply a variance-oriented, non-directional approach. Here, the aim is to quantify the interviewer variance, i.e. the interviewers' contribution to the total variation in a survey variable, resulting from the individual biases introduced by each interviewer. The more homogeneous the responses collected by individual interviewers, the higher the share of variance that is due to the interviewers. Most of the research builds on the concept of 'intra-class correlation' or 'intra-interviewer correlation' (ICC, ρ_{int}) initially proposed by Kish (1962):

$$\rho_{int} = \frac{s_a^2}{s_a^2 + s_b^2}$$

where s_a^2 is the between-interviewer component, s_b^2 is the within-interviewer component, and $(s_a^2 + s_b^2)$ is the total variance of the variable of interest. Thus, the larger the variation in responses collected across interviewers compared to the variation within interviewers, the larger the interviewers' contribution to the total variance in a survey measure. The precision of survey estimates decreases with growing intra-interviewer correlation as well as an increasing number of interviews per interviewer.

There is a large body of literature on measures of intra-interviewer correlation in survey variables (e.g. Groves 2004, p. 365; Schnell and Kreuter 2005; West and Olson 2010). They show that interviewer effects are present across all survey topics and question types. Studies investigating interviewer variance share a common analytic approach: The interviewers influence on a survey measure is captured within a single parameter. It is a convenient strategy in the sense that no information about the interviewer beyond a unique identifier is needed. However, the parameter provides no information on the amount of interviewer bias, i.e. a survey-measure's systematic deviation from the true score of a population parameter. Hence, by relying exclusively on the intra-interviewer correlation, researchers cannot make statements about whether an estimate is biassed. Moreover, the parameter does not provide any information about the underlying mechanism(s) causing the effects in the first place.

Additionally, scholars face problems of disentangling the targeted interviewer effects from confounding factors and other types of measurement error, especially in the case of face-to-face interviewing. In many (panel) surveys, interviewers are not randomly allocated to respondents or households. Rather, they are allocated to convenient geographic areas, sometimes defined by primary sampling units. Often, only a single interviewer is working in a given area and interviewers are allocated to only a single area. In these cases, estimated interviewer effects are confounded with area effects (Schnell and Kreuter 2005; Campanelli and O'Muricheartaigh 1999; Durrant and D'Arrigo 2014). Answers observed by a single interviewer may be more homogeneous not because of the interviewers' biasing effects on responses, but due to the homogeneity of individuals living in the same geographic area. So-called 'interpenetrated survey designs' are needed in order to unequivocally separate interviewer from area effects in face-to-face surveys. In a fully interpenetrated design, interviewers are randomly allocated to respondents (Mahalabonis 1946). Due to high travel costs, these designs are very rarely implemented in face-to-face surveys. Thus, some researchers have implemented partially interpenetrated designs that allow separation of interviewer and area effects but minimise the costs (O'Muircheartaigh and Campanelli 1998; Schnell and Kreuter 2005). In cases where deliberate interpenetrated designs cannot be implemented, researchers make use of a 'natural' crossing of interviewers and areas (e.g. Sturgis 2009) in order to separate interviewer from area effects. For instance, in many panel surveys, (some) interviewers work in multiple areas over time and multiple interviewers are working in a given area. Given such a cross-nesting of interviewers and areas, scholars can apply hierarchical regression models that include crossed random effects of interviewers and areas (see, Brunton-Smith et al. 2017, p. 5; West and Blom 2017).[5] In non-interpenetrated designs, it is common to also include interviewer- and area-level fixed effects in order to further 'mitigate endogeneity of area/interviewer selection effects' (Sturgis 2009). Including good covariates is crucial as a crossing of interviewer and areas is usually far from random, thereby potentially introducing biases due to selection effects.

13.2.4.2 Interviewer Bias

When taking a bias-oriented perspective, scholars usually investigate how interviewer characteristics (or their interaction with respondent characteristics)

5 Multiple membership models are an alternative to cross-classified models, e.g. Vassallo et al. 2015; Lynn et al. 2014.

systematically, i.e. directionally, affect respondent answers. Many examples of this strategy are found in the numerous studies on race-of-interviewer and gender-of-interviewer effects. For instance, average responses to a question about gender-equality are compared between male and female interviewers. In a regression setting, addressing interviewer bias usually involves estimating effects (coefficients) for specific interviewer characteristics such as ethnicity, gender, or age.

While such approaches are bias-oriented, they usually do not allow quantification of the size of the bias in a given estimate; a mean for instance. This is mainly for practical reasons: An estimation of bias usually requires information on the 'true' value of a survey parameter or at least specific assumptions about the (psychometric) reliability of a survey measure. For many survey measures, such as opinions and attitudes, external information about true values does not exist. Examples of studies on the size of interviewer-related bias are found in West et al. (2013) and Kroh (2005).[6] Moreover, estimated interviewer effects (regression coefficients) may be confounded with, again, area effects or correlated respondent and interviewer characteristics. Finally, the success of this method is highly dependent on the depth of information that is available about interviewers. In most cases, information is very limited, for instance to socio-demographics such as gender and age.

13.2.4.3 Using Panel Data to Identify Interviewer Effects

When investigating interviewer effects in panel surveys, scholars can use the same identification strategies as for cross-sectional studies. Thus, researchers may calculate intra-interviewer correlations or run regression analysis on interviewer characteristics for single panel waves or a pooled database.

We propose using the panel data structure itself to identify interviewer effects. Our main analytic argument states that changes in responses in accordance to a change in interviewer characteristics (opinions, traits) serve as an indicator of interviewer effects.[7] Thus, we propose the usage of both respondent and interviewer longitudinal data in order to identify and analyse interviewer effects. The

---

6 Other studies have focused on interviewer effects on indicators of data quality. A common indicator of (low) data quality is "item non-response," which refers to missing data for a survey question or item. Bilgen (2011) shows that inexperienced interviewers achieve lower rates of item non-response. Riphahn and Serfling (2003) find higher item non-response rates in female interviewers. Singer et al. (1983) found that interviewers' expectations about the difficulty of obtaining valid answers affect non-response rates. Other indicators of data quality used in the study of interviewer effects include rounding errors (indicating social desirability; Weinhardt et al. 2010), acquiescence ('yes-saying') and psychometric reliability (Hox et al. 2004), 'centering' and extreme responses (Lipps 2007; Pickery and Loosveldt 2004).

7 An and Winship (2017) apply the only comparable approach known to the author. In order to review popular parametric models for analysing panel data, they analyse effects of changes in interviewers' race in the U.S. General Social Survey as an application example.

upcoming Section 13.4 sets out each estimation method in detail. At this point, we want to illustrate the basic idea of using the panel data structure to identify interviewer effects.

In principle, there are two ways through which a respondent is exposed to changing interviewer characteristics (e.g. opinions or traits). First, interviewer changes occur regularly in panel surveys. For instance, in the SOEP, about 7% of all respondents experience a change in interviewer in a given wave. Thus, many respondents experience a change in interviewer during their time in the panel.[8] Second, the same interviewers change over time, for instance by changing their own opinions towards certain topics.

We see two major advantages of using respondent and interviewer longitudinal data to identify interviewer effects. First, it provides a stronger basis for causal inference, as it is based on within-variation at the respondent level that allows controlling for unobserved heterogeneity. We argue that panel analysis approaches are especially suited for inferring causal relationships. Second, the proposed panel approach may be especially useful for panel surveys lacking a sufficient cross-nesting of areas and interviewers. By applying a panel regression approach, the likely confounding through area effects is diminished as analysis is based on variation in interviewer characteristics within areas.

13.3 Data

13.3.1 The Socio-Economic Panel

For this research, we rely on data derived from the SOEP, an ongoing longitudinal survey of households in Germany (Goebel et al. 2018). Conducted annually since 1984, the study covers a variety of topics, including household composition, employment, family biography, health, education, personality, and attitudes. SOEP version v32.1, years 1984–2015 (doi: http://doi.org/10.5684/soep.v32, see Socio-Economic Panel (SOEP) 2017) is used for this analysis. In 2015, there were 27 183 individuals in 13 782 households participating in the study. Data collection in the SOEP is largely based on personal, face-to-face interviews. Since 1998, the SOEP has been gradually replacing paper-and-pencil personal interviewing (PAPI) with computer-assisted personal interviewing (CAPI) as the

8 Reasons for interviewer changes are diverse. In the SOEP, in most cases (67%), interviewer changes occur because interviewers drop-out of the staff either because they quit, are dismissed, retire, or are re-allocated to other studies operated by the fieldwork agency. Only in about 8% of cases do interviewer changes occur because a household moves. In 18%, interviewers have temporarily dropped out of the staff and in 7%, interviewers have been allocated to other geographic areas.

predominant mode of data collection. For instance, in 2015, 72% of the interviews were interviewer-administered, with 89% of those being conducted via CAPI.

In the SOEP, longitudinal survey data is available not only for respondents, but for interviewers as well. In 2006, 2012, and 2015, all SOEP interviewers were asked to take part in a mail survey. Many interviewers (about 66%) took part in more than one interviewer survey, thus allowing analysis of effects of changing opinions within individual interviewers over time. The questionnaires covered the interviewers' socio-demographics, personality traits, self-assessments, as well as work-related topics such as experience, motivation, and attitudes towards their job. In addition, the questionnaires were designed to overlap in large part with the main SOEP questionnaire. Thus, the database allows a 1 : 1 matching of respondent and interviewer answers to the same questions in a longitudinal perspective.

13.3.2 Variables

For the identification of potential interviewer effects and the comparison of strategies, we make use of three sets of non-factual questions about attitudes, opinions, and traits. First, we use eight sub-questions regarding the level of worries for different social and political issues. Second, we look at political orientation as a combination of party identification and strength of leaning towards this party. Thirdly, we analyse interviewer effects in questions about life satisfaction and subjective health status. All survey questions are shown to be susceptible to interviewer effects in existing empirical studies.[9] Table 13.1 displays descriptive statistics for the items for both respondents and interviewers. The exact question texts and answer options are presented in the online supplementary materials for this chapter.

13.4 The Size and Direction of Interviewer Effects in Panels

In this section we analyse interviewer effects in the data by means of different identification strategies. After providing details of the analytic approaches, we compare estimates of interviewer effects across the 11 variables described above.

13.4.1 Methods

A sub-sample of the SOEP is used for comparing the different interviewer effect identification strategies. First, we focus on the panel waves 2006, 2012, and 2015

9 See, Kühne (2018), Chadi (2013), Lipps (2007), and Warren and Halpern-Manners (2012).

Table 13.1 Descriptive statistics for items used in the analyses.

Variable	Scale	Respondent		Interviewer	
		Mean	SD (n_{obs})	Mean	SD (n_{obs})
Worries about …					
Economy in general	1 (not at all)	2.10	0.65 (20 942)	2.14	0.64 (664)
Own economic situation	2 (somewhat)	1.81	0.71 (20 949)	1.73	0.67 (664)
Own health	3 (very)	1.92	0.70 (20 948)	1.97	0.65 (664)
Environment		2.13	0.62 (20 943)	2.21	0.65 (663)
Peace		2.30	0.68 (20 937)	2.33	0.71 (663)
Crime		2.22	0.70 (20 929)	2.32	0.68 (662)
Immigration		1.96	0.76 (20 910)	2.04	0.60 (663)
Xenophobia		2.08	0.68 (20 901)	2.15	0.69 (662)
Political orientation (left/right)	−5 (far left) … +5 (far right)	−0.09	2.33 (20 829)	−0.09	2.90 (640)
Health	1 (very good) … 5 (bad)	2.67	0.94 (20 997)	2.46	0.83 (661)
Life satisfaction	0 (compl. dissatisfied) 10 (compl. satisfied)	7.19	1.69 (20 647)	7.68	1.41 (662)

Pooled information waves 2006, 2012, 2015.

as interviewer surveys were conducted in these years only. Second, we only include those respondents (and their years of observation) who participated in the SOEP via personal interviews (PAPI and CAPI). Third, as we apply panel data approaches, respondents needed to take part in at least two out of the three waves (either consecutive or non-consecutive). Finally, we use a subsample of interviewers and geographic areas that provides a sufficient cross-nested data structure in order to separate area from interviewer effects. Geographic areas are defined at the German county level. There are 401 counties in Germany. The subsample includes only those interviewers who were allocated to at least three counties and only those counties in which at least three interviewers conducted interviews over the course of the three panel waves.[10] This holds true for about 60% of the sub-sample.

10 By counting the cross-nesting over time and not separately within waves, interviewer effects may be, in principle, confounded with time (wave) effects. Nonetheless, there are two reasons this should be unproblematic. First, there are no systematic effects of time (specific waves) on

The final subsample amounts to 20 700 observations (person-years) of 9040 respondents in 5708 households interviewed by 361 interviewers in 351 areas (counties) in the years 2006, 2012, and 2015. All interviewer effect identification strategies are applied to this identical sub-sample. This allows direct comparison of the results.

Strategy 1, which we refer to as 'variance-oriented', is a well-established (cross-sectional) identification strategy. We estimate the intra-interviewer correlation coefficient (Kish's ρ_{int}), representing the share of variance in a survey measure that is due to the interviewer. For the estimation of interviewer variance in the items, we use hierarchical models that include crossed-random effects of interviewers and areas (as done by others to estimate interviewer effects, e.g. Brunton-Smith et al. 2017; West et al. 2013; Sturgis 2009).

The respondent's answer functions as the dependent variable. For each item, we estimate a multivariate hierarchical linear regression model (see Brunton-Smith et al. 2017, p. 5):

$$y_{ih(jk)} = x'_{ih(jk)}\beta + u_j + v_k + w_h + e_{ih(jk)}$$

where $y_{ih(jk)}$ represents the answer of respondent i in household h living in area k interviewed by interviewer j. The cross-classification of interviewers and areas is indicated by placing their indices in parentheses. $x'_{ih(jk)}$ then represents a vector of respondent, household, interviewer, and area covariates as well as corresponding coefficients β. Random intercepts for interviewers, areas, and households are represented by u_j, v_k, and w_h. The respondent-specific residual (error term) is $e_{ih(jk)}$. The intra-class correlation for the interviewers is then derived as the share of the interviewer-level variance compared to the total variance that is decomposed into interviewer-specific variance, the area-specific variance, the household-specific variance, and the individual respondent residual variance:

$$\rho_{int} = \frac{s^2_{u_j}}{s^2_{u_j} + s^2_{v_k} + s^2_{w_h} + s^2_{e_{ih(jk)}}}$$

We compare the interviewer variance estimates with the results in other existing studies in order to evaluate the strength of interviewer effects in the data.

Strategy 2 is an established (cross-sectional) strategy to identify the effect of specific interviewer characteristics. We refer to it as 'bias-oriented, pooled cross-sectional.' It involves adding the interviewer characteristic (the opinion or trait) as an independent variable into the multilevel model presented above. Now,

---

the level of cross-nesting of areas and interviewers. Thus, even though the area-interviewer allocation is subject to change over multiple waves, there are no general shifts in how the interviewer staff is organized and allocated across regions in Germany. Second, we estimated each model separately for each individual wave, showing similar results. Nonetheless, in order to address the potential confounding, we include wave identifiers as controls into the models.

interviewer effects are identified by analysing the regression coefficient β and related significance level for this particular interviewer characteristic rather than the overall interviewer variance.

Third, we apply the suggested panel data identification strategy that makes use of the longitudinal information available for both respondents and interviewers (Strategy 3, 'bias-oriented, panel fixed effects'). A multivariate linear panel fixed effects regression model is used to estimate the effect of changes in interviewer opinions/traits on respondent answers to the exact same questions:

$$y_{it} = x'\beta + a_i + e_{it}$$

where y_{it} is the response of individual i at time (wave) t and x' represents a vector of respondent, household, interviewer, and area time-variant covariates with their corresponding coefficients β. This includes the (changing) interviewer opinions/traits within respondents over time. We now assume a respondent-level fixed effect, a_i, representing all time-invariant individual variables affecting the outcome. Also referred to as 'unobserved heterogeneity', it denotes general differences between individuals. In the mixed models applied in strategies 1 and 2, this individual heterogeneity is part of the residual error term. Note that the previous mixed models include random intercepts for interviewers, areas, and households while the panel fixed effects regression does not.[11] Finally, e_{it} represents the error term of individual i at time t. Again, we analyse the regression coefficients and their standard errors in order to identify interviewer effects.

We control for the same set of covariates in all three identification strategies. First, rich information about areas is added to the models in order to counteract the potential endogeneity of area/interviewer selection effects. This includes structural data at the German county, municipality, and neighbourhood levels. A list of all the variables added as area-specific fixed effects is provided in the online supplementary materials for this chapter.

At the household level, we control for interviewer changes (yes/no) and life events that may both affect response behaviour as well as simultaneously influence the probability of a change in interviewer. Thus, we integrate two binary variables indicating whether the household membership of a person has changed (y/n), for instance because a couple moves in together, and whether a household has moved geographically (y/n).[12]

---

11 We did not include area- and household-level fixed effects as there is only very small variation within respondents over time, i.e. in most cases, area and household characteristics are constant. Moreover, adding interviewer-level fixed effects did not change any results.

12 We coded a household as having moved if there was a distance of at least 100 m between the new and the old address.

Third, a set of respondent-level socio-demographics is added including gender, age, schooling, and nationality (German, non-German). Moreover, the respondents' prospective number of years of participation is included in the models. This measure captures whether the responses provided for given interviewer characteristics are different for respondents who will refuse to participate in the following wave of the survey compared to those who will participate in the survey for several more years. Finally, the survey year is included in order to control for potential period effects.

13.4.2 Results

Table 13.2 displays the results of the three identification strategies across all 11 items. There are a number of interesting similarities and differences across strategies as well as items.

For strategy 1, the variance-oriented approach, the estimated interviewers' contribution to the total variation in responses ranges from 2% in the questions about health status and political orientation to 10% in the question on worries towards peace. Overall, stronger interviewer effects are observed for the item battery relating to worries compared to political orientation, health, and life satisfaction. Compared to other studies investigating the size of interviewer variance (see Groves and Magilavy 1986; Schnell and Kreuter 2005), the observed effects are quite strong, pointing to a substantial impact of interviewers on response behaviour. However, note that these are maximum estimates as it is unlikely that the model can explain 100% of the between-area variation in the true values.

Within the bias-oriented approaches (strategies 2 and 3) we focus on whether the interviewers' own opinions and traits are associated with respondent answers to the same questions. Based on the pooled data approach (strategy 2), positive effects of interviewer opinions and traits are observed for all 11 items with the majority of the regression coefficients being statistically significant ($p < .001$). Thus, the results point to an association of interviewers' own opinions/traits and the answers provided by their respondents. One plausible explanation would be that some respondents adjust their true answers towards anticipated interviewer opinions in order to appear in a good light in front of the interviewer. Turning to strategy 3, the proposed panel data approach using fixed effects regressions, highly similar results are observed with significant positive effects of interviewer characteristics on responses for the majority of items. While the coefficients in the pooled regressions (strategy 2) may be potentially confounded with time-invariant factors of respondents, households, and areas, the fixed effects regressions are controlling for this 'unobserved heterogeneity' by design. Thus, the fact that the effects do not disappear in the panel regressions indicates that the observed results in the pooled data approach are not (entirely) due to confounding factors.

Table 13.2 Comparison of interviewer effects identification strategies.

Strategy	1	2	3
	Variance-oriented (interviewer random effects)	Bias-oriented (pooled data)	Bias-oriented (panel fixed effects)
	Kish's ρ_{int}	Coeff. *b* (int. opinion)	Coeff. *b* (int. opinion)
Worries about ...			
Economy in general	0.09	0.05***	0.05***
Own economic situation	0.06	0.09***	0.09***
Own health	0.04	0.06***	0.05***
Environment	0.09	0.10***	0.10***
Peace	0.10	0.07***	0.07***
Crime	0.06	0.09***	0.04***
Immigration	0.07	0.13***	0.05***
Xenophobia	0.08	0.06***	0.07***
Political orientation	0.02	0.03**	0.01
Health	0.02	0.02	0.00
Life satisfaction	0.02	0.05**	0.02*

* $p < .05$, ** $p < .01$, *** $p < .001$.
Strategy 1: Hierarchical Linear Regression, Pooled Data.
Strategy 2: Model a + Interviewer opinion added as explanatory variable.
Strategy 3: Panel fixed effects regression.
21 015 observations, 5664 households, 9129 respondents, 359 interviewers, 343 counties.

Nonetheless, there are substantial differences between the two methods for some items. For example, estimated effect sizes for the worries about crime and immigration are larger in the pooled-data approach (strategy 2) compared to the panel fixed effects approach (strategy 3). A possible explanation could be that the association observed in the pooled regression is in part due to existing similarities of respondents and interviewers living in the same geographic areas. As the panel regression focuses on effects of changing interviewer characteristics within areas, the strong effect disappears when strategy 3 is applied.[13]

13 Hausman tests (Hausman 1978) point to the appropriateness of the panel fixed effects models compared to panel regressions with random effects; i.e. the individual-level effects would have not been adequately modelled in random effect models.

Overall, all three strategies point to the existence of interviewer effects in the data. While there are differences across strategies in effect sizes and the ranking of items in terms of effect sizes, the results are highly comparable. For instance, all three strategies find that the weakest interviewer effects are on responses to the questions relating to political orientation, health, and life satisfaction.

13.4.3 Effects on Precision

The analysis reveals a substantial clustering of responses within interviewers reflected in the comparatively large values of interviewer variance (Kish's ρ_{int}). What does this mean for data quality and reliability specifically?

The so-called interviewer design effect reflects the factor through which the variance of a survey statistic (a mean for instance) is increased due to the homogeneity of responses within interviewers. The design effect can be estimated as:

$$deff_{int} = 1 + \rho_{int} \times \left(\overline{n_{int}} - 1\right)$$

As can be seen, $deff_{int}$ is a function of the interviewer variance ρ_{int} and the interviewers' average workload $\overline{n_{int}}$, i.e. the average number of conducted interviews per interviewer. The design effect then can be related to the concept of effective sample size, which represents the size of the sample needed to achieve the same precision in case of the absence of any design effect. The effective sample size is calculated as a given sample size divided by the design effect.

Figure 13.1 visualises the potential loss in precision by means of effective samples sizes (y-axis) across the observed values of interviewer variance (x-axis) and average interviewer workloads (the three lines). In this example, 5000 respondents have been interviewed. As can be seen, the effective sample size decreases with an increasing clustering of responses within interviewers. Even small values of ρ_{int} can have severe effects on the precision of estimates. This becomes particularly evident when the average workload for each interviewer is comparatively large. While 10 interviews (dark solid line) per interviewer result in a comparatively small loss in precision for lower levels of interviewer variance, precision drops dramatically for 40 interviews (light, even dashed line, about the average in the SOEP), and 100 interviews (dark, uneven dashed line; a large but still not uncommon workload). Thus, when it comes to interviewer variance, it is in the researcher's best interest to have a fieldwork staff consisting of many interviewers conducting only a small number of interviews each. Obviously, this likely increases the survey costs and, consequently, fieldwork agencies usually aim for rather large interviewer workloads.[14]

---

14 Given a fixed budget for conducting a survey, researchers also need to take into account other interviewer error sources decreasing precision and accuracy of estimates. For instance, employing a large interviewer staff may come at the expense of quantity and quality of

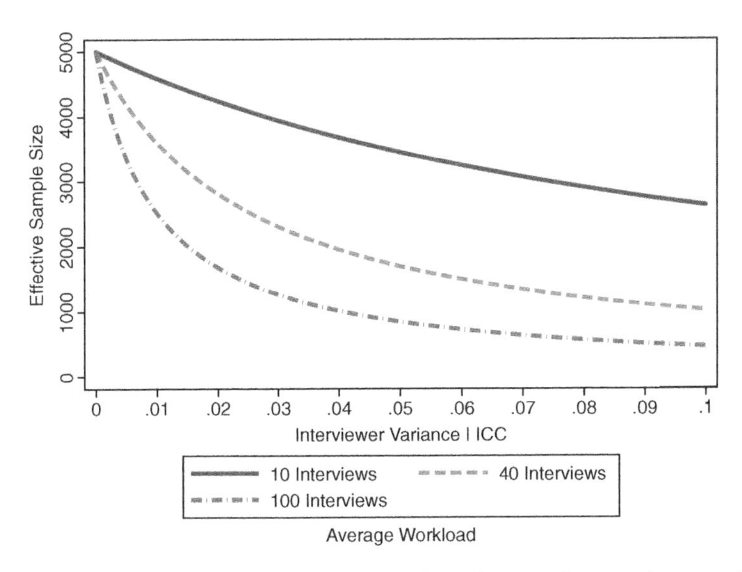

Figure 13.1 Effective sample size across interviewer variance and average interviewer workload.

13.4.4 Effects on Validity

Interviewer-related measurement error can also threaten the validity of survey measures, as estimates may be biased. The results above already showed that interviewer opinions significantly relate to respondent answers. What does this mean for overall estimates and substantial analyses of the data? As there is no external validation data or a 'gold standard' for the questions analysed in this chapter, there is no way of quantifying the bias introduced by interviewers. However, the potential of bias can be assessed by looking at marginal effects across interviewer characteristics.

Figure 13.2 displays margins across variable-values for each interviewer characteristic. The margins represent predictions of a regression model at fixed values of a covariate (the interviewer opinion/trait) controlling for all other remaining covariates. The estimates are based on applying strategy 2, the bias-oriented approach relying on the pooled database. In each of the 11 margin plots, a dot represents a linear prediction of the respondent answers given a specific interviewer characteristic and controlling for covariates. The estimates are accompanied by 95% confidence intervals. The horizontal line in each plot represents the overall mean

---

interviewer training procedures. Thus, employing more interviewers will not necessarily reduce overall interviewer effects.

of respondent answers in the sample. In line with the previous results, in the questions relating to worries, the responses collected by different subgroups of interviewers differ significantly from each other. Moreover, there is quite a clear positive linear relationship between interviewer worries and respondents reported worries.

The potential of bias lies in the composition of the interviewer staff. For instance, given the observed differences in margins, an interviewer staff entirely consisting of interviewers with major worries is expected to collect responses reflecting more worries compared to a staff consisting of interviewers with no worries at all. In both of these 'extreme' compositions of an interviewer staff, a large bias in estimates of a mean is expected. In contrast, in case the interviewer staff represents a random sample of the underlying population of interest (and given a constant workload), the individual errors of each interviewer should more or less cancel each other out: Some interviewers 'overestimate' the worries of their respondents while others 'underestimate' them.[15] However, it is rather unlikely that survey interviewers represent a random sample of the population. Thus, the potential of bias for a given survey variable increases with the deviation of the distribution of characteristics of a given interviewer staff from the distribution in the underlying population.[16]

13.5 Dynamics of Interviewer Effects in Panels

So far, we tested and compared different strategies in order to identify interviewer effects in panel survey data. However, we did not test whether there are effects of interviewers that are specific to longitudinal data. The nature and size of interviewer effects may be subject to change over the course of a panel survey. Thus far, little is known about these potential specifics of interviewer effects in longitudinal studies. In this section, we ask whether interviewer effects are stable in panel surveys or whether they vary 'over time'. Specifically, we analyse whether they are a function of the number of waves in which the respondent has participated, the number of waves in which the interviewer has participated, or the number of waves for which the same respondent-interviewing pairing has been maintained. For instance, respondents may be less likely to adjust their answers towards the interviewer if they have known their interviewer for many years and, thus, have gained trust and in turn feel more comfortable to answer truthfully.

15 But even in this situation, though estimates of means may be unbiased, estimates of measures of distribution or association are likely to be biased: for example, regression coefficients may be systematically attenuated.

16 Setting aside other sources of error, weighting factors may be used to decrease bias. Interviewers may be assigned with 'more importance' in the analyses if they hold characteristics that are underrepresented in a given interviewer staff composition. Thus far, the usage of interviewer weights is an almost entirely unexplored area in survey research.

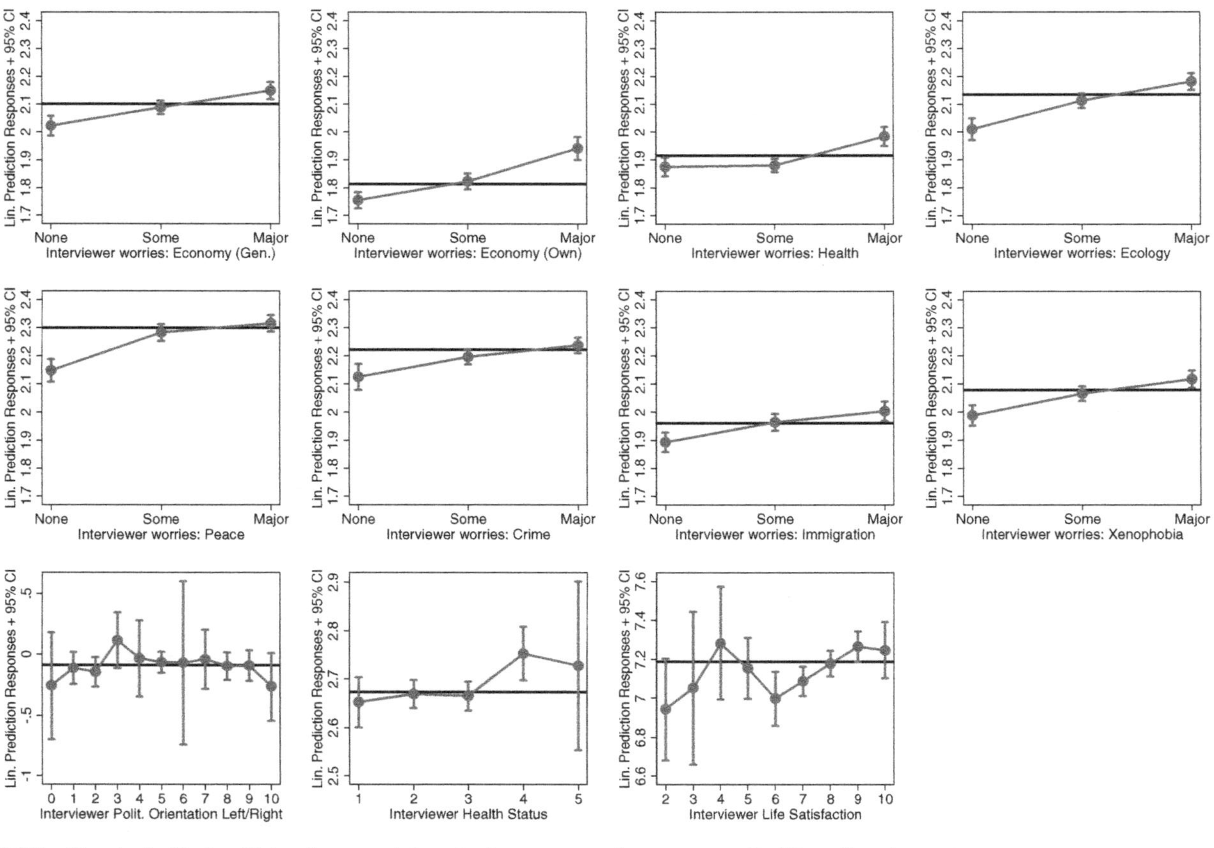

Figure 13.2 Marginal effects of interviewer opinions/traits on respondent answers. Red line: Grand sample mean.

13.5.1 Methods

We apply two analysis methods in order to investigate how interviewer effects develop over time in the SOEP.

First, we focus on the variable error component by analysing interviewer variance over time (years) in the panel. Rather than comparing interviewer variance estimates for each wave, we compare estimates across the number of participations, i.e. the first participation of a respondent, the second participation, etc., regardless of the exact year. In order to control for confounding through panel attrition and refreshment samples, a balanced panel dataset is constructed. Respondents were included if they participated in at least six consecutive waves (about the median with respect to the maximum number of participations in the panel) beginning with their first year of participation. We further restrict the database to the years 2000 through 2015 so that the balanced panel contains a high share of respondents and interviewers who are also part of the subsample constructed for the previous comparison of identification strategies. Thus, respondents who entered the panel in 1999 and earlier are not included in the analyses. As before, we estimate mixed models with individuals nested in households nested in a cross-classified structure of areas and interviewers. Again, the survey year, respondent socio-demographics, and area covariables are included as controls. For each of the 11 items, six models are estimated, each representing a rank of respondent panel participation ranging from 1 (the first time being interviewed) to 6 (the 6th time being interviewed). Each model incorporates data from about 9440 respondents. We then visually analyse the interviewer variance estimates over the six participations for each of the 11 items.

Second, we analyse biasing effects of interviewer opinions on related responses. For this, we replicate the panel fixed effect models using the same exact subsample (Strategy 3, years 2006, 2012, and 2015), adding three interaction terms: (i) interviewer opinion × interviewer work experience (in years); (ii) interviewer opinion × respondent panel experience (in years); and (iii) interviewer opinion × familiarity between respondent and interviewers (in years). Systematic effects of the three variables on the impact of interviewer opinions on responses are reflected by significant interaction terms. In the subsample, on average, respondents' panel experience amounts to $\bar{x} = 12.22$ years (SD = 8.79, Min = 1, Max = 32), interviewers' work experience amounts to $\bar{x} = 14.26$ years (SD = 8.71, Min = 1, Max = 32), and the average familiarity between respondents and interviewers amounts to $\bar{x} = 7.50$ years (SD = 6.19, Min = 1, Max = 30).

13.5.2 Results

13.5.2.1 Interviewer Variance

Figure 13.3 displays intra-interviewer correlation coefficients (Kish's ρ_{int}) for each of the 11 items for the balanced panel of respondents participating in six

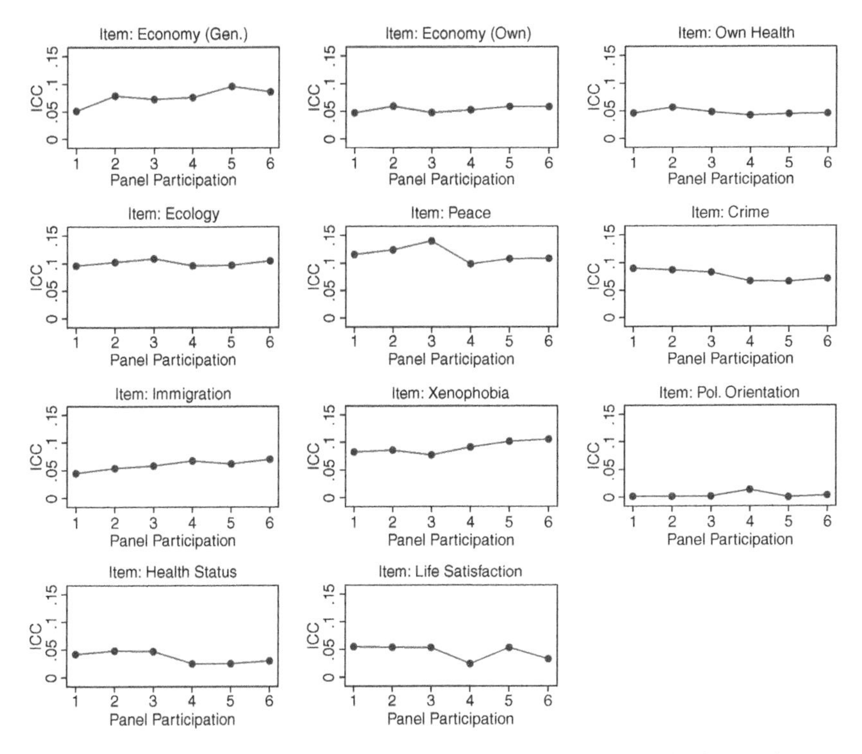

Figure 13.3 Estimates of interviewer variance (Kish's ρ_{int}) over years in the panel, waves, 2000–2015, balanced panel.

consecutive waves. As seen in the connected dot plots, there are no systematic patterns in estimates of interviewer variance over consecutive panel participations. To put it differently: Answers collected from respondents in their first year(s) of participation are not more or less homogeneous within interviewers compared to answers collected in later panel rounds. This is quite a positive result: Even though interviewer effects do not vanish over time, there is no evidence for an increase in interviewer-related error in panel surveys.

13.5.2.2 Interviewer Bias

Next, we assess whether directional interviewer effects in panel surveys depend on respondents' panel experience, interviewers' panel (work) experience, or the familiarity between both.

Table 13.3 displays the results of the panel fixed effects regressions. A positive interaction effect of interviewer work experience (in years) and interviewer opinion is observed for worries towards own economic situation, health, and environment as well as political orientation. This suggests that more experienced interviewers are more likely to affect their respondents through their own opinions on the topics.[17] A possible explanation might be that interviewers are decreasingly motivated to follow the standardised interviewing protocol but instead tend to develop own interviewing routines, for instance by applying a more conversational style of interviewing.

No clear pattern of interactions is observed for respondent panel experience or for familiarity between respondent and interviewer. Comparing the results across items, even the direction of the effects is rather ambiguous. However, 3 of the 11 interaction effects of familiarity reach statistical significance. For worries related to the general economic situation and immigration as well as political orientation, the interviewer's own opinions show a stronger influence on responses with greater familiarity between respondent and interviewer.

13.6 Summary and Discussion

This chapter addresses interviewer effects in panel surveys, a largely unexplored area in survey research. First, we discussed how to identify interviewer effects in panel surveys using existing (cross-sectional) identification strategies and a novel approach relying on the panel data structure itself. Second, we investigated how interviewer effects develop over time in panel surveys. For the analyses, we made use of rich longitudinal data about respondents and interviewers derived from the SOEP.

All three identification strategies tested point to (strong) interviewer effects in the items. Moreover, while there are differences in the ranking of effects, the three strategies agree on the items associated with the smallest effects. A possible explanation for this result could be that the estimates across strategies all represent the same underlying mechanisms, causing the effects in the first place. Comparing the two bias-oriented strategies, both the cross-sectional and the panel regression approach point to a positive effect of an interviewer's opinion on the respondent answer to the same question. For instance, respondents are more likely to state that they are worried about immigration if interviewed by an interviewer who does so. Some respondents may adjust their responses towards anticipated interviewer opinion in order to appear in a good light in front of the interviewer. Moreover, interviewers may 'prime' their respondents through their appearance and behaviour that relates to their own opinions about topics.

Estimated effects are generally smaller in the panel regression approach compared to the models using pooled data. This suggests an overestimation of

17 Similar results are observed when using the cumulative number of interviews conducted by each interviewer in a given year.

Table 13.3 Interviewer effects over time.

	1	2	3	4	5	6	7	8	9	10	11
	Econ. gen.	Econ. own	Health	Environment	Peace	Crime	Immigration	Xenophobia	Political orient.	Health status	Life satisf.
Interviewer opinion	0.056*	0.035	0.017	0.050*	0.041*	−0.005	0.041*	0.076**	0.029	0.014	−0.033
Interviewer experience	−0.004*	−0.003	−0.002	−0.007***	−0.003*	0.002	0.001	−0.002	−0.009*	−0.001	−0.005
I-opinion * I-experience	−0.002	0.005***	0.004*	0.004**	0.001	0.001	−0.001	−0.000	0.004***	0.001	0.003
Respondent experience	−0.115	0.035	−0.191	−0.200*	−0.209	−0.151	−0.057	0.004	0.581*	0.315*	−1.150***
I-opinion * R-experience	−0.000	0.001	0.001	−0.001	0.002	0.002	−0.001	0.001	0.001	−0.004*	−0.002
Familiarity	−0.002	0.003	0.003	−0.006**	−0.005*	−0.002	−0.006**	−0.006**	−0.002	0.002	−0.001
I-Opinion * familiarity	0.005*	−0.003	−0.002	−0.002	−0.002	−0.001	0.005**	−0.003	0.004*	0.002	−0.002
n_{resp}	18 848	18 853	18 848	18 843	18 829	18 829	18 792	18 787	17 982	18 852	18 678
n_{int}	360	360	360	360	359	360	360	359	355	359	359

* $p < .05$, ** $p < .01$, *** $p < .001$.

Note: Panel fixed effects regressions. Coefficients b.

Controls: Area covariates, wave dummies, prospective panel participation, hh-moves, hh-changes, respondent age & age^2, interviewer age & age^2.

interviewer effects due to a confounding with area effects when applying the two 'cross-sectional' approaches. While all identification strategies reveal evidence for the existence of interviewer effects in the data, using the panel data structure and (longitudinal) information about interviewers seems worthwhile and beneficial. We see two main advantages. First, using respondent fixed-effects models diminishes the likely confounding of interviewer effects through area effects. This may be of special importance in cases where there is no (fully) interpenetrated survey design (the default in the majority of existing large-scale face-to-face panel surveys). In these cases, it is often impossible to estimate cross-classified multilevel models as the data are not cross-nested sufficiently. Second, panel data analysis provides a stronger basis for causal inference, as it is based on within-variation at the respondent level and thus controls for unobserved heterogeneity. However, the proposed panel data approach is associated with comparatively high demands regarding the data, as (longitudinal) information about interviewers is needed.

As for the second objective in this chapter, we analysed how interviewer effects develop in panel surveys over time. Analysing interviewer variance over different years of respondent participation in a balanced panel did not reveal any systematic pattern. With respect to potential directional effects of interviewer characteristics on responses, we replicated some analyses and added three interaction terms: the interviewer's opinion interacting with (i) the interviewers' panel experience (in years); (ii) the respondents' panel experience (in years); and (iii) the familiarity between both (in years). We observe some evidence for an increase in interviewer effects with growing interviewer work experience and familiarity between respondent and interviewer. However, the results do not show a clear and coherent pattern. Thus, overall, the results do not point to a substantial increase or decrease in interviewer-related measurement error in panel surveys compared to cross-sectional studies. This is good news, as there is no evidence for another panel conditioning effect threatening the quality of data collected in longitudinal surveys.

The study faces some limitations and leaves room for further research. First, in many of the analyses, we made use of three SOEP panel waves (2006, 2012, and 2015) for which interviewer survey data is available. Moreover, we rely on a sub-sample of SOEP respondents who took part in at least two of these three waves as well as a sub-sample of interviewers, who took part in at least one interviewer survey. Even though we evaluate it as rather unlikely, this could have introduced selection bias into the estimates. Second, interviewer effect estimates are maximum effects as we cannot rule out area confounding. We addressed this by selecting a subsample of interviewers and areas that represent a sufficient cross-nesting as well as adding a set of area- and respondent-level controls into the models. Third, changes in interviewer opinions – both due to changing interviewers (between) as well as changes in opinions of the same interviewers (within) – occur non-randomly in the SOEP. While we control for potential confounding factors being responsible for both interviewer changes as well as changes in response behaviour, results may still be confounded to some degree. Experimental (and unfortunately expensive) research designs, in which interviewers are randomly assigned to respondents, may solve these issues in future studies.

We see three main implications of the results for panel survey practitioners and methodologists. First, we highly recommend the collection of interviewer data; for instance, by means of regular (annual) interviewer surveys. Detailed information about interviewers allows for an in-depth investigation of interviewer effects and the mechanisms underlying them. This is not limited to interviewer opinions, traits, and behaviour, but also includes work-related information such as motivation and performance indicators as well as paradata (timestamps, audio-recordings, etc.). Second, the results provide suggestions for staff management and interviewer-respondent allocation. Kühne (2018) shows that an ongoing relationship between respondents and interviewers decreases the occurrence of socially desirable responding. While the results in this chapter do not point to a reduction of the effects of interviewer opinions on the respondent answers, there is also no evidence of an increase in the effects. Thus, based on what we know thus far, there is no need to change the common practice of allocating the same interviewers to the same respondents in longitudinal studies. Third, more research is needed that addresses underlying mechanisms and, thus, allows researchers to actually explain the effects found. While the approaches in this chapter allow identification and analysis of the size and development of effects, further research is needed to isolate mechanisms such as socially desirable responding, priming, or expectancy effects.

References

Ackermann-Piek, D. and Massing, N. (2014). Interviewer behavior and interviewer characteristics in PIAAC Germany. *Methods, Data, Analyses* 8 (2): 199–222.

Ambady, N., Hallahan, M., and Rosenthal, R. (1995). On judging and being judged accurately in zero-acquaintance situations. *Journal of Personality and Social Psychology* 69 (3): 518–529.

An, W. and Winship, C. (2017). Causal inference in panel data with application to estimating race-of-interviewer effects in the general social survey. *Sociological Methods & Research* 46 (1): 68–102.

Athey, K.R., Coleman, J.E., Reitman, A.P. et al. (1960). Two experiments showing the effect of the interviewer's racial background on responses to questionnaires concerning racial issues. *Journal of Applied Psychology* 44 (4): 244–246.

Biemer, P.B. and Lyberg, L.E. (2003). *Introduction to Survey Quality*. Wiley.

Bilgen, I. (2011). Is less more & more less …? The effect of two types of interviewer experience on 'don't know' responses in calendar and standardized interviews. PhD thesis. University of Nebraska.

Blaydes, L. and Gillum, R.M. (2013). Religiosity-of-interviewer effects: assessing the impact of veiled enumerators on survey response in Egypt. *Politics and Religion* 6 (3): 459–482.

Blom, A.G. and Korbmacher, J.M. (2013). Measuring interviewer characteristics pertinent to social surveys: a conceptual framework. *Survey Methods: Insights from the Field* (23 January). http://surveyinsights.org/?p=817 [Accessed December 20th 2018].

Borkenau, P. and Liebler, A. (1993). Convergence of stranger ratings of personality and intelligence with self-ratings, partner ratings, and measured intelligence. *Journal of Personality and Social Psychology* 65 (3): 546–553.

Börsch-Supan, A., Brandt, M., Hunkler, C. et al. (2013). Data resource profile: the survey of health, ageing and retirement in Europe (SHARE). *International Journal of Epidemiology* 42 (4): 991–1001.

Brunswik, E. (1956). *Perception and the Representative Design of Psychological Experiments*, 2e. Berkeley: University of California Press.

Brunton-Smith, I., Sturgis, P., and Leckie, G. (2017). Detecting and understanding interviewer effects on survey data by using a cross-classified mixed effects location-scale model. *Journal of the Royal Statistical Society Series A* 180 (2): 551–568.

Buck, N. and McFall, S. (2012). Understanding society: design overview. *Longitudinal and Life Course Studies* 3 (1): 5–17.

Campanelli, P. and O'Muircheartaigh, C. (1999). Interviewers, interviewer continuity, and panel survey nonresponse. *Quality & Quantity* 33 (1): 59–76.

Campanelli, P., Sturgis, P., and Purdon, S. (1997). *Can You Hear Me Knocking? An Investigation into the Impact of Interviewers on Survey Response Rates*. London: The Survey Methods Centre, SCPR.

Cannell, C., Miller, P., and Oksenberg, L. (1981). Research on interviewing techniques. In: *Sociological Methodology* (ed. S. Leinhardt), 389–437. Jossey-Bass.

Catania, J.A. and Binson, D. (1996). Effects of interviewer gender, interviewer choice, and item wording on responses to questions concerning sexual behavior. *Public Opinion Quarterly* 60 (3): 345–375.

Chadi, A. (2013). The role of interviewer encounters in panel responses on life satisfaction. *Economics Letters* 121 (3): 550–554.

De Leeuw, E.D., Hox, J.J., and Kef, S. (2003). Computer-assisted self-interviewing tailored for special populations and topics. *Field Methods* 15: 223–251.

DeMaio, T.J. (1984). Social desirability and survey measurement. A review. In: *Surveying Subjective Phenomena*, vol. 2 (eds. C. Turner and E. Martin), 257–281. New York: Russell Sage Foundation.

DeYoung, C.G. and Peterson, J.B. (2002). Higher-order factors of the big five predict conformity: are there neuroses of health? *Personality and Individual Differences* 33 (4): 33–45.

Dohrenwend, B.S., Colombotos, J., and Dohrenwend, B.P. (1968). Social distance and interviewer effects. *Public Opinion Quarterly* 32 (3): 410–422.

Durrant, G.B. and D'Arrigo, J. (2014). Doorstep interactions and interviewer effects on the process leading to cooperation or refusal. *Sociological Methods & Research* 43 (3): 490–518.

Durrant, G.B., Groves, R.M., Staetsky, L. et al. (2010). Effects of interviewer attitudes and behavior on refusal in household surveys. *Public Opinion Quarterly* 74 (1): 1–36.

Esbensen, F.-A. and Menard, S. (1991). Interviewer-related measurement error in attitudinal research: a nonexperimental study. *Quality & Quantity* 25 (2): 151–165.

Essig, L. and Winter, J. (2009). Item nonresponse to financial questions in household surveys: an experimental study of interviewer and mode effects. *Fiscal Studies* 30 (4): 367–390.

Fendrich, M., Johnson, T., Wislar, J.S. et al. (1999). The impact of interviewer characteristics on cocaine use underreporting by male juvenile arrestees. *Journal of Drug Issues* 29 (1): 37–58.

Fisher, P. (2019). Does repeated measurement improve income data quality? *Oxford Bulletin of Economics and Statistics* 81 (5): 989–1011.

Fowler, F.J. and Mangione, T.W. (1990). *Standardized Survey Interviewing: Minimizing Interviewer-Related Error*. Newbury Park, CA: Sage.

Goebel, J., Grabka, M.M., Liebig, S. et al. (2018). The German Socio-Economic Panel (SOEP). *Journal of Economics and Statistics* Online first. https://doi.org/10.1515/jbnst-2018-0022 [Accessed December 20th 2018].

Groves, R.M. (2004). *Survey Error and Survey Costs*. Hoboken, NJ: Wiley.

Groves, R.M. and Lyberg, L. (2010). Total survey error: past, present, and future. *Public Opinion Quarterly* 74 (5): 849–879.

Groves, R. and Magilavy, L. (1986). Measuring and explaining interviewer effects in centralized telephone surveys. *Public Opinion Quarterly* 50 (2): 251–266.

Groves, R.M., Fowler, F.J. Jr., Couper, M.P. et al. (2009). *Survey Methodology*, 2nd edition. Hoboken, NJ: Wiley.

Hajek, K., Schumann, N. (2014). Continuity trumps? The impact of interviewer change on item nonresponse. Handout prepared for the Panel Survey Methodology Workshop 2014, Ann Arbor, Michigan, U.S.

Halpern-Manners, A., Warren, J., and Torche, F. (2014). Panel conditioning in a longitudinal study of illicit behaviors. *Public Opinion Quarterly* 78 (3): 565–590.

Hausman, J. (1978). Specification tests in econometrics. *Econometrica* 46: 1251–1271.

Healy, A. and Malhotra, N. (2014). Partisan bias among interviewers. *Public Opinion Quarterly* 78 (2): 485–499.

Hilgert, L., Kroh, M., and Richter, D. (2016). The effect of face-to-face interviewing on personality measurement. *Journal of Research in Personality* 63: 133–136.

Himelein, K. (2016). Interviewer effects in subjective survey questions: evidence from Timor-Leste. *International Journal of Public Opinion Research* 28 (4): 511–533.

Hogg, M.A. (2010). Influence and leadership. In: *Handbook of Social Psychology*, 5e (eds. S. Fiske, D. Gilbert and L. Gardner), 1166–1207. New York: Wiley.

Holbrook, A.L., Green, M.C., and Krosnick, J.A. (2003). Telephone versus face-to-face interviewing of national probability samples with long questionnaires: comparisons of respondent satisficing and social desirability response bias. *Public Opinion Quarterly* 67 (1): 79–125.

Hox, J.J., De Leeuw, E.D., and Kreft, I.G.G. (2004). The effect of interviewer and respondent characteristics on the quality of survey data: a multilevel model. In: *Measurement Errors in Surveys* (eds. P.P. Biemer, R.M. Groves, L.E. Lyberg, et al.), 439–461. Hoboken, NJ: Wiley.

Jäckle, A., Lynn, P., Sinibaldi, J. et al. (2013). The effect of interviewer experience, attitudes, personality and skills on respondent co-operation with face-to-face surveys. *Survey Research Methods* 7 (1): 1–15.

Kane, E.W. and Macaulay, J. (1993). Interviewer gender and gender attitudes. *Public Opinion Quarterly* 57 (1): 1–28.

Katz, D. (1942). Do interviewers bias poll results? *Public Opinion Quarterly* 6 (2): 248–268.

Kelley, J. (2017). The accuracy of using paradata to detect and understand interviewer question-reading deviations. Presentation at the Understanding Society Scientific Conference 2017, July 11–13. University of Essex.

Kelman, H.C. (1958). Compliance, identification, and internalization: three processes of attitude change. *Journal of Conflict Resolution* 2 (1): 51–60.

Kish, L. (1962). Studies of interviewer variance for attitudinal variables. *Journal of the American Statistical Association* 57 (297): 92–115.

Kroh, M. (2005). Interviewereffekte bei der Erhebung des Körpergewichts in Bevölkerungsumfragen. *Gesundheitswesen* 67 (8/9): 646–655.

Kroh, M., Winter, F., and Schupp, J. (2016). Using person-fit measures to assess the impact of panel conditioning on reliability. *Public Opinion Quarterly* 80 (4): 914–942.

Krumpal, I. (2013). Determinants of social desirability bias in sensitive surveys: a literature review. *Quality & Quantity* 47 (4): 2025–2047.

Kühne, S. (2018). From strangers to acquaintances? Interviewer continuity and socially desirable responses in panel surveys. *Survey Research Methods* 12 (2): 121–146.

Kühne, S., 2020: Interpersonal Perceptions and Interviewer Effects in Face-to-Face Surveys. Sociological Methods & Research. June 2020, Online first. doi:10.1177/0049124120926215

Levesque, M. and Kenny, D. (1993). Accuracy of behavioral predictions at zero acquaintance: a social relations analysis. *Journal of Personality and Social Psychology* 65 (6): 1178–1187.

Lipps, O. (2007). Interviewer and respondent survey quality effects in a CATI panel. *Bulletin de méthodologie sociologique* 95: 1–16.

Lipps, O. and Lutz, G. (2010). How answers on political attitudes are shaped by interviewers: evidence from a panel survey. *Swiss Journal of Sociology* 36 (2): 345–358.

Lynn, P. and Lugtig, P. (2017). Total survey error for longitudinal surveys. In: *Total Survey Error in Practice: Improving Quality in the Era of Big Data* (eds. P.P. Biemer, E.D. De Leeuw, S. Eckman, et al.), 279–298. Wiley.

Lynn, P., Kaminska, O., and Goldstein, H. (2014). Panel attrition: how important is interviewer continuity? *Journal of Official Statistics* 30 (3): 443–457.

Mahalabonis, P.C. (1946). Recent experiments in statistical sampling in the Indian statistical institute. *Journal of the Royal Statistical Society* 109: 325–370.

McGonagle, K.A., Schoeni, R.F., Sastry, N. et al. (2012). The panel study of income dynamics: overview, recent innovations, and potential for life course research. *Longitudinal and Life Course Studies* 3 (2): 268–284.

Mensch, B.S. and Kandel, D.B. (1988). Underreporting of substance use in a national longitudinal youth cohort: individual and interviewer effects. *Public Opinion Quarterly* 52 (1): 100–124.

Mitchell, M., Collins, D., and Brown, A. (2015). *Factors Affecting Participation in Understanding Society: Qualitative Study with Panel Members*, Understanding Society Working Paper Series No. 2015-04. UK: ISER Institute for Social and Economic Research, University of Essex.

Nestler, S. and Back, M. (2013). Applications and extensions of the lens model to understand interpersonal judgments at zero acquaintance. *Current Directions in Psychological Science* 22: 374–379.

Nestler, S., Egloff, B., Küfner, A. et al. (2012). An integrative lens model approach to bias and accuracy in human inferences: hindsight effects and knowledge updating in personality judgments. *Journal of Personality and Social Psychology* 103 (4): 689–717.

Nybo Andersen, A.-M. and Olsen, J. (2002). Do interviewers' health beliefs and habits modify responses to sensitive questions? A study using data collected from pregnant women by means of computer-assisted telephone interviews. *American Journal of Epidemiology* 155 (1): 95–100.

Olson, K. and Bilgen, I. (2011). The role of interviewer experience on acquiescence. *Public Opinion Quarterly* 75 (1): 99–114.

O'Muircheartaigh, C. and Campanelli, P. (1998). The relative impact of interviewer effects and sample design effects on survey precision. *Journal of the Royal Statistical Society Series A* 161 (1): 63–77.

Pickery, J. and Loosveldt, G. (2004). A simultaneous analysis of interviewer effects on various data quality indicators with identification of exceptional interviewers. *Journal of Official Statistics* 20 (1): 77–89.

Rendtel, U. (1995). *Lebenslagen im Wandel: Panelausfälle und Panelrepräsentativität*. Campus Verlag.

Riphahn, R.T. and Serfling, O. (2003). Heterogeneity in item non-response on income and wealth questions. *Schmollers Jahrbuch* 123: 95–108.

Rosenthal, R. and Rubin, D.B. (1978). Interpersonal expectancy effects: the first 345 studies. *Behavioral and Brain Sciences* 1 (3): 377–386.

Samochowiec, J., Wänke, M., and Fiedler, K. (2010). Political ideology at face value. *Social Psychology and Personality Science* 1 (3): 206–212.

Schaeffer, N.C. (2018). Interviewer deviations from scripts. In: *The Palgrave Handbook of Survey Research* (eds. D.L. Vannette and J.A. Krosnick), 465–472. London: Palgrave Macmillan.

Schaeffer, N.C., Dykema, J., and Maynard, D.W. (2010). Interviewers and Interviewing. In: *Handbook of Survey Research*, 2nd edition (eds. J.D. Wright and P.V. Marsden), 437–470. Bingley, UK: Emerald Group Publishing Limited.

Schnell, R. and Kreuter, F. (2005). Separating interviewer and sampling-point effects. *Journal of Official Statistics* 21 (3): 389–410.

Schräpler, J.-P. (2004). Respondent behavior in panel studies – a case study for income- nonresponse by means of the German socio-economic panel (SOEP). *Sociological Methods & Research* 33 (1): 118–156.

Schuman, H. and Converse, J.M. (1971). The effects of black and white interviewers on black responses in 1968. *Public Opinion Quarterly* 35 (1): 44–68.

Singer, E. and Kohnke-Aguirre, L. (1979). Interviewer expectation effects: a replication and extension. *Public Opinion Quarterly* 43 (2): 245–260.

Singer, E., Frankel, M.R., and Glassman, M.B. (1983). The effect of interviewer characteristics and expectations on response. *Public Opinion Quarterly* 47: 68–83.

Smit, J.H., Dijkstra, W., and Van der Zouwen, J. (1997). Suggestive interviewer behavior in surveys: an experimental study. *Journal of Official Statistics* 13 (1): 19–28.

Smith, H.L. and Hyman, H. (1950). The biasing effect of interviewer expectations on survey results. *Public Opinion Quarterly* 14 (3): 491–506.

Socio-Economic Panel (SOEP) (2017). *soep.v32 – Sozio-oekonomisches Panel (SOEP), Daten der Jahre 1984-2015*. DOI: http://doi.org/10.5684/soep.v32.SOEP/DIW Berlin.

Stocké, V. (2004). Entstehungsbedingungen Von Antwortverzerrungen Durch Soziale Erwünschtheit. *Zeitschrift für Soziologie* 33 (4): 303–320.

Stoop, I., Billiet, J., and Koch, A. (2010). *Improving Survey Response: Lessons Learned from the European Social Survey*. Wiley.

Strickland, B.R. and Crowne, D.P. (1962). Conformity under conditions of simulated group pressure as a function of the need for social approval. *The Journal of Social Psychology* 58 (1): 171–181.

Sturgis, P. (2009). Using Cross-Classified Multi-level Models to Separate Interviewer from area variability in face-to-face interview surveys. Presentation held at the University of Jyväskylä, Finland, 2009.

Sturgis, P., Allum, N., and Brunton-Smith, I. (2009). Attitudes over time: the psychology of panel conditioning. In: *Methodology of Longitudinal Surveys* (ed. P. Lynn), 113–126. Chichester: Wiley.

Sudman, S., Bradburn, N.M., Blair, E.D. et al. (1977). Modest expectations. The effects of interviewers' prior expectations on responses. *Sociological Methods & Research* 6 (2): 171–182.

Tourangeau, R. and Yan, T. (2007). Sensitive questions in surveys. *Psychological Bulletin* 133 (5): 859–883.

Tourangeau, R., Rips, L.J., and Rasinski, K.A. (2000). *The Psychology of Survey Response*. Cambridge: Cambridge University Press.

Tulving, E. and Schacter, D.L. (1990). Priming and human memory systems. *Science* 247: 301–306.

Turner, M., Sturgis, P., Martin, D. et al. (2015). Can interviewer personality, attitudes and experience explain the design effect in face-to-face surveys? In: *Improving Survey Methods. Lessons from Recent Research* (eds. U. Engel, B. Jann, P. Lynn, et al.), 72–85. New York/London: Routledge.

Uhrig, S. C. N. and Lynn, P. (2008). The Effect of Interviewer Continuity on Measurement Error in Panel Surveys. Institute for Social and Economic Research (ISER), University of Essex. http://panelsurveymethods.files.wordpress.com/2013/03/2008uhriglynn.pdf [Accessed December 20th 2018].

Uhrig, S.C.N. and Sala, E. (2011). When change matters: an analysis of survey interaction in dependent interviewing on the British Household Panel Study. *Sociological Methods & Research* 40 (2): 333–366.

Vassallo, R., Durrant, G.B., Smith, P.W.F. et al. (2015). Interviewer effects on non-response propensity in longitudinal surveys: a multilevel modeling approach. *Journal of the Royal Statistical Society Series A* 178: 83–99.

Voorpostel, M., Tillmann, R., Lebert, F. et al. (2017). *Swiss Household Panel User Guide (1999-2016), Wave 18, December 2017*. Lausanne: FORS.

Warren, J.R. and Halpern-Manners, A. (2012). Panel conditioning in longitudinal social science surveys. *Sociological Methods & Research* 41 (4): 491–534.

Watson, N. and Wooden, M. (2009). Identifying factors affecting longitudinal survey response. In: *Methodology of Longitudinal Surveys* (ed. P. Lynn), 157–181. Chichester: Wiley.

Webster, C. (1996). Hispanic and Anglo interviewer and respondent ethnicity and gender: the impact on survey response quality. *Journal of Marketing Research* 33 (1): 62–72.

Weeks, M.F. and Moore, P.R. (1981). Ethnicity-of-interviewer effects on ethnic respondents. *Public Opinion Quarterly* 45 (2): 245–249.

Weinhardt, M., Schupp, J., and Kreuter, F. (2010). *The Impact of Interviewers' Personality on Measurement Error. Findings from the SOEP 'Interviewer Survey'*. Berlin: DIW. https://www.diw.de/documents/dokumentenarchiv/17/diw_01.c.97230.de/soep_po2009_interviewer.pdf.

West, B.T. and Blom, A.G. (2017). Explaining interviewer effects: a research synthesis. *Journal of Survey Statistics and Methodology* 5 (2): 175–211.

West, B.T. and Olson, K. (2010). How much interviewer variance is really nonresponse error variance? *Public Opinion Quarterly* 74 (5): 1004–1026.

West, B.T., Kreuter, F., and Jaenichen, U. (2013). 'Interviewer' effects in face-to-face surveys: a function of sampling, measurement error, or nonresponse? *Journal of Official Statistics* 29 (2): 277–297.

Williams, J.A. Jr., (1964). Interviewer-respondent interaction: a sthudy of bias in the information interview. *Sociometry* 27 (3): 338–352.

Williams, J.A. Jr., (1968). Interviewer role performance: a further note on bias in the information interview. *Public Opinion Quarterly* 32 (2): 287–294.

Wooden, M., Freidin, S., and Watson, N. (2002). The Household, Income and Labour Dynamics in Australia (HILDA) survey: wave 1. *The Australian Economic Review* 35 (3): 339–348.

14

Improving Survey Measurement of Household Finances: A Review of New Data Sources and Technologies

Annette Jäckle[1], Mick P. Couper[2], Alessandra Gaia[3] and Carli Lessof[4]

[1] *Institute for Social and Economic Research, University of Essex, Colchester, UK*
[2] *Institute for Social Research, University of Michigan, Ann Arbor, MI, USA*
[3] *Department of Sociology and Social Research, University of Milano-Bicocca, Milan, Italy*
[4] *Department of Social Statistics and Demography, University of Southampton, Southampton, UK*

14.1 Introduction

There is widespread interest in the potential of process-generated data and new technologies to provide new data sources for research, whether to replace or supplement traditional surveys. These include data organically generated by social media (e.g. Facebook or Twitter), administrative processes of private companies (e.g. credit rating data), or local and national government (e.g. health, education, or benefit records), and data collected with new technologies (e.g. smartphone apps or sensors). These new data sources already exist or are typically considered cheap to collect; often contain large volumes of data; may provide good quality objective data; may be measured passively; may measure concepts that cannot easily be captured with survey questions; or measure concepts in greater detail. For longitudinal surveys, such data can be particularly valuable, as they can provide more detailed information about changes over time. In reality, there are also several limitations that can affect the suitability of such data for research, most notably access restrictions; coverage of the population of interest; limited covariates; and data that are often designed for a different purpose than is needed for research.

In this chapter we review different process-generated data sources and new technologies that could be used to enhance the measurement of household finances in longitudinal surveys. We examine financial aggregators, loyalty cards, credit and debit cards, credit ratings, barcode scanning, receipt scanning, and mobile apps. Our aim is to contribute to a greater understanding of errors that may arise at different stages of the data generating mechanism (with process-generated data), or

Advances in Longitudinal Survey Methodology, First Edition. Edited by Peter Lynn.

Companion website: www.wiley.com/go/lynn/advancesinlongitudinalsurvey

data collection (with new technologies), and how resulting errors may affect data quality. This will inform research and development into methods to reduce the likelihood and impact of errors. We propose an expanded version of the total survey error (TSE) framework for evaluating these new data sources. We conclude with a discussion of implications for survey practice and research needs.

14.1.1 Why Is Good Financial Data Important for Longitudinal Surveys?

Policy development and evaluation in a number of areas relies on survey data about individual and household financial behaviours and circumstances and, specifically, draws on longitudinal data which captures the dynamics of income, expenditure, debts, and savings. As a result, many of the world's key panel studies – for example, the Panel Study of Income Dynamics (PSID), *Understanding Society*: the UK Household Longitudinal Study (UKHLS), the German Socio-Economic Panel (SOEP), the Canadian Longitudinal and International Study of Adults (LISA), and the Household, Income and Labour Dynamics in Australia survey (HILDA) – have a strong focus on measuring changing financial circumstances of households and how they relate to a range of outcomes.

Alongside the recognition of the importance of measuring individual and household financial data, there is widespread acknowledgement of the difficulties of doing so in surveys. Reporting on finances is a burdensome, tedious, and error-prone task for respondents. Information on finances may be fragmented (e.g. respondents may have multiple bank accounts), and knowledge of income and expenditures may vary between household members (e.g. joint versus separate bank accounts, division of responsibilities for shopping or paying bills, single versus multiple earners in a family). Survey questions about finances are typically considered sensitive; some respondents may not be willing to answer such questions or may misreport their answers, potentially leading to biased estimates. Further, detailed financial information may not be encoded into memory and/or may be difficult to recall; for example, respondents may be unable to report on finances that are only intermittently reviewed (e.g. their mortgage balance or interest payments), may forget small and frequent purchases (e.g. supermarket spending), and/or may find it difficult to locate in a precise time frame infrequent and/or irregular purchases. Given the challenges of asking questions about finances, the number of questions is often limited, both in time (e.g. asked at relatively infrequent intervals) or in terms of content (e.g. high-level detail or aggregated information). Few if any surveys are thus able to collect information simultaneously about the entire budget constraint (i.e. income, expenditure, and changes in assets and debts) (see Hurd and Rohwedder 2009).

These obstacles severely limit the usefulness of financial survey data and, as a result, many important policy questions remain unanswered.

While many enhancements have been made within surveys to improve the quality and completeness of financial data (e.g. electronic diaries, budget reconciliation exercises, feeding forward data, etc.), further improvements are necessary. There is a need to find alternative sources of financial data or ways to measure these important domains while reducing respondent burden and improving data quality.

14.1.2 Why New Data Sources and Technologies for Longitudinal Surveys?

While the need for better financial measurement applies to all kinds of surveys, the issue is particularly critical for longitudinal surveys. The following are some of the reasons that these new measurement tools may be especially important for longitudinal surveys:

1. Longitudinal surveys are often focused on dynamics (i.e. change over time) and new technologies and data sources may permit measurement that fills in the gaps between waves of data collection. That is, there is the promise of greater granularity, in terms of both the time period covered and in the level of detail captured.
2. Interview time in longitudinal surveys is scarce, so gathering data from external sources has great advantages and gathering more detailed information from intensive supplementary activities such as app surveys could enhance the limited data that can be collected in interview.
3. Longitudinal studies have the opportunity to ask committed participants to provide additional types of data because of the trust already established though an ongoing relationship.
4. Longitudinal studies already have a rich set of covariates to evaluate the success of alternative approaches and assess the additional survey errors that may be introduced.
5. Longitudinal studies operate over long time frames, creating the opportunity for the development of innovative approaches, whether through innovation panels (IPs), supplemental data collection, or embedded sub-samples.

Longitudinal surveys, however, also present challenges for the adoption of new sources of measurement. Some of these include:

1. The possible additional burden of participating in between-wave studies and concerns that requests for consents to link to external data sources may affect ongoing participation in the panel.

2. The additional measurement may produce panel conditioning, with respondents changing their behaviour as a result of their participation in the supplemental data collection activity.
3. The need for consistent measures over time constrains longitudinal surveys in ways that are different from cross-sectional surveys.
4. The volatile nature of the emerging technologies and alternative data sources may not be suitable for stable measurement over a long period of time. Changes in technology and in the status of commercial enterprises holding the data, along with changing legal frameworks regarding access to or use of government or commercial data, may limit the longer-term viability of these approaches. In other words, these new data sources may be more suited to shorter periods of measurement.

These opportunities and challenges illustrate the many gaps in research knowledge about the adoption of new methods in longitudinal surveys.

14.1.3 How Can New Technologies Change the Measurement Landscape?

Our interest is in both new data sources and new technologies. The widespread adoption of mobile computing (specifically, smartphones and tablets) and the development of apps for these devices have given consumers a wide array of tools to manage their financial transactions. The way people spend money and the way that they keep track of their expenses are being fundamentally altered by the introduction of these new technologies (see Wang and Wolman 2016). On the one hand, these shifts may hurt traditional survey measurement: problems with recalling expenditures may be exacerbated when the transaction is less memorable – a swipe or tap may be easier to forget than counting out cash or writing a check. On the other hand, the shift from cash to digital spending means that an increasing proportion of transactions can be traced and a digital record of all such payments exists – if consumers would give researchers access to those records, recall and reporting errors could be eliminated and respondent burden reduced.

Our primary focus here is on technologies to supplement, rather than replace, survey data collection (see Groves 2011; Japec et al. 2015). The distinction between these two approaches is important. We illustrate this with a brief example. Gelman et al. (2014) base their research on an existing source of 'organic' data, produced by Check (now part of Mint.com), a financial aggregation and service app (See Section 14.3.1). They analyse data from a sample of about 75 000 Check users, selected at random from the pool of over 1.5 million active US-based users. While the income distribution of the sample approximates that of the US population, the

Check users cannot be viewed as representative of the US population. Nonetheless, the analysis yields important insights into how expenditures match income. A contrasting example comes from Angrisani et al. (2018). Rather than using existing data collected by a financial aggregator, they used the technology to collect detailed additional information on a population representative sample: they invited members of the probability-based internet panel Understanding America Study (UAS) (see https://uasdata.usc.edu) to sign up with a financial aggregator, link data from financial institutions, and then make those data available to the researchers. This approach will yield a smaller data set than the first, and it may suffer from selection biases due to survey non-response or non-participation in the aggregator study, but it has the potential to link detailed financial data to the existing wealth of survey data available on panel members. Both approaches provide opportunities for longitudinal researchers, but they present very different types of errors.

14.2 The Total Survey Error Framework

The TSE framework is widely used to conceptualise and categorise the error sources that may arise in surveys (see Groves and Lyberg 2010). While we frame our discussion within the TSE perspective, we also need to address the question whether that framework is sufficient for understanding the errors that may arise when new data sources and technologies are used to supplement survey data collection. Given that our focus is on the use of such data to supplement, rather than replace, surveys, we believe an expansion of the existing TSE framework is appropriate for understanding the contribution of new technologies and data sources to survey errors.

There is already considerable variation in how the TSE framework is used, with some types of errors (e.g. specification error, processing errors) receiving less attention. Still others have elaborated the framework to incorporate special types of errors common to longitudinal surveys. For example, Lepkowski and Couper (2001) and Couper and Ofstedal (2009) discuss location as a distinct element in longitudinal non-response, related to panel members' propensities to move during waves of data collection. Similarly, Lynn and Lugtig (2017) view non-response error in a longitudinal survey as the cumulative result of initial non-response, item- and wave-non-response, and attrition.

Other recent developments have been to propose alternatives to the TSE framework to accommodate so-called 'Big Data,' 'organic,' or 'found' data (Biemer 2014; Japec et al. 2015; Baker 2017), Twitter data (Hsieh and Murphy 2017), mobile web surveys (Couper et al. 2017), and the linkage of survey and administrative data (Sakshaug and Antoni 2017). Lyberg and Stukel (2017) review the evolution of the TSE framework over time.

A key area where elaboration of the TSE framework is needed relates to non-response. The relatively straightforward notion of cooperation (given contact) that is used in cross-sectional surveys or waves of longitudinal surveys no longer suffices when considering the complex processes involved in using new technologies. With some technologies, participants are actively involved in the data collection process; with others, data are collected passively, after initial consent and compliance has been obtained.

For tasks that require active participation, the process of gaining cooperation and participation consists of several steps. These may or may not be applicable to all new technologies:

1. Initial *willingness*. In some settings, respondents are asked if they are willing to participate in the measurement activities (e.g. Scherpenzeel 2017) before being sent consent materials, instructions, or equipment. Several researchers are exploring stated willingness in the context of new technologies (e.g. Wenz et al. 2019; Révilla et al. 2019; Keusch et al. 2017).
2. *Consent*. At some point, the panel member is asked to consent to the measurement request, either through an explicit request or implicitly by inviting participation. While a lot of research has focused on consent to administrative record linkage (e.g. Sakshaug et al. 2012; Sala et al. 2012), there is as yet little known about the process of consent to new data forms such as transaction data or social media data – especially, there is limited knowledge on how well-informed participants are about the consent they provide.
3. *Initial compliance*. This too may involve different steps depending on the tasks and technologies used. For example, compliance may involve installing an app, accepting and setting up a device, logging into a portal, registering an account, etc.
4. *Ongoing adherence*. This refers to continued use of the technology over the course of data collection. Both item missing data (missed episodes or transactions) and drop-out are issues here, whether due to respondent actions or technical failures. Defining what is viewed as sufficient participation for inclusion in the analytic dataset may vary from study to study.

The last of these can be viewed as a source of measurement error due to missing items and episodes. The process of moving from the initial request to full and complete measurement involves many more steps than is typical in surveys, and selection errors can arise at all steps in this process.

The use of technologies to capture data unobtrusively is often viewed as 'passive' or 'non-reactive' measurement (e.g. Fritsche and Linneweber 2006; Swan 2013). Even if the measurement itself is passive, some action is required on the part of the respondent, whether initially (consenting to data collection, setting up a system, downloading an app) or on an ongoing basis (e.g. continuing to wear or carry the device, keeping the app switched on, etc.). In this context, Wenz et al. (2019) identify some of the characteristics of mobile data collection tasks that may affect willingness to participate in data collection activities using mobile devices. This expands on the conceptual model of causes and correlates of non-response (see Groves and Couper 1998), and illustrates the many ways in which missing observations may come about in such studies.

Measurement error also requires some elaboration in the context of new technologies and data sources. Specification error is often ignored in surveys where the questions are designed to measure the concepts of interest. But in the context of direct measurement or data from other sources, the mismatch between the data available and the concepts of interest takes on added importance. This may occur at the unit level (e.g. whether an expenditure is associated with an individual or household), and at the item level (whether a particular transaction or event represents the intended measurement concept). Further, given that the data often come from a different source over which the researcher has less control, processing or coding errors may become more salient. Finally, apps or tools that provide feedback to respondents about their spending (often used as an incentive for participation) may change their behaviour. This may result in panel conditioning as a source of measurement error (see Lynn and Lugtig 2017; Struminskaya and Bosnjak 2021, Chapter 12 in this book).

With this brief elaboration of the TSE framework in mind, we turn next to a review of selected types of new data sources and technologies for obtaining financial data. We then review what is known about these data sources in the context of the TSE framework.

14.3 Review of New Data Sources and Technologies

Table 14.1 provides a broad overview of the different types of financial data we review, with examples of the potential sources of error for each type. The first four types of data are process generated where the data are collected passively: financial aggregator, loyalty card, credit card, and credit rating data. The latter three are technologies where the respondent is more actively engaged in the data collection task: barcode scanning, receipt scanning, and mobile applications. We elaborate on each data source or new technology briefly in turn below, before reviewing the potential errors. We do not attempt an exhaustive review, but use examples to illustrate the challenges of using these sources in longitudinal surveys and identify gaps in knowledge. More detailed description of each source of data can be found in Jäckle et al. (2019b).

Table 14.1 Comparison of technologies and methods – sources of total survey error.

Type of error		Financial aggregators	Loyalty cards	Credit/debit cards	Credit rating	Barcode and price data	Receipt scanning
Representation/selectivity							
Coverage	Designed (linked to survey)	Has bank account; has internet access	Has loyalty card or is willing to join a loyalty scheme	Has credit card (applied and not been refused); different card issuers attract different types of clients	Has applied for credit with an organisation that reports to credit rating agencies	Has own device; device loaned	Has own device; device loaned
	Organic	Has financial aggregator account	Has loyalty card; different shops attract different types of consumers			Volunteers; often internet users only	Volunteers
Participation	Designed (linked to survey)	Consent and set up account	Consent	Consent	Consent	Set up and use for period of time	Set up and use for period of time
	Organic				N/A		
Measurement							
Specification	Concepts not covered	Payments deducted from gross income; cash income not deposited; categories of cash spending	Spending with other shops or service providers	Non-credit card spending; spending with other credit card	Income; cash not deposited; spending	Non barcode or non-standard barcode items; price information separately	Non-receipt purchases

Unit of analysis	Unclear if individual or joint accounts	Unclear if individual or joint	Credit card account	Usually individual	Unclear if individual or joint	Unclear if individual or joint
Missing or duplicate items/episodes (conditional on specification)	Transactions in accounts that are not linked	Forgot to use the card	None	Credit not reported to agencies; records that are unscorable	Items or entire shops not scanned; duplicates	No receipt; forgot to scan; scan not codeable; duplicate receipt scans
Errors in data captured	Misclassification of categories of income and spending	Barcode error; point of sale (scanning) error	Missing or misclassified categories of spending	Linkage errors; processing errors	Barcode error; error in reported price; error in matched price if price not reported; misclassification of items	Scanning errors
Coding and processing errors	Coding of income and spending categories by account holder or aggregator service	None	Coding error in assigned categories	Coding error in assigned categories	Non-unique barcodes	Data entry errors; coding errors: misclassification of spending categories
Conditioning	Changes in spending behaviour	Vouchers may influence the purchasing behaviour	No	No	Changes in purchase behaviour	Changes in purchase behaviour

14.3.1 Financial Aggregators

Data from financial aggregators provide detailed measures of income and spending, based on the financial transactions recorded in bank statements. Financial aggregators, or account aggregators, are online platforms designed to help consumers manage their finances. Registered users upload the login details for their bank accounts, permitting the platform to automatically retrieve data on transactions. The platforms classify individual transactions in broad categories (e.g. groceries or electricity), provide users with summary information on their income, debts, and investments, and offer tools for budgeting and financial goal setting. Examples of financial aggregators include Money Dashboard (www.moneydashboard.com), OnTrees (www.ontrees.com), Mint (www.mint.com), and Bankin' (www.bankin.com). Some aggregators (e.g. Yodlee, www.yodlee.com) primarily provide services to financial institutions, which in turn offer these services to their customers.

Examples of analyses using existing financial aggregator data from a single provider include Gelman et al. (2014), Olaffson and Pagel (2018), Baugh et al. (2014), Kuchler (2015), and Baker (2015). An example of combining financial data with survey data is The Vanguard Research Initiative (see http://ebp-projects.isr.umich.edu/VRI/index.html), which focuses on older Americans who are clients of the Vanguard Group, an investment firm. The only known effort to start with survey panel members is the UAS (https://uasdata.usc.edu), which invited panellists to create an account with a financial aggregator (Yodlee), with the data then being passed to UAS (see Angrisani et al. 2018).

While the number of users of financial aggregator services (and indeed, the number of such services) is growing, market penetration is still far from universal and likely to be uneven across the population. Initial efforts to enrol survey panel members show the challenge of this approach, but will also provide useful data on selection biases and measurement errors. The regulatory climate may also affect adoption of these services. For example, while the UK Financial Conduct Authority (2015) is working towards reducing restrictions on financial aggregators, many UK banking institutions are discouraging the use of financial aggregators or developing their own tools, officially on the grounds of security concerns and technical issues, but potentially also to limit competition (Sidel 2015; Crosman 2015).

14.3.2 Loyalty Card Data

Data from loyalty cards provide detailed measures of spending at a defined set of outlets, based on purchases recorded by the company's loyalty scheme. Loyalty cards (also called rewards cards, points cards, or club cards) are personalised cards that are usually issued by a retail chain or travel-related company (air, rail, hotels, etc.). Loyalty cards are used for research purposes by matching the card owner's demographic data with transactional data (point-of-sale scanner data, flight booking information, etc.), thereby creating consumer panel data on expenditures linked to the loyalty card (see Tin et al. 2007). Most major supermarket chains have loyalty card schemes, and their use by consumers is widespread. For example, the population of loyalty card users in the largest UK shopping retailer (Tesco) amounts to 1.4 million shoppers (Hornibrook et al. 2015), accounting for about 80% of Tesco shoppers (Worthington and Fear 2009). Similarly, Andreyeva et al. (2012) estimate that 90% of transactions in a large supermarket chain in New England are made using a loyalty card.

While much of the research using loyalty card data is proprietary or internal research, some examples are found in the literature. For example, Andreyeva et al. (2012) used loyalty card data to track beverage purchases among participants in federal assistance programs, while Hornibrook et al. (2015) studied consumer responses to carbon footprint labels. Panzone et al. (2016) combined data from Tesco Clubcard holders with survey responses from Tesco's Shopper Thoughts opt-in panel to study environmental attitudes and food purchases.

Given the specific nature of loyalty cards, they may be more useful for studying behaviour over time within a limited range of goods that are likely to be purchased at a single chain. We know of no studies that have explored the use of generic cards (e.g. loyalty cards that work across networks of stores or credit cards that award points for all types of purchases), but suspect that this still suffers from high coverage error.

14.3.3 Credit and Debit Card Data

Data from credit and debit cards provide detailed measures of all spending, based on purchases made using card payments. The content of credit card data and the frequency with which the information is collected varies between credit card issuers. The information collected can be grouped into four main types: (i) data from the monthly billing statement for the account (i.e. balances, payments, credit limits, interest charges, and interest rates); (ii) data obtained by the issuers from credit bureaus (e.g. external credit scores, number of other credit cards, and combined balances on other credit cards); (iii) administrative data related to each account (e.g. internal credit scores, changes in credit limits, and interest rates); and (iv) socio-demographic data from the credit application such as age, income, and marital status (see Agarwal et al. 2007; Gross and Souleles 2002).

Credit and debit card data have been used extensively for research purposes. Researchers typically use random samples of credit card data, either from a single card issuer (e.g. Agarwal et al. 2007) or combining data from more than one issuer (e.g. Gross and Souleles 2002). One example of the range of analyses that are possible with credit card data comes from the JP Morgan Chase Institute (www.jpmorganchaseinstitute.com), where researchers select samples of credit records from a universe of over 37 million Chase account holders for analysis.

14.3.4 Credit Rating Data

Credit rating data provide detailed measures of credit applications, loan payments, and credit scores, based on information provided to potential lenders (see Avery et al. 2003). Credit rating data are created by credit reporting or credit reference agencies, which collect data on credit applications and credit payments, derive credit scores, and report these to potential lenders. There are several credit reporting agencies in the UK, providing records on over half of the economically active population (Jentzsch 2007). In the UK, Equifax, Experian and Callcredit are the main agencies, while in the US, three nationwide credit reporting agencies (Equifax, Experian, and TransUnion) collect credit data.

Credit rating data have been used for research purposes on their own (see, e.g. Bhardwaj and Sengupta 2011; Brevoort and Cooper 2013; Brevoort and Kambala 2014), or as sampling frames for surveys of consumers (see, e.g. Bucks and Couper 2015; Carroll 2015). For example, credit rating data have been used to create anonymised administrative data panels and a database of mortgages. The New York Federal Reserve Board's Consumer Credit Panel (CCP) is a panel containing quarterly data on 5% of US consumers with a social security number and a credit history, from 1999 to the present (Lee and van der Klaauw 2010). The database includes information on all mortgage instalments and revolving credit accounts, as well as information on all other loan types, including car purchase loans, bankcard loans, student loans, and consumer finance loans (Lee and van der Klaauw 2010). It also includes information from public records, such as records of bankruptcy and tax liens, information reported by collection agencies and individual characteristics such as the year of birth and individual credit score at the end of each quarter. The Consumer Financial Protection Bureau's (CFPB) CCP and the Home Mortgage Disclosure Act (HMDA; see http://www.consumerfinance.gov/data-research/hmda) are other examples (Essene and Byrne 2014; Avery et al. 2014). The CFPB's CCP has been used as a sampling frame for surveys, for example for the Survey of Consumer's View on Debt (see CFPB 2017) and the National Survey of Mortgage Originations (see FHFA n.d.).

As is the case with other proprietary data, access to credit rating data is restricted. For example, access to the New York Fed's CCP and the CFPB's CCP are limited to staff at those agencies. Access to credit rating data for the purposes of sampling is presumably even more restricted, unless the survey is administered by the credit agency. Further, as with other existing data sources, the consistency of reporting and coverage of the data may change over time or in response to external events (such as the 2017 Equifax data breach), adding additional risks for longitudinal measurement.

Having reviewed four potential sources of existing financial data, we next turn to an examination of technologies that require more active engagement on the part of survey respondents in gathering the data.

14.3.5 In-Home Scanning of Barcodes

Data from in-home barcode scanning provides detailed measures of products purchased; associated spending data are derived from receipts or prices reported by the consumer. In-home or household-based barcode scanner data are collected by asking a sample of households to scan the barcodes of purchased items over an extended period of time (Leicester 2013; Zhen et al. 2009). The products scanned are generally food and grocery items and the scanning is normally carried out after each shopping trip, using a barcode reader installed in the sample member's home (Leicester 2013). Non-barcoded items such as loose fruit and vegetables need to be entered manually. New developments include the use of smartphone apps, to reduce the costs of devices and to encourage householders to scan on-the-go purchases such as snacks (Kantar Worldpanel, personal communication). Shoppers are asked to provide additional information, such as when and where the products were bought, as well as transmitting the associated receipts using a digital camera or scanner and email. Some panels, such as the Consumer Network panel, provide scanner cards that can be presented at certain shops at the point-of-sale so that barcode data from purchases are uploaded automatically. Examples of in-home scanner databases are the National Consumer Panel (which in the US combined the Nielsen's Homescan panel and IRI's Consumer Network panel), ShopandScan (run by and commonly referred to as Kantar Worldpanel), and the Nielsen panel in the UK. Commercial panels of this kind grew rapidly to meet the demand for market research into shopper behaviour and, in time, these datasets became available for economic and social research.

As well as the substantial body of research that uses existing household-based barcode scanner datasets to examine individual and household purchasing behaviours, researchers have explored the possibility of introducing barcode scanning methods within government-funded expenditure or food consumption surveys, either to supplement or replace existing surveys (see the Consumer Expenditure Survey Project Gemini, http://www.bls.gov/cex/geminiproject.htm; Leicester 2013; Westat 2011). Scanner data has also been proposed for imputation: respondent burden could be reduced by asking limited questions about aggregate category-level expenditure in budget surveys and supplementing this with more detailed information from scanners (Tucker 2011). However, given the high level of investment needed, dedicated social science projects using scanning technology remain scarce.

14.3.6 Scanning of Receipts

Data from scanned receipts provide detailed spending data on purchases of goods and services (Feenstra and Shapiro 2003).

Receipt scanning involves individuals or households retaining receipts that have been given to them at a point of sale and transmitting them to a central database. Receipt scanning provides information about purchases of a broad range of goods and services and basic information about each item purchased, though not the supplementary detail provided by barcodes. In the past, paper receipts were returned by post, or scanners were used to capture images, which were then sent by email. More recent examples depend on smartphone apps that use the mobile device's camera to photograph receipts and automatically transmit the images collected. Supplementary activities may be required such as annotating receipts before they are scanned or providing information about non-receipted purchases. The scanned receipts are then data-entered or machine-read and coded for analysis.

The use of receipt scanning appears to be a relatively recent growth area in the market research industry and, similar to barcode scanning, appears to be used to obtain detailed expenditure data to understand shopping behaviours. As with barcode scanning, this is usually done with opt-in panels. Examples include ReceiptPal (www.receiptpalapp.com, used by NPD Group), Receipt Hog (see www.receipthog.com), and Ibotta (https://ibotta.com). Until recently, the main academic research applications of receipt scanning were studies using receipts to supplement or replace dietary surveys (e.g. Turrini et al. 2001; Rankin et al. 1998; Ransley et al. 2001). As far as we know, the *Understanding Society* Innovation Panel is the first population representative survey to carry out a receipt scanning exercise to measure household spending on goods and services. Participants were asked to download an app onto their smartphone and scan receipts, directly enter spending without a receipt, or record 'no spend' days over a one month period (Jäckle et al. 2019a).

14.3.7 Mobile Applications and Expenditure Diaries

The final set of new technologies we discuss are not included in the summary table, as they mostly enable methods or approaches we have already discussed or potentially enhance tools that have already been in use for some time.

Smartphone, tablet, and PC applications are an emerging methodology for social research (Volkova et al. 2016; Lessof and Sturgis 2017), especially in the fields of health- and travel-related research. The built-in capabilities of mobile devices to measure location and movement (General Population Sample [GPS]) and activity (accelerometry) make them ideal for passive measurement of activity in these fields. Financial behaviour is harder to track passively, and requires more active involvement from participants, either directly (entering data) or indirectly (providing access to transaction data used for other purposes).

The market for mobile apps for managing and/or monitoring household finances is relatively recent but has grown rapidly and is already very extensive, although the published literature on these developments inevitably lags behind. Apps can broadly be divided into four main categories:

1. *Mobile wallets.* Apps are loaded on phones or tablets to store credit and debit cards, coupons and loyalty cards, and are used for in-store payments. Examples of mobile wallets are Apple Pay, PayPal, and Google Wallet.
2. *Spending diaries and budgeting apps.* These address a similar need to the financial aggregators, but may focus on an element of the household budget such as spending and rely on manually entered information. Examples include Dollarbird, Fudget, and Goodbudget (see Sharf 2016).
3. *Mobile versions of financial or account aggregators.* These apps scrape data from registered users' bank accounts and summarise income and expenses.
4. *Mobile apps that allow uploading and sharing of receipts.* Examples include Receipt Bank, Expensify, Concur, Wally, as well as *cashback* apps (e.g. Top Cashbacks), which allows the upload of till receipts in exchange for cashback on certain products purchased (Sharf 2016).

We discussed the latter two in the context of financial aggregators and receipt scanning. Mobile wallets include data on expenditures, shop check-in, coupons, and accumulated loyalty points (Fundinger 2016). We are not aware of any research studies using mobile wallets, so we focus on spending diaries here.

There is a long history of diary use for measuring time use, travel, food consumption, media use (TV and radio), income and expenditures, and a variety of other topics (Bolger et al. 2003). For several years, organisations have been exploring the replacement of paper-and-pencil diaries with electronic versions. Examples include the UK Living Cost and Food Surveys (Bulman and Kubascikova-Mullen 2015) and the US Consumer Expenditure Survey (National Research Council 2013). Mobile apps are making these electronic diaries more portable and, potentially, better able to measure in-the-moment behaviours. The introduction of mobile apps allows the use of device-contingent protocols, in which participants are prompted by the device to enter data under particular situations such as when they enter a known location for a given supermarket chain using geocoding or beacons (see Iida et al. 2012).

14.4 New Data Sources and Technologies and TSE

In this section we review common sources of error across the new approaches and provide examples of possible errors. More details of errors specific to each source are provided in Table 14.1 (see also Jäckle et al. (2019b)).

14.4.1 Errors of Representation

Financial aggregator, loyalty card, credit/debit card, and credit rating data can be used as stand-alone data sources, linked to survey data, or as sampling frames for surveys. When the data are used as a stand-alone source, coverage error is a key concern. When linked to surveys or used as sampling frames, participation – in the form of survey response and consent to linkage – are instead key concerns. For methods requiring active input from respondents, non-participation is the key source of error affecting representation.

14.4.1.1 Coverage Error

The use of *aggregators* is still relatively scarce. Only about 9% of UAS panellists surveyed in 2014 reported using an online personal financial management service. Data from the *Understanding Society* Innovation Panel in 2016 suggest that less than 1% of respondents in the UK use any financial aggregator. Those who do not have bank accounts are (by definition) excluded, as are those who do not do their banking online. Given this, financial aggregator data are likely to underrepresent those at the low end of the socio-economic scale.

While the use of loyalty cards and credit/debit cards is more widespread, there are still concerns about those who do not use these products, and how they differ from those who do. For instance, *loyalty card* schemes may attract frequent shoppers and/or customers who are more sensitive to prices and discounts (Cortiñas et al. 2008). In a survey in Australia, Worthington and Fear (2009) found that 83% of adults reported having at least one loyalty card, with the most common card being held by 59% of respondents. Women held an average of 2.02 loyalty cards, compared with 1.29 for men. Loyalty card possession also increased with age.

In similar fashion, *credit and debit cards* are designed to serve different segments of the market. For example, the JP Morgan Chase Institute notes that its card holders differ from the general US population on several key demographics, including income, age, and gender. While they adjust the samples to account for these differences (see, e.g. Farrell and Greig 2016), other differences, such as the under-representation of those who use paper instruments (checks, money orders, or cash) for payments (see Farrell and Greig 2017) may be harder to correct through post-stratification.

Credit rating data are collected on consumers, often without their choice (or knowledge, as the 2017 Equifax breach revealed), but still suffer from differential coverage in that they are limited to consumers with a credit history. Brevoort, Grimm and Kambara (2015) estimated that the CFPB's CCP did not cover 11% of the US population because of credit invisibility, and 8.3% because credit records were considered 'unscorable,' given insufficient information. Minorities are under-represented in the CCP, as are low-income communities. There is also differential coverage by age: while over 80% of young adults aged 18–19 are credit invisible or have unscored credit, this percentage drops to 40% for adults aged 20–24 years old, and further decreases until age 60–64.

A related coverage concern for longitudinal surveys is the stability of these data sources. In many cases, these are competitive markets that change in response to shifting demand, the introduction of new products, and efforts to attract new customers. Customer data are viewed as proprietary, so we know of no information on the volatility of these financial products, nor of the extent of customer turnover within and across different financial products.

14.4.1.2 Non-Participation Error

When new technologies and existing data sources are used to augment survey data, non-response error becomes a primary focus. As noted earlier, there are many sources of non-response when adding data linkages or new technologies to surveys. Angrisani et al. (2018) report on preliminary results from their study asking panel participants to sign up for a *financial aggregator*. Of those invited, 65% consented to the study. Of those 68% (or 32% of those invited) signed up for the financial aggregator, of whom 38% (or 12% of those initially invited) added one or more financial institutions to their account. They found significant differences by age and education in each of these stages of response, and differences in initial consent by income. In addition, those who already used the internet for banking were more likely to sign up.

Similar evidence on non-response error comes from Jäckle et al.'s (2019a) study asking *Understanding Society* Innovation Panel (IP) respondents to download an app and *scan receipts* for a month. They report that 17% of IP respondents completed the registration survey, with 13% downloading the app and using it at least once, and 10% doing so for the month of the study. Participation in the spending study was significantly associated with use of the internet, access to a suitable device (smartphone or tablet) and frequency of smartphone use. Further, women were more likely to participate than men, and participation decreased monotonically with age. Jäckle et al. (2019a) report significant participation biases with respect to some financial behaviours (e.g. using an app for banking, keeping a budget on a computer, using a store loyalty card) but not on several key income and spending-related indicators.

We know of no attempt reported in the literature to ask survey respondents for permission to link to *loyalty card* or *credit/debit card* data. However, in a study starting with a sample of loyalty card holders, only 19.2% of those invited to the survey completed it (Panzone et al. 2016). Similarly, a study using the CCP *credit rating data* as a frame to survey those at risk of debt collection achieved a response rate just under 20% (see CFPB 2017); a survey of mortgage holders from the same data source yielded a slightly higher response rate (see FHFA). However, one advantage of using process generated data as a sampling frame is that the administrative records are available for both survey respondents and non-respondents. The administrative data can be used to evaluate non-response bias and construct adjustment weights (Bucks and Couper 2015).

There is little evidence on non-response error for *in-home barcode scanner* panels, as these typically use quota sampling methods to invite participants (Tucker 2011; Zhen et al. 2009; USDA 2009). They have been criticised for providing limited information about sampling methods, response rates, and attrition (Westat 2011; Perloff and Denbaly 2007; Leicester 2012). Some differences have been observed between scanner panels and budget surveys although these are somewhat inconsistent. For example, in the US, Homescan households were on average smaller compared to the Consumer Expenditure Survey (CE) (Huffman and Jensen, 2004) but in the UK, Kantar Worldpanel households had more members on average than the Living Costs and Food Survey (Leicester and Oldfield 2009). Similarly, there are differences between panels with regard to income. While panel companies derive weights to ensure that the households are demographically representative (see Leicester 2013), there are likely to be unobservable differences in the characteristics of households in different data sources. For example, households who agree to collect scanner data are observed to be more price conscious than the general population (Leicester 2012).

With the exception of Jäckle et al. (2019a) discussed above, *receipt-based studies* have typically been based on relatively small sample sizes, often in localised geographical areas (French et al. 2008, 2009) based on purposive samples of population sub-groups. Further, the receipt scanning activities carried out for market research are typically based on volunteers and so also do not provide a good basis from which to assess response rates or non-response error. For example, Ransley et al. (2001) recruited a sample of Tesco Clubcard holders in one city in the UK to participate in a study involving the collection of receipts and completion of a shopping diary for non-receipt purchases over 28 days. Of those invited, 52% expressed initial interest, of whom 63% (or 34% of those invited) reported spending the majority of their food expenses at Tesco and were willing to take part in the study. Of those meeting the criteria, 75% (or 27% of invitees) participated in the study, but the degree of participation is not specified. Ransley et al. (2001) concluded that the responding sample was broadly representative of their consumer base but noted that people aged 30–59, women, and Social Class II were over-represented. Similarly, Smith et al. (2012) reported a 71% participation rate over four weeks in a study involving the collection of receipts and reporting of purchases with a group of targeted volunteers from low-income households. They also reported that low food security households, where food expenditure would likely have been lower, were less likely to participate.

While there is a long history of research on using paper *expenditure diaries*, the non-response challenges associated with their electronic equivalents are only now starting to be explored. Initial work on correlates of willingness to perform a variety of tasks using mobile apps and online tools (see, e.g. Wenz et al. 2019; Révilla et al. 2019) is a useful first step in understanding the process of compliance.

14.4.2 Measurement Error

Measurement error can stem from many sources. Some errors occur by design, such as if what is measured does not match the concept or unit of analysis of interest to the researcher (specification error). Other errors are the result of respondent behaviours (such as missing or duplicate transactions or panel conditioning), while yet other errors occur during the data generating process or at the hand of the researcher preparing data for analysis (processing and coding errors). The various sources of measurement error apply to all data sources and technologies reviewed.

14.4.2.1 Specification Error

Specification errors, or the mismatch between the constructs of interest and the available measures, are of particular concern for existing data sources. For example, the income measured by *financial aggregators* (money deposited into accounts) may not match the definition of interest for research purposes (e.g. gross income). Cash income not deposited into an account is excluded. Correspondingly, spending only includes payments made from a bank account and cash spending cannot be classified. Similarly, not all credit accounts are represented in *credit agency reporting* (Avery et al. 2003). Small retail, mortgage, finance companies, and some government agencies may not report to credit reporting agencies. Specification error may also be present in the more active technologies. For example, there may be a lack of alignment between *receipts* and total spending.

Differences in the unit of interest between the existing data and the analytic goals may also lead to measurement error. For *loyalty* and *credit/debit card* data, the unit of observation is usually the account, rather than the individual or household. Within households, multiple people may share the same account, multiple cards may be issued for the same account, and individuals may have multiple accounts (Agarwal et al. 2007; Gross and Souleles 2002). The unit of observation for *credit rating data* is the credit record, which again may not coincide with either individuals or households. While joint accounts are identified as such, co-signers of loans may or may not be part of the same household. The more active methods may suffer from similar problems. Separating individual from household income and expenditures, and separating these from spending for or by persons outside the household may be difficult. Little is known about such mismatches.

14.4.2.2 Missing or Duplicate Items/Episodes

Missed transactions or events may be of two types. First, the financial behaviour of interest may not cover all transactions (e.g. not all spending is done using credit/debit cards). Second, the survey participant may forget to record or deliberately misreport certain activities. Such errors may be systematic (e.g. exclusion of certain types of transactions or purchases) or more stochastic (e.g. forgetting).

For *credit/debit cards*, payments made through means other than these cards (or with cards held by other banks) will be missed. For example, comparing data to national estimates, Farrell and Greig (2016) estimate that about 71% percent of gasoline (petrol) spending in the US occurs on debit and credit cards, while only 58% of non-gas spending occurs on such cards. Similar problems exist for *loyalty cards*. In one study in Switzerland (Hauser et al. 2013), more than two-thirds of survey respondents indicated that they spent more than half of their total food expenditure in the retail chain whose loyalty card data they studied. That chain, in turn, claimed a food market share of around 27% in Switzerland. This leaves a large proportion of expenditures not covered by the loyalty card.

Existing *in-home barcode scanning* methods are typically limited to food and grocery products including cleaning products and personal care items. Leicester (2012) estimated that the set of products contained in the Worldpanel data make up about 18% of all non-housing expenditure (falling from around 24% in 1996) and that 'at best just over a third of total expenditures (by CPI weight) appear to be readily amenable to in-home scanner technology'. Duly et al. (2003), as reported in Leicester (2013), compared Nielsen Homescan data in the US to Consumer Expenditure Survey (CE) diary data from 2000, and found that only 45% of the CE diary items (but 83% of food items) were covered in the Homescan data (see Leicester 2012).

Similarly, for *receipt-scanning* apps, certain types of payments do not generate receipts. Spending outside structured shopping environments, for example at market stalls, and informal expenditures often occur without a receipt being provided. A growing number of shops only give receipts on request. Relying solely on receipts will result in gaps in spending records, and the accuracy of estimates of expenditure will depend on how much effort is put into gathering information about other types of expenditures. In the *Understanding Society* spending study, 62% of all reports of purchases over a one-month study period were based on scanned receipts with the remainder being entered in summary directly into the app, though it is not possible to distinguish the different reasons why receipts were not always used (Lessof et al. 2016). Receipts may also be produced but not collected in-store, or are lost or forgotten before scanning (Smith et al. 2013). Participants may fail to scan single receipts or may miss days or even whole weeks of reporting. Minor purchases at convenience stores or consumed 'on the go' (Leicester and Oldfield 2009) may be more likely to be missed. Participants may also be less likely to scan receipts that reveal spending on potentially sensitive items such as alcohol or parking fines relative to summary reports, which only identify the category of spending (such as food and drink, or travel).

14.4.2.3 Data Capture Error

The quality of financial transaction data is largely unknown. Reporting or data capture errors for individual transactions in *linked bank accounts* may be less likely than in survey data (Baker 2015), as both individuals and institutions have a vested interest in correcting errors. The number of errors in *credit reports* varies widely, from 0.2% to 70% (see Consumer Federation of America 2002; Jentzsch 2007). More recently, Smith et al. (2013) found that in a sample of US consumers, 26% had at least one error in their credit reports. While some of these errors are minor, others are sufficiently large to lead to the denial of credit or to less favourable credit terms.

14.4.2.4 Processing or Coding Error

There are also several sources of potential processing or coding error. For example for *credit rating data*, Avery et al. (2014) found both duplicate and missing records in the FRB's CCP due to linkage or processing errors. *Barcode scanner data* are subject to many of the same food composition database errors as traditional dietary assessment methods (Eyles et al. 2010). The process of linking price data to product information from barcodes may also lead to errors. Current *receipt-scanning* technology requires transmission, transcription, and coding, potentially introducing processing errors. A receipt may not have sufficient detail to allow for an item to be correctly categorised, or may be unclear or incomplete, leading to coding errors, whether human coders or machine coding is used. We are not aware of any literature on the challenges of consistently coding receipts and validating these activities.

14.4.2.5 Conditioning Error

Both *barcode scanning* and *expenditure diaries* also provide evidence of decreasing compliance over the course of participation, resulting in declines in reported expenditures (Weerts and Amoran 2011; Smith et al. 2013). For example, Leicester and Oldfield (2009) found that expenditures in a scanning study were highest in the first few weeks of participation but fell away slightly such that after about six months households spent, on average, about 5% less than in their first week. That said, this decline is slight relative to a 9% reduction between the first and second week of the Canadian Food Expenditure Survey (Ahmed et al. 2006; see also Silberstein and Scott 1991). Evidence from the *Understanding Society* receipt scanning study also showed that the amount of spending reported over a four-week period declined (Lessof et al. 2016). In addition, panel conditioning may occur when study participants change their spending behaviour as a result of reflecting on their spending patterns, although it is hard to disentangle changes in spending behaviour from changes in reporting behaviour.

14.5 Challenges and Opportunities

Process-generated data and new technologies present exciting new opportunities to rethink the way data about household finances are collected in longitudinal surveys. Depending on the data source or technology, they can provide a broad range of financial information (such as income and spending data from bank accounts), or a narrow view of particular aspects (such as spending in a particular outlet from loyalty card data). In all cases, they can provide more granular detail over longer periods of time than can be collected with survey questions or paper diaries. The financial transactions that are recorded by a particular process or technology are also likely to be recorded more accurately than when respondents are asked to recall their transactions.

Our review highlights emerging examples of how these new opportunities can be used to supplement survey data: from linking survey data with process generated data, to using process generated data as sampling frames, to asking sample members to use new technologies to collect supplementary data. In all cases, the implications for TSE require careful thought and examination.

A key challenge for representation is that of access to or use of the technologies or data sources described here. In contrast to health surveys, significantly fewer people use mobile devices to keep track of their expenditures than use health tracking apps or devices. In addition, people use a wide variety of instruments to make financial transactions, and measuring financial behaviour is more complex than, say, measuring physical activity or travel. Further, financial behaviour is viewed as more sensitive than many other things we ask about (Couper et al. 2010), presenting challenges for gaining consent to measure these activities, whether using active or passive approaches.

Additional challenges of particular concern for longitudinal surveys, relate to the stability of long- term measurement:

- *Changing environments.* Changing legal, financial, and commercial environments affect data sharing and access arrangements and costs, stability of existing data sources, and the like.
- *Changing financial behaviours.* For instance, rising levels of interest in financial aspects of the 'quantified self' may lead to greater use of financial aggregators or spending diaries. Similarly, the way people manage their finances is changing, e.g. increased direct debits and standing orders for regular spending, more spending online, greater use of contactless, or mobile payment methods.
- *Increasing capability of technology.* Examples include the ability to scrape online and mobile payments into aggregators, increasing scope to remind people in store to enter data or to provide transactional prompts.
- *Possible changes in willingness to share data.* Potentially, this could occur in response to external events like identity theft, data breaches, hacking, and the like.

As the digital economy continues to grow, and as every aspect of peoples' financial lives is increasingly leaving a digital trace, some of these concerns may diminish. While these new data sources and technologies may not be ready to supplement or replace designed data collection in longitudinal surveys in the short run, continued research on the approaches we reviewed here will help identify the magnitude of the errors that may arise, and identify ways to mitigate those errors.

We have identified a number of key gaps in the literature. Our knowledge of the various error sources associated with these alternative approaches to measuring financial behaviour is very limited. Only recently have there been attempts to try new ways of gathering such data (e.g. Angrisani et al. 2018; Jäckle et al. 2019a), while in other areas (e.g. in-home scanning) a lot of the data is proprietary and little is known about the error sources. There is much work that remains to be done to understand how those who are willing to use these technologies may differ from those who are not, how to increase the consent compliance, and adherence rates for using these tools, and how participating in such additional measurement activities may affect ongoing panel participation (in terms of both attrition and panel conditioning). Finally, the measurement properties of these new tools and technologies are not yet well understood, and careful comparisons of the data to alternative measures or benchmarks is needed. Nonetheless, there are many potential uses of such data other than fully replacing survey estimates for general populations. It is important to evaluate these data sources in terms of their fitness for use (Biemer and Lyberg 2003), rather than against some absolute standard.

As these technologies continue to evolve and are used by larger segments of the population, and as our knowledge of how best to implement these methods grows, we are likely to see increased adoption of these new data sources and technologies to enhance and extend the measurement of a range of financial behaviours in longitudinal surveys.

Acknowledgements

This work was funded by the UK Economic and Social Research Council Transformative Research scheme and the National Centre for Research Methods (grant number ES/N006534/1), as well as by an NCRM International Visitor Exchange Scheme grant funding research visits by Mick Couper to the University of Essex.

References

Agarwal, S., Liu, C., and Souleles, N.S. (2007). The reaction of consumer spending and debt to tax rebates – evidence from consumer credit data. *Journal of Political Economy* 115 (6): 986–1019.

Ahmed, N., Brzozowski, M., and Crossley, T. (2006). Measurement errors in recall food consumption data. In: *IFS Working Paper 06/21*. London: Institute for Fiscal Studies www.ifs.org.uk/publications/3752.

Andreyeva, T., Luedicke, J., Henderson, K.E., and Tripp, A.S. (2012). Grocery store beverage choices by participants in federal food assistance and nutrition programs. *American Journal of Preventive Medicine* 43 (4): 411–418.

Angrisani, M., Kapteyn, A., and Samek, S. (2018). Real time measurement of household electronic financial transactions in a population representative panel. Paper prepared for the 35th IARIW General Conference, Copenhagen. http://www.iariw.org/copenhagen/angrisani.pdf.

Avery, R.B., Calem, P.S., Canner, G. B., and Bostic, R. W. (2003). An overview of consumer data and credit reporting. Federal Reserve Bulletin 2003, February. Washington: The Federal Reserve Board. https://www.federalreserve.gov/pubs/bulletin/2003/0203lead.pdf.

Avery, R.B., Courchane, M.L., and Zorn, P. (2014). The creation of the National Mortgage Database. In: *What Counts: Harnessing Data for America's Communities* (ed. Federal Reserve Bank of San Francisco and Urban Institute), 124–136. San Francisco: Federal Reserve Bank of San Francisco.

Baker, S.R. (2015). Debt and the consumption response to household income shocks. SSRN: https://ssrn.com/abstract=2541142.

Baker, R. (2017). Big data: a survey research perspective. In: *Total Survey Error in Practice* (eds. P.P. Biemer, E. de Leeuw, S. Eckman, et al.), 47–70. New York: Wiley.

Baugh, B., Ben-David, I., and Park, H. (2014). Disentangling financial constraints, precautionary savings, and myopia: household behaviour surrounding federal tax returns. In: *NBER Working Paper No. 19783*. Cambridge: National Bureau of Economic Research http://www.nber.org/papers/w19783.pdf.

Bhardwaj, G. and Sengupta, R. (2011). Subprime loan quality. In: *Working Paper 2008-036E*. St. Louis, MO: Federal Reserve Bank of St. Louis https://files.stlouisfed.org/files/htdocs/wp/2008/2008-036.pdf.

Biemer, P.P. (2014). Toward a total error framework for big data. Paper presented at the Annual Meeting of the American Association for Public Opinion Research, Anaheim, CA.

Biemer, P.P. and Lyberg, L.E. (2003). *Introduction to Survey Quality*. New York: Wiley.

Bolger, N., Davis, A., and Rafaeli, E. (2003). Diary methods: capturing life as it is lived. *Annual Review of Psychology* 54 (1): 579–616.

Brevoort, K.P. and Cooper, C.R. (2013). Foreclosure's wake: the credit experiences of individuals following foreclosure. *Real Estate Economics* 41 (4): 747–792.

Brevoort, K.P. and Kambala, M. (2014). Data point: medical debt and credit scores. Washington: Consumer Financial Protection Bureau, Office of Research. https://www.consumerfinance.gov/data-research/research-reports/data-point-medical-debt-and-credit-scores.

Brevoort, K.P., Grimm, P., and Kambara, M. (2015). *Data Point: Credit Invisibles*. Washington: Consumer Financial Protection Bureau, Office of Research https://files.consumerfinance.gov/f/201505_cfpb_data-point-credit-invisibles.pdf.

Bucks, B. and Couper, M.P. (2015). Using information from credit records to improve survey data. Paper presented at the 6th conference of the European Survey Research Association, Reykjavik.

Carroll, C. (2015). The CFPB consumer credit panel: direct use and as a sampling frame. Paper presented at the Federal Economic Statistics Advisory Committee Meeting, June, Washington. https://www2.census.gov/about/partners/fesac/2015-06-12/Carroll_Presentation.pdf

Consumer Federation of America (2002). *Credit Score Accuracy and Implications for Consumers*. Washington, DC: Consumer Federation of America National Credit Reporting Association https://consumerfed.org/pdfs/121702CFA_NCRA_Credit_Score_Report_Final.pdf.

Consumer Financial Protection Bureau (CFPB) (2017). Consumer Experiences with Debt Collection: Findings from the CFPB's Survey of Consumer Views on Debt. Washington, DC: Consumer Financial Protection Bureau http://files.consumerfinance.gov/f/documents/201701_cfpb_Debt-Collection-Survey-Report.pdf.

Cortiñas, M., Elorz, M., and Múgica, J.M. (2008). The use of loyalty-cards databases: differences in regular price and discount sensitivity in the brand choice decision between card and non-card holders. *Journal of Retailing and Consumer Services* 15 (1): 52–62.

Couper, M.P. and Ofstedal, M.B. (2009). Keeping in contact with mobile sample members. In: *Methodology of Longitudinal Surveys* (ed. P. Lynn), 183–203. New York: Wiley.

Couper, M.P., Singer, E., Conrad, F.G., and Groves, R.M. (2010). Experimental studies of disclosure risk, disclosure harm, incentives, and survey participation. *Journal of Official Statistics* 26 (2): 287–300.

Couper, M.P., Antoun, C., and Mavletova, A. (2017). Mobile web surveys: a total survey error perspective. In: *Total Survey Error in Practice* (eds. P.P. Biemer, E. de Leeuw, S. Eckman, et al.), 133–154. New York: Wiley.

Crosman, P. (2015). The truth behind the hubbub over screen scraping. *American Banker*. 21 November. https://www.americanbanker.com/news/the-truth-behind-the-hubbub-over-screen-scraping.

Duly, A., Garner, T., Keil, E. et al. (2003). The Consumer Expenditure Survey and AC Nielsen Survey: a data comparison study. Internal document, Bureau of Labor Statistics, Washington, DC.

Essene, R. and Byrne, M. (2014). Applying technology advances to improve public access to mortgage data. In: *What Counts: Harnessing Data for America's Communities* (ed. Federal Reserve Bank of San Francisco and Urban Institute), 138–147. San Francisco: Federal Reserve Bank of San Francisco.

Eyles, H., Jiang, Y., and Mhurchu, C.N. (2010). Innovations in dietary assessment technology. *Journal of the American Dietetic Association* 110 (1): 106–110.

Farrell, D. and Greig, F. (2016). *The Consumer Response to a Year of Low Gas Prices: Evidence from 1 Million People*. New York: JP Morgan Chase Institute https://www.jpmorganchase.com/corporate/institute/report-consumer-response-low-gas-prices.htm.

Farrell, D. and Greig, F. (2017). *Coping with Costs: Big Data on Expense Volatility and Medical Payments*. New York: JP Morgan Chase Institute https://www.jpmorganchase.com/corporate/institute/report-coping-with-costs.htm.

Federal Housing Financial Agency (FHFA) (n.d.). National survey of mortgage originations. Washington: Federal Housing Financial Agency. https://www.fhfa.gov/PolicyProgramsResearch/Programs/Pages/National-Survey-of-Mortgage-Originations.aspx.

Feenstra, R.C. and Shapiro, M.D. (eds.) (2003). *Scanner Data and Price Indexes*. Chicago: University of Chicago Press.

Financial Conduct Authority (2015). Cash Savings Market Study Report: Part I: Final Findings Part II: Proposed Remedies. Market Study Report No. MS14/2.3. London: Financial Conduct Authority www.fca.org.uk/publication/market-studies/cash-savings-market-study-final-findings.pdf.

French, S.A., Shimotsu, S.T., Wall, M., and Gerlach, A.F. (2008). Capturing the spectrum of household food and beverage purchasing behavior: a review. *Journal of the American Dietetic Association* 108 (12): 2051–2058.

French, S.A., Wall, M., Mitchell, N.R. et al. (2009). Annotated receipts capture household food purchases from a broad range of sources. *International Journal of Behavioral Nutrition and Physical Activity* 6: 37.

Fritsche, I. and Linneweber, V. (2006). Nonreactive methods in psychological research. In: *Handbook of Multimethod Measurement in Psychology* (eds. M. Eid and E. Diener), 189–203. Washington, DC: American Psychological Association.

Fundinger, D. (2016). Mobile wallet analytics and personalization: the value of data. *Mobile Business Insights*. 10 August. https://mobilebusinessinsights.com/2016/08/mobile-wallet-analytics-personalization-value-data.

Gelman, M., Kariv, S., Shapiro, M.D. et al. (2014). Harnessing naturally-occurring data to measure the response of spending to income. *Science* 345 (6193): 212–215.

Gross, D.B. and Souleles, D.S. (2002). Do liquidity constraints and interest rates matter for consumer behavior? Evidence from Credit Card Data. *Quarterly Journal of Economics* 117 (1): 149–185.

Groves, R.M. (2011). Three eras of survey research. *Public Opinion Quarterly* 75 (5): 861–871.

Groves, R.M. and Couper, M.P. (1998). *Nonresponse in Household Interview Surveys*. New York: Wiley.

Groves, R.M. and Lyberg, L.E. (2010). Total survey error: past, present, and future. *Public Opinion Quarterly* 74 (5): 849–879.

Hauser, M., Nussbeck, F.W., and Jonas, K. (2013). The impact of food-related values on food purchase behavior and the mediating role of attitudes: a Swiss study. *Psychology & Marketing* 30 (9): 765–778.

Hornibrook, S., May, C., and Fearne, A. (2015). Sustainable development and the consumer: exploring the role of carbon labelling in retail supply chains. *Business Strategy and the Environment* 24 (4): 266–276.

Hsieh, Y.P. and Murphy, J. (2017). Total Twitter error: decomposing public opinion measurement on Twitter from a Total Survey Error perspective. In: *Total Survey Error in Practice* (eds. P.P. Biemer, E. de Leeuw, S. Eckman, et al.), 23–46. New York: Wiley.

Huffman, S. and Jensen, H. (2004). Demand for enhanced foods and the value of nutritional enhancements of food: the case of margarines. Paper presented at the American Agricultural Economics Association Annual Meeting, Denver, Colorado (August 1-4). http://ageconsearch.umn.edu/bitstream/20205/1/sp04hu05.pdf.

Hurd, M. and Rohwedder, S. (2009). Methodological innovations in collecting spending data: the HRS consumption and activities mail survey. *Fiscal Studies* 30 (3/4): 435–459.

Iida, M., Shrout, P.E., Laurenceau, J.-P. et al. (2012). Using diary methods in psychological research. In: *APA Handbook of Research Methods in Psychology: Vol 1 Foundations, Planning, Measures, and Psychometrics* (ed. H. Cooper), 277–305. Washington: American Psychological Association.

Jäckle, A., Burton, J., Couper, M.P., and Lessof, C. (2019a). Participation in a mobile app survey to collect expenditure data as part of a large-scale probability household panel: coverage and participation rates and biases. *Survey Research Methods* 13 (1): 23–44.

Jäckle, A., Gaia, A., Lessof, C., and Couper, M.P. (2019b). A review of new technologies and data sources for measuring household finances: Implications for total survey error. In: *Understanding Society Working Paper 2019-02*. Colchester: University of Essex www.understandingsociety.ac.uk/research/publications/525666.

Japec, L., Kreuter, F., Berg, M. et al. (2015). Big data in survey research: AAPOR task force report. *Public Opinion Quarterly* 79 (4): 839–880.

Jentzsch, N. (2007). *Financial Privacy: An International Comparison of Credit Reporting Systems*. New York: Springer-Verlag.

Keusch, F., Antoun, C., Couper, M.P. et al. (2017). Willingness to participate in passive mobile data collection. Paper presented at the annual meeting of the American Association for Public Opinion Research, New Orleans.

Kuchler, T. (2015). Sticking to your plan: empirical evidence on the role of present bias for credit card paydown. In: *SIEPR Discussion PaperNo. 12-025*. Stanford, CA: Stanford Institute for Economic Policy Research. https://ssrn.com/abstract=2629158.

Lee, D. and van der Klaauw, W. (2010). An Introduction to the FRBNY Consumer Credit Panel. In: Federal Reserve Bank of New York Staff Report no. 479. New York: Federal Reserve Bank of New York. https://papers.ssrn.com/sol3/papers.cfm?abstract_id=1719116.

Leicester, A. (2012). How might in-home scanner technology be used in budget surveys? In: *IFS Working Paper W12/01*. London: Institute for Fiscal Studies www.ifs.org.uk/wps/wp1201.pdf.

Leicester, A. (2013). The potential use of in-home scanner technology for budget surveys. In: *NBER Working Paper No. 19536*. Cambridge, MA: National Bureau of Economic Research. Available at: http://www.nber.org/papers/w19536.

Leicester, A. and Oldfield, Z. (2009). Using scanner technology to collect expenditure data. *Fiscal Studies* 30 (3/4): 309–337.

Lepkowski, J.M. and Couper, M.P. (2001). Nonresponse in longitudinal household surveys. In: *Survey Nonresponse* (eds. R.M. Groves, D. Dillman, J. Eltinge, et al.), 259–272. New York: Wiley.

Lessof, C. and Sturgis, P. (2017). New kinds of survey measurement. In: *The Palgrave Handbook of Survey Research* (eds. D.L. Vannette and J.A. Krosnick), 165–173. New York: Palgrave Macmillan.

Lessof, C., Jäckle, A., and Couper, M.P. (2016). Data quality from a mobile app survey to collect expenditure data as part of a large-scale probability household panel survey. Paper presented at the 7th Conference of the European Survey Research Association, Lisbon.

Lyberg, L.E. and Stukel, D.M. (2017). The roots and evolution of the total survey error concept. In: *Total Survey Error in Practice* (eds. P.P. Biemer, E.D. de Leeuw, S. Eckman, et al.), 3–22. New York: Wiley.

Lynn, P. and Lugtig, P.J. (2017). Total survey error for longitudinal surveys. In: *Total Survey Error in Practice* (eds. P.P. Biemer, E.D. de Leeuw, S. Eckman, et al.), 279–298. New York: Wiley.

National Research Council (2013). *Measuring What We Spend: Toward a New Consumer Expenditure Survey*. Washington, DC: National Academies Press.

Olaffson, A. and Pagel, M. (2018). The liquid hand-to-mouth: evidence from a personal finance management software. *Review of Financial Studies* 31 (11): 4398–4446.

Bulman, K.G. and Kubascikova-Mullen, J. (eds.) (2015). *Living Costs and Food Survey Technical Report for survey year: January – December 2014, Great Britain and Northern Ireland*. Newport: Office for National Statistics Social Surveys.

Panzone, L., Hilton, D., Sale, L., and Cohen, D. (2016). Socio-demographics, implicit attitudes, explicit attitudes, and sustainable consumption in supermarket shopping. *Journal of Economic Psychology* 55: 77–95.

Perloff, J.M. and Denbaly, M. (2007). Data needs for consumer and retail firm studies. *American Journal of Agricultural Economics* 89 (5): 1282–1287.

Rankin, J.W., Winett, R.A., Anderson, E.S. et al. (1998). Food purchase patterns at the supermarket and their relationship to family characteristics. *Journal of Nutrition Education* 30 (2): 81–88.

Ransley, J.K., Donnelly, J.K., Khara, T.N. et al. (2001). The use of supermarket till receipts to determine the fat and energy intake in a UK population. *Public Health Nutrition* 4 (6): 1279–1286.

Révilla, M., Couper, M.P., and Ochoa, C. (2019). Willingness of online panelists to perform additional tasks. *Methods, Data, and Analysis* 13 (2): 29. https://doi.org/10.12758/mda.2018.01.

Sakshaug, J.W. and Antoni, M. (2017). Errors in linking survey and administrative data. In: *Total Survey Error in Practice* (eds. P.P. Biemer, E.D. de Leeuw, S. Eckman, et al.), 557–573. New York: Wiley.

Sakshaug, J.W., Couper, M.P., Ofstedal, M.B., and Weir, D.L. (2012). Linking survey and administrative records: mechanisms of consent. *Sociological Methods and Research* 41 (4): 535–569.

Sala, E., Burton, J., and Knies, G. (2012). Correlates of obtaining informed consent to data linkage: respondent, interview and interviewer characteristics. *Sociological Methods and Research* 41 (3): 414–439.

Scherpenzeel, A. (2017) Mixing Online Panel Data Collection with Innovative Methods. Pp. 27-49 in Methodische Probleme Von Mixed-Mode-Ansätzen in Der Umfrageforschung, edited by S. Eifler and F. Faulbaum. Wiesbaden: Springer.

Sharf, S. (2016). 12 free apps to track your spending and how to pick the best one for you. *Forbes* 2 March. https://www.forbes.com/sites/samanthasharf/2016/03/02/12-free-apps-to-track-your-spending-and-how-to-pick-the-best-one-for-you/#15e812a54453.

Sidel, R. (2015). Big banks lock horns with personal-finance web portals. J.P. Morgan, Wells Fargo are snarling the flow of data to popular websites that help consumers manage their finances. *The Wall Street Journal*. 4 November. https://www.wsj.com/articles/big-banks-lock-horns-with-personal-finance-web-portals-1446683450 [Accessed 25/01/2018].

Silberstein, A.R. and Scott, S. (1991). Expenditure diary surveys and their associated errors. In: *Measurement Errors in Surveys* (eds. P.P. Biemer, R.M. Groves, L.E. Lyberg, et al.), 303–326. New York: Wiley.

Smith, C., Parnell, W.R., Brown, R.C., and Gray, A.R. (2012). Providing additional money to food-insecure households and its effect on food expenditure: a randomized controlled trial. *Public Health Nutrition* 16 (8): 1507–1515.

Smith, L.D., Staten, M., Eyssell, T. et al. (2013). Accuracy of information maintained by US Credit Bureaus: frequency of errors and effects on consumers' credit scores. *Journal of Consumer Affairs* 47 (3): 588–601.

Struminskaya, B. and Bosnjak, M. (2021). Panel conditioning: Types, causes and empirical evidence of what we know so far. In: *Advances in Longitudinal Survey Methodology* (ed. P. Lynn). Chichester: Wiley.

Swan, M. (2013). The quantified self: fundamental disruption in big data science and biological discovery. *Big Data* 1 (2): 85–99.

Tin, S.T., Mhurchu, C.N., and Bullen, C. (2007). Supermarket sales data: feasibility and applicability in population food and nutrition monitoring. *Nutrition Reviews* 65 (1): 20–30.

Tucker, C. (2011). Using multiple data sources and methods to improve estimates in surveys. Paper presented at BLS Household Survey Producers Workshop, Washington. http://www.bls.gov/cex/hhsrvywrkshp_tucker.pdf.

Turrini, A., Saba, A., Perrone, D. et al. (2001). Food consumption patterns in Italy: the INN-CA study 1994-1996. *European Journal of Clinical Nutrition* 55 (7): 571–588.

USDA (2009). The consumer data and information program: sowing the seeds of research. In: *Administrative Publication No. (AP-041)*. Washington, DC: USDA Economic Research Service. http://www.ers.usda.gov/publications/ap/ap041/ap041.pdf.

Volkova, E., Li, N., Dunford, E. et al. (2016). 'Smart' RCTs: development of a smartphone app for fully automated nutrition-labeling intervention trials. *JMIR mHealth and uHealth* 4 (1): e23. http://doi.org/10.2196/mhealth.5219.

Wang, Z. and Wolman, A.L. (2016). Payment choice and currency use: insights from two billion retail transactions. *Journal of Monetary Economics* 84: 94–115.

Weerts, S.E. and Amoran, A. (2011). Pass the fruits and vegetables! A community-university-industry partnership promotes weight loss in African American women. *Health Promotion Practice* 12 (2): 252–260.

Wenz, A., Jäckle, A., and Couper, M.P. (2019). Willingness to use mobile technologies for data collection in a probability household panel. *Survey Research Methods* 13 (1): 1–22.

Westat (2011). *Redesign Options for the Consumer Expenditure Survey*. Rockville, MD: Westat Available at: http://www.bls.gov/cex/redwrkshp_pap_westatrecommend .pdf.

Worthington, S. and Fear, J. (2009). *The hidden side of loyalty card programs. Retail Therapy*. Clayton, Victoria: The Australian Centre for Retail Studies http:// tapchibanle.org/retail-lib/hidden-side-of-loyalty.pdf.

Zhen, C., Taylor, J.L., Muth, M.K., and Leibtag, E. (2009). Understanding differences in self-reported expenditures between household scanner data and diary survey data: a comparison of Homescan and Consumer Expenditure Survey. *Review of Agricultural Economics* 31 (3): 470–492.

15

How to Pop the Question? Interviewer and Respondent Behaviours When Measuring Change with Proactive Dependent Interviewing

Annette Jäckle[1], Tarek Al Baghal[1], Stephanie Eckman[2] and Emanuela Sala[3]

[1] *Institute for Social and Economic Research, University of Essex, Colchester, UK*
[2] *Survey Research Division, RTI International, Washington, DC, USA*
[3] *Department of Sociology and Social Research, University of Milano-Bicocca, Milan, Italy*

15.1 Introduction

Dependent interviewing (DI) is a technique used in longitudinal surveys, whereby answers given in an interview are used to determine question routing or wording in the following interview. There are two different ways in which previous answers are used. With proactive dependent interviewing, respondents are reminded of their previous answer before being asked to update their status. For example, if at wave 1 a respondent reported being self-employed, the wave 2 question might read: 'Last time we interviewed you, you said you were self-employed. Is this still the case?' With reactive dependent interviewing, the respondent is asked about their status without reference to their prior answer (e.g. 'Are you an employee or self-employed?'). If the answer indicates a change in status, a follow-up question is triggered to check that the respondent's status has truly changed.

There is evidence that dependent interviewing questions improve the quality of longitudinal data, by reducing spurious changes in responses over time (e.g. Hoogendoorn 2004; Jäckle 2009; Jäckle and Lynn 2007; Lemaitre 1992; Lynn and Sala 2006; Murray et al. 1991; Perales 2014). As a result, DI questions are now routinely used in longitudinal surveys worldwide, such as the UK Household Longitudinal Study (UKHLS), the US Survey of Income and Program Participation (SIPP), and the German Panel Study Labour Market and Social Security (PASS). As documented in Lugtig and Lensvelt-Mulders (2014), DI has mainly been used to measure domains such as employment status, family situation, and income sources.

Examining the dependent interviewing questions used in different surveys shows that there are many different ways in which the questions can be worded.

Advances in Longitudinal Survey Methodology, First Edition. Edited by Peter Lynn.

Companion website: www.wiley.com/go/lynn/advancesinlongitudinalsurvey

Focusing just on proactive DI questions, the topic of this chapter, respondents can be asked to update their status in different ways. After being reminded of their previous status, they might be asked, 'Is this still the case?' as in the opening example. Alternatively, they might be asked 'Has this changed?' Or they might be asked a balanced question: 'Is this still the case or has it changed?' Sometimes respondents are just asked the original independent question after being reminded of their previous response: 'Are you an employee or self-employed?' (Jäckle 2009; Schoeni et al. 2013). In fact, even within a single survey, different wordings are sometimes used, although it is mostly unclear why the wording is varied. For example, in the English Longitudinal Study of Ageing, 'five different styles of dependent interviewing were used [...], no particular research guided those design decisions' (Pascale and Mcgee 2008, p. 144). Similarly, the UKHLS uses the 'Has this changed?' format for some questions and 'Is this still the same?' for others (Jäckle and Eckman 2019).

There is emerging evidence that how a proactive DI question is worded can affect the answers that respondents give. Al Baghal (2017) reports that respondents seem more likely to report a change when they are asked, 'Has this changed?' or, 'Has this changed or is it still the same?' than when they are asked, 'Is this still the same?'. Jäckle and Eckman (2019) report that re-asking the original independent question after reminding respondents of their previous response produces higher rates of change than other proactive DI wordings. There is, however, little research on the mechanisms through which question wording affects reports of change. As a result, there is little empirically based guidance on how best to word proactive DI questions.

In this chapter, we examine which question format is best by studying interviewer and respondent behaviours, and how the question wording affects these. The premise of this study is that departures from standardised interviewing are likely to affect data quality: if interviewers change question wording, this might influence the answers respondents give. We use audio-recordings of experimental questions in the Innovation Panel (IP) of the UKHLS to examine how different versions of proactive DI questions function, and why some wording versions are problematic. The specific research questions we address are:

1. Does the DI wording affect how interviewers and respondents behave?
2. Does the wording of DI questions affect the sequences of interviewer and respondent interactions?
3. Which interviewer behaviours lead to respondents giving codeable answers?
4. Are the different rates of change measured with different DI wordings explained by differences in interviewer and respondent behaviours?

15.2 Background

In 2000, Mathiowetz and McGonagle (2000) published a paper in which they assessed the state of the art of dependent interviewing in household surveys. After reviewing the relevant literature, they concluded that 'with respect to dependent interviewing the empirical literature is virtually non-existent' (p. 416) and proposed a research agenda with the core aims of assessing how DI affects data quality, respondent and interviewer burden, and interviewer–respondent interaction. Since the publication of this review, the survey methodology community has engaged in research on DI and an extensive body of research has been published.

Table 15.1 summarises the existing literature assessing the effects of dependent interviewing. The studies are of three types: (i) observational studies, where data collected with dependent interviewing are examined without comparison to other ways of asking the questions, (ii) pre/post studies, where data collected with dependent interviewing are compared to data collected before dependent interviewing was introduced in the survey, or (iii) experimental, where different question wordings are randomly allocated to sample members. Within each of these types of study designs, different evaluation criteria have been used to assess the effects of dependent interviewing.

Across these studies, and across the different evaluation criteria, the general picture that emerges is that dependent interviewing improves the quality of panel data. Dependent interviewing increases the rate of reporting unearned income sources where under-reporting is a known problem (Fisher 2019; Hale and Michaud 1995; Lugtig and Jäckle 2014; Lynn et al. 2006) and can reduce the rate of item non-response (Bates and Okon 2003). As a result, dependent interviewing reduces spurious changes in reported wealth (Hill 2006), assets (Hoogendoorn 2004), net income (Lugtig and Lensvelt-Mulders 2014), receipt of unearned income sources (Eggs and Jäckle 2015), employment characteristics including industry and occupation (Hill 1994; Lynn and Sala 2006; Perales 2014; Polivka and Rothgeb 1993), and in business surveys (Holmberg 2004). Dependent interviewing has further been shown to reduce spurious changes in monthly transition rates in receipt of unearned income sources (Moore et al. 2009) and labour market activities (Jäckle and Lynn 2007) and to increase the implied duration of unemployment spells (Busch 2012). Other studies have shown that the internal validity of panel data is improved. With dependent interviewing, predictors of industry and occupation changes (Hill 1994), of earnings and job satisfaction (Perales 2014), and of net wealth and wealth dynamics (Hill 2006) are more in line with theoretical expectations, and less affected by artefacts of the data collection such as the interview month.

Table 15.1 Summary of studies examining the effects of dependent interviewing.

Evaluation	Study Design		
Criteria	**Observational**	**Pre/post introduction of DI**	**Experimental**
Rate of reporting / item non-response	• Fisher (2019): RDI • Hale and Michaud (1995): PDI for some, RDI for other items	• Lugtig and Jäckle (2014): RDI • Bates and Okon (2003): item non-response, combination of PDI and RDI	• Lynn et al. (2006): RDI, PDI, INDI
Rate of change between interviews	• Hill (2006): RDI (call back) changes in wealth	• Hill (1994): PDI • Hoogendoorn (2004): PDI • Perales (2014): PDI • Polivka and Rothgeb (1993): PDI	• Al Baghal (2017): different PDI wordings • Holmberg (2004): PDI, INDI • Jäckle and Eckman (2019): different PDI wordings and INDI • Lugtig and Lensvelt-Mulders (2014): PDI, RDI, INDI • Lynn and Sala (2006): PDI, RDI, INDI
Seam effects in monthly changes / spell durations		• Busch (2012): PDI • Moore et al. (2009): enhanced PDI	• Jäckle and Lynn (2007): RDI, PDI, INDI
Internal validity / model based estimates of measurement error	• Hill (2006): RDI (call back), predictors of wealth and wealth dynamics	• Hill (1994): PDI, predictors of occupation and industry change • Perales (2014): PDI, predictors of wages and job satisfaction	• Lugtig and Lensvelt-Mulders (2014): PDI, RDI, INDI

Table 15.1 (Continued)

Evaluation	Study Design		
Criteria	**Observational**	**Pre/post introduction of DI**	**Experimental**
Record check (individual link unless stated)	• Dibbs et al. (1995): RDI comparison of amount of unemployment insurance • Eggs and Jäckle (2015): error in PDI preload of unemployment benefit • Murray et al. (1991): PDI, comparison with aggregate external data		• Jäckle and Eckman (2019): rate of change with different PDI wordings and INDI • Lugtig and Jäckle (2014): RDI, unearned income, amounts, durations, transitions • Lynn et al. (2012): rate of reporting with RDI, PDI, INDI
Interviewer / respondent burden	• Sala et al. (2011): PDI and RDI for different items	• Hoogendoorn (2004): PDI, administration time, subjective burden	• Jäckle (2008): PDI, RDI, INDI
Standardised interviewer / respondent behaviours	• Pascale and McGee (2008): different PDI wordings for different items • Uhrig and Sala (2011): PDI		• This chapter: different PDI wordings

Notes: PDI: proactive DI, RDI: reactive DI, INDI: independent interviewing. Unless stated, comparisons involved only one version of PDI/RDI wording each.

The interpretation that increased reporting, reduced rates of change, and smoothing of monthly transitions represent improvements in data quality has been borne out by several record check studies. In observational studies, monthly labour-market transitions collected with proactive DI followed the trends in aggregates derived from benchmark data (Murray et al. 1991) and reactive DI brought reported amounts of unemployment insurance closer to the individual's tax record (Dibbs et al. 1995). In experimental studies linked to individual administrative records, both proactive and reactive DI have been shown to reduce under-reporting of unearned income sources (Lugtig and Jäckle 2014; Lynn et al. 2012) and to reduce the under-reporting of spell durations spanning multiple interviews (Jäckle and Eckman 2019; Lugtig and Jäckle 2014).

Several studies have examined the risk that proactive DI might lead respondents to simply confirm the response they had given previously, leading to spurious stability and over-reporting of spell durations. However, there is little evidence of correlated measurement error with proactive or reactive DI (Lugtig and Lensvelt-Mulders 2014), and little evidence of satisficing in response to proactive DI questions (Eggs and Jäckle 2015).

A few studies have examined the effects of dependent interviewing on burden and found that it can reduce interview durations and that interviewers and respondents welcome the use of previous responses within the interview (Hoogendoorn 2004; Jäckle 2008; Sala et al. 2011).

Although the existing literature clearly shows that dependent interviewing improves data quality, there is little evidence on how best to word dependent interviewing questions. While some of the studies have used both reactive and proactive DI, either in observational studies or experimentally, no study has examined different ways in which a reactive question might be worded. Only Al Baghal (2017) and Jäckle and Eckman (2019) have examined different possible ways of wording proactive DI questions. Their results suggest that the way in which dependent interviewing questions are worded matters: depending on the wording, respondents are more or less likely to report a change in their status, affecting the accuracy of transitions compared to individual administrative records. To date, it is unclear through which mechanisms the wording affects reporting of change, but it is likely that the wording affects how the DI questions are administered within the interview.

Two observational studies have examined how dependent interviewing questions are administered and shown that interviewers frequently depart from the scripted question wording in ways that could influence the respondent's answer. Interviewers have, for example, been found to turn the reminder of what the respondent said previously into a question, or to turn questions into statements (Pascale and Mcgee 2008; Sala et al. 2011). When respondents report a change in response to the dependent interviewing question, standardisation frequently breaks down as they attempt to describe their current situation, effectively answering the follow-up questions (Uhrig and Sala 2011). If interviewers depart from standardised interviewing in different ways depending on the wording of the DI question, then that could be one channel through which the wording affects reporting of changes. The present study is the first experimental study to examine how the behaviours of interviewers and respondents vary depending on the wording of proactive DI questions.

15.3 Data

To address our research questions, we analyse audio-recordings of interviewer and respondent interactions from randomised experiments with dependent interviewing in waves 3 and 7 of the Innovation Panel of the UKHLS. The Innovation Panel (IP) is a separate panel set aside for methodological testing. It consists of a sample of respondents in Great Britain, interviewed every year since 2008. The IP sample design uses a clustered and stratified sample of households, with periodic refreshments. Selected households complete a household interview and then each person 16 years or older completes an individual interview. More details about the initial design of the survey can be found in Lynn (2009). Details on waves 3 and 7 of the IP – hereafter referred to as IP3 and IP7, respectively – are available in the Innovation Panel user guide.[1]

Data collection for IP3 was from April to June 2010. Interviews were achieved with 1621 individuals in 1022 households. Data collection for IP7 was from June to November 2014. IP7 included a refreshment sample, which is excluded from our analyses. Because these sample members were interviewed for the first time, it was not possible to ask them any DI questions. In addition, in IP7 households were randomly allocated to two mode conditions: face-to-face or mixed-mode (web with non-respondents followed up by face-to-face interviewers). We exclude all web respondents, because no interviewer–respondent interaction is possible for interviews completed via the web. Face-to-face interviews were achieved with 924 individuals in 593 non-refreshment households.

Both IP3 and IP7 included experiments with the wording of proactive DI questions. In both cases, the information fed forward and used in the question wording came from the prior interview. For most respondents this was the interview in the previous year, however, for respondents who had temporarily dropped out of the survey, the feed-forward information was from the last time they had been interviewed.

1 Available at www.understandingsociety.ac.uk/documentation/innovation-panel/user-guide.

The IP7 experiment included 4 treatment groups. In each case, the respondent was first reminded of the status they had reported in the previous interview, often including a reference to the date of their last interview. As an example, one of the questions read 'The last time we interviewed you on *[June 5th 2013]*, you said that, in general, your health was *[excellent / very good / good / fair / poor]*'. The experimental variation was in how respondents were asked to update their status. The four question versions were:

(1) Is that still the case? (Yes/No)
(2) Has that changed? (Yes/No)
(3) Is that still the case or has it changed? (Still the case/Has changed)
(4) Has that changed or is it still the case? (Has changed/Still the case)

If the respondent's answer indicated that her situation had changed, she was asked an independent follow-up question to ascertain her current status. In the example above, this was 'In general, would you say your health is excellent, very good, good, fair, poor?'

The IP3 experiment was set up in the same way, but used only the first two question versions. In both waves, the allocation to experimental groups was at the household level. This meant that all respondents within a household were asked all experimental DI questions in the same version. Interviewers, however, administered questions in all versions.

The IP3 experiment used four items in the individual questionnaire: general health, job permanence, hours worked as an employee, and hours worked as self-employed. The IP7 experiment involved four questions in the household questionnaire (number of rooms in the home, whether the house was owned, and the amount of the last mortgage or rent payment) and 15 questions from the individual questionnaire, some of which were asked separately of employees and self-employed people (likelihood of moving house, type of school attended (for those in education), job permanence, industry, occupation, employment status (employee or self-employed), size of firm, hours worked, gross pay, net pay, pay arrangement (salaried or hourly), commuting method). The wording of all experimental DI questions is documented in Appendices 15.A and 15.B.

The analysis sample is restricted to IP3 and IP7 respondents who (i) gave a full computer-assisted personal interview (CAPI), (ii) had been interviewed in a previous wave such that feed-forward data were available and they were routed into the DI questions, (iii) consented to audio recording, and (iv) for whom audio-recordings are available that could be coded.

The final sample sizes for our study are documented in Table 15.2. The IP3 data consist of 1145 question administrations (i.e. questions x respondents), nested in 583 respondents, nested in 79 interviewers; the IP7 data consist of 3235 question administrations, nested in 564 respondents, nested in 88 interviewers. There are

Table 15.2 Analysis sample sizes.

		IP3	IP7
Interviewers		79	88
Respondents	Full CAPI interview	1621	924
	Interviewed previously	1310	821
	Consent to audio recording	959	641
	Codeable audio file	583	564
Item administrations	(respondents x items)	1145	3235
	Still group	622	701
	Changed group	523	806
	Still/changed group	–	869
	Changed/still group	–	859

121 respondents who are in both the IP3 and IP7 data and 905 respondents who are only in one of the waves. The fieldwork agency changed between IP3 and IP7, which means that different interviewers administered the survey in the two waves. There were several reasons for missing audio-recordings: the recordings of some interviewers were not retrieved (in particular in IP3 recordings are missing for 33 interviewers), some recordings were not audible, and sometimes specific items were missing in the recordings.

15.4 Behaviour Coding Interviewer and Respondent Interactions

The next step in the analysis process was to code the interviewer-respondent interactions captured in the digital audio-recordings. In coding the interviewer and respondent behaviours, the unit of coding was each turn taken by the interviewer or respondent. Turns are defined as ending when the other speaker begins to speak.

The coding frame was developed in several stages. The initial coding frame was drafted based on our research goals and prior studies, which also behaviour coded DI questions (Pascale and McGee 2008; Uhrig and Sala 2011). The initial frame was tested on a sub-sample of respondents; three coders independently coded all questions recorded for 50 randomly selected respondents, equalling 118 items. Discrepancies in outcomes were identified and resolved through discussion between the coders. To clarify any final discrepancies, another randomly selected

22 respondents provided 53 additional test items which were coded independently by the three coders. During resolution of these discrepancies the coding scheme was expanded and refined. The main change was to code all possible behaviours of interest as to whether or not they occurred. This was a departure from the initial coding frame, which had included definitions for which codes to assign if multiple behaviours of interest occurred. Coding all possible behaviours of interest, instead of trying to agree a priority of codes, increased the robustness of the coding frame. In addition, codes were added to capture additional interviewer and respondent behaviours, which might affect outcomes. The full coding frame for both interviewers and respondents can be found in Appendix 15.C.

Interviewer codes focused on the reading of the questions and deviations from standardisation. Codes included whether the recording was not audible or the question was omitted entirely. If the question was asked and audible, the interviewer could either have read the question exactly as written or deviated in some way from the script. Deviations included whether the interviewer asked a DI question but used a version different from what was in the script (referred to as 'version error' in the following text), or some other change. These other changes included whether the interviewer omitted the reference to the date of the last interview, made a major or minor change in wording, reworded the question as a statement, provided only the reminder of the previous wave's response, or omitted the reminder text, asking only the question. A minor change in wording was one where a small number of words were changed that did not alter question meaning. A major change entailed rephrasing the question or omitting phrases. Additional codes were included for whether the interviewer provided additional information or said things not directly related to the question, indicated a problem with the question or respondent's answer and whether the interviewer answered the question instead of the respondent.

Respondent codes included whether the recording was not audible or a response was not applicable due to interviewer omission of the question. The respondent could provide an answer that corresponded to one of the scripted response categories, the goal of standardised interviewing. In the following, we refer to this outcome as a 'codeable answer'. The respondent could answer the question, but using phrasing other than the offered response categories, or could provide additional elaboration in the response. Respondents might also interrupt the interviewer when the question was being read, ask the interviewer for an explanation, or ask the interviewer to repeat the question. Additionally, the respondent could express uncertainty in the response, note that the preloaded reminder contained incorrect information, or indicate some other problem with the question or the answer, including correcting the previous response.

Some codes were mutually exclusive, while others could be used in combination. For example, an interviewer could omit the date of the last interview and read the

Current job permanent or temporary (JBTERM1) – IP7, Changed or still condition.

Question Wording
Last time we interviewed you on [ff_IntDate] you said that, leaving aside your own personal intentions and circumstances, your job was a permanent job [if ff_jbterm1 = 1] / was not a permanent job in some way [if ff_jbterm1 = 2]. Has that changed or is it still the case?

Response Options
1 Has Changed
2 Still the case

Transcription and Behaviour Coding

I1. "When we interviewed you last year, you said that, leaving aside your personal intentions –right, it was a permanent job, is that still the case?"

Coded: Date omitted, majorchange, version change to 'still'

R1. "Yeah, it is permanent, yeah"
Coded: Answered, using different wording than questionnaire options

Figure 15.1 Example question transcription and behaviour coding.

wrong DI version. A respondent could give a codeable answer and provide additional elaboration on the same turn. However, an interviewer code of reading the question as scripted could not be combined with other codes. Similarly, a respondent could not simultaneously provide a codeable response and a response with other wording.

An example of the coding scheme is provided in Figure 15.1, with transcribed text for the question on job permanency status in the 'changed or still' format. The scripted question wording and response options are documented in the upper section of Figure 15.1. On the first turn (I1), the interviewer paraphrased the question (coded as 'major change' in question wording), omitted the reference to the date of the previous interview, and made a version error by asking '… is that still the case?' instead of 'Has that changed or is it still the case?' On the next turn (R1) the respondent answered the question, but using wording other than the scripted response categories.

After practicing and finalizing the coding frame using the test set, the interrater agreement between the three coders was checked, before one coder went on to code the rest of the data. Interrater agreement was computed as the proportion of all possible turns where coders had assigned the same codes. Agreement was generally good on all codes and turns, ranging from 88.7% to 100% for all pairwise comparison, with two related exceptions. The code for exact reading the question was lower, with the lowest agreement of 64.2% for one comparison. The reason identified was that the continuing coder identified far more small changes to questions (for example, not including words like 'is') than the other coders. These discrepancies were discussed in depth and we clarified how to better code these. An additional test set was generated with 30 new questions from a new random selection of 8 respondents. Each coder initially coded these separately and we then met to compare and discuss each outcome in depth. It was agreed that concordance was acceptably high, and the remaining coding was completed by the one coder.

15.5 Methods

For the analyses, we pool all DI questions from a given wave and compare differences in the average effects across items between question versions. All analyses are adjusted for clustering of items within respondents.

To address research question 1, whether the DI version affected how interviewers and respondents behaved, we examine all behaviours that occurred at the interviewer and respondent's first turns. For most items, answering the question required just two turns: the interviewer read the question and the respondent gave an answer. There were only 37 item administrations in IP3, and 56 in IP7, where the interviewer and respondent each had a second turn, and 2 items in IP3 and 1 in IP7, where they each had a third turn. The only behaviour that is coded across all turns is whether the respondent gave a codeable answer. We use χ^2 tests, accounting for clustering of items within respondents, to determine whether each coded behaviour was significantly more or less likely in the different DI versions.

To address research question 2, whether the DI version affected the sequences of respondent-interviewer interactions, we also focus on just the first two turns. However, we priority coded the detailed codes of which behaviours occurred into just one code for the respondent and another for the interviewer at each turn. Interviewer behaviours were priority coded as 1: the interviewer read the question exactly as worded, 2: read a different version of the DI question, or 3: changed the question wording in some other way. Respondent behaviours were priority coded as 1: the respondent gave a codeable answer, i.e. using one of the scripted response categories, 2: interrupted the interviewer, 3: showed signs of uncertainty about the question, requested an explanation, or asked for the question to be repeated, 4: answered the question in words other than the scripted response categories, 5: gave additional information to explain their answer, or 6: noticed a problem with the preload or the question. We then created sequence-index plots using the Stata sqindexplot ado-file by Brzinsky-Fay et al. (2006) to examine the patterns of interviewer and respondent behaviour across the different question versions. These plots show graphically the different behaviours taken by interviewers and respondents and how they relate, but do not permit significance testing (see a similar use of sequence analysis by Kreuter and Kohler 2009). Thus, our results from the sequence analysis are more descriptive than inferential.

To address research question 3, which interviewer behaviours lead to respondents giving codeable answers, we used logit models at the item administration level, accounting for clustering of items in respondents. The dependent variable in the models is whether the respondent gave a codeable response (1) or did not (0). The explanatory variables are binary indicators of interviewer behaviours, each coded as 1 if it occurred, and 0 otherwise: whether the interviewer read the question as scripted, made a version error, omitted the date of the previous interview, or omitted a major or minor change of wording. The results are presented as average marginal effects because these can be interpreted as the percentage point change in the probability of giving a codeable answer associated with a change in the explanatory variable from 0 to 1. Six separate models are estimated, one for each question version and year. To test whether the associations between interviewer behaviours and the probability that respondents give codeable answers vary between question versions, we test for interactions. We rerun the logit model for each wave, including interactions between the interviewer behaviours and the question version. We test for differences in the effect of each interviewer behaviour between question versions using the 'margins' and 'lincom' commands in Stata 15 to test for differences-in-differences in the predicted probability of codeable answers.

To address research question 4, whether the different rates of change measured with different DI versions are explained by differences in interviewer and respondent behaviours, we estimate logit models at the item administration level, accounting for clustering of items in respondents. The dependent variable is coded as 1 if the respondent's answer implied a change, and 0 if it implied no change. The explanatory variables include binary indicators for the DI question version, for which interviewer behaviours occurred, and for respondent behaviours. Interviewer and respondent behaviours are coded as 1 if they occurred, and 0 otherwise. We estimate separate models for each wave and again present the results as average marginal effects. We first estimate models where the only explanatory variable is the DI question version. We then add the indicators of interviewer and respondent behaviours to the models to test whether the effect of the question version is explained by differences in interviewer and respondent behaviours.

15.6 Results

As found in other research (Al Baghal 2017; Jäckle and Eckman 2019), the DI question version clearly influenced the answers respondents gave (Figure 15.2):

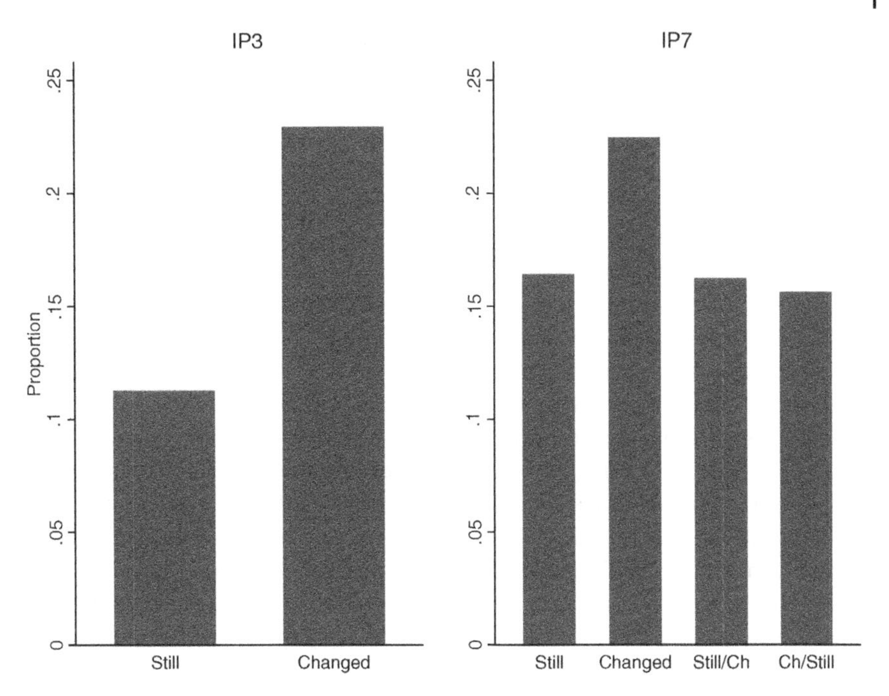

Figure 15.2 Proportion of respondents reporting a change in status, by DI version.

in the IP3 data 22.9% of respondents in the 'changed' group reported a change, compared to just 11.3% in the 'still' group (χ^2 test, $p < .001$). Similarly, in the IP7 data, 22.5% of respondents in the 'changed' group reported a change, compared to around 16% in each of the other three groups ($p < .05$ for each pairwise comparison with the 'changed' group).

One possible explanation for the differences in reporting of changes could be that interviewers and respondents behave differently, when administering different question versions. We now turn to answering our research questions on how interviewer-respondent interactions vary with the question wording, and what effect they have on responses to DI questions.

15.6.1 Does the DI Wording Affect how Interviewers and Respondents Behave? (RQ1)

Table 15.3 reports the prevalence of each of the behaviours coded from the audio transcripts and shows how they differ between the question formats and waves. The top of the table gives results for interviewer behaviour and the bottom part for respondent behaviour. If the respondent answered the question using the scripted response options at any turn, the code 'codeable answer' is applied. All other behaviours are coded only for the first interviewer and respondent turn, respectively. The last column in each wave indicates (with stars) whether the p-value from a χ^2 test for differences in the distributions of behaviours between question formats is significant. The superscripts on some results indicate which of the pairwise comparisons of the IP7 formats were significant at $p \leq .05$. For example, at 59.6% the probability of reading the question exactly as worded was significantly higher in group 1 ('still'), than in groups 2 ('changed'), 3 ('still/changed/'), or 4 ('changed/still').

Some behaviours were rare across all versions, such as interviewers omitting or answering the question, or turning it into a statement (0.2–3.3%), or respondents showing signs of uncertainty (1.5–2.3%).

In the IP3 data, interviewers in the 'still' version were more likely to read the question as worded and conversely less likely to read out the wrong format, make a major change in wording, or omit reference dates than those in the 'changed' version. Respondent behaviours did not differ between the two versions in IP3.

In the IP7 data, the probability of reading the question as worded was significantly higher in the 'still' version (59.6%) than in each of the other three versions (42.5–50.9%). Conversely, the probability of version errors ranged from 2.0% in the 'still' version to 39.4% in the 'still/changed' version, with significant differences between all versions. The probability of omitting the reference date differed between the 'still/changed' version (6.4%) and both the 'still' format (9.7%) and the 'changed' version (11.0%). The probability that the interviewer answered the question was significantly lower in the 'still' version than in each of the other three versions, although low overall.

Examining respondent behaviours, the probability of interrupting the interviewer was significantly lower in both the 'still' and 'changed' groups than the 'still/changed' and 'changed/still' groups. Conversely, the probability that the respondent gave a codeable answer at any turn was significantly higher in the 'still' and 'changed' groups than the 'still/changed' and 'changed/still' groups.

15.6.2 Does the Wording of DI Questions Affect the Sequences of Interviewer and Respondent Interactions? (RQ2)

Before addressing this research question, we exclude 87 item administrations where the interviewer did not read out the question text or answered the question themselves and 69 item administrations where the respondent behaviour was not audible or not coded. This reduces the number of sequences analysed by 3.8% in IP3 (to 1101 sequences) and by 3.5% in IP7 (to 3123 sequences).

Figure 15.3 shows the sequence-index plots of the interviewer and respondent behaviours, for each wave and question wording. The left block indicates the interviewer's behaviour, and the right block indicates the respondent's reaction. The

Table 15.3 Interviewer and respondent behaviours.

	IP3		IP7			
	1-Still	2-Changed	1-Still	2-Changed	3-Still/Ch	4-Ch/still
Interviewer behaviours (1st turn)						
Read exactly as worded	47.9	31.9***	$59.6^{2,3,4}$	50.9^{1}	42.5^{1}	49.4^{1}**
Omitted question	0.2	1.3*	0.3	1.0	0.6	1.3
I answered question	1.6	2.1	$0.4^{2,3,4}$	3.3^{1}	2.0^{1}	2.1*
Turned question into statement	2.7	1.1	0.9	2.2	1.6	1.6
Major change in wording	27.2	33.5*	22.0	23.7	20.5	22.6
Minor change in wording	11.6	14.9	12.1	12.9	9.8	10.4
Version error	0.8	9.0***	$2.0^{2,3,4}$	10.0^{134}	$39.4^{1,2,4}$	$24.7^{1,2,3}$***
Date omitted	24.6	32.5***	9.7^{3}	11.0^{3}	$6.4^{1,2}$	8.6*
Respondent behaviours (1st turn)						
Interrupted interviewer	1.9	2.3	$3.3^{3,4}$	$2.2^{3,4}$	$8.1^{1,2}$	$8.1^{1,2}$***
Signs of uncertainty	2.3	1.7	1.7	1.6	1.7	1.5
Stray talk	10.8	11.3	12.1	12.4	10.1	13.0
Other answer	9.0	9.8	$13.8^{3,4}$	$14.4^{3,4}$	$54.2^{1,2,4}$	$35.5^{1,2,3}$***
Codeable answer (all turns)	87.3	85.9	$83.2^{3,4}$	$80.9^{3,4}$	$40.5^{1,2,4}$	$59.7^{1,2,3}$***
N item administrations	622	523	701	806	869	859

Notes: *p*-values from χ^2 tests. *** $p \leq .001$, ** $.001 < p \leq .01$, * $.01 < p \leq .05$. Superscripts identify pairwise comparisons that are significantly different at $p \leq .05$ in χ^2 tests.

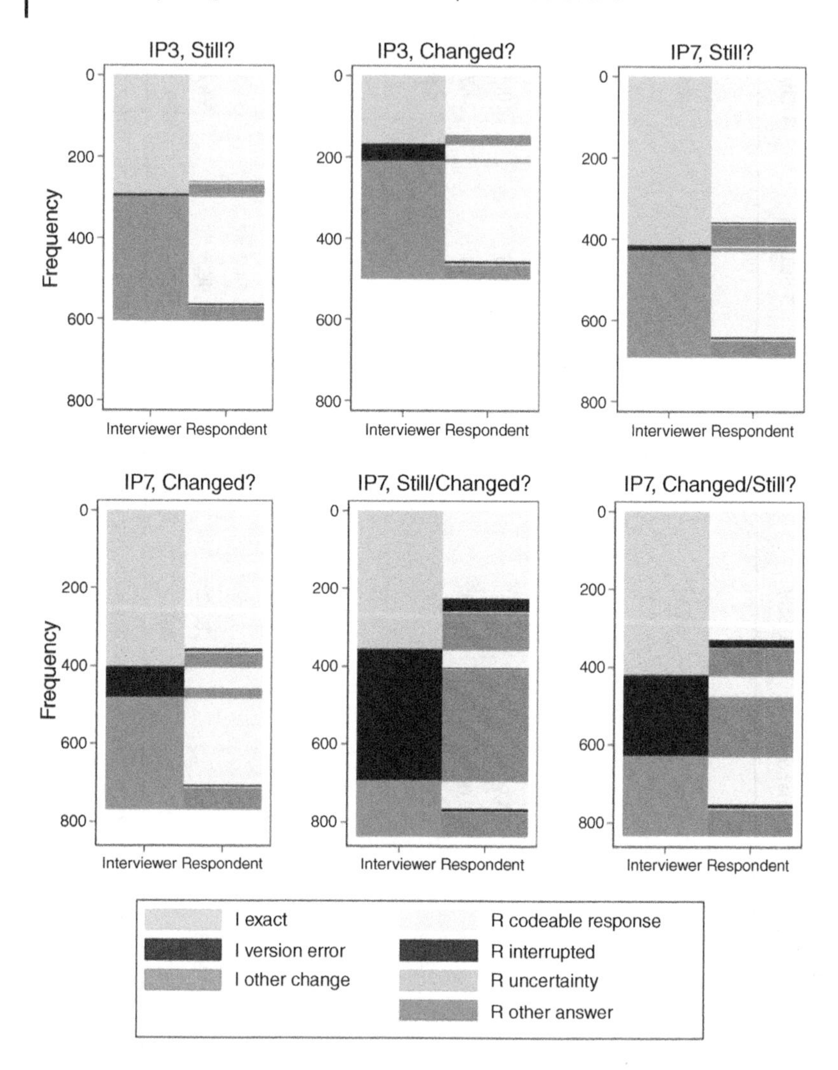

Figure 15.3 Sequence of interviewer and respondent behaviours at first interaction, by DI format.

height of the blocks corresponds to the frequency with which different behaviours occurred. The respondent behaviours 'other answer' and 'stray talk' are combined, to increase readability of the graph. For example, in the first panel, for the still version in IP3, we see that when interviewers read the question exactly as worded (left side of panel), most respondents gave a codeable response (right side of panel). A small share of respondents instead expressed uncertainty, interrupted the interviewer, or gave another answer or stray talk.

The figure shows that across all question versions, when interviewers read items exactly as scripted, in the majority of cases, respondents gave a codeable answer. With the 'still/changed' and 'changed/still' versions, respondents were, however, more likely than with other question formats to interrupt or give an answer that did not correspond to the response categories.

When interviewers misread the question as a different version, respondents in the 'changed' version usually nonetheless gave a codeable response. This was because nearly 70% of version errors in the 'changed' group consisted of the interviewer reading the question as 'still' instead of 'changed'. The answer categories remained the same (yes/no), although the meaning was reversed. In contrast, in the 'still/changed' and 'changed/still' formats, the most common version errors were to turn the question into a yes/no question. In the 'still/changed' group, 92.4% of version errors were that the question was read as 'still', while 3.2% were read as 'changed'. Similarly, in the 'changed/still' group, 60.4% of version errors were that the question was read as 'still', while 28.8% were read as 'changed'. In these cases, misreading the question lead most respondents to give a yes/no answer, which did not correspond to the response categories (code 'R other answer').

When interviewers made some other change to the question wording (including omitting reference dates, turning the question into a statement, or other large or small changes to wording), the vast majority of respondents in the 'still' and 'changed' formats nonetheless gave a codeable answer. In the 'still/changed' and the 'changed/still' versions, other answers were again more likely.

15.6.3 Which Interviewer Behaviours Lead to Respondents Giving Codeable Answers? (RQ3)

Table 15.4 shows the average marginal effects of interviewer behaviours on the probability that the respondent gave an answer using one of the response categories, by wave and question version. The estimates are derived from logit models of the probability that the respondent gave a codeable answer, as described in the Methods section.

In the IP3 'still' version, and the IP7 'changed' version, there are no significant effects. In the other versions, there are some marked effects of the interviewers' behaviours. Version errors reduced the probability of a codeable answer by 24–32 percentage points in the still, still/changed and changed/still versions at IP7. Reading the question as worded increased the probability of a codeable response by 13 and 15 percentage points in the IP7 'still/changed' and 'changed/still' formats. Curiously, omitting the reference date and making a small change to question wording increased the probability of a codeable answer by 9 percentage points

Table 15.4 Average marginal effects for the probability that respondent gave a codeable answer.

	IP3		IP7			
	Still	Changed	Still	Changed	Still/changed	Changed/still
I: exact wording	0.023	0.086	0.042	0.077	0.146**	0.134*
I: version error	−0.156	0.009	−0.233**	−0.062	−0.322***	−0.291***
I: date omitted	0.043	0.092*	0.047	−0.035	−0.009	0.045
I: large wording change	−0.013	0.026	−0.040	−0.025	0.044	−0.051
I: small wording change	−0.020	0.043	0.042	0.008	0.121*	0.010
N	611	505	696	775	848	836

Notes: Average marginal effects estimated from logistic regressions. *** $p \leq .001$, ** $.001 < p \leq .01$, * $.01 < p \leq .05$.

in the IP3 'changed' format and 12 percentage points in the IP7 'still/changed' format.

Testing for interactions in the IP3 data shows no differences in the effects of interviewer behaviours between question versions. In the IP7 data there is one significant interaction confirming the pattern in Table 15.4: whether or not the interviewer made a version error had no effect on the predicted probability of a codeable answer in the 'changed' group, but lowered the probability in each of the other groups. The differences in the effect of version errors on the predicted probability of codeable answers between the 'change group' and each of the other three groups are significant at $p \leq .05$.

15.6.4 Are the Different Rates of Change Measured with Different DI Wordings Explained by Differences in I and R Behaviours? (RQ4)

Table 15.5 shows the average marginal effects estimated from logit models of the probability of reporting a change. Models 1 and 3 indicate the effect of question wording on the reporting of change already documented in Figure 15.2: in IP3 respondents asked the 'changed' question were 11.5 percentage points more likely to report a change than respondents asked the 'still' question. Similarly, in IP7 respondents asked the 'changed' questions were 5.6 percentage points more likely to report a change. There were no differences between the 'still/changed' or the 'changed/still' and the 'still' version.

Models 2 and 4 show that several interviewer and respondent behaviours are associated with the probability of reporting a change. When interviewers omitted the question, the probability of reporting a change increased by 17.1 percentage

Table 15.5 Average marginal effects from logit models of probability of reporting a change, by wave.

	IP3		IP7	
	m1	m2	m3	m4
DI Version Reference: Still				
Changed	0.115***	0.106***	0.056*	−0.055*
Still/changed	n/a	n/a	−0.002	−0.003
Changed/still	n/a	n/a	−0.009	−0.020
Interviewer Behaviours[a)]				
I: omitted Q		0.055		0.171***
I: answered Q		0.059		0.050
I: exact wording		0.028		0.034
I: version error		0.131**		−0.071***
I: date omitted		−0.061		0.022
I: major change		0.021		0.066**
I: minor change		0.022		0.041
Respondent Behaviours[a)]				
R: interrupted		−0.040		−0.088**
R: uncertain		0.132		0.099**
R: stray talk		0.204***		0.247***
R: codeable answer		−0.109***		−0.110***
N	1145	1145	3235	3235

Notes: Average marginal effects estimated from logistic regressions. *** $p \leq 0.001$, ** $0.001 < p \leq 0.01$, * $0.01 < p \leq 0.05$.
a) More than one behaviour could be coded.

points; when they made a major change to the question wording, it increased by 6.6 percentage points (both IP7 only). Version errors also significantly affected the reporting of change.

If the respondent provided additional information and engaged in stray talk on their first turn, the probability that they reported a change increased by 20/25 percentage points in IP3/IP7. Similarly, if she showed signs of uncertainty about how to answer the question, the probability of reporting a change increased by 13/10 percentage points in IP3/IP7 (IP3 estimate marginally significant at p = 0.051). Conversely, if she gave a codeable answer, the probability of reporting change decreased by 11 percentages points in both waves. If she interrupted the interviewer, the probability of reporting a change decreased by 9 percentage points (IP7 only).

However, the effects of question wording in models 1 and 3 remain after controlling for interviewer and respondent behaviours in models 2 and 4, suggesting that these effects are not mediated by differences in interviewer and respondent behaviours between the question formats.

The associations between interviewer and respondent behaviours and the reporting of change cannot, however, be interpreted as causal effects. It may be that respondents who interrupt the interviewer are less likely to report a change because they have not heard the full question before answering. However, the opposite is also possible: the study by Uhrig and Sala (2011), for example, showed that when the respondent's situation has in fact changed, she is more likely to engage in stray talk describing her current situation than when her situation is still the same as in the previous interview. In this way she is more likely to, in effect, answer later questions, leading interviewers to omit the relevant questions later on.

15.7 Conclusion

As in previous studies (Al Baghal 2017; Jäckle and Eckman 2019), the findings in this chapter show that respondents can give markedly different answers to proactive dependent interviewing questions, depending on how these are worded. In this study, respondents were more likely to report a change in their status when they were asked, 'Has this changed?' than with the other question versions.

The premise of this chapter is that departures from standardised interviewing can affect responses, and therefore the best way of wording dependent interviewing questions is such that interviewers stick to the wording.

The results suggest that the 'Is that still the case?' wording was most likely to lead to standardised behaviours: interviewers were more likely to read questions as scripted than in any of the other versions, and respondents were more likely to give a codeable answer than in any of the other versions. As a result, standardised sequences of behaviours were more likely, with interviewers reading the question as scripted and respondents giving a codeable answer. Finally, in the 'still' version, interviewer departures from standardised interviewing did not appear to influence whether the respondent gave a codeable response.

With the 'Has that changed?' wording, interviewers were more likely to make version errors than with the 'still' wording, and they were more likely to omit questions or answer them themselves. Respondents however behaved no differently with the 'changed' version than the 'still' version. As a result, non-standardised question sequences were more likely than with the 'still' version, especially due to the interviewer's behaviour. The interviewer behaviour again had very little effect on whether the respondent gave a codeable answer. Overall, however, the 'changed' version worked less well in terms of standardisation than the 'still' version.

With the two balanced versions – 'Is that still the case or has it changed?' and 'Has that changed or is it still the case?' – interviewers were more likely to make version errors than with the 'still' or 'changed' versions, and respondents were more likely to interrupt the interviewer and were less likely to give a codeable answer. In these formats, interviewer behaviours such as reading the question as worded or making version errors had large effects on the probability that the respondent gave a codeable answer. In terms of standardisation, the 'still' version therefore seemed to work best, the two balanced formats least well.

Since the different question wordings lead to different interviewer and respondent behaviours, and since some of those behaviours are associated with the reporting of change, we expected that the differences in reports of change between question versions might in part be because the question version leads to different behaviours. This does, however, not seem to be the case; the interviewer and respondent behaviours do not explain differences in reporting of change between question versions. That is, non-standardised behaviours do not seem to produce the differences in reporting of change between formats.

So if interviewer and respondent behaviours do not explain higher reporting of change in the 'changed' group, what does? Anecdotally, from listening to the recordings, it seems that the 'changed' version prompts respondents to think of changes that have occurred since the previous interview, even if current status is again the same as in the previous interview. For example, when asked (last year your health was very good, has that changed?), one respondent described having had an operation, which the interviewer coded as a change in health status. In response to the follow-up question, however, the respondent said that their health was 'very good', i.e. the same as it had been at the previous interview.

One limitation of our study is the assumption that standardised interviewing produces more accurate responses. There are examples in the literature where conversational interviewing produces more accurate data than standardised interviewing (e.g. Schober and Conrad 1997). Ideally, we would have external data against which to validate the responses to the different DI question formats, as in the study by Jäckle and Eckman (2019). Another caveat is that the Innovation Panel mainly uses 'still' wording in other non-experimental dependent interviewing questions, although 'changed' is also used. Therefore, the interviewer default error was to ask the 'still' question. It is possible that interviewers might make fewer version errors if a single question version was used throughout the survey.

Acknowledgements

This research was funded by the UK Economic and Social Research Council grants for *Understanding Society* (RES-586-47-0001, ES/K005146/1, ES/N00812X/1).

References

Al Baghal, T. (2017). Last year your answer was … the impact of dependent interviewing wording and survey factors on reporting of change. *Field Methods* 29 (1): 61–78.

Bates, N. and Okon, A. (2003). Improving quality in the collection of earnings: The survey of Income and Program Participation 2004 panel. In: *Proceedings of the American Statistical Association, Government Statistics Section*. Alexandria, VA: American Statistical Association.

Brzinsky-Fay, C., Kohler, U., and Luniak, M. (2006). Sequence analysis with Stata. *Stata Journal* 6 (4): 435.

Busch, M.E. (2012). The CPS redesign's effects on measured unemployment duration in the great recession. *The BE Journal of Economic Analysis & Policy* 12 (1) (Conributions), Article 19.

Dibbs, R., Hale, A., Loverock, R. et al. (1995). Some effects of computer assisted interviewing on the data quality of the Survey of Labour and Income Dynamics. In: *SLID Research Paper Series*. Ottawa: Statistics Canada.

Eggs, J. and Jäckle, A. (2015). Dependent interviewing and sub-optimal responding. *Survey Research Methods* 9 (1): 15–29.

Fisher, P. (2019). Does repeated measurement improve income data quality? *Oxford Bulletin of Economics and Statistics* 81 (5): 989–1011.

Hale, A. and Michaud, S. (1995). Dependent interviewing: Impace on recall and on labour market transitions. In: *SLID Research Paper Series*. Ottawa: Statistics Canada.

Hill, D.H. (1994). The relative empirical validity of dependent and independent data collection in a panel survey. *Journal of Official Statistics* 10 (4): 359–380.

Hill, D.H. (2006). Wealth dynamics: reducing noise in panel data. *Journal of Applied Econometrics* 21 (6): 845–860.

Holmberg, A. (2004). Pre-printing effects in official statistics: an Experimental Study. *Journal of Official Statistics* 20 (2): 341–355.

Hoogendoorn, A.W. (2004). A questionnaire design for dependent interviewing that addresses the problem of cognitive satisficing. *Journal of Official Statistics* 20 (2): 219–232.

Jäckle, A. (2008). Dependent interviewing: effects on respondent burden and efficiency of data collection. *Journal of Official Statistics* 24 (3): 1–21.

Jäckle, A. (2009). Dependent interviewing: a framework and application to current research. In: *Methodology of Longitudinal Surveys* (ed. P. Lynn), 93–111. Chichester: Wiley.

Jäckle, A. and Eckman, S. (2019). Is that still the same? Has that changed? On the accuracy of measuring change with dependent interviewing. *Journal of Survey Statistics and Methodology* 8 (4): 706–725. https://doi.org/10.1093/jssam/smz021.

Jäckle, A. and Lynn, P. (2007). Dependent interviewing and seam effects in work history data. *Journal of Official Statistics* 23 (4): 529–551.

Kreuter, F. and Kohler, U. (2009). Analyzing contact sequences in call record data. Potential and limitations of sequence indicators for nonresponse adjustments in the European Social Survey. *Journal of Official Statistics* 25 (2): 203–226.

Lemaitre, G. (1992). Dealing with the seam problem for the Survey of Labour and Income Dynamics. *SLID Research Paper Series*. Ottawa: Statistics Canada.

Lugtig, P. and Jäckle, A. (2014). Can I just check...? Effects of edit check questions on measurement error and survey estimates. *Journal of Official Statistics* 30 (1): 45–62.

Lugtig, P. and Lensvelt-Mulders, G. (2014). Evaluating the effect of dependent interviewing on the quality of measures of change. *Field Methods* 26 (2): 172–190.

Lynn, P. (2009). Sample design for Understanding Society. *Understanding Society Working Paper* 2009-01. Colchester: University of Essex.

Lynn, P. and Sala, E. (2006). Measuring change in employment characteristics: the effects of dependent interviewing. *International Journal of Public Opinion Research* 18 (4): 500–509.

Lynn, P., Jäckle, A., Jenkins, S.P. et al. (2006). The effects of dependent interviewing on responses to questions on income sources. *Journal of Official Statistics* 22 (3): 357–384.

Lynn, P., Jäckle, A., Jenkins, S.P. et al. (2012). The impact of interviewing method on measurement error in panel survey measures of benefit receipt: evidence from a validation study. *Journal of the Royal Statistical Society, Series A* 175 (1): 289–308.

Mathiowetz, N.A. and McGonagle, K.A. (2000). An assessment of the current state of dependent interviewing in household surveys. *Journal of Official Statistics* 16 (4): 401–418.

Moore, J., Bates, N., Pascale, J. et al. (2009). Tackling seam bias through questionnaire design. In: *Methodology of Longitudinal Surveys* (ed. P. Lynn), 73–92. Chichester: Wiley.

Murray, T.S., Michaud, S., Egan, M. et al. (1991). Invisible seams? The experience with the Canadian Labour Market Activity survey. In: *Proceedings of the US Bureau of the Census Annual Research Conference*. Washington, DC: US Census Bureau.

Pascale, J. and McGee, A. (2008). Using behavior coding to evaluate the effectiveness of dependent interviewing. *Survey Methodology* 34 (2): 143–151.

Perales, F. (2014). How wrong were we? Dependent interviewing, self-reports and measurement error in occupational mobility in panel surveys. *Longitudinal and Life Course Studies* 4 (3): 299–316.

Polivka, A.E. and Rothgeb, J.M. (1993). Redesigning the CPS questionnaire. *Monthly Labor Review*: 10–28.

Sala, E., Uhrig, S.C.N., and Lynn, P. (2011). "It is time computers do clever things!": the impact of dependent interviewing on interviewer burden. *Field Methods* 23 (1): 3–23.

Schober, M.F. and Conrad, F.G. (1997). Does conversational interviewing reduce survey measurement error? *Public Opinion Quarterly* 61 (4): 576–602.

Schoeni, R.F., Stafford, F., Mcgonagle, K.A. et al. (2013). Response rates in national panel surveys. *The Annals of the American Academy of Political and Social Science* 645 (1): 60–87.

Uhrig, S.C.N. and Sala, E. (2011). When change matters: an analysis of survey interaction in dependent interviewing on the British Household Panel Study. *Sociological Methods & Research* 40 (2): 333–366.

15.A IP3 Stems of Experimental Dependent Interviewing Questions

General Health (SF1):

The last time we interviewed you on *[ff_INTDATE]*, you said that, in general, your health was *[excellent/very good /good/fair/poor]*.

Job Permanent (JBTERM1):

Last time we interviewed you on *[ff_INTDATE]*, you said that leaving aside your own personal intentions and circumstances, your job was (a permanent job/was not a permanent job in some way).

Employee – hours worked (JBHRS):

Last time we interviewed you, you said that in your (main) job, you were expected to work *[ff_jbhrs]* hours in a normal week, excluding overtime and meal breaks.

Self-employed – hours worked (JSHRS):

Last time we interviewed you, you said that you usually work *[ff_jshrs]* in total each week in your job.

15.B IP7 Stems of Experimental Dependent Interviewing Questions

Hsroomchk

When we interviewed you on *[ff_Idate]*, you said you had *[ff_HsBeds]* bedroom(s), excluding any you may sublet and *[ff_HsRooms]* other rooms, excluding kitchens and bathrooms.

Hsowndchk

Last time you said that this accommodation was (Owned outright/Owned or being bought on mortgage/Shared ownership [part-owned part-rented] /Rented/Rent free).

Xpmg

Last time your total monthly instalment on all mortgages or loans for this property was *[ff_xpmg]*.

Rentchk

Last time you paid *[ff_rent] [ff_rentperiod]*.

Lkmove

Last time we interviewed you on *[ff_IntDate]* you said that if you could choose, you would (stay here in your present home/prefer to move somewhere else).

Edtype

The last time we interviewed you on *[ff_IntDate]* you said that you were (At School/ At Sixth Form College /At Further Education (FE) College/At Higher Education (HE) College/at University).

Jbterm 1

Last time we interviewed you on *[ff_IntDate]* you said that, leaving aside your own personal intentions and circumstances, your job (was a permanent job / was not a permanent job in some way).

Jbsic07

Last time you said that the firm or organisation where you work, makes or does *[ff_jbsic07]*.

Jbsoc00

Last time you described your occupation in your **main** job as *[ff_jbsoc00]*.

Jbsemp

Last time you said that you were (an employee / self-employed).

Jbsizechk

Last time, you said that there were *[1–2 / 3–9 / 10–24 / 25–49 / 50–99 / 100–199 / 200–499 / 500–999 / 1000 or more]* people employed at the place you work.

Jbhrschk

Last time we interviewed you, you said that in your **main** job, you were expected to work *[ff_jbhrs]* hours in a normal week, excluding overtime and meal breaks.

Paygl

When we interviewed you on *[ff_IntDate]*, you said that last time you were paid, your gross pay – that is including any overtime, bonuses, commission, tips or tax refund but before any deductions for tax, National Insurance or pension contributions, union dues, and so on – was £*[ff_paygl] [ff_paygperiod]*.

Paynl

And when we interviewed you on *[ff_IntDate]*, you said that last time you were paid, your net pay – that is after any deductions were made for tax, National Insurance, pensions, union dues and so on – was £ *[ff_paynl] [ff_paynperiod]*.

Paytypchk

Last time you said that (you were salaried / you received a basic salary plus commission / you were paid by the hour).

Wktravchk

Last time you said that you usually (travel to work by *motorcycle, moped, scooter / taxi, minicab / bus, coach, train / Underground, Metro, Tram, Light railway [if Region = GB]*) / (*drive a car or van / get a lift with someone from the household / get a lift with someone outside the household / cycle/ walk* to work).

Jshrschk

Last time we interviewed you, you said that you usually work *[ff_jshrs]* hours, in total each week, in your job.

Jspartchk

Last time we interviewed you, you said that you were working (on your own account [sole owner] / in partnership with someone else).

Jstravchk

Last time you said that you usually (travel to work by *car or van / motorcycle, moped, scooter / taxi, minicab / bus, coach / train / Underground, Metro, Tram, Light railway [if Region = GB]*) / (travel to work by getting a *lift with someone from the household / lift with someone outside the household*) / (*cycle / walk* to work).

15.C Behaviour Coding Frame

Column	Code	Label	Definition	Examples
Interviewer	X	Not audible	Question is not audible.	
	O	Omitted	Int. did not read the question.	
(If code E, codes S-M must not apply).	E	Exact	Int. read question exactly as worded, including the response options.	Including if interviewer reads 'is this still the case?' instead of 'is that…?' and vice versa Including if interviewer adds 'And…' at the beginning of the question.
(Codes S-U mutually exclusive)	S	Version error: Int. read 'Is that still the case?'	Int. read 'Is that still the case?' instead of allocated version.	
	C	Version error: Int. read 'Has that changed?'	Int. read 'Has that changed?' instead of allocated version.	
	I	Version error: Int. read independent question	Int. did not remind R of the answer they had given in the previous interview.	'Is your job permanent?'
	T	Version error: Int. read 'Is that still the case or has it changed?'	Int. read 'Is that still the case or has it changed?' instead of allocated version.	
	U	Version error: Int. read 'Has that changed or is it still the case?'	Int. read 'Has that changed or is it still the case?' instead of allocated version.	
	F	Fact / statement	Instead of reminding R of the answer they had given in the previous interview, the Int. turned the reminder into a statement about the respondent's current situation.	'Your job is (still) permanent.'

Column	Code	Label	Definition	Examples
	W	Date omitted	Int. skipped the date of the last interview, but read everything else as scripted.	Code only if interviewer mentions neither day, month or year of last interview.
	B	Big change of wording	Int. rephrased the question using her own words instead of the scripted words, or omitted phrases.	If only omitted a single word, code as 'L'
	L	Little change of wording	Int. made a small change to the question wording that did not alter the question; usually only change in one word.	
Code in combination with codes E-L	R	Reminder only	Int. read the reminder part of the question.	'Last time we interviewed you, you said your job was permanent' (code as R, W).
	Q	Question only	Int. read the question part of the question.	'Is that still the case?' / 'Has that changed?'
	A	Additional information, elaboration, stray talk	Int. says things that are not directly related to the question.	Int. remembering something about what respondent had said last year, other unrelated talk.
	M	Mistake / indicates problem	Int. noticed a problem with the question, the preload or queried the respondent's answer.	
	Y	Int. answered	Int. answered the question instead of letting R answer.	'you already said it's 38 hours'
Respondent	O	Omitted, not applicable	Int. didn't ask the question.	
	X	Not audible	Response is not audible.	

Column	Code	Label	Definition	Examples
	I	Interrupts	R spoke before interviewer had finished reading the question.	Exclude: if respondent interrupts to acknowledge the preload, do not code as 'I'
	E	Explanation	R asked for explanation / clarification about the meaning of the question.	
	R	Repeat	R asked for question to be repeated.	
	U	Uncertainty	R shows signs that she is uncertain about her answer (recall or judgement problems); R used hedging words ('perhaps'), express statements of uncertainty over the answer provided, 'don't know'.	
	C	Response category	R's answer matched one of the response categories.	Answer is one of the explicit response options, including Don't Know / Refused.
	D	Other answer	R answered using words other than the response categories.	'It is permanent' instead of 'yes/no'.
	A	Additional information, elaboration	R gave additional information to explain their answer.	e.g. I: 'last time you worked 22 hours… Is that still the case?' R: 'I now work 22.5 hours' – the respondent is answering the next question, which is 'how many hours do you work?' code as 'DA'. 'yes, my job is still permanent' –this does not include any additional information.

Column	Code	Label	Definition	Examples
	P	Preload wrong	R said preload was wrong.	
	M	Mistake / indicates problem	R indicated a problem with the question or their answer, corrected the answer they had already given.	'I just told you that'.

Abbreviations: R – respondent, Int. – interviewer.

16

Assessing Discontinuities and Rotation Group Bias in Rotating Panel Designs

Jan A. van den Brakel[1,2], Paul A. Smith[3], Duncan Elliott[4], Sabine Krieg[1], Timo Schmid[5] and Nikos Tzavidis[3]

[1] *Methodology Department, Statistics Netherlands, Heerlen, Netherlands*
[2] *School of Business and Economics, Maastricht University, Maastricht, Netherlands*
[3] *Department of Social Statistics and Demography, University of Southampton, Southampton, UK*
[4] *Methodology Division, Office for National Statistics, Newport, UK*
[5] *Institute for Statistics and Economics, Freie Universität Berlin, Berlin, Germany*

16.1 Introduction

Longitudinal surveys come in many forms, ranging from large-scale cohort studies to repeating surveys which have some common sample elements from period to period, with intermediates depending on how the population and sample are refreshed and followed up. *Rotating panel surveys* are a specific type of longitudinal survey where on each survey occasion a new panel is added to the sample, and followed for a number of periods according to a predetermined pattern, after which the panel is (normally) dropped and replaced by a new one. Thus, on each occasion, the sample consists of several panels, each previously surveyed a different number of times.

Rotating panel surveys follow particular rotation schemes, and Steel (1997) coded these according numbers of months in and out of the sample. So a simple monthly survey where households were interviewed for six consecutive months (such as the Canadian Labour Force Survey [LFS]) is known as an in-for-6 survey. An annual survey with continuous interviewing, where households are interviewed four times at yearly intervals (for example the UK Annual Population Survey) would have a 1-11-1 (4) design if considered on months, but could also be thought of as an in-for-4 yearly design. The Netherlands and UK LFSs discussed later both use a 1-2-1 (5) monthly design, where households are reinterviewed at three-monthly intervals with five interviews in total. The rotation pattern determines how many panels there are on any given survey occasion (the

Advances in Longitudinal Survey Methodology, First Edition. Edited by Peter Lynn.

Companion website: www.wiley.com/go/lynn/advancesinlongitudinalsurvey

cross-sectional sample). In a steady state, on each occasion one panel will be introduced, replacing a panel that had its last interview on the previous occasion.

Rotating panels are common in business surveys, and used more generally in surveys that are intended to measure changes well, because the overlap of units from period to period gives a large sampling covariance, so the estimated sampling variance of the change in the total of a variable *y* between periods 1 and 2, $var(\hat{t}_{y_1} - \hat{t}_{y_2}) = var(\hat{t}_{y_1}) + var(\hat{t}_{y_2}) - 2\text{cov}(\hat{t}_{y_1}, \hat{t}_{y_2})$ is substantially reduced over a design with independent samples in the two periods. There is a tension in design between retaining panels as long as possible to maximise the covariance so that estimates of changes are as accurate as possible, and the need to refresh the panel to avoid problems caused by selective non-response and respondent fatigue in order to make accurate cross-sectional estimates. Some of the considerations and methods for designing such samples are discussed in Smith et al. (2009, section 2.4).

Rotating panels and (pure) longitudinal surveys are designed to reinterview the same respondents multiple times. In rotating panels particularly, it is possible to see that respondents' answers depend on how many times they have previously been interviewed, because the responses can be compared with the average in each cross-sectional dataset. The average difference between the responses of one panel and the overall average at each wave is called the *rotation group bias* (RGB; Bailar 1975). It is a bias – in fact the average difference of several relative biases – because it is not accounted for by the sampling error, and is systematic. The size of the RGB depends on differences in fieldwork methods, the number of waves and the time between the waves. See Section 16.3.1 for more details on factors responsible for RGB. Although it is particularly visible in rotating panels, it is probably present in longitudinal surveys and even one-off surveys may be affected (if their single survey occasion is subject to a bias that would conceptually change with further collection occasions). We generally assume that RGB is fixed, or evolves slowly, so that we can consider its average over a number of time periods – so we would estimate RGB for the second interview occasion averaged over the difference between the second interview and the cross-sectional average from a number of time periods.

There is also a systematic difference in a particular panel's responses compared to the overall average that persists across the interview occasions, depending on which units are selected in that particular panel. This, however, *is* part of the sampling error, which manifests itself in rotating panel surveys over multiple periods – the sampling errors are correlated, and indeed this is exactly the property that we take advantage of to reduce the variance of estimates of changes.

A further challenge in repeating surveys arises when the survey procedures are updated. One of the most important properties of such surveys is their consistency from period to period, so that they can provide good estimates of changes (Van den Brakel et al. 2008). When methods are necessarily updated (which may be for a variety of reasons), it causes a change in the series, and this systematic difference is distinct from the sampling error and is known as a *discontinuity*. Discontinuities are due to changes in the survey process, not real changes in the underlying population parameter. However, any estimate of such a discontinuity must be disentangled from the evolution of the parameter of interest, and this raises some methodological challenges.

RGB and discontinuities are in fact similar phenomena – both are systematic differences in non-sampling errors in a survey. Therefore, they can be dealt with using common approaches, and in this chapter we review the methods for planning a transition to a new approach, and estimating and adjusting for the changes in the evolution of the series. Section 16.2 reviews the available methods to quantify discontinuities induced by redesigns of the survey process. Section 16.3 shows how time series models can be fitted to the outputs from rotating panels to account for RGB. Section 16.4 extends these times series models to estimate discontinuities. Section 16.5 presents some example applications, and some general conclusions are drawn in Section 16.6.

16.2 Methods for Quantifying Discontinuities

Quantifying discontinuities in repeating surveys is challenging because the underlying population parameter evolves at the same time the discontinuity is introduced. Therefore, if no special steps are taken, the discontinuity and the actual change in the series will be confounded (not separately estimable). We would like the series to reflect the changes in the underlying parameter, without any discontinuities. The different approaches to estimating discontinuities are based on what information is available to separate these two effects.

The simplest approach is to make an assumption about the evolution of the underlying parameter – either that it is zero, or that it follows some simple trend (for example, linear growth), so that the discontinuity introduced at time 2 is $\hat{t}_{y_1} - \hat{t}_{y_2} - c_{12}$ where c_{12} is the assumed change between times 1 and 2 (possibly $= 0$). As well as the need for an appropriate and untestable assumption, the whole of the variance of change will be attributed to the discontinuity, with the result that inference about it will be conservative.

A more structured approach is to have a gradual transition where the new methods are initially introduced on part of the sample while the remainder continues on the existing methods. The proportion of the sample moving to the new methods can be gradually increased if there is no evidence of a significant discontinuity. If the original and new methods are viewed as treatment and control in an embedded experiment (van den Brakel and Renssen 1998; van den Brakel 2008, 2013), then there is sufficient information to separately estimate the usual evolution of the series and the discontinuity, although possibly with large variances. In this approach, we obtain an estimate of the discontinuity as the difference between the estimates of period to period change in the control and treatment groups. If the transition is staged over several periods, then we may end up with several such estimates and be able to combine them in a suitable way, taking account of their estimated variances.

A third approach, which can be considered if there is no budget or capacity for an embedded experiment, is to use the time series of observations and improve the estimate of the discontinuity using the autocorrelation in the series – the extent of the improvement depends on the partial autocorrelation function. A convenient way to estimate a discontinuity in a time series is through a state space model fitted with the Kalman filter. The state–space approach can also combine information from an embedded experiment with the information from the time series observed before and after the change-over. RGB can be included in the model structure, and then the bias (measured relative to one or a combination of waves, since the true bias is unknown) can be adjusted in the same procedure. Assumptions about the outcomes of the different waves and the specific survey situation determine how RGB is measured and therefore whether it is benchmarked to one particular wave or the average over all waves. For more details see Section 16.3.2 and van den Brakel and Krieg (2015). Other features of the time series such as seasonality can also be included, and the model can additionally account for the sampling error in the survey estimates (Pfeffermann 1991). In addition, the time series modelling approach is a very powerful method for small area estimation to borrow strength across space as well as time (van den Brakel et al. 2016).

16.3 Time Series Models for Rotating Panel Designs

In this section, the structural time series model, originally proposed by Pfeffermann (1991), as a method for time series small area estimation that accounts for RGB, is introduced. In Section 16.4, the model is extended with an intervention component to account for discontinuities.

16.3.1 Rotating Panels and Rotation Group Bias

National statistical institutes often use rotating panel designs for their LFSs (Kalton 2009), including the Dutch and UK LFSs. The UK LFS has always been conducted as a rotating panel since it moved from annual to continuous in 1992. The Dutch LFS changed from a cross-sectional to a rotating panel design in 2000. Both surveys have the same 1-2-1 (5) rotation pattern where households are observed five times at quarterly intervals. Each month a probability sample of households is selected. All household members aged 16 years or older are included in the sample.

In the first wave of the Dutch LFS an extended questionnaire is used to establish the labour market position of the household members aged 16 years or older, and this was originally administered by computer-assisted personal interviewing (CAPI). In the follow-up waves, a condensed questionnaire is used that focusses on changes in the labour market position of these respondents, and this was collected using computer-assisted telephone interviewing (CATI). Around 2009, budget cuts forced Statistics Netherlands to reduce the administration costs of its surveys. Therefore, a sequential mixed-mode data collection strategy was introduced. Sampled persons are invited to complete a questionnaire online through the web (henceforth web Interviewing, [WI]). After three reminders, non-respondents are approached through CATI if a non-secret landline telephone number is available or through CAPI otherwise. In 2009, Statistics Netherlands did not have sufficient experience with WI for household surveys. Therefore, in 2010, the data collection in the first wave changed to a mixed-mode design based on CATI and CAPI. In 2012, the sequential mixed-mode data collection strategy based on WI, CATI, and CAPI was introduced in the first wave. The data collection in the follow-up waves was based on CATI throughout.

The data collection methods used for the UK LFS are similar to those of the Dutch LFS. Currently, the first wave is mostly based on CAPI, with follow-up waves based on CATI where a telephone number has been provided. Some interviews in the first wave are conducted by telephone (e.g. areas north of the Caledonian Canal in Scotland), and since January 2011, if a telephone number can be matched to a sampled address, the first approach is by telephone. Approximately 15% of first wave interviews are conducted by CATI.

Generally, design-based or model-assisted estimation techniques, known from classical sampling theory (Särndal et al. 1992), are applied to produce official statistics about the labour force. A major drawback of these estimators is that they have unacceptably large standard errors in the case of small sample sizes. Sample sizes of LFS's are often too small to produce sufficiently reliable monthly figures with these design-based estimators. Since there is a large demand for labour force indicators at a monthly frequency, many national statistical institutes publish a rolling quarterly figure each month as a pragmatic solution, and indeed the 1-2-1 (5) design is particularly suited to this (McLaren and Steel 2000).

For rotating panels, it is often accepted that level estimates are biased with RGB. A common strategy is to design a rotation group pattern that is balanced at each point in time such that estimates of period-to-period changes are unbiased (Bailar 1975). This assumes that the RGB is constant over time and does not have a periodic component. The change-over from a cross-sectional design to a rotating panel in the Dutch LFS, however, made the effects of RGB visible and compromised data quality, particularly for measures of change during the transition period (Van den Brakel and Krieg 2015).

In the Dutch LFS, the estimated unemployment rate in the first wave is systematically higher (10%) than in the CATI waves. There are also systematic differences between the seasonal patterns. This is the net result of the following strongly confounded factors:

- Selective non-response between the subsequent waves, i.e. panel attrition.
- Systematic differences between the populations reached with the different data collection modes used in the different waves (coverage errors).
- Systematic differences in measurement bias because different data collection modes are used in different waves.
- Differences in wording and questionnaire design used in the different waves generally have systematic effects on the outcomes.
- Panel conditioning effects, i.e. systematic changes in the behaviour of the respondents in the panel. For example, questions about activities to find a job in the first wave might increase the search activities of the unemployed respondents in the panel. Respondents might also adjust their answers in the subsequent waves systematically, since they learn how to keep the routing through the questionnaire as short as possible.

16.3.2 Structural Time Series Model for Rotating Panels

With a structural time series model, a series is decomposed into a trend component, a seasonal component, other cyclic components, a regression component and an irregular component. For each component, a stochastic model is assumed. The trend, seasonal, and cyclic components are usually time dependent, but this approach also allows the regression coefficients to be time dependent. If necessary, ARMA components can be added to capture the autocorrelation in the series beyond these structural components. See Harvey (1989) or Durbin and Koopman (2012) for details about structural time series modelling.

Consider a rotating panel, where monthly samples are observed M times according to a particular rotation pattern. In the Dutch and the UK LFS, $M = 5$, where monthly samples are observed at quarterly intervals. Let θ_t denote the unknown population parameter for month t. With the rotating panel, each time period's data are collected from M panels, each at a different wave. These data can be used to construct M direct estimates for θ_t, which are denoted $\hat{y}_t^j$, $j = 1, \dots, M$. Direct estimators are e.g. the Horvitz-Thompson estimator to account for complex sample designs or general regression (GREG) estimators to further reduce the design variance and correct, at least partially, for selective non-response with auxiliary information available from administrative sources (Särndal et al. 1992).

At each time period $t = 1, \dots, T$, the M direct estimates can be collected in a vector, say $\hat{\boldsymbol{y}}_t = (\hat{y}_t^1, \dots, \hat{y}_t^M)'$. Pfeffermann (1991) proposed the following M dimensional structural time series model as a method of small area estimation and to account for RGB:

$$\hat{\boldsymbol{y}}_t = \mathbf{1}_{[M]}\theta_t + \boldsymbol{\lambda}_t + \boldsymbol{e}_t, \tag{16.1}$$

with $\mathbf{1}_{[M]}$ an M dimensional vector with each element equal to one, $\boldsymbol{\lambda}_t = (\lambda_t^1, \dots, \lambda_t^M)'$ an M dimensional vector with components that model the RGB and $\boldsymbol{e}_t = (e_t^1, \dots, e_t^M)'$ an M dimensional vector with sampling errors. The population parameter is modelled as

$$\theta_t = L_t + S_t + \varepsilon_t, \tag{16.2}$$

with L_t a stochastic model for the trend/cycle, S_t a stochastic model for the seasonal component and ε_t a white noise for the unexplained variation of the population parameter. Frequently applied models for L_t in econometric time series modelling are the local level model, the smooth trend model and the local linear trend model, see Durbin and Koopman (2012, Ch. 3) for expressions. These are stochastic models that can change gradually over time and are therefore appropriate to model the low-frequency variation in the time series. These models have the flexibility to model business cycles, but it is also possible to include a separate component in addition to L_t and S_t for the business cycle. For S_t the so-called dummy seasonal model or trigonometric seasonal model can be used, see Durbin and Koopman (2012, Ch. 3) for expressions. Both are stochastic models and therefore the seasonal pattern can gradually change over time.

The RGB is modelled in (16.1) with $\boldsymbol{\lambda}_t$, which contains time-dependent components that model the systematic differences between the M input series $\hat{\boldsymbol{y}}_t$. Without external information, it is only possible to estimate differences between the input series, not the absolute bias, and an additional restriction is required to identify the model for the RGB. One possibility is to set one of the M components of $\boldsymbol{\lambda}_t$ equal to zero, typically the series expected to be the most reliable and with the smallest bias. Another way to identify the RGB component is to add the restriction that the sum over the elements of $\boldsymbol{\lambda}_t$ is equal to zero, i.e. $\sum_{j=1}^{M} \lambda_t^j = 0$ for all t. The chosen identification determines the level at which the RGB-adjusted series is benchmarked. If there is external information about the absolute bias, one of the λ_t^j's or the sum over the elements of $\boldsymbol{\lambda}_t$ could be set equal to this value, as an attempt to correct for the absolute bias.

The λ_t^j's are modelled with random walks (or local level trend models) to allow for time-varying RGB. In the Dutch LFS, the RGB component is identified by assuming that the first wave is observed without bias, i.e. $\lambda_t^1 = 0$. This choice implies that the time series estimates for θ_t are benchmarked to the level of the series observed in the first wave. This choice is based on two considerations.

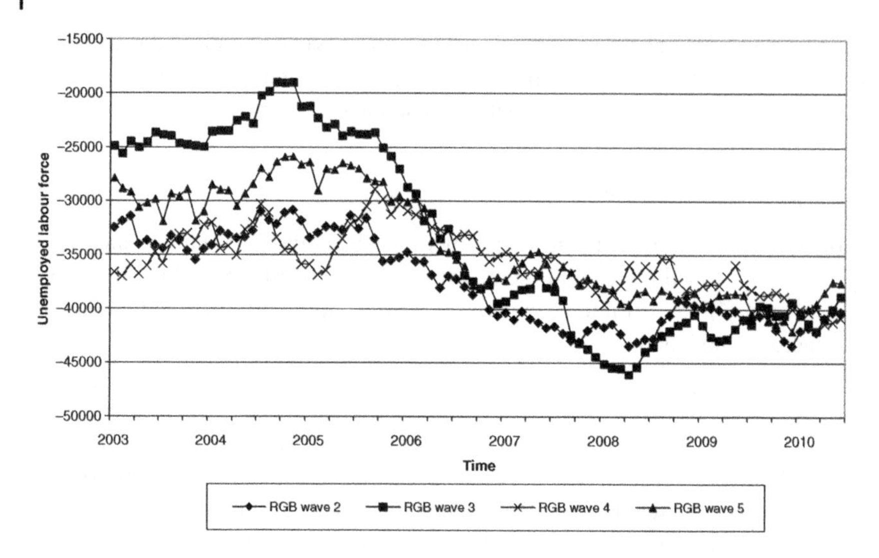

Figure 16.1 RGB estimates in the Dutch LFS for unemployed labour force at national level.

Firstly, based on the factors that contribute to the RGB (see Section 16.3.1), it is anticipated that measurement bias and selection effects are smallest in the first wave. Secondly, the Dutch LFS changed in 2000 from a cross-sectional design to a rotating panel. After the redesign, the RGB became visible as a discontinuity. Benchmarking the estimates for θ_t to the level of the observations in the first wave made the results under the panel design comparable with the results before the change-over under the cross-sectional design, since the questionnaire and data collection methods in the first wave were similar to the methods used under the cross-sectional design. With this choice, the discontinuity due to the introduction of a rotating panel was corrected. Figure 16.1 shows the filtered RGB estimates in the Dutch LFS for the four follow-up waves. RGB is clearly time dependent. In Figure 16.1 as well as the other figures in the results section, the first three years are ignored, since the Kalman filter requires some time to converge to proper values for the state variables.

In the UK LFS, RGB is modelled in a similar way to the Dutch LFS. The λ_t^j's are again modelled as random walks, but with the constraint that $\sum_{j=1}^{M} \lambda_t^j = 0$, which is used as there are clearly differences in the level of the wave-specific estimates, but there is no evidence that the current survey based estimates have a systematic bias in the level. By assuming that the RGB in wave-specific estimates sum to zero this maintains consistency in the level of model-based and design-based estimates.

The sampling errors are modelled in (16.1) with $\boldsymbol{e}_t$. This component accounts for serial correlation due to the partial sample overlap of the rotating panel design,

Table 16.1 Partial autocorrelation for wave specific unemployment (UK LFS).

Lag	Wave 2	Wave 3	Wave 4	Wave 5
3	0.63	0.52	0.47	0.61
6		0.31	−0.04	−0.17
9			0.30	0.05
12				0.25

Notes: Analysis uses the approach of Pfeffermann et al. (1998) over the period January 2002–November 2017. Estimation is based on respondents aged 16 and over. Shaded cells indicate significant partial autocorrelations based on an approximate standard error of 0.07 at the 5% significance level.

and for heteroscedasticity due to fluctuations in the sample size over time. A convenient approach is to estimate the sampling variance and the autocorrelation from the microdata of the LFS. The GREG estimates for the design variances of the survey errors are incorporated in the time series model using the survey error model $e_t^j = k_t^j \tilde{e}_t^j$ with

$$k_t^j = \sqrt{Var(\hat{y}_t^j)},$$

$Var(\hat{y}_t^j)$ an estimate for the design variance of $\hat{y}_t^j$ derived from the micro data and $\tilde{e}_t^j$ an appropriate stationary process for the scaled sampling error of the j-th wave.

The scaled sampling errors $\tilde{e}_t^j$ account for the serial autocorrelation induced by the sampling overlap of the rotating panel. An appropriate model can be derived by calculating the autocorrelation from the micro data or from the input series using the approach proposed by Pfeffermann et al. (1998). Based on these autocorrelations an appropriate AR model can be derived using the Yule-Walker equations.

In the case of the Dutch LFS, the autocorrelations for the four follow-up waves are modelled with an AR(1). More details about the state-space model developed for the Dutch LFS can be found in Van den Brakel and Krieg (2009a, 2015, 2016). For the UK LFS the partial autocorrelations for the variable total unemployment ages 16 and over (Table 16.1) suggest that an AR($M-1$) process may be required for modelling the sampling error $\tilde{e}_t^M$. Further details on the UK LFS models are given in Elliott (2017).

16.3.3 Fitting Structural Time Series Models

A widely applied approach to fit structural time series models is to write them in state-space form and analyse them with the Kalman filter. The Kalman filter is a recursive procedure that runs from period $t = 1$ to T and gives, for each time period, an optimal estimate for the state variables based on the information available up to and including period t. These estimates are referred to as the *filtered estimates*. The filtered estimates of past state vectors can be updated, if new data after period t become available. This procedure is referred to as smoothing and results in *smoothed estimates* that are based on the complete time series.

To start the filter at $t = 1$, the filter must be initialized with starting values for the state variables and its covariance matrix for $t = 0$. The non-stationary variables (i.e. the states for the trend, seasonal, and RGB), are initialised with a diffuse prior, i.e. the expectation of the initial states are equal to zero and the initial covariance matrix of the states is diagonal with elements diverging to infinity. The sampling errors are stationary and therefore initialised with a proper prior. The initial values for the sampling errors are equal to zero and the covariance matrix is available from the aforementioned model for the sampling errors.

Expressions for the state space representation of trend and seasonal components can be found in Durbin and Koopman (2012). Details of how to obtain the state space representation for the sampling error component are given by Pfeffermann (1991). The state space representation of the time series model for the Dutch LFS is given by Van den Brakel and Krieg (2009b), and for the UK LFS by Elliott (2017).

Several software packages are available to fit structural time series models. Most standard structural time series models can be fitted with STAMP (Koopman et al. 2007). For more advanced models like the one proposed in Section 16.3.2, more advanced software is required. One option is to implement these models in OxMetrics in combination with the subroutines of SsfPack 3.0, see Doornik (2009) and Koopman et al. (1999, 2008). Another possibility is to implement these models in R (R Core Team 2017) using packages like KFAS (Helske 2017) or DLM (Petris 2010).

16.4 Time Series Models for Discontinuities in Rotating Panel Designs

There are various strategies to roll out a new survey design in a rotating panel set-up. One approach is to start the change-over in the first wave and start the introduction of the new design with this panel. In this way the new design is gradually introduced according to the rotation pattern, i.e. the new design is introduced in the second wave, third wave, etc., at the moment that the first wave under the new design is observed for the second time, third time, etc. An alternative way is to introduce the new design at the same time in all M waves. The latter approach is less advisable but might be feasible if e.g. only the wording of the questions or the data collection mode changes. Changing a panel that is observed for the j-th time from the old to the new approach might result in different estimates for the discontinuity in this wave, compared to a comparison in a panel where the preceding waves are also conducted under the new design.

In Section 16.4.1 the time series model proposed in Section 16.3.2 is extended to account for discontinuities. A parallel run might precede the change-over to the new design. Some design issues of a parallel run are discussed in Section 16.4.2. Section 16.4.3 discusses how this information can be combined with the time series model from Section 16.4.1. In Section 16.4.4, the time series model is extended to include auxiliary series, which might be useful to better separate discontinuities from real evolution in the population parameters of interest.

16.4.1 Structural Time Series Model for Discontinuities

Rolling out a new survey design in a rotating panel, will likely introduce discontinuities in the series of the M waves. To avoid model-misspecification, (16.1) is extended with level interventions, which account for these sudden shocks, leading to the following model

$$\hat{\mathbf{y}}_t = \mathbf{1}_{[M]}\theta_t + \boldsymbol{\lambda}_t + \boldsymbol{\Delta}_t\boldsymbol{\beta} + \mathbf{e}_t \quad (16.3)$$

In (16.3), $\boldsymbol{\Delta}_t$ is an $M \times M$ diagonal matrix where the diagonal elements, say δ_t^j, contain dummy indicators that change from zero to one the moment the survey design for wave j changes from the old to the new design;

$$\delta_t^j = \begin{cases} 0 \ \textit{if} \ t < T^j \\ 1 \ \textit{if} \ t \geq T^j \end{cases},$$

where T^j is the moment that the new survey design is introduced in wave j. Furthermore, $\boldsymbol{\beta} = (\beta^1, \ldots, \beta^M)'$ denotes an M dimensional vector containing regression coefficients belonging to the dummy indicators δ_t^j. Under the assumption that the time series component (16.2) correctly models the evolution of the population variable, the regression coefficients in $\boldsymbol{\beta}$ can be interpreted as the systematic effect of the redesign on the level of the series observed in the five panels. With a step intervention it is assumed that the redesign only has a systematic effect on the level of the series. Alternative interventions, e.g. for the slope or the seasonal components are also possible, see Durbin and Koopman (2012, Ch. 3) and Van den Brakel and Roels (2010). More details about the application of model (16.3) in the Dutch LFS can be found in Van den Brakel and Krieg (2015).

A redesign might affect not only the point estimates but also the variance of the GREG estimates. Since the standard errors of the GREG estimates are used as prior information in the time series model for the sampling error, as explained in Section 16.3.2, systematic differences in the variance of the GREG estimates are automatically taken into account. An alternative possibility would be to allow for different values for the variance of the measurement equation disturbance terms before and after the survey redesign, which can be interpreted as an intervention on the variance hyperparameter of the survey error (Bollineni-Balabay et al. 2016).

16.4.2 Parallel Run

Rolling out a new survey design without any form of pre-testing or parallel data collection and relying solely on an intervention analysis to quantify discontinuities increases statistical risks for various reasons. If the new survey approach turns out to be a failure and it is decided to turn back to the old approach, a period without data for constructing official figures is created. Furthermore, there is no control over the minimum size of the discontinuities that can be detected. Directly after the change-over to the new design, the initial estimates for the discontinuities obtained with intervention model (16.3), are very inaccurate. As a result, the estimates for the discontinuities are subject to revisions in the period directly after the change-over and the final estimates are available only with a substantial delay. Finally, there is always the risk that part of the real evolution of the population parameter is incorrectly absorbed in the estimate for the discontinuity.

Based on these considerations, it can be useful to conduct a parallel run based on some kind of randomised experiment embedded in the ongoing sample. Planning and designing an experiment requires key decisions on several themes. A clear definition is needed about the treatments to be tested and the number of factors to be included in the experiment. For example, is the purpose of the experiment to estimate the difference between the old and new design as a whole, which can be done with a two-sample experiment, or to explain the effect of each change separately, which usually requires a factorial design?

An important design aspect is to decide in advance which difference should at least result in a rejection of the null hypotheses at a pre-specified significance and power level. If the budget and thus the sample size are fixed in advance, it can be calculated what the minimum observable differences are at pre-specified significance and power levels. This gives an indication of what can be expected from the experiment. The sample design provides a framework to design efficient randomised experiments. For details, see Fienberg and Tanur (1987, 1988), Van den Brakel and Renssen (2005), and Van den Brakel (2008). For the analysis of experiments embedded in sample surveys, Van den Brakel and Renssen (2005), Van den Brakel (2008, 2013), and Chipperfield and Bell (2010) developed design-based inference procedures that account for the sample design as well as the superimposition of the applied experimental design on the sampling design.

An important design choice is whether a parallel run of equal size and length is conducted in each wave or whether a major part of the sample is allocated to a parallel run in one of the waves. This choice is related to the way that the RGB component in model (16.1, 16.3) is identified; see Section 16.3.2. In the case of the

Dutch LFS, it was argued that the RGB in the first wave can be assumed to be equal to zero and that the outcomes of the follow-up waves are benchmarked to the level of the first wave. This choice naturally implies that the quality of the population parameter estimates is dominated by the data quality of the first wave. This might be an argument to quantify the discontinuities in the first wave as precisely as possible and to allocate a major part of the sample size for the parallel run to the first wave. In this case, an estimate for the discontinuity in the population parameter estimate is equal to the discontinuity observed in the first wave. An example of this approach is worked out in Section 16.5.1.

If the RGB is identified with the restriction that the sum over the elements of λ_t is equal to zero, then the M waves have similar impacts on the quality of the population parameter estimates. This can be an argument for a parallel run covering all M waves – for example, by starting a parallel run for K panels as they begin their first wave. Each panel follows the rotating pattern under the new design until it reaches the last wave. The advantage of this approach is that direct estimates for the discontinuities in all waves are available at the end of the parallel run. An estimate for the discontinuity for the population parameter is obtained by averaging the discontinuity estimates over the M waves. Since the discontinuity in each wave is estimated using many common respondents, there will be a strong positive correlation between the M wave-specific discontinuity estimates.

16.4.3 Combining Information from a Parallel Run with the Intervention Model

In many cases, the available budget for a parallel run will be insufficient to meet pre-specified precision requirements. It is therefore important to combine the information from the parallel run with all information observed in the time series before and after the change-over.

In the intervention model (16.3), the time independent regression coefficients β^j of the intervention variables are included in the state vector. In the absence of any prior information about these state variables, they are initialised in the Kalman filter with a diffuse prior, as described in Section 16.3. One way to incorporate the initial information about the discontinuities available from the parallel run in the model is to use an exact initialization for the regression coefficients β^j. This can be done by using the direct estimate for the discontinuity obtained from the parallel run in the initial state vector for β^j and the estimated variance of this direct estimate as an uncertainty measure for β^j in the covariance matrix of the initial state vector. In this way, the Kalman filter combines the information from the parallel run with the information available in the entirely observed series before and after the change-over to the new design.

16.4.4 Auxiliary Time Series

If auxiliary time series that are strongly related to the target variable measured with the rotating panel design are available, then this information can be incorporated in the model to better separate real developments from discontinuities in the intervention model and to improve the precision of the discontinuity estimates. For the unemployed labour force, the number of people formally registered at the employment office is a potential auxiliary variable to be included in the model.

There are different ways to incorporate auxiliary information in the model. One possibility is to extend the time series model (16.2) for the population parameter of the LFS with a regression component for the auxiliary series, i.e. $\theta_t = L_t + S_t + bx_t + \varepsilon_t$, where x_t denotes the auxiliary series and b the regression coefficient.

Another approach is to extend model (16.3) with an auxiliary series (Harvey and Chung 2000; Van den Brakel and Krieg 2015):

$$\begin{pmatrix} \hat{\mathbf{y}}_t \\ x_t \end{pmatrix} = \begin{pmatrix} \mathbf{1}_{[M]}\theta_t^y \\ \theta_t^x \end{pmatrix} + \begin{pmatrix} \boldsymbol{\lambda}_t \\ 0 \end{pmatrix} + \begin{pmatrix} \boldsymbol{\Delta}_t\boldsymbol{\beta} \\ 0 \end{pmatrix} + \begin{pmatrix} \mathbf{e}_t \\ 0 \end{pmatrix}. \tag{16.4}$$

In (16.4) θ_t^y and θ_t^x are the population parameters of the LFS and the auxiliary variable, respectively. Both series are modelled according to (16.2) with their own trend, seasonal component and white noise, i.e. $\theta_t^z = L_t^z + S_t^z + \varepsilon_t^z$ with $z = y, x$. The correlation between both series can be modelled, e.g. by specifying a full covariance matrix for the disturbance terms of the trend components in L_t^y and L_t^x. In this way, additional information from the auxiliary series is used to better separate discontinuities from real developments of the monthly unemployed labour force. The auxiliary information will also result in increased precision of the model estimates for the monthly unemployment figures. Details of how to fit multivariate state-space models with correlated disturbance terms can be found in e.g. Harvey (1989, Section 8.5) or Koopman et al. (2007, Section 9.1).

16.5 Examples

16.5.1 Redesigns in the Dutch LFS

Three major redesigns took place in the Dutch LFS. The first one was in 2000, when the LFS changed from a cross-sectional to a rotating panel design. The second was in 2010, when the data collection in the first wave changed from single mode CAPI to a mixed mode design using CATI for households with a non-secret landline telephone number and CAPI for the remaining households. To make the questionnaire of the first wave suitable for CATI, questions were reformulated and the length of the questionnaire needed to be reduced substantially. Therefore, several blocks were moved from the first wave to one of the follow-up waves. As a result, the questionnaire changed in each of the five waves. The third redesign took place in 2012, when the data collection in the first wave changed to a sequential mixed-mode design starting with WI, with a follow up using CATI or CAPI. The questionnaire was again revised in each of the five waves.

The introduction of the rotating panel design in 2000 resulted in a discontinuity in the labour force figures because of the RGB of the rotating panel. The model applied in the Dutch LFS assumes that the RGB in the first wave equals zero. This implies that the outcomes of the follow-up waves are benchmarked to the level of the first wave. In this way, the official figures are corrected for the discontinuity introduced by the change to the rotating panel. The time series model (16.1) was, however, implemented in June 2010 to produce official model-based monthly figures about the labour force. Until June 2010, rolling quarterly figures about the labour force were published each month, and a rigid correction was applied to correct for the RGB. For the most important parameters, the ratio between the estimates based on the first panel only and the estimates based on all panels was computed using the data of the 12 preceding quarters. Rolling quarterly estimates were multiplied by this ratio to correct for RGB.

For the change to mixed-mode data collection in 2010 and in 2012, the intervention approach of model (16.3) in combination with a parallel run was required. The assumption that the RGB in the first wave equals zero was an argument to use the major part of the budget for a parallel run in the first wave (Van den Brakel and Krieg 2015).

From now on, the initial survey design of the rotating panel used from 2000 until 2010 is denoted with Aj, with $j = 1, \ldots, 5$ referring to the five waves (i.e. the subsequent survey occasions of the rotating panel design). Similarly, the intermediate design used from 2010 through 2012 is denoted Bj and the final design used since 2012 is denoted Cj. The roll-out patterns are shown in Figure 16.2. For the first parallel run in 2010, there was a budget to conduct parallel collection for new panels for two calendar quarters (six months); both the old and new design were at the regular sample size. Due to the rotation pattern, the first new panel continues the parallel run for a further quarter in wave 2. During the parallel run, official publications were based on observations under the old design in all waves. When the parallel run finished after two quarters, the intermediate design was assessed as satisfactory, and time series model (16.1) was extended to (16.3) with an intervention component at the point of change in each panel to account for the difference in the input series. The intermediate design data then replaced the original design data for the parallel run periods, and were used for official publications going forward. Although there were small differences in the intermediate design, official monthly publications during the parallel run, based on data observed under the old design, were not revised. This method resulted in a smooth change-over to

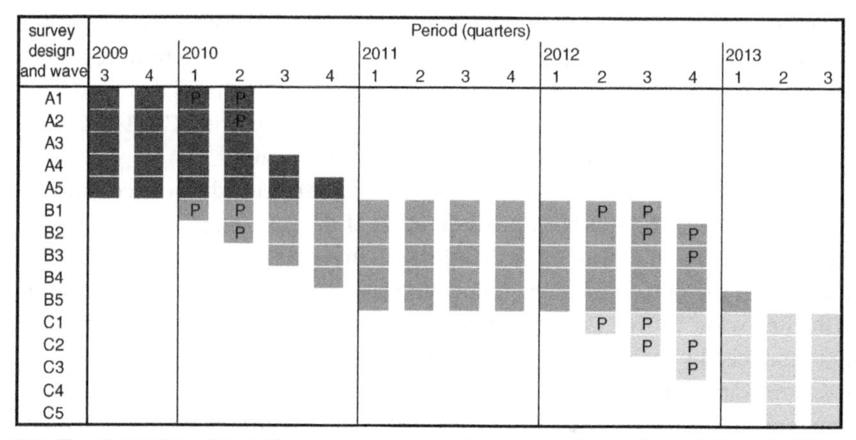

Note: The observations obtained from one sample (panel) are represented by diagonal elements of the same shading. For example, the panel first fielded in quarter 1 of 2010 under the intermediate design contributes the observations B1 (2010 Q1), B2 (2010 Q2), B3 (2010 Q3), B4 (2010 Q4) and B5 (2011 Q1)

Figure 16.2 Roll-out strategy for new data collection approaches for the Dutch LFS. 'P' indicates parallel running where waves on two approaches are undertaken at the same time.

the intermediate design. The panels that started under the old design during the parallel run were stopped (but those without a parallel sample under the intermediate design continued under the old design until their rotation was complete).

The parallel run for the second design change, from B to C in 2012, had a similar structure and outcome, but there was a budget to conduct parallel collection for new panels at the regular sample size, for three quarters (nine months). Due to the rotation pattern, this gave two quarters parallel run at wave 2, but none at wave 3 because the first parallel panel was stopped after wave 2. At the end of the parallel run, the final collection design was assessed as satisfactory for use in model (16.3) with a further intervention component for the new change. Then the final data replaced the intermediate data during the parallel run period for official publications, and the panels that started under the intermediate design during the parallel run were stopped.

Both parallel runs were designed as two-sample experiments where the strata of the sample design were used as block variables in a randomised block design. This design gives the most precise direct estimate for the discontinuity but cannot disentangle the effect of different data collection modes and different questionnaires. During the period that the intermediate design B was in place, the time series model (16.3) was used to produce labour force figures adjusted to the level of the initial design A. Since the implementation of Design C, publications have been based on this new design. In this way users were only confronted once with a discontinuity.

Table 16.2 Results of parallel runs in the Dutch LFS in 2010 and 2012.

Parameter	Parallel run 2010		Parallel run 2012		
	Discontinuity	Level (under A)	Discontinuity		Level (under A)
	B–A		C–B	C–A	
Unemployed	68 (17)	400	−19 (17)	49 (24)	460
Employed	−97 (35)	8350	16 (33)	−81 (48)	8400
Total labour force	−29 (34)	8750	−3 (32)	−32 (47)	8860

Notes: Cell entries are point estimates and standard errors (in brackets) of the discontinuities in thousands. The level refers to the average value of the trend during the parallel run under the initial design A.

Direct estimates for the discontinuities obtained with the parallel runs with a length of two quarters in the first wave of the redesign in 2010 and the first and second wave in 2012 were used as a priori values for the regression coefficients in the intervention components for these waves in model (16.3). The discontinuities for the other waves were obtained with the Kalman filter after diffuse initialization of the regression coefficients (it would be possible additionally to initialise with the first quarter of wave 2 in the first experiment and the first quarter of wave 3 in the second, but this resulted in less smooth transitions (Van den Brakel and Krieg 2015)).

Estimates for the discontinuities in the first wave are summarised in Table 16.2 at the national level for monthly unemployed, employed, and total labour force. For the unemployed labour force it follows that the implementation of the intermediate design B, resulted in an increase of 68 000 people, which is significantly different from zero at the two-sided 5% level. The change-over from the intermediate design B to the final design C resulted in an insignificant decrease of the unemployed labour force of 19 000 people. The net discontinuity between the initial design A and the final design C is an increase of 49 000 people, which is still significant at the 5% level.

In Figure 16.3 the filtered estimates for the discontinuities in the four follow-up waves due to the changeover to the intermediate design in 2010 are presented as an example. This graph illustrates that directly after the changeover the estimates for the discontinuities are unstable and subject to large revisions as new observations under the new design become available. It takes more than a year before the estimates converge to a more or less stable value. The standard errors converge to 12 000 for the second wave, 13 000 for the third and fourth wave, and 14 000 for the fifth wave. As an illustration the filtered trend under the initial design (A) and the final design (C) are compared with the time series of GREG estimates for the five waves for the unemployed labour force in Figure 16.4. The trend under design A is

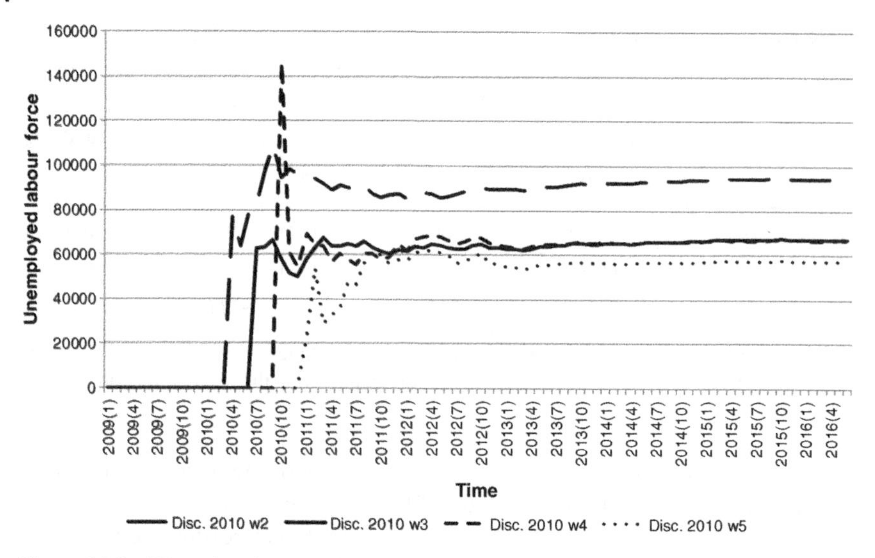

Figure 16.3 Filtered estimates of discontinuities for the follow-up waves for the redesign of the Dutch LFS 2010 for the unemployed labour force.

in fact the filtered estimate for L_t of the population parameter in (16.2). This trend is benchmarked to the level of the GREG estimates in the first wave under design A and is corrected for the discontinuities in 2010 and 2012 by level interventions in time series model (16.3). The trend under design C is in fact L_t plus the discontinuities in the first wave in 2010 and 2012. The thin solid line in black is the GREG estimate for the first wave. Before 2010, the trend under design A is at the level of the GREG of the first wave. After 2012, the trend under design C is at the level of the GREG of the first wave. Due to the positive discontinuities, the level of the trend under design A drops below the level of the input series after 2010.

In Figure 16.5 the standard errors of the trend under design A are shown. During the period 2003–2009, these standard errors gradually decrease since more information from preceding periods becomes available. In 2009, the standard errors increase, which is the result of a reduction in the monthly sample size. In 2010, the standard errors increase rapidly, because of the change to the intermediate design. During this period the model components for the discontinuities are activated and add additional uncertainty to the model, in particular to the level of the trend. After completing the change to the intermediate design in 2011, the standard errors of the trend gradually decrease since more information under the intermediate design becomes available. In 2012, the standard errors increase rapidly again due to the change to the final design, which adds additional uncertainty to the level of the trend. Once the change-over is completed in 2013, the standard errors gradually decrease as more and more information under the final design becomes

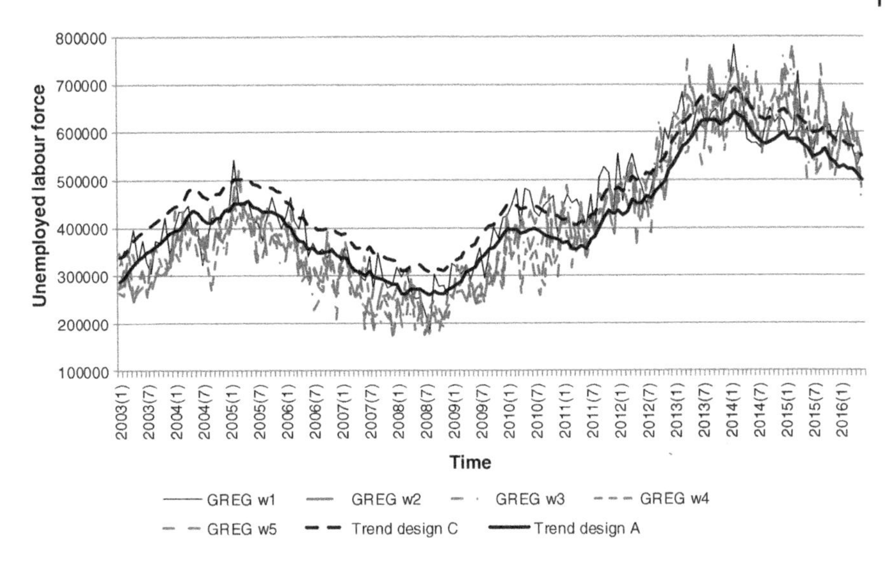

Figure 16.4 Filtered trend in the unemployed labour force in the Dutch LFS under initial design (A) and final design (C) with the observed series in the five panels.

available. The introduction of two intervention components to account for discontinuities results in a substantial increase of the uncertainty of the trend, which can be seen as the cost for modelling the change with intervention components.

16.5.2 Using a State Space Model to Assess Redesigns in the UK LFS

Over the period of time for which the LFS has been conducted in the UK, there have been numerous changes, details of which can be found in ONS (2016). The data available as inputs for state space modelling are from January 2002 onwards, and so this section focuses on changes since that date, none of which have involved parallel runs. The most notable changes that have the potential to affect the main LFS variables include an increase in the proportion of CATI in the first wave, a change to the way in which multiple occupancy addresses are dealt with and a change to interviewing of households where all occupants are over the age of 75. The development of a state space model for the UK LFS occurred after these potential discontinuities, so their effects are analysed, retrospectively, by introducing appropriate intervention variables into the model.

From January 2011 onwards, if a match between a telephone number and a sampled address was available, then the first approach to that address was by telephone rather than CAPI. This is in addition to a very small proportion of addresses north of the Caledonian Canal that receive only telephone interviews. This resulted in approximately 15% of first wave interviews as CATI, the remainder being CAPI. To

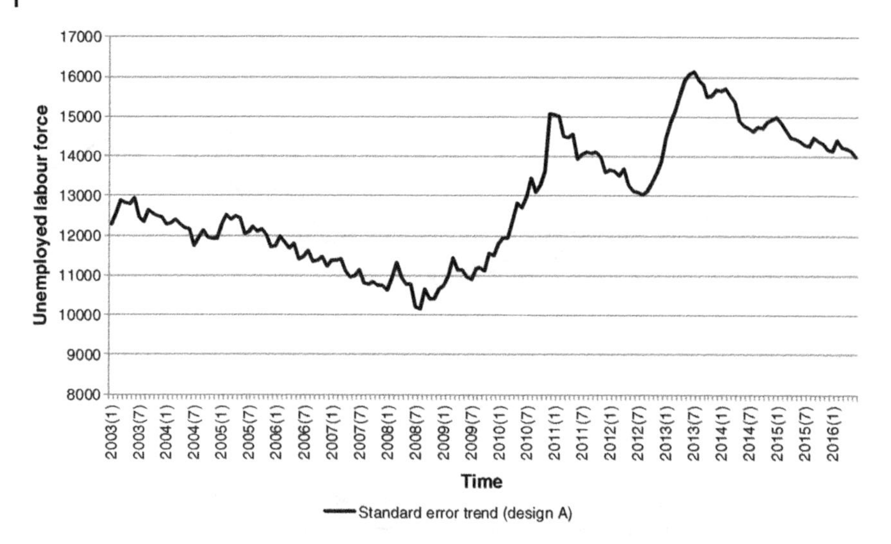

Figure 16.5 Standard error of the filtered trend in the unemployed labour force in the Dutch LFS under initial design (A).

model the effect of this discontinuity an intervention variable can be introduced to the model, for the wave one series only, from January 2011 onwards.

A change to the treatment of multiple occupancy addresses (addresses containing more than one household) began in July 2010 for wave one addresses, was introduced to wave two addresses from October 2010 onwards and so on as the change fed through the waves, with wave five only having the change in July 2011. Before the change all households at the sampled address were interviewed, whereas after these dates only one household from the address was sampled. Again, intervention variables are introduced into the model for each wave at the appropriate date for when the change occurred, as discussed in Section 16.4.1. It is assumed that the magnitude of the effect is the same for each wave, but that the impact starts at different periods based on when the change was introduced.

An additional change in July 2010 that could have had an impact on waves two to five only is the treatment of households where all members of the household are over the age of 75. From this date, the status of those members is effectively rolled forward as responses for all the subsequent waves. The model assumes that the discontinuity will have an equal effect on each wave.

Incorporating these variables into the model gives no evidence of a statistically significant discontinuity due to the changed treatment of addresses of multiple occupancy for any of the main UK LFS variables, whereas there is evidence of a discontinuity in wave 1 estimates due to the introduction of CATI and a discontinuity in waves 2–5 due to the changed treatment of households where all members

Table 16.3 Estimated coefficients of discontinuity variables for UK unemployment, employment, and inactivity ages 16 and over in thousands.

Discontinuity	Unemployment	Employment	Inactivity
CATI wave 1	−87 (45)	273 (75)	−186 (60)
Multiple occupancy	−54 (61)	41 (115)	13 (97)
Over 75	−74 (43)	196 (69)	−121 (54)

Note: Standard errors in parentheses.

are over the age of 75. Table 16.3 presents the *smoothed estimates* of these three discontinuities using the span of data from January 2002 to October 2017. There should be no discontinuity in the sum of total employment, unemployment, and inactivity, which should, by definition, be equal to the estimated population, and so the sum of the discontinuities over the three variables should be zero. To ensure that this constraint is met the models for the three variables based on Eq. (16.3) are combined into one model, stacking in a similar way to that described in (16.4) with each discontinuity for the unemployment variable equal to minus the sum of that discontinuity for the other two variables. Model (16.3) is extended using the superscripts U, E and I to denote unemployment, employment, and inactivity, respectively:

$$\begin{pmatrix} \hat{\mathbf{y}}_t^U \\ \hat{\mathbf{y}}_t^E \\ \hat{\mathbf{y}}_t^I \end{pmatrix} = \begin{pmatrix} \mathbf{1}_{[M]}\theta_t^{y^U} \\ \mathbf{1}_{[M]}\theta_t^{y^E} \\ \mathbf{1}_{[M]}\theta_t^{y^I} \end{pmatrix} + \begin{pmatrix} \lambda_t^U \\ \lambda_t^E \\ \lambda_t^I \end{pmatrix} + \begin{pmatrix} -\boldsymbol{\Delta}_t\boldsymbol{\beta}_t^E - \boldsymbol{\Delta}_t\boldsymbol{\beta}_t^I \\ \boldsymbol{\Delta}_t\boldsymbol{\beta}_t^E \\ \boldsymbol{\Delta}_t\boldsymbol{\beta}_t^I \end{pmatrix} + \begin{pmatrix} \mathbf{e}_t^U \\ \mathbf{e}_t^E \\ \mathbf{e}_t^I \end{pmatrix}. \tag{16.5}$$

Filtered estimates due to the discontinuities from the increased use of CATI in wave 1 and for the changes in waves 2–5 on the treatment of households over the age of 75 are presented in Figure 16.6 for UK employment ages 16 and over. The filtered estimates show that it is necessary to have a reasonable number of observations after the cause of the discontinuity before the estimates become stable, as discussed for the Dutch case in Section 16.5.1.

16.6 Discussion

In rotating panel designs, RGB can occur due to non-sampling errors that are visible as systematic differences between the multiple occasions that samples are observed. Major redesigns of the survey process generally also change the net effect of non-sampling errors, resulting in systematic differences in the outcomes of a survey that are called discontinuities. To avoid confounding real period-to-period

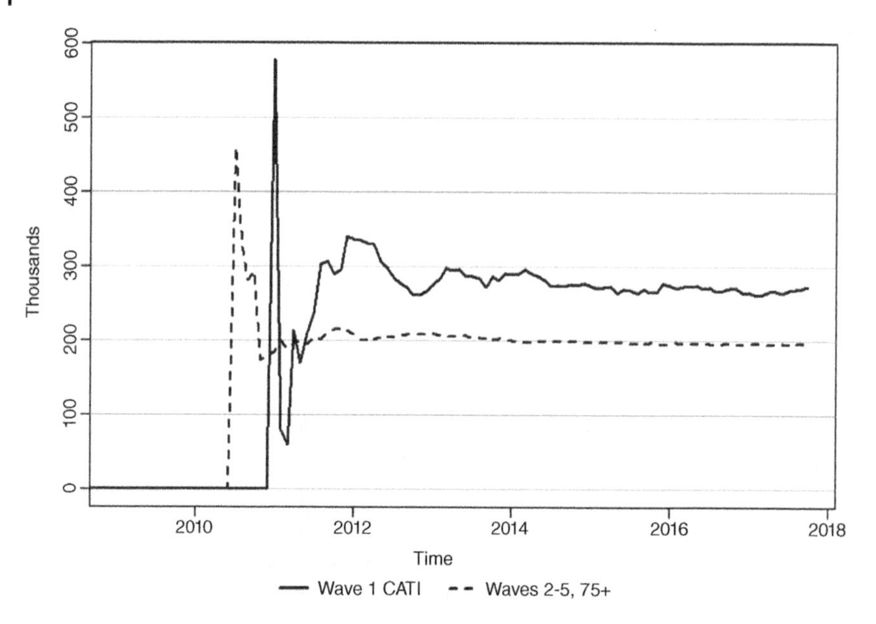

Figure 16.6 Filtered estimates of discontinuities for UK employment ages 16 and over.

change with systematic differences in measurement error due to a redesign, it is important that discontinuities are quantified. RGB in level estimates is often accepted if it does not affect estimates of period-to-period change. For the Dutch LFS this is not a satisfactory approach since the change-over from a cross-sectional design to a rotating panel made the RGB visible as a discontinuity.

In this chapter, a structural time series modelling approach is described that enables estimation of RGB and discontinuities. The time series model can also be used as a form of small area estimation since it increases the effective sample size of the direct estimator with sample information from preceding periods.

For the Dutch LFS, the time series model has been used to publish official monthly labour force figures since June 2010. The unemployed labour force estimate in the first wave is about 10% larger than in the follow-up waves. By benchmarking the outcomes to the level of the first wave, the discontinuity from introduction of a rotating panel is removed. Additionally, there are clear indications that the RGB is time dependent. This suggests that the common assumption that RGB can be ignored since it does not affect period-to-period change under a balanced rotation pattern is questionable.

Changing the data collection in the first wave from CAPI to sequential mixed-mode design using WI, CATI, and CAPI resulted in an increase in the unemployed labour force estimate of about 15%. The roll-out strategy in combination with the time series modelling approach allowed for a smooth transition to the new designs, without disturbing the official publications.

For the UK LFS, the model is not yet used for official publication purposes as it is still being developed and tested. Estimates from the model are evaluated each month as part of the publication schedule and compared to the current design-based estimators to help producers of the data understand the model-based outputs. The evidence of potential discontinuities is also being investigated further to test the conclusions from the model-based approach.

References

Bailar, B.A. (1975). The effects of rotation group bias on estimates from panel surveys. *Journal of the American Statistical Association* 70: 23–30.

Bollineni-Balabay, O., van den Brakel, J.A., and Palm, F. (2016). Multivariate state-space approach to variance reduction in series with level and variance breaks due to sampling redesigns. *Journal of the Royal Statistical Society, Series A* 179: 377–402.

Chipperfield, J. and Bell, P. (2010). Embedded experiments in repeated and overlapping surveys. *Journal of the Royal Statistical Society, Series A* 173: 51–66.

Doornik, J.A. (2009). *An Object-Oriented Matrix Programming Language Ox 6.* London: Timberlake Consultants Press.

Durbin, J. and Koopman, S.J. (2012). *Time Series Analysis by State Space Methods*, 2e. Oxford: Oxford University Press.

Elliott, D. (2017). Increasing frequency and improving timeliness of key variables in the UK Labour Force Survey. Masters thesis. University of Southampton.

Fienberg, S.E. and Tanur, J.M. (1987). Experimental and sampling structures: parallels diverging and meeting. *International Statistical Review* 55: 75–96.

Fienberg, S.E. and Tanur, J.M. (1988). From the inside out and the outside in: combining experimental and sampling structures. *Canadian Journal of Statistics* 16: 135–151.

Harvey, A.C. (1989). *Forecasting, Structural Time Series Models and the Kalman Filter.* Cambridge: Cambridge University Press.

Harvey, A.C. and Chung, C.H. (2000). Estimating the underlying change in unemployment in the UK. *Journal of the Royal Statistical Society, Series A* 163: 303–339.

Helske, J. (2017). KFAS: exponential family state space models in R. *Journal of Statistical Software* 78 (10) https://doi.org/10.18637/jss.v078.i10.

Kalton, G. (2009). Designs for surveys over time. In: *Handbook of Statistics, Sample Surveys: Design, Methods and Applications*, vol. 29A (eds. D. Pfeffermann and C.R. Rao), 89–108. Amsterdam: Elsevier.

Koopman, S.J., Harvey, A.C., Doornik, J.A. et al. (2007). *STAMP 8: Structural Time Series Analyser, Modeller and Predictor*. London: Timberlake Consultants Press.

Koopman, S.J., Shephard, N., and Doornik, J.A. (1999). Statistical algorithms for models in state space form using Ssfpack 2.2. *The Econometrics Journal* 2: 113–166.

Koopman, S.J., Shephard, N., and Doornik, J.A. (2008). *SsfPack 3.0: Statistical Algorithms for Models in State Space Form*. London: Timberlake Consultants Press.

McLaren, C.H. and Steel, D.G. (2000). The impact of different rotation patterns on the sampling variance of seasonally adjusted and trend estimates. *Survey Methodology* 26: 163–172.

ONS (2016). Labour Force Survey user guide: Volume 1 - LFS Background and Methodology 2016. www.ons.gov.uk/file?uri=/employmentandlabourmarket/peopleinwork/employmentandemployeetypes/methodologies/labourforcesurveyuserguidance/volume1nov17.pdf (accessed 12 October 2018).

Petris, G. (2010). An R package for dynamic linear models. *Journal of Statistical Software* 36: 1–16.

Pfeffermann, D. (1991). Estimation and seasonal adjustment of population means using data from repeated surveys. *Journal of Business & Economic Statistics* 9: 163–175.

Pfeffermann, D., Feder, M., and Signorelli, D. (1998). Estimation of autocorrelations of survey errors with application to trend estimation in small areas. *Journal of Business & Economic Statistics* 16: 339–348.

R Core Team (2017). *R: A Language and Environment for Statistical Computing*. Vienna, Austria: R Foundation for Statistical Computing.

Särndal, C.-E., Swensson, B., and Wretman, J. (1992). *Model Assisted Survey Sampling*. New York: Springer Verlag.

Smith, P., Lynn, P., and Elliot, D. (2009). Sample design for longitudinal surveys. In: *Methodology of Longitudinal Surveys* (ed. P. Lynn), 21–33. Chichester: Wiley.

Steel, D. (1997). Producing monthly estimates of unemployment and employment according to the International Labour Office definition (with discussion). *Journal of the Royal Statistical Society, Series A* 160: 5–46.

Van den Brakel, J.A. (2008). Design-based analysis of embedded experiments with applications in the Dutch labour force survey. *Journal of the Royal Statistical Society, Series A* 171: 581–613.

Van den Brakel, J.A. (2013). Design based analysis of factorial designs embedded in probability samples. *Survey Methodology* 39: 323–349.

Van den Brakel, J.A. and Krieg, S. (2009a). Estimation of the monthly unemployment rate through structural time series modelling in a rotating panel design. *Survey Methodology* 35: 177–190.

Van den Brakel, J.A. and Krieg, S. (2009b). Structural time series modelling of the monthly unemployment rate in a rotating panel design. Statistics Netherlands Discussion Paper 09031. https://www.cbs.nl/nl-nl/achtergrond/2009/24/

structural-time-series-modelling-of-the-monthly-unemployment-rate-in-a-rotating-panel (accessed 12 October 2018).
Van den Brakel, J.A. and Krieg, S. (2015). Dealing with small sample sizes, rotation group bias and discontinuities in a rotating panel design. *Survey Methodology* 41: 267–296.
Van den Brakel, J.A. and Krieg, S. (2016). Small area estimation with state-space common factor models for rotating panels. *Journal of the Royal Statistical Society, Series A* 179: 763–791.
Van den Brakel, J.A. and Renssen, R. (1998). Design and analysis of experiments embedded in sample surveys. *Journal of Official Statistics* 14: 277–295.
Van den Brakel, J.A. and Renssen, R. (2005). Analysis of experiments embedded in complex sample designs. *Survey Methodology* 31: 23–40.
Van den Brakel, J.A. and Roels, J. (2010). Intervention analysis with state-space models to estimate discontinuities due to a survey redesign. *Annals of Applied Statistics* 4: 1105–1138.
Van den Brakel, J.A., Smith, P.A., and Compton, S. (2008). Quality procedures for survey transitions – experiments, time series and discontinuities. *Survey Research Methods* 2: 123–141.
Van den Brakel, J.A., Buelens, B., and Boonstra, H.J. (2016). Small area estimation to quantify discontinuities in repeated sample surveys. *Journal of the Royal Statistical Society, Series A* 179: 229–250.

17

Proper Multiple Imputation of Clustered or Panel Data

Martin Spiess[1], Kristian Kleinke[2] and Jost Reinecke[3]

[1]*Department of Psychological Methods and Statistics, University of Hamburg, Hamburg, Germany*
[2]*Department of Education Studies and Psychology, University of Siegen, Siegen, Germany*
[3]*Faculty of Sociology, University of Bielefeld, Bielefeld, Germany*

17.1 Introduction

Most datasets are affected by missing values, i.e. values intended to be surveyed but are not observed. In contrast to these unintended missing values there may also be planned missingness like, e.g., in split questionnaire designs. These are examples of item nonresponse. An extreme form of item nonresponse is unit nonresponse where selected units are completely missing. In addition, panel data are affected by attrition, i.e. units observed in wave t are not observed in later waves.

In this chapter, we consider multiple imputation (MI) as a method to compensate for missing items. If the number of cases dropping out is large, however, compensating using weights might be more convenient in this situation but will not be treated here. Combining both methods is, for example, considered in Spiess (2006) and some promising results are presented in Seaman et al. (2012b).

If a dataset is affected by missing values then, in the analysis step, assumptions about the process leading to nonresponse are made either implicitly (implicit strategy), e.g. by accepting the default techniques of many software packages like analysing only the completely observed cases, or explicitly (explicit strategy) by overtly stating assumptions, e.g. about the process that generated the missing values. If imputations are generated at the analysis stage, then the imputation method can optimally be adapted to the assumptions underlying the planned analyses. On the other hand, if an analyst is planning to use imputations provided with a publicly available dataset, then it is necessary that the method, the models and variables used to generate imputations are sufficiently well documented so that the underlying assumptions are clear. If these assumptions, which can be debated, are not correct, then inferences are likely to be invalid: Estimators may

Advances in Longitudinal Survey Methodology, First Edition. Edited by Peter Lynn.

Companion website: www.wiley.com/go/lynn/advancesinlongitudinalsurvey

be biased and/or standard errors may be wrong, usually leading to confidence intervals being too short and true null hypotheses being rejected too often. As a consequence, non-existing effects may be reported more often than prespecified by α. Thus, the best strategy to minimise the risk of invalid inferences is to avoid missing values. As missing values are almost inevitable, the second best strategy is to adopt an explicit strategy because then inferences will be valid if the known modelling assumptions are (approximately) met. Usually these strategies are preferable over implicit models because the latter are derived based on criteria like simplicity of downstream analyses that do not necessarily support statistical validity of the inferences of interest.

17.2 Missing Data Mechanism and Ignorability

A nowadays standard classification of missing data mechanisms goes back to Rubin (1976) distinguishing missing values as being missing completely at random (MCAR), missing at random (MAR), or missing not at random (MNAR), where the event of observing or not observing a value is modelled via a binary variable.

If the probability for the observed pattern of missing and not missing values in the dataset at hand depends neither on variables whose values are observed nor on variables whose values are not observed, then the missing data are called MCAR. An example for missing values being MCAR is that, due to a poorly designed questionnaire but not depending on any variable surveyed, some units simply overlook an item in a questionnaire. If the probability of observing the pattern of the response indicators in the dataset at hand may depend on observed values but not (in addition) on variables whose values are not observed, then the missing data are called MAR. An example would be if younger individuals are more likely to provide their income than the older, independently from income itself. Finally, missing values are MNAR if the probability of the observed pattern of response indicators depends on variables whose values are not observed, even after conditioning on observed values. Missing income information would, e.g. be MNAR if in wave $t+1$ of a longitudinal dataset individuals are less likely to provide their income if it significantly decreased from wave t to wave $t+1$, which is not observed, even after controlling for their observed income in t and other relevant variables.

Missing data mechanisms are conceptualised as stochastic models governed by unknown parameters. To avoid invalid inferences due to a misspecification of this mechanism, it is important to know under which conditions the missing data mechanism can safely be ignored for downstream analysis.

Since analyses in applications are often based on the method of maximum likelihood (ML) and estimation results are evaluated from a frequentist point of view, e.g. in terms of (asymptotic) bias, rejection, or coverage rates, this is the situation considered in what follows. Within this framework, a missing mechanism is generally ignorable in large enough samples, if the missing values are MCAR or MAR for all possible patterns of the response indicators. A further but often not crucial technical condition if missing values are MAR is that the parameters of the missing data mechanism and the parameters of the scientifically interesting model are not linked to each other. For further discussion, including Bayesian and non-likelihood approaches, see Rubin (1976, 1987), Seaman et al. (2013), or Kleinke et al. (2020b). Generally, a missing data mechanism is not ignorable if the missing values are MNAR.

There are, however, exceptions to the above results. One such example is a regression model where only the dependent variable is affected by missing values with probability depending solely on values of the predictor variables considered as fixed. Then the incompletely observed part of the dataset can be ignored for valid inferences as long as the regression model is correctly specified for the dataset if there were no missing values (complete-dataset). Up to this point, it is only required that the missing values are MAR. Additionally, predictors may be unobserved for those units for which the values of the dependent variable are missing, and thus could even be MNAR. Analysing the completely observed part of the dataset only would still allow valid inferences (Kleinke et al. 2020b). Apart from this specific situation, however, if missing values are MNAR, then in general the regression relationship in the observed dataset will differ from the corresponding relationship in the complete dataset.

Unfortunately, ignorability of the missing mechanism does not necessarily mean that intended analyses for the incompletely observed dataset can proceed in the same way as for the complete dataset. This would be possible in the above-mentioned regression case or if units drop out in wave $t > 1$ completely at random. In the latter case, analysis of a balanced panel with standard methods would be possible, if there were no additional item non-response. In general, however, estimation of incomplete datasets turns out to be rather difficult if all available information in a dataset is intended to be used because the estimation procedure has to deal with possibly many different patterns of observed variables. In addition, with the exception of cases like the above regression example, ignoring incompletely observed units often leads to ignoring information and thus to less precise inferences. Thus, if missing values are MCAR or MAR, it may be beneficial to apply the method of MI.

17.3 Multiple Imputation (MI)

17.3.1 Theory and Basic Approaches

MI is an attractive method to compensate for missing values. Most software implementations presuppose that the missing data mechanism is ignorable, an assumption we adopt in what follows, although the method is much more general and allows imputation also under non-ignorability, which, however, requires strong external information. The assumption of ignorability can be made more plausible by including all available variables, relevant functions, and interactions thereof into the imputation model.

The basic idea of MI is to replace each missing value with several ($m = 1, \ldots, M$) predictions, called imputations. Each of the M versions of the same dataset is created differing only in the predictions for the unobserved values. The M completed datasets are analysed using standard software for completely observed datasets and the final inferences are based on combining the estimation results using simple rules. Let us denote the estimator of interest, e.g. a regression parameter, using the mth imputed dataset as $\hat{\theta}_m$ and its estimated variance as $\widehat{Var}(\hat{\theta}_m)$. Then the combining rules are

$$\overline{\theta} = M^{-1} \sum_{m=1}^{M} \hat{\theta}_m \quad \text{and} \quad \widehat{Var}(\overline{\theta}) = \overline{\mathbf{W}} + (1 + M^{-1})\mathbf{B},$$

where $\overline{\mathbf{W}} = M^{-1} \sum_{m=1}^{M} \widehat{Var}(\hat{\theta}_m)$ is the so-called *within variance* and $\mathbf{B} = (M-1)^{-1} \sum_{m=1}^{M} (\hat{\theta}_m - \overline{\theta})(\hat{\theta}_m - \overline{\theta})^T$ is the *between variance* (Rubin 1987).

Inferences are (confidence) valid if for $M = \infty$ the estimator $\overline{\theta}$ is approximately unbiased and if the variance estimator $\widehat{Var}(\overline{\theta})$ is approximately equal to (or larger than) the variance of $\overline{\theta}$. Confidence validity implies that the actual coverage of confidence intervals is equal to or larger than (this is the 'confidence' part) the nominal level of $1 - \alpha$. Note that, at least in large samples, many estimators are approximately normally distributed.

For valid inferences using the above combining rules the process to create MIs cannot be arbitrary. In fact, imputations should be (confidence) proper for the complete-data estimates as given in Rubin (1987, 1996), meaning that for $M = \infty$ and given the realised sample, the expectations of $\overline{\theta}$ and $\overline{\mathbf{W}}$ are approximately equal to the corresponding complete-data estimates and that the expectation of $\mathbf{B}$ is, over repeated samples and $M = \infty$, at least as large as the difference between the variance of $\overline{\theta}$ and the variance of the complete-data estimator $\hat{\theta}$. If in addition the complete-data inference is (confidence) valid assuming $\hat{\theta}$ to be normally distributed, then for sufficiently large M and sample sizes, inferences assuming $\overline{\theta}$ to be normally distributed tend to be (confidence) valid (Rubin 1987, 1996). In what follows, we will understand the term *valid* to mean 'confidence valid' and the term *proper* to mean 'confidence proper'.

The theory of MI is derived from a Bayesian framework, but the procedures are evaluated from a frequentist point of view (Rubin 1987, 1996). At the core of an imputation procedure is the posterior predictive distribution of the variables whose values are missing given all observed values. Drawing M independent values for each missing value from this distribution generates M imputations. Note that for (confidence) proper imputation, this needs not be the 'true' distribution, because only expectations and variances are subject of the definition.

In general, the posterior predictive distribution is difficult to handle because in most datasets different types of variables are present, e.g. continuous, truncated, categorical, and count variables. In addition, the observed and missing values do usually not show a well-behaved pattern, like e.g. a monotone missing data pattern. A missing pattern is monotone, if the columns (variables) of a dataset can be arranged such that for all rows (units) a missing value is preceded only by columns with observed values and followed only by columns with missing values. For this definition to work, a longitudinal dataset would have to be converted into wide format, i.e. all collected variables at all time points for a given unit are written in one line. If a missing pattern is monotone, then imputation is considerably simplified and well justified. If the missing pattern is arbitrary, generating imputations is complicated but possible if Markov Chain Monte Carlo (MCMC) techniques are adopted (see Schafer 1997).

There are two basic approaches to generate imputations in datasets with arbitrary missing patterns. The first approach, joint modelling (JM), assumes a particular joint distribution for all variables with missing values conditional on the observed values and then applies a MCMC technique to generate the imputations under this model. The existence of such a joint distribution is a condition for the algorithm to converge and for interpreting the selected values as realisations from the posterior predictive distribution. A popular assumption is the joint normal distribution even if discrete variables need to be imputed. After having generated imputations under the normal model, it is proposed for discrete variables to round each imputed value to the closest possible value. If the distribution is not normal, then the suggestion is to transform the observed values of the variable accordingly. However, rounding cannot be endorsed (see e.g. Horton et al. 2003) and transforming observed values is hazardous, because it may be difficult to justify a transformation such that the conditional distribution of variables with missing values given the observed values is jointly normal based on the, possibly selective, marginal distributions of observed values of the incompletely observed variables. The classic software for JM is the program norm (Schafer 1997).

An alternative technique only assumes that a common conditional distribution exists and specifies, for each incompletely observed variable, an appropriate univariate model conditional on the observed variables ('fully conditional specification', FCS; van Buuren and Groothuis-Oudshoorn 2011; Raghunathan et al. 2016). Thus, for a continuous variable a linear and for a binary variable a logistic regression model could be specified. Assuming the modelling assumptions are appropriate and given a monotone missing data pattern FCS would be a well justified strategy. This does not hold if the missing data pattern is arbitrary. One way to generate imputations in this case consists in, firstly, imputing only so many values until the pattern of the remaining missing values is monotone and, in a second step, successively imputing all incompletely observed variables by iteratively adopting appropriate regression models. As predictors in the imputation models of the second step, the values of all other variables, observed or imputed, can be used. This procedure approximates a MCMC technique, is very flexible and allows to generate imputations for many different types of variables. The drawback is that it is not always clear whether a joint probability distribution even exists and if it does not, what the consequences are. Some discussion may be found, e.g. in van Buuren et al. (2006). The classic software for FCS are the programs Iveware (Raghunathan et al. 2016) and mice (van Buuren and Groothuis-Oudshoorn 2011).

Because application of the corresponding software may still be inconvenient in some cases, imputations could also be generated using a simpler but improper method – and several software packages provide such techniques, like imputing the mean of observed values or nearest neighbour imputation (see next Section). Although inferences may be valid in exceptional situations, inferences should be expected to be invalid if not otherwise justified. In some cases, improper imputation methods may allow valid inferences but not using the simple combining rules (Robins and Wang 2000).

17.3.2 Single Versus Multiple Imputation

Standard software usually requires units to be completely observed with respect to the variables that enter the analysis. Single imputation methods replace each missing value by a single value that is either explicitly or implicitly assumed to be a plausible replacement for the unobserved value. The imputed values are then treated just as any other observation. This means that, in contrast to multiple imputation, the uncertainty regarding the estimate of each missing value is usually ignored by these techniques.

Unconditional mean imputation, regression imputation (conditional mean imputation), and the 'last observation carried forward' (LOCF) technique are common single imputation methods that are described briefly in the following section (see also Enders 2010). Specific to panel data is the row-and-column (RAC) imputation technique proposed by Little and Su (1989), an improved mean imputation procedure that combines cross-sectional and longitudinal information. An example applying LOCF and RAC to juvenile delinquency data of the CrimoC study (e.g. Boers et al. 2010)[1] demonstrates the consequences of

1 The panel study 'Crime in the modern City' (CrimoC) is located at two German universities (Bielefeld and Muenster), for details see www.crimoc.org. The CrimoC study examines the emergence and development of deviant and delinquent behaviours throughout the phase of adolescence. Data from adolescents were collected annually from the age of 13 with classroom interviews from 2002 to 2009 following a panel design. From the age of 20 and after leaving

these single imputation techniques for panel data. Naturally, the study is affected by missing data and panel attrition (see Reinecke and Weins 2013; Kleinke et al. 2020a).

17.3.2.1 Unconditional Mean Imputation and Regression Imputation

Figure 17.1a displays a three-dimensional scatterplot of a variable that has been measured at three panel waves. Unconditional mean imputation replaces the missing values of an incompletely observed variable by the mean of the available cases of this variable. This maintains the original sample size but the variability in the data is reduced (see Figure 17.1b), which leads to standard errors being systematically too small. The magnitude of covariances and correlations also decreases, which often causes biased parameter estimators of the substantive model, irrespective of the underlying missing data mechanism. Although this method is still available in most of the common statistical packages, it is not recommended.

Regression imputation approaches use predicted values from a regression equation to impute the missing data values. The imputed values are points on the regression plane (Figure 17.1c), which leads to an infusion of the data with perfectly correlated observations. For multivariate data, correlations and R^2 statistics are overestimated. Standard errors are typically underestimated. There are ad hoc adjustments available in some statistical packages (e.g. SPSS): By adding residuals to the imputed values, these so-called stochastic versions of regression imputation try to ensure an adequate variability of the data, but may lead to biased inferences. Unless justified correction procedures based on bootstrap (Efron 1994) or jackknife (Rao and Shao 1992) techniques are applied, standard errors are usually too small. The most common procedure today to obtain correct standard errors is to generate multiple imputations following Rubin's (1987) theory (Figure 17.1d). Here, the standard errors and thus the widths of confidence intervals directly depend on how well missing information and the associated uncertainty can be represented by the imputations.

17.3.2.2 Last Observation Carried Forward

The LOCF method is a very simple and popular method to fix the drop out problem in panel studies. The method imputes the last observed value of an individual for subsequent unobserved panel waves. LOCF thus assumes no change in the outcome variable for these individuals after they dropped out of the study. The major drawback of this method is that the assumption of stability of the outcome variable after drop-out is usually quite unrealistic and thus this technique will underestimate the dynamic in the variable of interest.

---

school, data collection was proceeded biannually with mail surveys. The last panel wave will be proceeded in 2019 at the age of 29.

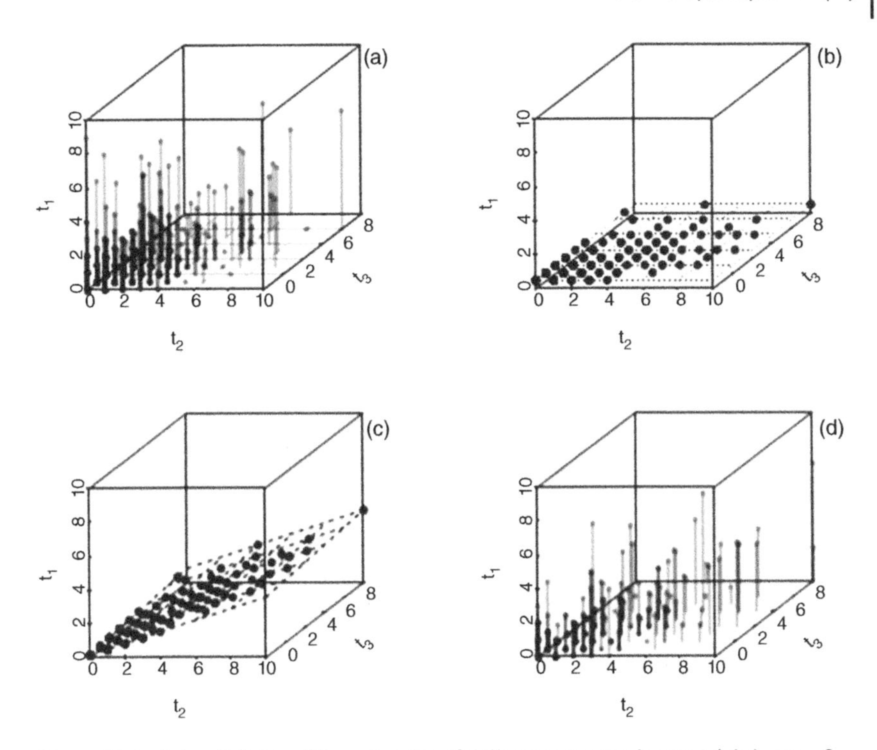

Figure 17.1 CrimoC-Data – 3D-scatterplot of delinquency rates in wave 1 (t_1), wave 2 (t_2), and wave 3 (t_3). *Note.* (a) original incomplete data; (b) unconditional mean imputation of missing data at t_1; (c) regression imputation of missing data at t_1; (d) multiple imputation of missing data at t_1.

For example, applying LOCF to the juvenile delinquency data of the CrimoC study would artificially stabilise the individual trajectories and would bias the amount of change across time towards zero. Furthermore, depending on how many participants drop out at a particular panel wave, LOCF could lead to both overestimation and underestimation of mean delinquency rates over time.

Table 17.1 shows parameter estimates of an unconditional quadratic growth curve model fitted to four waves ($t = 1, \ldots, 4$) of the CrimoC-data:

$$y_i = \Lambda\eta_i + \varepsilon_i$$

$$\begin{bmatrix} y_{1i} \\ y_{2i} \\ y_{3i} \\ y_{4i} \end{bmatrix} = \begin{bmatrix} 1 & 0 & 0 \\ 1 & 1 & 1 \\ 1 & 2 & 4 \\ 1 & 3 & 9 \end{bmatrix} \begin{bmatrix} \eta_{1i} \\ \eta_{2i} \\ \eta_{3i} \end{bmatrix} + \begin{bmatrix} \varepsilon_{1i} \\ \varepsilon_{2i} \\ \varepsilon_{3i} \\ \varepsilon_{4i} \end{bmatrix}, \quad E(\varepsilon_{ti}) = E(\eta_{ji}\varepsilon_{t'i}) = 0 \quad \forall t, t', j, i, \tag{17.1}$$

Table 17.1 CrimoC-data – parameter estimates of an unconditional growth curve model.

	α_I	α_S	α_Q	ψ_I	ψ_S	ψ_Q	ψ_{IS}	ψ_{IQ}	ψ_{SQ}
MI	0.69	0.63	−0.16	1.67	2.34	0.22	0.12	−0.13	−0.66
LOCF	0.60	0.39	−0.08	1.55	1.66	0.11	−0.18	−0.01	−0.39
RAC	0.65	0.63	−0.16	1.25	0.87	0.09	0.90	−0.24	−0.27

Note. α are the means, ψ the variances and covariances of the growth factors with subscripts I, S, Q denoting the latent intercept, slope, and quadratic slope, respectively. MI is multiple imputation, LOCF is last observation carried forward, RAC the row-and-column imputation method proposed by Little and Su (1989).

where y_{ti} is the repeatedly measured delinquency score for case i, Λ the matrix of loadings (which were constrained in such a way that they represent the starting level, linear growth and quadratic growth respectively), η_{ji} are the latent growth factors and ε_{ti} are random measurement errors. For the latent growth factors, we estimated means α, variances and covariances ψ of errors ζ:

$$\begin{bmatrix}\eta_{1i}\\ \eta_{2i}\\ \eta_{3i}\end{bmatrix} = \begin{bmatrix}\alpha_1\\ \alpha_2\\ \alpha_3\end{bmatrix} + \begin{bmatrix}\zeta_{1i}\\ \zeta_{2i}\\ \zeta_{3i}\end{bmatrix}, \quad E(\zeta_{ji}) = E(\eta_{ji}\zeta_{j'i}) = 0 \quad \forall j, j', i. \tag{17.2}$$

Table 17.1 contains results based on LOCF imputation, and on multiple imputation using FCS package mice in R with default imputation methods and models.

As can be seen from Table 17.1, LOCF predicted less change over time in comparison to MI. The predicted mean delinquency rates obtained by MI were 0.69, 1.16, 1.30, and 1.12 for waves 1–4 (i.e. age 13–16). The delinquency rates reflect the typical age–crime curve – an inverted U-shaped development of delinquency over time: on average, delinquency rates increase until around the age of 15, and decrease thereafter. In comparison, LOCF yielded 0.60, 0.91, 1.07, and 1.06. Especially noteworthy is the fact that the turning point and decline of delinquency from wave 3 to wave 4 is no longer visible. Furthermore, LOCF imputation reduced the estimated intercept and slope variability quite noticeably.

We discourage the use of LOCF because the probability to draw wrong substantive conclusions is very high.

17.3.2.3 Row-and-Column Imputation

A characteristic of panel datasets is the large number of covariates from both the same and other panel waves that could be used to predict missing information in certain items. State-of-the-art missing data procedures make use of this information. Identification of suitable predictors is usually a time-consuming effort, and proper imputation models usually have to be fairly complex to represent all aspects of interest (e.g. multilevel structure, interaction terms, higher-order associations, auxiliary variables). In the late-1980s, the complexity of the imputation models was furthermore restricted by limited computational capacities. Missing data researchers proposed pragmatic procedures that had to trade off what is proper against what is computationally feasible. These methods often ignored auxiliary information. One of these methods is the RAC single imputation approach by Little and Su (1989), which is still used in applications with panel data (see the recent overview in the field of longitudinal wealth data in Westermeier and Grabka 2016).

The RAC method combines cross-sectional and longitudinal information of a certain variable of interest to create the imputations. Basically, the imputed value is a combination of a column effect, a row effect, and a stochastic component, in which the incomplete case is matched to a complete one.

The column effect c_t shows trend information over time and is calculated as the mean of the observed part of the variable of interest $\bar{y}_t$ at wave t divided by the overall mean of the observed values of that variable across the T panel waves – the denominator in (17.3):

$$c_t = \frac{\bar{y}_t}{T^{-1} \sum_{t=1}^{T} \bar{y}_t}. \tag{17.3}$$

A value of 0.8 at wave 1, for example, would indicate that the mean of wave 1 is 20% lower than the overall mean.

The row effect r_i for case i is

$$r_i = m_i^{-1} \sum_t \frac{y_{it}}{c_t}, \tag{17.4}$$

where m_i is the number of available observations of the variable of interest y for case i. The sum is calculated over the m_i available observations for case i. The row effect gives the individual departures from the overall trend.

Incomplete cases are then matched to a nearest complete case d (the so-called donor) based on the individual row effects of the complete and incomplete cases. The matching itself works analogous to the matching strategy in predictive mean matching (Little 1988; Rubin 1986). Assuming a multiplicative model, the imputation is finally obtained by

$$y_{it}^{\text{imp}} = \frac{r_i}{r_d} y_{dt}, \tag{17.5}$$

where r_i is the row effect of the incomplete case, r_d is the row effect of the donor, and y_{dt} is the observed value of the donor at timepoint t.

Table 17.1 displays the application of the RAC method to the CrimoC data. While the means of the growth factors are very similar to those obtained by MI, the variances and covariances of the growth factors are not: RAC yielded a lower between-person variability of intercepts, slopes, and quadratic slopes. Furthermore, RAC yielded a much larger estimate of the covariance between the intercept and slope. A disadvantage in comparison to MI is that RAC cannot use auxiliary information like gender or school type to predict incomplete delinquency scores (cf. Reinecke and Weins 2013). Instead of RAC effects, MI can consider more information in form of auxiliary variables. It is therefore not astonishing that RAC and MI yielded different estimates for the variances and covariances of the latent growth factors.

17.4 Issues in the Longitudinal Context

In longitudinal or panel studies usually samples of n units (or clusters) are selected and several variables are surveyed at each of $t = 1, \ldots, T$ time points or waves. Observations between the waves are usually not independent and these dependencies have to be accounted for not only in the analyses of interest, but even more so when generating the imputations. It may yet be more demanding if the selected units are, for example, households introducing additional cluster structures. In fact, properly accounting for the complicated structure of dependencies is a challenging task, because for each unit, many time-invariant and time varying variables are intended to be observed over T waves. To yield proper imputations in complex survey settings, information regarding stratification and cluster membership needs to be taken into account (Rubin 1996). Effects of misspecifications of the imputation model in the clustered or longitudinal context are discussed, for example, in Taljaard et al. (2008), Andridge (2011), and Drechsler (2015).

Another important issue facing most longitudinal surveys is whether and how to update imputations when a new wave is completed. However, this depends on the pattern of missing values. If the pattern of missing values is monotone even if the dataset is extended by one more wave, then generating new imputations for the former waves is not necessary. Imputations for the new wave can be generated based on the observed and already imputed former waves. If the missing pattern is non-monotone, then a new wave may carry new information and former waves should be imputed in the light of this new information using an MCMC technique which might change results based on former waves only. Possible differences should, however, be small in large datasets if appropriate imputation models are adopted and a large number of imputations are generated (100 or more; cf. Graham et al. 2007). The latter reduces variation due to the simulation error inherent in the imputations.

Usually, applied researchers have two options to impute missing data in clustered or panel datasets: (i) to use single-level imputation provided by standard statistical software and (ii) to use multilevel imputation provided by special MI packages.

17.4.1 Single-Level Imputation

Panel data are a special case of clustered data, where repeated measurements of certain variables of interest (level 1) are clustered in persons (level 2). Unlike multilevel data in cross-sectional designs (e.g. students clustered in schools), the number of level 2 units is usually very large (about 3400 participants in the CrimoC dataset) in comparison to the number of level 1 units (e.g. 10 panel waves). We therefore need to discuss whether the simple strategies proposed for cross-sectional multilevel data can be recommended for the analysis of panel data, as well.

Most statistical software packages only include basic MI implementations like imputation under the joint multivariate normal model (Schafer 1997) or using approaches like k-nearest neighbour imputation based on predictive mean matching (van Buuren and Groothuis-Oudshoorn 2011).[2] Multilevel models are supported only by a few packages (see the next section for details). Due to the sparseness of adequate MI software, three simple strategies using standard MI techniques and software to impute clustered data have been proposed (e.g. Eddings and Marchenko n.d.; Graham 2009):

1. Apply single-level MI to the clustered dataset.
2. Include cluster indicator variables into the imputation model.
3. Impute each cluster separately.

Typically, these strategies are applied to *long-format* data, i.e. the repeated measurements of a certain variable of interest are stacked on another and stored in a single variable. The first and most simple strategy ignores the fact that the data are clustered and applies single-level MI (e.g. MI under the multivariate normal model, Schafer 1997) to the panel dataset. Therefore, it oversimplifies the imputation model, implicitly assuming independence between observations at different waves. We illustrate this strategy with the example of the latent growth curve model from the previous section, specifying a random intercept, a random linear, and a random quadratic slope (see (17.1) and (17.2)). A single-level imputation algorithm ignores the fact that participants can differ in intercepts, and in linear and quadratic slopes, thus simplifying (17.2) to

$$\begin{bmatrix} \eta_{1i} \\ \eta_{2i} \\ \eta_{3i} \end{bmatrix} = \begin{bmatrix} \alpha_1 \\ \alpha_2 \\ \alpha_3 \end{bmatrix}. \tag{17.6}$$

2 For a discussion of whether predictive mean matching is a proper MI method, see Schenker and Taylor (1996), or Gaffert et al. (2016).

The imputation procedure thus assumes that all participants have the same delinquency rate at wave 1 and the same trajectory across time. As imputations are created under the assumption that the variance of the random effects is zero, single-level imputation of clustered data biases the estimators of random effects towards zero. The severity of this effect obviously depends on the number of missing values to be imputed. Van Buuren (2011) has shown by means of Monte Carlo simulations that ignoring the hierarchical structure of the data can result in biased estimators of random effects. But fixed effects estimators can also be biased when the model includes incomplete predictors.

The second strategy using cluster indicators tries to overcome the problem of assuming a common baseline level. By including cluster dummy variables into the imputation model, cluster-specific intercepts are estimated. A drawback of this strategy is that it makes the imputation model unnecessarily complex. Even in a study with 'only' 500 participants, including 499 cluster dummy variables into the imputation model would cause severe convergence problems of any MI algorithm. To avoid estimation problems, Graham (2009), for example, recommends a maximum number of variables of around 100 for the JM approach. In addition to these practical problems, research by Andridge (2011), Drechsler (2015), Speidel et al. (2018b), and van Buuren (2011) suggests that this strategy also yields biased estimators and biased standard errors.

The third strategy (separate imputations per cluster), is also of limited use to panel researchers. The advantage is that it allows parameters of the imputation models to differ among clusters. The major disadvantage is that the method requires a substantial number of observations within each cluster (Graham 2009). Unless there are at least 20 waves per unit, the separate imputations strategy should not be used. In addition to the described strategies imputing panel data in long format, single-level MI procedures can also be applied to panel data in wide format. Kleinke et al. (2011) have pursued this strategy. The reasoning behind this approach is that repeated measurements – especially the adjacent panel waves – are usually correlated, and missing information in one or several panel waves can be predicted by observed information from other panel waves. Kleinke et al. (2011) have compared the performance of JM (R package norm) and FCS by nearest-neighbour imputation based on predictive mean matching (pmm, R package mice) with multilevel MI based on a linear mixed effects model (R package pan). The single-level MI procedures did quite well (in terms of bias regarding parameter estimators and standard errors). In a Monte Carlo simulation based on empirical panel data, pan was neither superior to norm nor pmm. However, the analysis model was fairly simple, and the missing data problem with only 20% drop-outs after five waves was not too severe. Future research needs to establish how this strategy fares in comparison to multilevel imputation strategies in more complex scenarios with more missing data. For limitations of predictive mean

Table 17.2 Overview of multiple imputation software for clustered or panel data.

Package/macro	Software	Framework
pan	R	JM
MMI_impute	SAS	JM
	REALCOM-IMPUTE	JM
jomo	R	JM
mice	R	FCS
hmi	R	FCS
	Mplus	JM/FCS

Note. JM = joint modelling, FCS = fully conditional specification.

matching MI, see furthermore Schenker and Taylor (1996) or Kleinke (2017). For newer and more adaptive matching strategies, see Gaffert et al. (2016).

17.4.2 Multilevel Multiple Imputation

Single-level imputation strategies usually oversimplify the imputation model and can thus not be expected to be proper. The most appropriate strategy to fill-in incomplete clustered or panel data would be to use an MI method that explicitly allows for clustering. Multilevel MI methods are available for both MI frameworks – joint and conditional modelling (for an overview, see Table 17.2).

One of the earliest joint modelling solutions has been proposed by Schafer and Yucel (2002) who generate multiple imputations of continuous incomplete variables from a multivariate linear mixed-effects model. The method appears to work fine when all model assumptions (like normality of level-1 errors, normality of random effects, homoscedasticity) are met, the imputation model is correctly specified, and the data are missing at random (Enders et al. 2016). The approach is implemented in R package pan. Note that pan does not allow for incomplete predictors. These have to be included among the outcome variables in the multivariate linear mixed effects model. Package mitml provides a user-friendly way to specify the pan imputation model (instructions are given in Grund et al. 2016).

Mistler (2013) introduced a SAS macro called MMI_impute that also creates imputations of incomplete level-1 variables based on a Bayesian linear mixed model. Furthermore, the macro can also handle incomplete level-2 variables. In fact, MMI_impute is a combination of three existing multiple imputation algorithms: Level-2 variables are imputed using an adaptation of Schafer's (1997) JM approach (norm). Level-1 variables are imputed using an adaptation of the pan algorithm (Schafer and Yucel 2002). The two algorithms are combined using techniques described in Yucel (2008). MMI_impute thus makes the same assumptions as the methods implemented in norm and pan. If these assumptions are not met, then the imputation method may not be proper and inferences may be invalid.

The JM software REALCOM-IMPUTE (Carpenter et al. 2011) also builds on the pan approach and allows incomplete predictors and missing data at all levels. Furthermore, the program can handle incomplete categorical variables through latent normal variables. REALCOM-IMPUTE is especially interesting for researchers who use either MLwiN or Stata for multilevel modelling, as REALCOM-IMPUTE works hand in hand with both statistical packages. An R adaptation is available as package jomo (i.e. joint modelling, Quartagno and Carpenter 2017).

Finally, the structural equation modelling software M*plus* allows to create multiple imputations based on a joint multilevel model (for details, see Asparouhov and Muthén 2010).

The FCS R package mice contains three functions to impute continuous two-level data based on a 'normal' linear mixed-effects model: mice.impute.2l.lmer, mice.impute.2l.pan, and mice.impute.2l.norm. The first function calls the lmer function from package lme4, which is one of the standard functions in R to fit normal linear mixed-effects models. The second function calls the pan package to fit the univariate normal linear mixed-effects model. Both functions assume homogenous within-group variances. The 2l.norm method, on the other hand, allows for heteroscedastic data.[3] Furthermore, the countimp package (Kleinke and Reinecke 2019) provides mice addon-functions that allow to create imputations based on various two-level count models (e.g. Poisson, negative binomial, and hurdle models). The miceadds package (Robitzsch et al. 2017) has also additional mice imputation functions: imputations can be created based on a two-level logistic regression model (mice.impute.2l.binary) and based on two-level predictive mean matching (mice.impute.2l.pmm). Additionally, R package hmi (Speidel et al. 2018a) provides functions to create imputations based on single-level and multilevel models under the FCS framework. It has been built to be compatible with mice, so that convenient mice functions like plot.mids (for convergence checks) can be used. One helpful feature of hmi is that based on the model formula supplied by the user, the program automatically tries to identify the variable types and the imputation routines needed for those variables. Supported variable types are continuous, binary, or categorical, but also Poisson

3 Note that the current implementation of mice.impute.2l.norm in mice version 2.46.0 only includes predictors into the imputation model when they are specified as fixed *and* random effects. See help("mice.impute.2l.norm") for details.

distributed data, semi-continuous, or rounded continuous data (for details, see hmi's package vignette). However, the user may override these default options.

Most of the above-mentioned packages are based on rather restrictive parametric models focussing on linear relationships. If all model assumptions including associations over time are (approximately) met, then the imputation functions discussed in this section allow valid inferences. However, datasets usually contain different types of variables like categorical, continuous, or count variables. They often have non-normal distributions (e.g. hourly income). Hence, the overly simplistic normality assumption is often not correct. The FCS packages offer more modelling flexibility in that regard. In addition, the assumption of homoscedasticity in most of the discussed packages is often violated, as well – for example, if different age cohorts are considered. Unfortunately, due to scarce empirical results, it is not yet clear if or to what extent these methods are robust against violations of their distributional assumptions. Finally, there may also be non-linear relationships between variables, e.g. if one variable depends on another variable via a quadratic term or an interaction of two other variables.

17.4.3 Interactions and Non-Linear Associations

Imputation of the outcome variable(s) in (multivariate) linear mixed-effects models has received the most attention so far in the MI literature. Unfortunately, creating proper imputations of incomplete predictors is not straightforward, especially when these predictors are used to create interaction terms and non-linear associations.

One popular and widely used approach to handle this is passive imputation (PI). Passive imputation omits all non-linear terms and interactions from the imputation model and forms them passively from the imputed values afterwards. For example: We want to impute variable x_1 (age) and also have the quadratic term $x_2 = x_1^2$ (age^2) in the analysis model. Let us furthermore assume that the imputed value of x_1 for a certain case is 12. Then x_2 would be obtained by computing $12^2 = 144$. This ensures that values in variables x_1 and x_2 are consistent for all cases in the data file, i.e. that x_2 is always x_1^2.

The main problem of PI is that it usually leads to a misspecified imputation model (Seaman et al. 2012a). Let us assume a standard linear regression model:

$$y = \beta_0 + \beta_1 x_1 + \beta_2 x_2 + \varepsilon, \qquad E(\varepsilon) = 0, \quad var(\varepsilon) = \sigma^2. \tag{17.7}$$

Dependent variable y is regressed on predictor x_1 and its quadratic term x_2. Let us furthermore assume that y is completely observed, and x_1 has missing values. We could impute missing values in x_1 by a standard regression approach:

$$x_1 = \gamma_0 + \gamma_1 y + \nu. \tag{17.8}$$

Following the passive imputation idea, we would obtain x_2 based on the imputed x_1 values. The problem, however, is that (17.8) is 'correct' only when the 'true' coefficient of β_2 in (17.7) is zero (Seaman et al. 2012a) and (y, x) is normally distributed. If $\beta_2 \neq 0$, bias is to be expected because imputed values of x_1 only reflect the linear relationship between x_1 and y.

An alternative to PI is the 'just another variable' (JAV) method. It includes interaction terms and non-linear terms into the imputation model as 'just another variable'. In the example above, we would impute both x_1 and x_2. This, however, could lead to imputations where $x_2 \neq x_1^2$, which might not be problematic, since multiple imputation does not aim to make plausible predictions on the person level, but instead aims at making valid statistical inferences (see Section 17.3). Such an approach would thus be justified if it is a (confidence) proper imputation method. But this seems to be not the case. Research by Seaman et al. (2012a) suggests that JAV only works sufficiently well when the data are MCAR – an assumption that is often violated.

Recently, various promising solutions have been suggested to overcome the problems of both PI and JAV: Bartlett et al. (2015) have proposed a modification to the traditional chained equations MI framework so that covariates could be imputed from the appropriate model – i.e. from a model that is compatible to the assumed data generating model (that includes interactions and higher-order relationships).

Bartlett et al. (2015) implemented their approach as R and Stata packages denoted as smcfcs (i.e. substantive model compatible fully conditional specification).[4] Unfortunately, their method currently only supports standard linear or generalised linear models. Extensions to multilevel models are not yet available. We are also not aware of any Monte Carlo simulations that have evaluated their method in a multilevel scenario. Using their solution to handle interactions and non-linear associations in multilevel models therefore might currently only be an option under very restrictive assumptions – for example, that no cluster effects have to be assumed for the covariates in question.

An alternative is the Bayesian JM procedure proposed by Goldstein et al. (2014). Through a latent normal formulation, the method can handle variables of different types (e.g. discrete or non-normal continuous) in either the outcome variables or the predictors (see Goldstein et al. 2014 and the references therein). Missing data may furthermore occur in both outcome variables and predictors. The advantage over smcfcs is that it explicitly models the multilevel structure of the data. The method ensures compatibility of the imputation model and the data analyst's

---

4 The R package is available from https://CRAN.R-project.org/package=smcfcs, the Stata version from github: https://github.com/jwb133/Stata-smcfcs.

model by defining the imputation model as the conditional distribution of the outcome variables given the covariates (i.e. the data analyst's model) multiplied by the joint distribution of the covariates.

Their procedure is implemented in the Stat-JR programme (written in Python, see www.bristol.ac.uk/cmm/software/statjr), which is distributed with the MLwiN software.

R package jomo (Quartagno and Carpenter 2017) also creates multiple imputations of clustered or panel data under the JM framework. jomo uses either a linear mixed-effects model or a binomial generalised linear mixed-effects model to generate the imputations. Like smcfcs and the approach by Goldstein et al. (2014), jomo supports substantive model compatible imputations. Furthermore, jomo can also handle binary and categorical data through latent normal variables. So far, only very few Monte Carlo simulations have evaluated package jomo (e.g. Grund et al. 2018).

17.5 Discussion

In panel research, substantive models are typically rather complex and missing data methods usually also have to be fairly complex to represent all relevant aspects of interest. Usually, if a missing data method fails to capture these aspects, affected parameters will be biased and inferences will be invalid.

Typically, simple ad hoc adjustments or single imputation methods cannot be recommended for most social science scenarios that involve analysis of clustered or panel data. Instead, a state-of-the-art procedure that allows valid inferences like multiple imputation should be used.

In comparison to simple ad hoc methods like case deletion, MI typically involves more modelling efforts: Under the FCS framework, users not only need to specify an imputation model for each incompletely observed variable. They also should identify additional auxiliary variables (that are not part of the substantive model, but that could additionally predict missing information) to end up with an appropriate imputation model for each incomplete variable in the data file. The JM framework seems to have the advantage that only one common imputation model has to be specified. However, these models are far more restrictive. In case of 'mixed' datasets that contain variables of many different types (like skewed continuous, ordered and unordered categorical, or count variables), it is hard to specify joint models for these variables.

Furthermore, in comparison to simple ad hoc methods, all iterative MI algorithms make it necessary to monitor convergence if the missing pattern is non-monotone (e.g. Schafer 1997; van Buuren and Groothuis-Oudshoorn 2011). If users fail to recognise nonconvergence, they can end up with implausible imputations and finally with biased statistical inferences. However, in terms of unbiased statistical inferences it is typically worthwhile making this extra effort.

Secondly, standard single-level multiple imputation methods that are implemented in virtually any statistical package make too restrictive assumptions and are often unable to produce valid statistical inferences in complex scenarios, where data have a multilevel (panel) structure.

This problem can be solved by using an imputation procedure that can adequately account for the clustered structure of the data. This, however, also means increased computational complexity in comparison to single-level MI, resulting in longer computing times, which depend on the number of observations, on the complexity of the model, and upon whether or not the respective MI package uses compiled code to make the computations faster. Some ideas have already been proposed to lessen the computational burden in very large data files (e.g. Vink et al. 2015). This could be an interesting avenue for future research.

Furthermore, many of the currently available multilevel MI procedures also make overly restrictive assumptions like normality or homoscedasticity. Extensions to the mice software like the countimp package (Kleinke and Reinecke 2013) or the miceadds package (Robitzsch et al. 2017) provide more modelling flexibility regarding the outcome variable beyond the normal model. However, all standard joint modelling functions like pan or all FCS multilevel imputation functions in mice assume that all predictors enter the imputation model linearly and lead to invalid inferences if this assumption does not hold. In this regard, a promising step towards greater modelling flexibility are R packages smcfcs and jomo, as well as the approach by Goldstein et al. (2014).

Unfortunately, Monte Carlo simulations that evaluate these recently proposed methods are still scarce. Additionally, most evaluation studies that tested multilevel multiple imputation procedures used fairly simple models (e.g. only random intercept models; only very few predictors, only missing data in one variable), and usually, model assumptions like the MAR assumption, homoscedasticity, or normality of the errors were fully met (e.g. Andridge 2011; Drechsler 2015; Enders et al. 2016; Lüdtke et al. 2017; Taljaard et al. 2008; van Buuren 2011).

Thus, although research so far suggests that multilevel MI yields adequate results when model assumptions are met and that multilevel MI has advantages over more simple strategies, we need to extend this body of research to more complex and more realistic models, and also to scenarios in which model assumptions are violated, before practical guidelines can be established regarding which method(s) to prefer in a given scenario. For example, it cannot be expected that every non-normal problem can be handled by JM approaches through latent normal variables. Future research needs to establish, under which scenarios these methods yield valid inferences.

For situations where variables are highly skewed or heavy tailed, semi-parametric imputation models (as proposed by Salfrán and Spiess (2018)) can be a suitable alternative. These models allow heteroscedasticity and non-linear relationships and can be applied, when sample size is sufficiently large. However, so far, these methods have not yet been generalised to clustered data structures.

References

Andridge, R.R. (2011). Quantifying the impact of fixed effects modeling of clusters in multiple imputation for cluster randomized trials. *Biometrical Journal* 53 (1): 57–74. https://doi.org/10.1002/bimj.201000140.

Asparouhov, T. and Muthén, B. (2010). Multiple imputation with *Mplus*. *M*plus *web notes*. https://www.statmodel.com/download/Imputations7.pdf.

Bartlett, J.W., Seaman, S.R., White, I.R. et al. (2015). Multiple imputation of covariates by fully conditional specification: Accommodating the substantive model. *Statistical Methods in Medical Research* 24 (4): 462–487.

Boers, K., Reinecke, J., Seddig, D. et al. (2010). Explaining the development of adolescent violent delinquency. *European Journal of Criminology* 7 (6): 499–520.

van Buuren, S. (2011). Multiple imputation of multilevel data. In: *Handbook of Advanced Multilevel Analysis* (eds. J.J. Hox and J.K. Roberts), 173–196. New York, NY: Taylor & Francis.

van Buuren, S. and Groothuis-Oudshoorn, K. (2011). MICE: Multivariate imputation by chained equations in R. *Journal of Statistical Software* 45 (3): 1–67.

van Buuren, S., Brand, J.P.L., Groothuis-Oudshoorn, C.G.M. et al. (2006). Fully conditional specification in multivariate imputation. *Journal of Statistical Computation and Simulation* 76 (12): 1049–1064.

Carpenter, J., Goldstein, H., and Kenward, M. (2011). REALCOM-IMPUTE software for multilevel multiple imputation with mixed response types. *Journal of Statistical Software* 45 (5): 1–14.

Drechsler, J. (2015). Multiple imputation of missing multilevel data – rigor versus simplicity. *Journal of Educational and Behavioral Statistics* 40 (1): 69–95.

Eddings, W. and Marchenko, Y. (n.d.). *Accounting for clustering with MI impute*. https://www.stata.com/support/faqs/statistics/clustering-and-mi-impute.

Efron, B. (1994). Missing data, imputation, and the bootstrap. *Journal of the American Statistical Association* 89 (426): 463–475.

Enders, C.K. (2010). *Applied Missing Data Analysis*. New York, NY: Guilford.

Enders, C.K., Mistler, S.A., and Keller, B.T. (2016). Multilevel multiple imputation: A review and evaluation of joint modeling and chained equations imputation. *Psychological Methods* 21 (2): 222–240.

Gaffert, P., Meinfelder, F., and Bosch, V. (2016). Midastouch: Towards an MI-proper predictive mean matching. Discussion paper. https://www.uni-bamberg.de/fileadmin/uni/fakultaeten/sowi_lehrstuehle/statistik/Personen/Dateien_Florian/properPMM.pdf (accessed 21 April 2016).

Goldstein, H., Carpenter, J.R., and Browne, W.J. (2014). Fitting multilevel multivariate models with missing data in responses and covariates that may include interactions and non-linear terms. *Journal of the Royal Statistical Society: Series A (Statistics in Society)* 177 (2): 553–564.

Graham, J.W. (2009). Missing data analysis: Making it work in the real world. *Annual Review of Psychology* 60: 549–576.

Graham, J.W., Olchowski, A., and Gilreath, T. (2007). How many imputations are really needed? Some practical clarifications of multiple imputation theory. *Prevention Science* 8 (3): 206–213.

Grund, S., Lüdtke, O., and Robitzsch, A. (2016). Multiple imputation of multilevel missing data: An introduction to the R package pan. *SAGE Open* 6 (4): 1–17. https://doi.org/10.1177/2158244016668220.

Grund, S., Lüdtke, O., and Robitzsch, A. (2018). Multiple imputation of missing data at level 2: A comparison of fully conditional and joint modeling in multilevel designs. *Journal of Educational and Behavioral Statistics* 43 (3): 316–353. https://doi.org/10.3102/1076998617738087.

Horton, N.J., Lipsitz, S.R., and Parzen, M. (2003). A potential for bias when rounding in multiple imputation. *The American Statistician* 57 (4): 229–232. https://doi.org/10.1198/0003130032314.

Kleinke, K. (2017). Multiple imputation under violated distributional assumptions: A systematic evaluation of the assumed robustness of predictive mean matching. *Journal of Educational and Behavioral Statistics* 42 (4): 371–404. https://doi.org/10.3102/1076998616687084.

Kleinke, K. and Reinecke, J. (2019). countimp version 2: A Multiple Imputation Package for Incomplete Count Data. In: *Technical Report*. Siegen, Germany: University of Siegen, Department of Education Studies and Psychology https://countimp.kkleinke.com.

Kleinke, K., Stemmler, M., Reinecke, J. et al. (2011). Efficient ways to impute incomplete panel data. *Advances in Statistical Analysis* 95 (4): 351–373.

Kleinke, K., Reinecke, J., and Weins, C. (2020a). The development of delinquency during adolescence: A comparison of missing data techniques revisited. *Quality and Quantity, Advance Online Publication*. https://doi.org/10.1007/s11135-020-01030-5.

Kleinke, K., Reinecke, J., Salfrán, D. et al. (2020b). *Applied Multiple Imputation. Advantages, Pitfalls, New Developments and Applications in R*. Cham, Switzerland: Springer Nature.

Little, R.J.A. (1988). Missing-data adjustments in large surveys. *Journal of Business & Economic Statistics* 6 (3): 287–296.

Little, R.J.A. and Su, H.L. (1989). Item non-response in panel surveys. In: *Panel Surveys* (eds. D. Kasprzyk, G. Duncan and M.P. Singh), 400–425. New York, NY: John Wiley.

Lüdtke, O., Robitzsch, A., and Grund, S. (2017). Multiple imputation of missing data in multilevel designs: A comparison of different strategies. *Psychological Methods* 22 (1): 141–165.

Mistler, S.A. (2013). A SAS® Macro for Applying Multiple Imputation to Multilevel Data. Proceedings of the SAS® Global Forum 2013 Conference. Paper 438-2013. https://support.sas.com/resources/papers/proceedings13/438-2013.pdf.

Quartagno, M. and Carpenter, J. (2017). jomo: A package for multilevel joint modelling multiple imputation. Computer software manual. https://CRAN.R-project.org/package=jomo.

Raghunathan, T., Solenberger, P., Berglund, P. et al. (2016). IVEware: Imputation and variance estimation software (version 0.3). Computer software manual. http://www.src.isr.umich.edu/software.

Rao, J.N. and Shao, J. (1992). Jackknife variance estimation with survey data under hot deck imputation. *Biometrika* 79 (4): 811–822.

Reinecke, J. and Weins, C. (2013). The development of delinquency during adolescence: A comparison of missing data techniques. *Quality & Quantity* 47 (6): 3319–3334.

Robins, J.M. and Wang, N. (2000). Inference for imputation estimators. *Biometrika* 87 (1): 113–124.

Robitzsch, A., Grund, S., and Henke, T. (2017). miceadds: Some additional multiple imputation functions, especially for mice. R package version 2.8-24. Computer software manual. https://CRAN.R-project.org/package=miceadds.

Rubin, D.B. (1976). Inference and missing data. *Biometrika* 63 (3): 581–592.

Rubin, D.B. (1986). Statistical matching using file concatenation with adjusted weights and multiple imputations. *Journal of Business & Economic Statistics* 4 (1): 87–94.

Rubin, D.B. (1987). *Multiple Imputation for Nonresponse in Surveys*. New York, NY: Wiley.

Rubin, D.B. (1996). Multiple imputation after 18+ years. *Journal of the American Statistical Association* 91 (434): 473–489.

Salfrán, D. and Spiess, M. (2018). Generalized additive model multiple imputation by chained equations with package ImputeRobust. *The R Journal* 10 (1): 61–72.

Schafer, J.L. (1997). *Analysis of Incomplete Multivariate Data*. London: Chapman & Hall.

Schafer, J.L. and Yucel, R.M. (2002). Computational strategies for multivariate linear mixed-effects models with missing values. *Journal of Computational and Graphical Statistics* 11 (2): 437–457.

Schenker, N. and Taylor, J.M. (1996). Partially parametric techniques for multiple imputation. *Computational Statistics & Data Analysis* 22 (4): 425–446.

Seaman, S.R., Bartlett, J.W., and White, I.R. (2012a). Multiple imputation of missing covariates with non-linear effects and interactions: An evaluation of statistical methods. *BMC Medical Research Methodology* 12 (1): 46.

Seaman, S.R., White, I.R., Copas, A.J. et al. (2012b). Combining multiple imputation and inverse-probability weighting. *Biometrics* 68 (1): 129–137.

Seaman, S.R., Galati, J., Jackson, D. et al. (2013). What is meant by 'missing at random'? *Statistical Science* 28 (2): 257–268.

Speidel, M., Drechsler, J., and Jolani, S. (2018a). hmi: Hierarchical multiple imputation. R package version 0.8.2. Computer software manual. https://CRAN.R-project.org/package=hmi.

Speidel, M., Drechsler, J., and Sakshaug, J.W. (2018b). Biases in multilevel analyses caused by cluster-specific fixed-effects imputation. *Behavior Research Methods* 50 (5): 1824–1840. https://doi.org/10.3758/s13428-017-0951-1.

Spiess, M. (2006). Estimation of a two-equation panel model with mixed continuous and ordered categorical outcomes and missing data. *Journal of the Royal Statistical Society: Series C (Applied Statistics)* 55 (4): 525–538.

Taljaard, M., Donner, A., and Klar, N. (2008). Imputation strategies for missing continuous outcomes in cluster randomized trials. *Biometrical Journal* 50 (3): 329–345.

Vink, G., Lazendic, G., and van Buuren, S. (2015). Partitioned predictive mean matching as a multilevel imputation technique. *Psychological Test and Assessment Modeling* 57 (4): 577–594.

Westermeier, C. and Grabka, M.M. (2016). Longitudinal wealth data and multiple imputation – An evaluation study. *Survey Research Methods* 10 (3): 237–252.

Yucel, R.M. (2008). Multiple imputation inference for multivariate multilevel continuous data with ignorable non-response. *Philosophical Transactions of the Royal Society, Series A* 366 (1874): 2389–2403.

18

Issues in Weighting for Longitudinal Surveys

Peter Lynn[1] *and Nicole Watson*[2]

[1] *Institute for Social and Economic Research, University of Essex, Colchester, UK*
[2] *Melbourne Institute, University of Melbourne, Melbourne, Australia*

18.1 Introduction: The Longitudinal Context

The production and application of survey weights is rather more complex in the context of a longitudinal survey than that of a cross-sectional survey. And it can be particularly complex for some types of longitudinal surveys, due to certain common features of design, such as sample units and survey instruments that are hierarchically related or cross-classified (individuals within households; pupils within both schools and households) and dynamic sampling. We note that general issues in survey weighting – that apply to any survey – apply also to longitudinal surveys. We do not discuss those issues here but instead refer the interested reader to the excellent texts by Bethlehem and Keller (1987), Biemer and Christ (2008), Brick (2013), Elliot (1992), Kalton and Flores-Cervantes (2003), Little and Vartivarian (2005), Särndal (2011), and Valliant et al. (2013). In this chapter we outline some of the aspects of weighting that are unique to the longitudinal context and discuss some of the advantages and disadvantages of different possible solutions to some of the issues that are encountered.

18.1.1 Dynamic Study Population

Longitudinal survey data relates to a population defined in time in addition to other dimensions such as geography and age. During the period of time to which the data refer, there may be new entrants to the population and there will almost certainly be exits from the population. Exits will occur due to mortality and, for most longitudinal surveys, due to emigration (as data collection is typically restricted to a national or sub-national territory). If the study population is

Advances in Longitudinal Survey Methodology, First Edition. Edited by Peter Lynn.

Companion website: www.wiley.com/go/lynn/advancesinlongitudinalsurvey

additionally defined by other restrictions such as residence in the household population (i.e. excluding those residing in institutions), or membership of a limited age group, then movements between the household and institutionalised population, and ageing, will constitute entrants and exits. In general, the process of defining the target population (with a time dimension) determines some of the challenges for weighting the sample.

It is unlikely that population reference data exist for a dynamic longitudinal population (Smith et al. 2009) such as 'all people who were continually resident in the country for the duration of the survey reference period' (the intersection of all cross-sectional populations during the period) or 'all people who were ever resident in the country during the survey reference period' (the union of all cross-sectional populations). Available population data generally refers to a cross-sectional population at a point in time (though comprehensive administrative data, where they exist, could be used to derive longitudinal population statistics). This rules out routine application of post-stratification or calibration techniques (Smith 1991; Zhang 2000). We discuss alternatives in Section 18.2.1 below.

Furthermore, all population exits amongst sample members should be identified in order to be able to create a base for non-response adjustment. Without this information, it will be unclear which sample elements constitute the eligible sample with respect to a given longitudinal population. In practice, identification of population exits is not always possible. For example, the field outcome at the survey wave after a sample member has died may be 'non-contact.' This may be particularly likely when the sample member lived alone. Such an outcome leaves uncertainty as to whether the sample member is still alive. Even if the dwelling can be observed to be unoccupied, or to have new occupants, this could be simply because the sample member has moved elsewhere (such as into a nursing home), so the uncertainty remains. And in some cases, a sample member may die several years after losing contact with the survey. It becomes necessary to develop methods to deal with the under-identification of mortality (Sadig 2014; Watson 2016) and we return to this in Section 18.2.3 below. Otherwise, if all cases of undetermined eligibility are assumed (still) eligible, respondents with similar characteristics (in terms of the auxiliary variables used in the weighting process) to those who have died will tend to be overweighted, leading to biased estimation.

18.1.2 Wave Non-Response Patterns

The set of sample units available for any particular longitudinal analysis depends on wave non-response patterns. The set of responding units can differ markedly between different combinations of waves, both numerically and in terms of composition. For example, using data from a 10-wave survey, analyst A may want to construct an estimate that requires data collected at each wave. Thus, the estimation sample will consist solely of sample members who responded in all 10 waves. But analyst B may want to construct an estimate that only requires data collected at waves 1 and 10. The estimation sample for analyst B will consist of all sample members who responded at waves 1 and 10, regardless of whether they also responded at all or any of the intervening waves. It is likely that analyst B's estimation sample will be considerably larger than analyst A's. A single weighting adjustment based on sample members who responded at all waves (i.e. using 'response at all waves' as the longitudinal analogue to the cross-sectional concept of 'unit respondent') would result in many units in analyst B's estimation sample receiving a weight of zero. This is clearly suboptimal for analyst B, both because information is lost due to unnecessarily dropped cases and because the weights were not designed specifically for analysis of the remaining cases, which may have different characteristics to analyst A's estimation sample.

To overcome this limitation, a set of weights would be required for respondents to each combination of waves (waves 1 and 10 is the combination relevant to analyst B; waves 1, 2, 3, ... and 10 is the combination relevant to analyst A). After n waves, there are $2^n - 1$ such combinations. As n increases, it soon becomes impractical to produce such a large number of sets of weights. In Section 18.3 we discuss practical solutions.

18.1.3 Auxiliary Variables

In developing weighting adjustments for any components of wave non-response or dropout subsequent to the first wave of participation, the wealth of previously collected survey data can be used as auxiliary variables. This gives longitudinal surveys the potential to make powerful weighting adjustments. However, it raises the question of how best to select and utilise auxiliary variables from amongst the very large number typically available. In particular, compromises may be needed between maximising the information in the auxiliary variables and other aspects such as parsimony and variance. To return to the example of the previous paragraph, developing a non-response adjustment for analyst B may involve modelling participation at wave 10 conditional on participation at wave 1. This could be done in a straightforward way using those wave 1 respondents who were still eligible at the time of wave 10 as the base and fitting a single-step model involving a set of covariates from the wave 1 questionnaire data. This is a parsimonious approach, but it ignores all the data collected at waves 2 through to 9. Some, if not most, of these data will be available for most of the wave 10 participants and, being more recent than the wave 1 data, these data from the intervening waves may be more strongly predictive of participation at wave 10 and hence may have the potential for greater bias reduction. But the patterns of available data will differ greatly between sample members, so rather more complex approaches to weighting will be needed, such as those involving chained series of non-response models. Section 18.4 discusses alternative ways of using auxiliary variables in non-response modelling for wave non-response and attrition.

18.1.4 Longitudinal Surveys as a Multi-Purpose Research Resource

Longitudinal surveys are often designed as multi-disciplinary, multi-purpose, research resources. They tend to attract a large and heterogeneous body of users, with different analysis objectives and different degrees of analysis expertise and sophistication. Some users will be willing and able to develop their own analysis-specific weights or to use other analytical methods to account for errors of nonobservation (sampling and non-response). But many will not. They will expect to be provided with an analysis weight, in the form of a variable on the survey dataset (or a very simple means of deriving one, e.g. by combining multiple variables), which they can then use to run weighted analysis using generic software. In the absence of such a weight, these users may either run unweighted analysis or not use the data at all. Provision of general-purpose weights (or the equivalent thereof) in public-access datasets is therefore essential to ensuring usability of the data resource and guarding against grossly misleading analysis. To support these aims, appropriate documentation, guidance, and user support are also needed.

18.1.5 Multiple Samples

Although most of the complications in longitudinal weighting arise in the context of adjustment for non-response, there are also some complexities related to design weights. Longitudinal surveys may include multiple samples, selected at different points in time. This could be due to a regular pattern of refreshment samples or the result of ad-hoc opportunities to boost certain subgroups or simply increase the sample size (see Chapter 1 in this book, Watson and Lynn 2021). This may result in some sample units having multiple chances of selection. Thus, for each sample unit it becomes necessary to know, or at least to estimate, their probability of selection for each sample, and not just for the sample in which they were actually selected. In some contexts, it may be necessary to include additional survey questions in order to establish selection probabilities for earlier samples. For example, all members of the Immigrant and Ethnic Minority boost sample added to *Understanding Society* at wave 6 (Lynn et al. 2018) were asked questions to establish where they were living at each of the points in time in which previous component samples had been selected, viz. 1991 (England, Scotland and Wales sample), 1999 (Scotland and Wales samples), 2001 (Northern Ireland sample), and 2009 (England, Scotland, Wales, Northern Ireland and ethnic minority boost samples).

Furthermore, the same sample unit may have different selection probabilities relating to different longitudinal study populations. For example, suppose that a longitudinal survey has an initial sample, first fielded at wave 1, plus two refreshment samples of some kind first fielded at waves 5 and 9 respectively. A particular individual i may have selection probabilities of p_{1i}, p_{2i}, and p_{3i}, respectively, for the three samples. Then, for longitudinal analysis of data from the first four waves of the survey (or any analysis involving any subset of those waves), the appropriate design weight for this individual would be $1/p_{1i}$. But for any analysis of data from waves 5–8, the appropriate design weight would be $1/(p_{1i}+p_{2i})$, and for analysis of data from wave 9 onwards, the design weight would become $1/(p_{1i}+p_{2i}+p_{3i})$.

18.2 Population Dynamics

18.2.1 Post-Stratification

As described in Section 18.1 above, standard approaches to post-stratification or calibration require modification when the study population is dynamic. Four approaches can be considered:

1. Post-stratify the wave 1 responding sample to the cross-sectional population at the time of wave 1, and thereafter rely on non-response adjustment.
2. Post-stratify the combined responding and out of scope sample at each wave to the wave 1 population (based on wave 1 characteristics of sample members).
3. Post-stratify the responding sample at each wave to the latest cross-sectional population estimates (based on current wave characteristics of sample members).
4. Use administrative data to construct longitudinal population targets.

For survey designs in which there are no sample entrants after wave 1, the first approach might be effective. In this situation, the analysis weight for data from waves 1 and 2 might take the form:

$$w_{i2} = dw_{i1} \times nr_{i1} \times ps_{i1} \times nr_{i2} \quad (18.1)$$

where

dw_{i1} is the design weight for element i at wave 1;

nr_{i1} is a non-response adjustment for element i at wave 1 (based on adjusting the wave 1 responding sample to the characteristics of the eligible wave 1 gross sample);

ps_{i1} is a post-stratification adjustment for element i at wave 1 (based on adjusting the preliminarily weighted wave 1 responding sample to the characteristics of the population from an external source, where the preliminary weight is $dw_{i1} \times nr_{i1}$); and

nr_{i2} is a conditional non-response adjustment for element i at wave 2 (based on adjusting the wave 1 to wave 2 balanced sample [elements that participated in both waves] to the characteristics of the wave 1 responding sample after removing those that had become ineligible by the time of wave 2).

The second approach has a potential advantage over the first approach in that the post-stratification step could adjust for any distortions due to non-response at wave 2 or later that are not captured by the non-response adjustments, whereas the first approach relies exclusively on the non-response adjustments to correct all distortions due to non-response. However, as variables available at population level are typically limited to socio-demographics that are also available for (wave 1) survey respondents, this advantage may be more theoretical than real, so long as these variables are included in the models on which the non-response adjustments are based (in the same form as used in the wave 1 post-stratification). However, if there is only imperfect information about eligibility at waves since a sample member last responded (see Section 18.2.3), then the advantage of approach 2 might be very real. The second approach differs from the first in that for analysis of data up to and including wave n, ps_{i1} is replaced by ps_{in}, a post-stratification adjustment for element i at wave n, obtained by adjusting the preliminarily weighted wave n combined responding and ineligible sample (classified by their wave 1 characteristics) to the characteristics of the wave 1 population from an external source, where the preliminary weight is $(dw_{i1} \times nr_{i1} \times \ldots \times nr_{in})$; and nr_{in} is a non-response adjustment for element i at wave n, based on adjusting the wave $(n-1)$ – wave n balanced combined responding and ineligible sample (elements that at both waves were either responding or ineligible) to the characteristics of the wave $(n-1)$ combined responding and ineligible sample.[1]

Approach 3 post-stratifies to the most up-to-date population data available, but it ignores the distinction between longitudinal and cross-sectional populations and, therefore, implicitly assumes that the sample is designed to remain cross-sectionally representative. This may be true for certain specific survey designs but is not generally true. For example, most longitudinal surveys do not attempt to achieve coverage at every wave of new population entrants through immigration since the previous wave. New immigrant samples are either added periodically (every several waves) or not at all. If recent immigrants (or at least a sizeable proportion of them) are included in the population statistics there will be a mismatch with the sample coverage, with the potential implication that rather than reducing estimation error, post-stratification could actually introduce a new source of bias. Attempts can be made to remove immigrants from the population

1 This assumes that nonresponse adjustment is based on modelling each step in the attrition process (wave, instrument) sequentially and applying a series of multiplicative adjustments. Other ways of making the adjustment are discussed in Section 18.4 below.

benchmark statistics, but this requires underlying population micro-data for each cross-sectional population (wave) that includes date of entering the country.

The only application of the fourth approach of which we know is on the Israel Longitudinal Survey, but it is a tantalising possibility for other surveys. In situations where a population register or other administrative data with complete coverage exists and is maintained continually, it would in principle be possible to derive population statistics for any longitudinally defined population. An example of such a population would be all persons who were alive and resident in the country at each of a number of time points, corresponding to the times of survey waves. However, this would depend on the register recording accurately when people exit the population (emigration or death) and on this having been done consistently for the entire period of time during which the survey data were collected. In principle, though, this could be a useful approach for longitudinal surveys in countries/territories where such continuous population data exist.

18.2.2 Population Entrants

In situations where the survey sample design includes features intended to regularly add representative samples of new population entrants to the sample, the weighting procedures must appropriately reflect these features. Two issues typically arise. The first is how to derive a base weight for these sample entrants. The second is how to combine a non-response adjustment for the new entrants with one for the continuing sample, given than the adjustment for the continuing sample is likely to depend on auxiliary variables collected at waves before the new entrants were observed.

For example, household panel surveys typically add to their sample at wave n all new births between waves $(n - 1)$ and n where the mother (or possibly either parent) is a sample member. In the absence of non-response and immigration this would, notwithstanding marginal issues about the definition of a mother/parent, maintain the cross-sectional representativeness of the survey sample. The probability of a new birth entering the sample is the product of the selection probability of the mother (which may itself be a combination of multiple probabilities; see Section 18.1.5 above) and the conditional probability of continuing to observe the mother until wave n. An estimate of the latter will usually be required as a component of the longitudinal weight of the mother, so this same estimate can be used for the newborn child. This effectively leads to a situation where the longitudinal weight of the mother is assigned as the inclusion (base) weight of the child. In a design with multiple samples and therefore multiple longitudinal weights (Section 18.1.5 above) it is the most recent longitudinal weight that should be used, i.e. the one that applies to analysis of data collected since the most recent sample was added.

The method for applying a non-response adjustment to new entrants (from their second wave of observation onwards) will depend on the strategy being used to produce longitudinal weights for continuing sample members (see Section 18.4 below). If the strategy is based on a chain of wave-on-wave conditional non-response models, then the non-response adjustment for new entrants can be made in the same way as for other sample members, but starting at wave $n+1$ (as an adjustment for non-response at wave n is already included in the base weight). But if the strategy involves conditional non-response models for larger numbers of waves in combination, or for pairs of waves that are more than one wave apart, a dual-estimation procedure will be needed, in which separate models are developed for new entrants and the adjustments derived from the two parallel sets of models are combined at a later step.

18.2.3 Uncertain Eligibility

For reasons introduced in Section 18.1.1, some uncertainty may exist regarding the continuing eligibility of sample members who are no longer participating in the survey. The proportion of sample units that are in fact ineligible but at risk of being assumed eligible will tend to accelerate over time, because of the combination of two factors. First, there tends to be higher numbers of dropouts per wave in early waves. Second, the chance of identifying a move into ineligibility (e.g. death or emigration) will tend to be lower the longer it occurs after the sample unit was last observed. The longer the survey has been running, then, the greater the risk of substantial estimation bias due to under-identified ineligibility.

Watson (2016) uses four different methods to estimate the extent of under-identification of deaths at each of 12 annual waves of the Household, Income and Labour Dynamics in Australia (HILDA) survey (wave 2 to wave 13). The methods gave similar results after several waves/years, though in early waves the method based on life tables tended to over-state sample mortality due to the initial sample being healthier than average (as it excluded those in institutions such as nursing homes, and those who were unable to participate in wave 1 due to ill health). A variant of Watson's life table method was used to estimate the proportion of deaths identified over various periods for the British Household Panel Survey (BHPS). Estimated identification rates using life table methods were similar between the two surveys. For example, after six years (at wave 7), the estimated proportion of deaths identified was 0.83 (HILDA) and 0.84 (BHPS), while after 12 years (at wave 13) it was 0.77 (HILDA) and 0.76 (BHPS). The BHPS estimates extend over a longer period and show a continuing decline to 0.73 at 19 years and 0.64 at 24 years. Of course, the estimated number of deaths amongst wave 1 sample members increases over time, so even a modest decline in the *proportion* identified represents a large increase in the *number* not identified (and

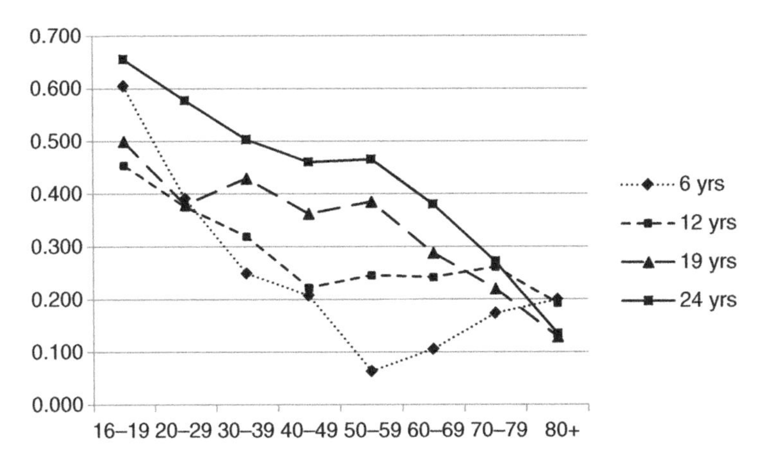

Figure 18.1 Proportion of deaths not identified, by age at wave 1 and elapsed years, BHPS.

therefore in the proportion of the total sample whose eligibility status may be mis-assigned). Amongst the BHPS sample, the estimated proportion of sample members who were dead, but not identified as such, rose from 1.2% after 6 years to 3.6% after 12 years, 6.3% after 19 years, and 10.4% after 24 years.

Furthermore, these proportions vary greatly over age groups. The chance of a death being successfully identified by the survey is greater the older the sample member (Figure 18.1). For example, after 24 years the proportion of deaths identified is only 34% amongst sample members who were aged 16–19 at wave 1, but rose to 54% amongst those initially aged 40–49 and 87% amongst those initially aged 80 or older. However, mortality rates are of course much higher in the older age groups, so this leads to the overall proportion of sample members incorrectly assumed to be alive being much higher in the older age groups (Figure 18.2), while the rate of increase over time in this proportion is greater in the middle age groups. For example, amongst those aged 50–59 at wave 1, the proportion incorrectly assumed to be alive was 0.3% after 6 years, 3.1% after 12 years, 9.5% after 19 years, and 17.0% after 24 years.

The strong association between under-identified mortality and age (and other important factors such as health status) means that it would be dangerous to define eligibility based solely on mortality identified in the course of the survey and to develop weighting adjustments for non-response on the assumption that sample members not explicitly identified as having died (or emigrated) remain eligible. Arguably, the best way to establish mortality would be regularly to link the survey sample to death registers or other equivalent administrative data. Linkage would have to be attempted at the time of each survey wave for all sample members who are neither identified in the field as being alive at that wave (e.g.

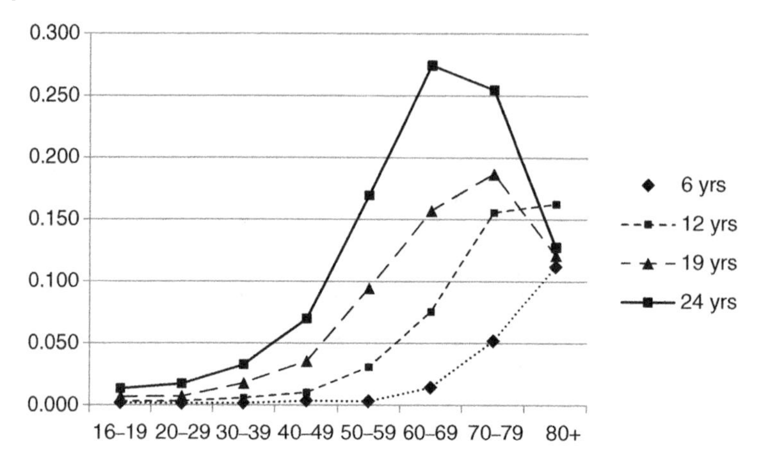

Figure 18.2 Deaths not identified as a proportion of sample, by age at wave 1 and elapsed years, BHPS.

survey participants and explicit refusals) nor already identified as having died. To rely solely on information from the linkage, linkage error rates should ideally be very low (thus, probably ruling out linkage that is restricted to sample members who have given explicit consent, for example). Where such linked data are either unavailable or do not have sufficiently high coverage, additional estimation methods are recommended.

Implementing a weighting adjustment based on estimated mortality has two conceptual steps. The first step is to produce estimates of mortality probabilities or mortality rates for subgroups. The second step is to use these estimates to adjust the weights. Each of the steps can be done in a number of different ways, the relative merits of which are under-researched.

The estimation of mortality can be based on external mortality indices, such as those typically published annually by national statistical offices, or on internal (to the sample) modelling. An example of the former is the life table method (method 2) of Watson (2016), in which the probability of death is estimated for each sample member whose status is uncertain as an appropriately weighted product of the published annual mortality rates for the period since the wave, m_i, at which the sample member was last observed alive, up until each subsequent survey wave n, thus:

$$\widehat{p_{ijn}} = 1 - \left(\prod_{t=m_i}^{n} (1 - w_{ijt} d_{jt})\right) \tag{18.2}$$

where:

sample member i in subgroup j (e.g. age x gender) was last observed (alive) in year m_i;

d_{jt} is annual mortality rate for individuals in subgroup j in year t;
w_{ijt} are weights reflecting the relevant proportion of a calendar year (for example, if the sample member was last observed on 31 March of year m and the reference date for wave n is also 31 March then $w_{ijm} = 0.75$ and $w_{ijn} = 0.25$).

An alternative example, introduced in Sadig (2014) and used to produce the BHPS estimates in Figures 18.1 and 18.2, is to estimate the number of expected deaths in each age–gender sample subgroup between wave 1 and wave n and to subtract the number of identified deaths in the subgroup to produce an estimated number of unidentified deaths. The ratio of this number to the number of sample members in the subgroup whose status is uncertain provides an estimate of the probability of death by wave n for each member in the latter group, i.e.:

$$\widehat{p_{ijn}} = \frac{n_j(1 - \prod_{t=1}^{n}(1 - w_{jt}d_{jt})) - o_{jn}}{n_j - c_{jn} - o_{jn}} \tag{18.3}$$

where:

n_j is initial sample size in group j, of which c_{jn} known alive in year n;
o_{jn} is observed number of deaths in group j by year n (year of current wave).

Watson (2016) outlines two methods of using sample-based modelling to estimate mortality probabilities. The first (method 3) involves building an explicit model of time to death based on characteristics observed at wave 1, where the dependent variable is based on survey observations of mortality at each wave (alive, dead, not observed). The model is then applied to each sample member whose mortality status is uncertain to produce a predicted value based on their characteristics at the last wave at which they were observed alive. The second (method 4) involves building a model of the probability of mortality status being observed at the time of a particular survey wave (known to be alive or known to be dead) and using the model-predicted probabilities as weighting adjustments, in the same way that probabilities of survey response would usually be used as adjustments for non-response (if eligibility status were known for all sample members). Comparison of the weighted and unweighted sample for whom mortality status is known provides an estimate of the number of unidentified deaths and hence the probability of death amongst those of unknown status.

Any of the above methods will produce an estimate, $\widehat{p_{in}}$, of the probability of individual i having died by wave n. Regardless of the estimation method, there are a number of ways in which the $\{\widehat{p_{in}}\}$ can be used to adjust the analysis weights (the second step). These fall into three broad categories.

The first approach is to impute a dichotomous indicator of eligibility for each individual in the sample whose mortality status is unknown. This could be done, for example, by generating for each individual a random number, r_{in}, from the uniform distribution (0,1) and imputing the status 'eligible/alive' if $r_{in} > \widehat{p_{in}}$. The second approach is to estimate eligibility *rates* for sample *subgroups* and to then apply a uniform adjustment to the weights for each member of the subgroup. The third approach is to make the non-response adjustment in the usual way, based on model-based estimates of response propensity, where the model is based on all sample members who are either known to be alive or whose mortality status is unknown, but where the model is weighted with each sample member assigned the weight $w_{in} = (1 - \widehat{p_{in}})$. This third method was proposed by Kaminska and Lynn (2012) and has been applied to *Understanding Society* to deal with uncertainty in the eligibility status of addresses that were selected for inclusion in the ethnic minority boost sample (Berthoud et al. 2009), but where no screen interview could be carried out. (Note that for all sample members who respond at wave n, $\widehat{p_{in}} = 0$.)

18.3 Sample Participation Dynamics

In Section 18.1.2 we introduced the issues that are raised for weighting by the complex non-response patterns observed in longitudinal surveys. The complexity of the patterns depend both on sample members' response behaviour and on survey data collection policy (whether attempts are made to interview people who were nonrespondents at previous waves; whether secondary instruments such as self-completion questionnaires or biomarker collections are only administered to people who also completed the main interview/questionnaire at that wave, and so on). Every combination of survey instruments is a potential analysis base for which weights might be desirable. If there are, say, two instruments at each wave and no restrictions in data collection, then after n waves, there are $2^{2n} - 1$ combinations. After 10 waves, this would imply production of more than half a million sets of weights! For data producers, some possible approaches to this logistical challenge are:

- *Provide weights for a (necessarily small) subset of the possible combinations of instruments.* With this approach, it is helpful to additionally provide guidance on how to proceed when the analysis base is something other than one of the combinations for which weights have been provided.
- *Provide weights for each pair of instruments.* For any analysis base, the user can construct their own analysis weight as the product of two or more of these component weights.
- *Provide the means for users to produce their own analysis-specific weights.* This could range from software that produces weights based on a set of input parameters specified by the user to simple guidance on how to calculate appropriate weights.

Each of these three approaches is now discussed in turn.

18.3.1 Subsets of Instrument Combinations

This is the approach adopted by, for example, *Understanding Society*. Consider, for simplicity, only the four individual-level instruments that are administered at every wave to collect information about each sample member aged 16 or over:

A. *Household grid*. Basic information about each person in the household.
B. *Core individual questionnaire*. Questions that are included in both the individual interview and the proxy interview (the latter is administered if a personal interview is not possible for health or availability reasons).
C. *Personal questionnaire*. Questions that are included only in the individual interview, i.e. if the individual is interviewed in person rather than by proxy.
D. *Self-completion questionnaire*. Additional personal questions. This was administered as a paper document at waves 1 and 2, but incorporated into the individual interview script from wave 3 onwards.

As mentioned earlier, the data collection policy influences the number of possible instrument combinations observable in the data. At each wave, instruments B and D are only administered conditional on instrument A being administered (a household grid must always be completed before any other instruments can be completed), and instrument C is only administered conditional on instrument B also being administered (it is not possible to complete the individual interview omitting the core instrument questions). At waves 3 onwards, instrument D is only administered conditional on instrument C also being administered. Thus, after seven waves, the number of possible instrument combinations is $(7^2 \times 5^5) - 1 = 153{,}124$. However, the number of sets of weights actually produced for analysis of combinations of these three instruments was just 52. For the vast majority of possible instrument combinations, no bespoke weight has been produced. The choice of instrument combinations for which to produce weights is therefore important, as it affects the usability of the data.

Lynn and Kaminska (2010) suggested five considerations that should guide the choice of instrument combinations for which weights will be produced. The first was to ignore any combinations that are ruled out by survey design. These combinations are already excluded from the 153 124 calculated above. Here are the other considerations:

- *Analytic use*. Some combinations are more likely than others to be required by analysts. For example, if a particular module of questions is included only in the individual interview and only at waves 1, 4, and 7, then some researchers interested in the topic addressed by that module may wish to carry out analysis that only draws on data from the individual interview in those three waves.
- *Levels of non-response*. If the samples responding to two instrument-combinations differ only by a very small number of cases, it is unlikely

that using weights defined for one combination will make much difference to analysis for the other combination. In other words, the sub-optimality will be trivial. If combinations with minimal differences in respondent sets could be identified then just a single combination from each such set need be included in the set of combinations for which weights are to be produced. At the first two waves of Understanding Society, it was very rare for a sample member to complete the proxy interview and the self-completion questionnaire (and from wave 3 onwards, it is impossible). Thus, no weights are produced for combinations of instruments B and D (without C).

- *Correlates of non-response.* The covariates predictive of non-response may differ notably between certain instruments or instrument combinations. For example, correlates of attrition may be quite distinctive (wave-specific) at waves 2 and 3, but are thereafter similar across waves. Thus, amongst the set of cases responding to waves 1 and 2, weights derived for response to waves 1, 2, and 3 might be quite different from weights derived specifically for response to waves 1 and 2, whereas amongst the set of cases responding to waves 7 and 8, weights derived for response to waves 7, 8, and 9 might be quite similar to weights derived specifically for response to waves 7 and 8, suggesting that it may not be necessary to produce both sets of weights.
- *Impact on estimates.* Empirical comparison of exemplar estimates using weights derived for alternative instrument combinations could reveal sets of combinations which appear equivalent. It would then only be necessary to produce one set of weights from each such set. Within the set, choice of the combination that maximises the sample size would tend to maximise the precision of estimates. The results of this approach could, however, be somewhat dependent on the choice of exemplar estimates. Furthermore, it is necessary to complete the full weighting procedure for all instrument combinations in order to make the comparisons, so there is no saving in the effort needed to produce weights, only in the number of weight variables included in the datasets.

However decisions are made about which instrument combinations to provide weights for, data users would benefit from guidance on how to proceed in the event that their desired analysis base is not one of those combinations. Options include the following:

1. *Use the weight from the hierarchically superior instrument combination that is closest in sample size to the analysis base.* For example, if the analysis base is the combination of instruments (E, F, G, H) – a combination for which weights have not been produced – then the analyst should use the weight for whichever combination of three out of those four instruments has the smallest sample size. In practice, it may be that a weight has been produced for only one of the four possible three-instrument combinations, say (E, F, G), in which case

that weight should be used. This solution will result in all sample units in the analysis sample having a non-zero weight. The weight will only be sub-optimal in so far as response to instrument H conditional on response to (E, F, G) is informative (systematic) with respect to the analysis being carried out.

2. *Adjust the weight from a hierarchically superior instrument combination.* Continuing the example of the previous paragraph, the user could fit their own model of response to instrument H conditional on response to (E, F, G) and use the predicted values from this model as adjustments to the weight for (E, F, G). Users should be advised on how to select covariates for the model and how to save the predicted values and use them as weight adjustments.
3. *Use the weight from a hierarchically inferior instrument combination that is close in sample size to the analysis base.* If a weight is available for the combination (E, F, G, H, I) and there are very few sample members who responded to (E, F, G, H) but not to I, then it may be appropriate to use the weight produced for (E, F, G, H, I). This will result in a small number of potential members of the analysis base being excluded (as they will have a weight of zero), but the non-response correction will be appropriate.

The difference between options 1 and 3 might be thought of as a bias-variance trade-off, with option 3 offering better bias reduction (bespoke non-response adjustment) at the price of slightly increased variance (smaller analysis sample size). Option 2 is clearly preferable and may even result in greater bias reduction than would have been achieved by a standard weight adjustment for the same analysis base derived by the data producers, because the user can include covariates in the adjustment model that are appropriate for their particular analysis. But it is of course also rather more work for the data user.

18.3.2 Weights for Each Pair of Instruments

In this approach, the data producer provides weights for the baseline participating sample (recruitment or wave 1) and for every subsequent combination of two instruments (not only 'consecutive' pairs). If there are N instruments (for example, N waves with one instrument per wave, or $N/2$ waves with two instruments per wave), then the total number of sets of weights to produce will be $1 + \sum_{n=1}^{N}(n-1)$, which is very considerably fewer than the $2^N - 1$ possible combinations of instruments. For example, if $N = 10$, then $1 + \sum_{n=1}^{N}(n-1) = 46$, whereas $2^N - 1 = 1{,}023$. Nevertheless, it should be noted that at each wave, the number of new sets of weights to produce will be Q^2 greater than the number produced at the previous wave, where Q is the number of instruments administered at each wave.

For each pair of instruments, the weight should correspond to the inverse of the probability of responding to one of the instruments conditional on having responded to the other, e.g. $w_{i(D|A)} = 1/\hat{P}(I_{iD} = 1 \mid I_{iA} = 1)$, where I_{in} is a 0/1 indicator of whether sample member i responded to instrument n. The direction of the conditionality should be consistent, in the sense that the instruments should be treated as ordinal. However, the nature of the ordinality is not greatly important (the numerator for the non-response model will be the same in both cases; only the denominator differs), though it is important that it is clearly documented. In many situations the logical ordinality of instruments corresponds to the temporal ordering of the instrument administration. This also tends to provide the larger of the two possible denominators for modelling of conditional response and hence potentially more powerful non-response models. Thus, for example, in the one-instrument-per-wave scenario there is no reason why we should not model response to wave 3 conditional on response to wave 6, but it is more natural to model response to wave 6 conditional on response to wave 3. Proceeding in this way is also a logical consequence of the typical need to produce weights (and release data) after each wave of data collection, rather than wait until multiple waves have been completed.

Once all the weights for pairs of instruments have been produced, users can derive the appropriate weight for any combination of instruments by taking the product of component weights. For example, the weight for analysing data from the combination (A, B, F, M) could be computed as:

$$w_{i(ABFM)} = w_{i(A)} \times w_{i(B|A)} \times w_{i(F|B)} \times w_{i(M|F)} \tag{18.4}$$

If the data structure is strictly hierarchical (e.g. a data collection protocol that results in instrument B only being administered if A is completed, F only being administered if B is completed, and so on) then this approach is equivalent to explicitly producing weights for each combination of instruments using chained conditional non-response models (as there are only N combinations in the hierarchical case rather then $2^N - 1$ in the non-hierarchical case). But if the structure is non-hierarchical, the weights from this simplified approach will be suboptimal to the extent that some of the conditional non-response models differ from the ones that would have been obtained had the optimal conditionality been modelled. In the example above, the optimal derivation of the analysis weight would be:

$$w_{i(ABFM)} = w_{i(A)} \times w_{i(B|A)} \times w_{i(F|AB)} \times w_{i(M|ABF)} \tag{18.5}$$

so the suboptimality arises from the difference between $w_{i(F \mid B)}$ and $w_{i(F \mid AB)}$ and between $w_{i(M \mid F)}$ and $w_{i(M \mid ABF)}$. Whether this approach is appropriate therefore depends both on the survey's data collection policies and on response outcomes.

18.3.3 Analysis-Specific Weights

Some surveys provide systems that produce weights on demand for any analysis base specified by the user. An example is the custom weighting program of the

National Longitudinal Surveys of Youth (NLSY) in the USA (http://www.nlsinfo.org/weights). The user can specify any combination of waves and the program will return a set of weights derived specifically for the set of sample members who participated in that combination of waves. For NLSY1979, 26 waves of data were available as of mid-2018, as waves were annual from 1979 to 1994 and thereafter biennial, so over 67 million combinations are possible. However, the program applies a fairly simple algorithm that calculates a post-stratification adjustment to the wave 1 weight, based on population information from the time the sample was selected in 1978. There is therefore an implicit assumption that every longitudinal population has the same characteristics as the initial cross-sectional population. The issues raised in Sections 18.2.1 and 18.2.3 of this chapter are ignored. Furthermore, the weighting does not use any of the survey data as auxiliary variables, relying solely on variables that are available at population level for 1978. It seems quite likely, then, that this approach might be less powerful at reducing bias than explicit non-response weighting.

It would be a big challenge to develop a system to produce on-demand weights that is sufficiently sophisticated to deal with all, or even most, of the issues relevant to weighting for longitudinal surveys. It seems likely that such systems will remain rather simplistic for the foreseeable future, or otherwise will be based on big and untestable assumptions.

18.4 Combining Multiple Non-Response Models

In Section 18.1.3 we emphasised the value of previous-wave survey data (or, more generally, data from any hierarchically superior instrument) as auxiliary variables in non-response models. We also pointed out that there are trade-offs to be made between maximising the power of the auxiliary variables and keeping the weighting process manageable and parsimonious. Possible approaches can be characterised as follows:

A. *Explicitly model the non-response pattern pertinent to each wave/instrument combination in a single step.* This approach requires auxiliary variables that are available for all sample members, including those who did not participate in any waves. The choice is therefore restricted to sampling frame, observation, and otherwise externally linked variables. With this approach, each time weights are needed for a new wave/instrument combination (for example, because a new wave has been carried out), all previously produced weights are ignored and development of a weighting model begins again from scratch.
B. *Model each step in the attrition process (wave, instrument) sequentially and apply a series of multiplicative adjustments.* This approach maximises the extent and

relevance of auxiliary variables that can inform the non-response weights, as these can include all survey measures from all previous steps in the process. With this approach, each time weights are needed for a new wave/instrument combination, the starting point is to identify a previously produced weight as a base weight and to develop a new adjustment to be made to that weight.

C. *Adopt a hybrid approach, between approaches A and C, in which some steps in the attrition process may be modelled separately, while others may be combined into a single model.* For example, one might model wave 1 non-response using frame data as auxiliary variables, but then model the combination of wave 2, 3, and 4 non-response in a single step, using wave 1 survey data as auxiliary variables.

Each of these approaches has advantages and disadvantages, the importance of which may depend on the survey design and on the missing data patterns. The more steps that are modelled separately, the more recent and relevant the auxiliary variables will be in each non-response model (and hence, possibly, the more powerful the bias reduction will be). However, the more steps that are modelled separately, the more random variance is likely to be introduced into the weights (as additional random variance is added at each step, and non-response models do not tend to be strongly predictive of either non-response or substantive *y*-variables, let alone both).

Also, for any 'incomplete' instrument/wave combinations of interest, approach B will tend to result in some members of the potential analysis base receiving a weight of zero (because they did not respond to one of the instruments that is not part of the combination of interest but is upstream, in terms of the sequence of non-response models, to one of the relevant instruments). The extent of zero-weights will tend to be greater the higher the proportion of steps that are modelled separately, and the greater the overall number of steps. Zero weights potentially lead to inefficiency in estimation, as not all sample information is used in the estimation. For example, in the one-instrument-per-wave scenario, where waves are denoted by sequential letters of the alphabet, after 13 waves an analyst may be interested in the combination A, E, I, M (waves 1, 5, 9, and 13). Under approach B, the analyst would have to use the weight designed for the combination of all 13 waves, so any sample member who was a non-respondent at any of the nine waves other than A, E, I, and M would have a zero weight even if they responded at each of the four waves of interest to the analyst. But approach C might involve a step of modelling response at wave 13 conditional on response at wave 9, in a single step. In that case it is only non-respondents to one of the six waves prior to wave 9 that are not part of the analysts's combination of interest that would receive a zero weight. Of course, if the data producers had followed the 'analytic use' criterion outlined in Section 18.3.1 in deciding which sets of weights to produce, they may have already produced bespoke weights for the combination A, E, I, M, in which case there would be no relevant sample members with a zero weight.

A common feature of the three approaches discussed here is that each analysis weight is based on the same set of models for all members of the analysis sample. An alternative way to make fuller use of the available covariates would be to allow the chain of models to differ between sample subsets. For example, suppose the aim is to produce weights for the sample that responded at waves 1 and 4. For sample members who responded at waves 1, 2, and 4, this could be based on a chain of three models: response at wave 1, response at wave 2 conditional on response at wave 1, and response at wave 4 conditional on response at waves 1 and 2. For sample members who responded at waves 1 and 4 but not wave 2, the weight could be based on a shorter chain of two models: response at wave 1, and response at wave 4 conditional on response at wave 1. This allows the wave 2 survey data to be used as auxilliary information for the wave 2 respondents, something that would not otherwise be possible (as the shorter chain of two models would have to be used for all sample members). Note that with this approach, the models within each chain must still be comprehensive in their sample coverage and hierarchical. So, in the above example, all wave 1 respondents must be included in the model of response at wave 4 conditional on response at wave 1 (second chain), not only those who would receive their weight from that chain. Producing separate chains of models for large numbers of sample subsets would be rather resource-intensive, but there may be situations in which doing so for a modest number of subsets (say, 2 to 5) could bring useful gains. Research in this area is lacking, however, so decisions in the meanwhile must continue to be intuition-led.

18.5 Discussion

As should be clear from our discussion in this chapter, in implementing a weighting strategy for a longitudinal survey there are many decisions to be made, many options available for each, and therefore a large number of combinations of those decisions. Many of the decisions made by longitudinal surveys have been based on intuition rather than extensive research. The overall implications of choosing one combination of approaches rather than another cannot currently said to be understood based on sound scientific theory or research. Some aspects of the weighting process are particularly lacking in research.

As regards post-stratification (Section 18.2.1), we are not aware of any research either comparing any of the methods we outline or exploring the properties of any of the methods. The fourth method, using administrative data to construct longitudinal population targets, is currently of considerable interest, as the possibility of doing this is becoming more realistic in several countries. We hope to see some evaluative research in this area before long.

As regards uncertain eligibility (Section 18.2.3 above), there has been some research into the relative performance of different methods of estimating the $\{p_{in}\}$, though this could usefully be extended to other variants of the methods and to datasets with different characteristics in terms of the available information. However, to our knowledge there has been no research into the relative merits of the three alternative ways – that we outline in the final paragraph of that section – of using these estimates to adjust weights.

In choosing which sets of waves or instruments to provide weights for (Section 18.3), or to provide the means for users to produce their own weights, there is little or no research that provides guidance on the decision-making process. In particular, clearer insights into the factors that determine the extent and nature of sub-optimality in using weights for hierarchically superior or hierarchically inferior instrument combinations (Section 18.3.1) or from non-hierarchical sequences of instrument-pairs (Section 18.3.2) would be helpful.

Regarding how to combine models for attrition (Section 18.4), we believe that best practice currently is to model separately steps in the attrition process that may have distinct characteristics (causes, correlates) and to combine those which are likely to have similar characteristics (or which have such low levels of non-response that it is hard to develop a predictive model) into single-step models. However, we can provide no precise guidance on what is meant by 'distinct' or 'similar' as this is another area in which research knowledge is lacking. Also, we are not aware of any evaluative studies of the precision loss due to weighting approaches which result in zero weights, or that compare this to the extra resources needed to implement more comprehensive approaches to weighting.

In sum, weighting for longitudinal surveys is a field that is ripe for further research. Given the large number of longitudinal surveys that are carried out, the considerable resources that are expended on the collection and promotion of longitudinal data, and the rapidly growing use of longitudinal survey data by researchers, investment in research in this field has the potential to have a major impact on the quality of research.

Acknowledgements

Peter Lynn's contribution to this research was funded by the UK Economic and Social Research Council grant for *Understanding Society* (ES/N00812X/1).

References

Berthoud, R., Fumagalli, L., Lynn, P., and Platt, L. (2009). Design of the Understanding Society ethnic minority boost sample. In: *Understanding Society Working Paper 2009–02*. Colchester: University of Essex.

Bethlehem, J.G. and Keller, W.J. (1987). Linear weighting of sample survey data. *Journal of Official Statistics* 3: 141–153.

Biemer, P. and Christ, S. (2008). Constructing the survey weights. In: *Sampling of Populations: Methods and Applications* (eds. P. Levy and S. Lemeshow), 489–516. Hoboken: Wiley.

Brick, J.M. (2013). Unit nonresponse and weighting adjustments: a critical review. *Journal of Official Statistics* 29 (3): 329–353.

Elliot, D. (1992). *Weighting for Non-Response: A Survey Researcher's Guide*. London: OPCS.

Kalton, G. and Flores-Cervantes, I. (2003). Weighting methods. *Journal of Official Statistics* 18 (1): 81–97.

Kaminska, O. and Lynn, P. (2012). Taking into account unknown eligibility in nonresponse correction. Paper presented at the *23rd International Workshop on Household Survey Nonresponse*, Ottawa, September.

Little, R.J.A. and Vartivarian, S. (2005). Does weighting for nonresponse increase the variance of survey means? *Survey Methodology* 31 (2): 161–168.

Lynn, P. and Kaminska, O. (2010). Criteria for developing non-response weight adjustments for secondary users of complex longitudinal surveys. Paper presented at the 21st International Workshop on Household Survey Nonresponse, Nürnberg, August.

Lynn, P., Nandi, A., Parutis, V., and Platt, L. (2018). Design and implementation of a high quality probability sample of immigrants and ethnic minorities: lessons learnt. *Demographic Research* 38: 513–548.

Sadig, H. (2014). Unknown eligibility whilst weighting for nonresponse: the puzzle of wo has died and who is still alive. In: *Institute for Social and Economic Research Working Paper 2014–35*. Colchester: University of Essex.

Särndal, C.-E. (2011). Morris Hansen lecture: dealing with survey nonresponse in data collection, in estimation. *Journal of Official Statistics* 27 (1): 1–21.

Smith, T.M.F. (1991). Post-stratification. *Journal of the Royal Statistical Society Series D (The Statistician)* 40 (3): 315–323.

Smith, P., Lynn, P., and Elliot, D. (2009). Sample design for longitudinal surveys, Chapter 2. In: *Methodology of Longitudinal Surveys* (ed. P. Lynn), 21–33. Chichester: Wiley.

Valliant, R., Dever, J., and Kreuter, F. (2013). *Practical Tools for Designing and Weighting Survey Samples*. New York: Springer.

Watson, N. (2016). Dead or alive? Dealing with unknown eligibility in longitudinal surveys. *Journal of Official Statistics* 32 (4): 987–1010.

Watson, N. and Lynn, P. (2021). Refreshment sampling for longitudinal surveys. In: *Advances in Longitudinal Survey Methodology* (ed. P. Lynn). Chichester: Wiley.

Zhang, L.-C. (2000). Post-stratification and calibration – a synthesis. *The American Statistician* 54 (3): 178–184.

19

Small-Area Estimation of Cross-Classified Gross Flows Using Longitudinal Survey Data

Yves Thibaudeau[1], Eric Slud[1,2] and Yang Cheng[3]

[1] *Center for Statistical Research and Methodology, US Census Bureau, MD, USA*
[2] *Mathematics Department, University of Maryland, MD, USA*
[3] *National Agricultural Statistics Service, US Department of Agriculture, Washington, DC, USA*

19.1 Introduction

Our objective in this chapter is to review and contribute to the methodology of small-area model-assisted estimation in complex longitudinal surveys. Genuinely longitudinal survey analyses, which address the pattern of behaviour of individuals over time, are distinguished from the more common methods based on temporal and spatio-temporal models of marginal outcomes for data from repeated surveys or repeated cross-sectional waves from longitudinal surveys. We review the estimation of gross flows in longitudinal surveys, dating from early efforts (Fienberg 1980) to define conditional probabilities to parameterise transitions in status. Progress in this area includes the partial model-based approach of Pfeffermann et al. (1998) using logistic regression to discount measurement error. The approach of Thibaudeau et al. (2017) is based on loglinear models with suppressed high-order interactions, parameterised in such a way that design-based estimates suffice for the large-population contingency-table cells but that take full advantage of the model to estimate small cell populations related to uncommon changes in status. The main novelty in the chapter is to extend the loglinear methodology of Thibaudeau et al. (2017) to small-area estimation from longitudinal surveys. The history of loglinear models applied to small area estimation (Purcell and Kish 1979, 1980; Marker 1999) and of random effect models in longitudinal surveys (Feder et al. 2000) are recalled, to extend the loglinear models of Thibaudeau et al. (2017) to incorporate geographic-area random effects. We present illustrative data analyses from the Current Population Survey (CPS), the official labour force statistical survey in the United States, estimating gross flows related to changes in labour

Advances in Longitudinal Survey Methodology, First Edition. Edited by Peter Lynn.

Companion website: www.wiley.com/go/lynn/advancesinlongitudinalsurvey

force status over two successive months further classified by demographic variables at the regional or state level.

19.2 Role of Model-Assisted Estimation in Small Area Estimation

National statistical data-collection agencies generally attempt to publish results from national surveys under minimal modelling assumptions, using *design-based* or *model-assisted* estimates whenever possible. Sometimes, as in *generalised regression* estimation, the objects of interest are statistical parameters expressing the influence of some survey predictor variables on others, which may be estimated and applied predictively using population totals about the predictor variables that are known from other censuses or surveys. When the target parameters θ, such as the regression coefficients in the model Eqs. (19.6–19.7) below, can be viewed as the solution of a population-wide estimating equation, they are descriptors of a frame population and do not, without stringent assumptions, apply to the prediction of relationships between survey variables at the level of individual units or small subpopulations. For this reason, although model-assisted survey theory (Binder 1983, Särndal et al. 1992, and many later papers) provides general mathematical conditions on non-random arrays of frame-population data under which population-level target parameters are design-consistently estimated using the solutions of survey-estimated score equations such as (19.8), those parameters may not be useful in survey prediction without strong assumptions to relate them to unit-level variables.

However, the well-recognised method of small area estimation (Rao and Molina 2015) draws inferences about domains too small to be well handled by design-based methods, through the use of shared statistical parameters fitted under model assumptions to survey variables at the level of individual units or of aggregates. There are many instances, elaborated by Rao and Molina (2015) and others, of random-effect models fitted to (small-area aggregates of) survey data, where estimated parameters have been found very useful in the reporting of survey outcomes for small domains. However, once the underlying small area estimation models are assumed, there may be no advantage in estimating their parameters through model-assisted techniques. There may, indeed, be a considerable disadvantage. This is true in the context of estimation within mixed-effect models such as the generalised logistic models ((19.11)–(19.12)) considered in the current chapter. These are models (on the frame population, in the survey context) in which population units have outcomes modelled predictively in terms of observable regression variables (the fixed-effect part of the model), but in which the population is also partitioned into numerous subpopulations (strata or clusters), on each of which there is a subpopulation-specific unobserved random-effect variable affecting all units of the subpopulation. These random-effect variables are often modelled as independent and identically distributed, with unknown distributional parameters called variance components that must be estimated along with unknown regression coefficients. Within mixed-effect models on survey populations, it is still a largely unsolved practical problem how to estimate in a general design-consistent survey-weighted manner the fixed- and variance-component parameters. There are partial solutions by Rao et al. 2013 and by Korn and Graubard 2003 using second- or higher-order inclusion probabilities. However, despite attempts by Pfeffermann et al. 1998b and Rabe-Hesketh and Skrondal 2006, the problem of how to accomplish design-consistent estimation in a mixed-model unequally weighted complex survey is still unsolved.

The reasoning of the foregoing paragraphs leads to the conclusion that, while survey-weighted fixed-effect estimators in the generalised-logistic setting of this chapter are still relatively easy to fit and well worth fitting for the purposes of comparing with mixed-model point estimators and variances obtained without taking survey weights into account, there is no real benefit in performing mixed-model analyses using survey weights. Accordingly, the methodological comparisons that we make and recommend in this chapter employ mixed-model analyses only in the unweighted case.

19.3 Data and Methods

19.3.1 Data

Our data are two consecutive months of the U.S. Current Population Survey (CPS) microdata (not publicly available) for selected US states. The CPS is a rotating panel survey collecting information on many aspects of the labour force. We focus on estimating uncommon changes in labour force status between the two months, and on the classification of the gross flows generated by those changes. Although the methods of this chapter could be extended readily to multi-period longitudinal models, that could also be applied to surveys like the CPS, the small-area aspects of the illustrative analyses would be less well motivated with more parameters and smaller cells, so we confine ourselves to a two-period data illustration.

Assume that, for each of a large number of observed records on the set S of sampled subjects with a survey frame population U, we have data

$$(\overset{\circ}{\mathbf{X}}_i, K_i, R_i, R_i \cdot \mathbf{Y}_i, w_i \ : \ i = 1, \dots, n) \tag{19.1}$$

Here K_i is a geographic label that will be used as the cluster level for fixed or random effect models, and is assumed available for all i, as is the indicator R_i that the (later-stage) outcome variable $\boldsymbol{Y}_i$ is observed. We will continue to use the boldfaced letter $\boldsymbol{Y}$ to represent the outcome, as our setting generalises to multi-dimensional outcome variables. However, our application will focus on the case where $\boldsymbol{Y}$ is one-dimensional. The covariate vector $\boldsymbol{X}_i^{\circ}$ is assumed to be observed for all survey respondents. In the longitudinal setting, we might view $\boldsymbol{X}_i = (\boldsymbol{X}_i^{\circ}, K_i)$ as the observed data from an initial or early wave of the collected data, while $(R_i, R_i \cdot \boldsymbol{Y}_i)$ is the indicator of observation and the value for the same individual i of the later-wave outcome category vector $\boldsymbol{Y}_i$. Within mixed-effect models, K_i is treated as a cluster label, different from the remaining covariates. Also associated with each individual i, indexed according to the early wave of data collection, is a survey weight w_i ideally viewed as the reciprocal $\frac{1}{\pi_i}$ of the probability $\pi_i = P(i \in S)$ that individual i was sampled in the survey at the early wave. In the superpopulation view of survey data, where all variables are viewed as random (but not necessarily independent across indices i), the weights w_i will generally depend on some of the components of $\boldsymbol{X}_i^{\circ}$

Our goal is to estimate population totals associated with *gross flows* and related proportions measuring changes in labour force status cross-classified by demographic and geographical characteristics across time. We centre attention on measuring gross flows associated with uncommon or small changes taking place over small areas. In this context, we represent population totals of interest by $T_{\boldsymbol{y},\boldsymbol{x}}$, where $\boldsymbol{y}$ is a vector whose components represent specific attributes of an individual that are subject to change over time; $\boldsymbol{x}$ includes the same attributes as $\boldsymbol{y}$ measured one month earlier, and also includes information on basic demographic or geographical baseline descriptors that may be constant over time.

In the extended example of the paper, $\boldsymbol{y}$ is a categorical variable representing labour force status at the time of the July 2017 CPS interview. We retain the boldface notation $\boldsymbol{y}$ as it would more generally denote a vector outcome. We have

$$\boldsymbol{y} = \text{status07} = \begin{cases} 1 : \text{Employed in July 2017} \\ 2 : \text{Unemployed in July 2017} \\ 3 : \text{Not in Labor Force in July 2017} \end{cases}$$

The predictor vector $\boldsymbol{x}$ is four-dimensional:

$$\boldsymbol{x} = (\text{status06, age06, educ06, state})$$

Here "status06" is the three-category labour force status obtained in the June CPS interview, with same levels as "status07 ($\boldsymbol{y}$)." The categorical variables educ06, age06 partition the population and sample by education and age as of the

June 2017 data collection, with levels:

$$\text{educ06} = \begin{cases} 1 : \text{no college degree in June 2017} \\ 2 : \text{college degree in June 2017} \end{cases}$$

$$\text{age06} = \begin{cases} 1 : \text{if} \geq 65 \text{ in June 2017} \\ 2 : \text{if} \geq 55 \text{ and} < 65 \text{ in June 2017} \\ 3 : \text{if} \geq 35 \text{ and} < 55 \text{ in June 2017} \\ 4 : \text{if} < 35 \text{ in June 2017} \end{cases}$$

The geographic label 'state' represents one of 16 small or medium-sized US states from New England, the Southwest (excluding Texas), or the South. We anticipate reductions in sampling error using a small-area estimation model that borrows strength between states, relative to direct estimators (Horvitz-Thompson, or HT). Large states do not stand to benefit as much from small-area modelling since they involve sufficiently large sample size and can be analysed on their own (e.g. *California* in Thibaudeau et al. 2017). We have state levels 1 to 16 associated with the standard state abbreviations as follows:

Level	1	2	3	4	5	6	7	8	9	10	11	12	13	14	15	16
State	AL	AZ	CT	GA	LA	ME	MA	MS	NV	NH	NM	NC	RI	SC	UT	VT

In our example, each state has $3 \times 3 \times 4 \times 2 = 72$ cells for a total of 1152 cells cross-classifying approximately 18 000 sampling units, or individuals, in the 16 states.

As in any longitudinal household survey, there is a non-trivial issue of matching persons across time-periods. In our CPS data application, the monthly person-level CPS files for the two consecutive months, June and July 2017, were filtered in order to implement our data analysis on the restricted data universe of sampled CPS persons who responded in the two successive months. We checked first that the person's household identifier was scheduled under the rotational pattern to be sampled in both months, second that (post-editing) the main employment category field was not blank for the sampled person record in either month, and third, that the person-level demographic (race and ethnicity) variables were not missing and were identical for both months, and that the age in the second month was either equal to or one larger than the age in the first month.

19.3.2 Estimate and Variance Comparisons

The specific models and estimates we consider in this chapter are loglinear fixed-effect models that can alternatively be presented as generalised logistic model (19.7–19.8) with a geographic categorical predictor, and the extension to the mixed-effect generalised logistic model (19.12–19.13) in which the geographic units are associated with a random intercept, a formulation frequently adopted in small-area estimation. We will compare the weighted versus unweighted fixed-effect coefficient estimates, including the geographic effects within our two-period CPS data example, and will also compare these with the fixed-effect coefficients and random intercepts predicted in the mixed-model setting. For reasons given in Section 1.1, we do not also undertake the survey-weighted (model-assisted) estimation of coefficients and geographic effects within the mixed model. For all three sets of estimates (unweighted and weighted from fixed-effect models and unweighted from mixed-effect model), we provide the linearisation-based estimates of standard errors and compare these with standard-error estimates generated through a weight-replication method (balanced repeated replication, or BRR). Also, for all three methods of estimation we provide and compare point and standard-error estimates for a particular set of small cells in the data example.

19.4 Estimating Gross Flows

The gross flows $\{T_{y,x}\}$ are the population totals characterised by $\boldsymbol{y}$ and $\boldsymbol{x}$, which jointly identify the gross flows, that is the transitions from the labour force status in June to the labour force status in July for each unit. We have:

$$T_{y,x} = \sum_{i \in U} I_{((Y_i,X_i)=(y,x))} \tag{19.2}$$

The summation counts all units i in the population such that the value of $\boldsymbol{Y}_i$, the $\boldsymbol{y}$ -characteristic of unit i, and $\boldsymbol{X}_i$, the $\boldsymbol{x}$ –characteristic, are equal to specified values of $\boldsymbol{y}$ and $\boldsymbol{x}$, respectively.

The traditional Horvitz-Thompson (HT) estimator for the gross flows between June and July ($\boldsymbol{y}$ and $\boldsymbol{x}$), subject to the geographical and qualitative contingencies in $\boldsymbol{x}$, is

$$\hat{T}_{y,x} = \sum_{i \in U} w_i \ I_{(i \in S,\ (Y_i,X_i)=(y,x))} \tag{19.3}$$

where the w_i's are the sampling weights. The summation is over the entire population and $I_{(i \in S)}$ is the sample inclusion indicator.

Pfeffermann et al. (1998) proposed a partial model-based alternative in the context of distinguishing measurement error. Thibaudeau et al. (2017) borrow from their approach to derive small-domain estimators that are more stable than the traditional Horvitz-Thompson. Let

$$\hat{T}_{x} = \sum_{i \in U} w_i \ I_{(i \in S,\ X_i=x)} \tag{19.4}$$

The model-assisted estimator is

$$\hat{\hat{T}}_{y,x} = \hat{P}(Y_i = y \mid X_i = x) \cdot \hat{T}_x \tag{19.5}$$

Here $\hat{P}(Y_i = y \mid X_i = x)$ is the model-based estimate of the conditional probability, or transition probability, $P(Y_i = y | X_i = x)$, where x represents all the covariates including the geographical information. $\hat{P}(Y_i = y \mid X_i = x)$ is computed from the sample based on specific partial ignorability assumptions. The ignorability is characterised as follows: first, population is modelled as a sample from a iid random 'super-population' (Fuller 2009) that can be described by a loglinear model, and second, conditionally given x, y is independent of the design, i.e. of the weights w_i. Under these assumptions, estimation by HT is design-consistent, as is estimation of cell totals by plug-in estimates (19.5) for conditional probabilities based on parameters estimated by weighted likelihood (pseudo-ML) score equations (Binder 1983).

We give instances of the HT estimator and both unweighted and weighted instances of the model-assisted estimator below. We also give instances of an estimator based on a mixed effect model. The geographical information K_i is treated differently from other covariates, indexing a random intercept. Based on this mixed effect model, the estimator of gross flows is:

$$\tilde{T}_{y,x} = \tilde{P}(Y_i = y \mid X_i = x) \cdot \hat{T}_x \tag{19.6}$$

Recall that $X_i = (\overset{\circ}{X_i}, K_i)$ and so $x = (\overset{\circ}{x}, k)$ is an observed value for X_i. The conceptual difference between $\hat{P}(Y_i = y \mid X_i = x)$ and $\tilde{P}(Y_i = y \mid X_i = x)$ is that in the latter, the cluster label $K_i = k$ indexes a random effect vector $\epsilon_{y,k}$, where $k = 1, \ldots, 16$ identifies the clusters, or states. Then $\tilde{P}(Y_i = y \mid X_i = x)$ is define by integrating out the (unobserved) value of $\epsilon_{y,k}$.

19.5 Models

19.5.1 Generalised Logistic Fixed Effect Models

As discussed in the previous section, the estimation of the conditional probabilities $P(Y_i = y | X_i = x)$ for the fixed effect model can be performed either through unweighted or weighted multinomial-logit or generalised *(multicategory) logistic regression*. We present the model in some generality. Let the last or H'th level of y, 3 in our example, be designated as a reference category. Then the conditional probability $f(y | X_i, \theta)$ for $Y_i = y$ given X_i with unknown statistical parameter $\theta = (\beta^{(z)},$

z = 1,...,H-1) is expressed as

$$P(\boldsymbol{Y}_i = \boldsymbol{y} \mid \boldsymbol{X}_i = \boldsymbol{x}) = \exp(\boldsymbol{x}'\boldsymbol{\beta}^{(y)}) \Big/ \left(1 + \sum_{z=1}^{H-1} \exp(\boldsymbol{x}'\boldsymbol{\beta}^{(z)})\right), \quad \boldsymbol{y} = 1, 2, \ldots, H-1 \tag{19.7}$$

$$P(\boldsymbol{Y}_i = H \mid \boldsymbol{X}_i = \boldsymbol{x}) = 1 \Big/ \left(1 + \sum_{z=1}^{H-1} \exp(\boldsymbol{x}'\boldsymbol{\beta}^{(z)})\right) \tag{19.8}$$

This model is descriptive at the unit level (person level). When $H = 3$, this is the simplest fixed-effect model generalizing logistic regression (from 2 to 3 outcome categories) for the behaviour of $\{\boldsymbol{Y}_t\}_{t=1}^n$ conditional on $\{\boldsymbol{X}_t\}_{t=1}^n$, where the $\boldsymbol{Y}_i$'s are assumed conditionally independent across i given $\{\boldsymbol{X}_t\}_{t=1}^n$ and each $\boldsymbol{Y}_i$ depends on $\{\boldsymbol{X}_t\}_{t=1}^n$ only through $\boldsymbol{X}_i$, viewed as a row of the regression design matrix. Intercepts are handled by assuming that the first component of each vector $\boldsymbol{X}_t$ is 1. The model leads to the following conditional ($\boldsymbol{Y}$ given $\boldsymbol{X}$) likelihood factor:

$$L(\boldsymbol{n} \mid \{x\}, \{\boldsymbol{\beta}^{(z)}\}) = \prod_{x \in \Xi} \left\{ \prod_{y=1}^{H-1} \left(\frac{\exp(\boldsymbol{x}'\boldsymbol{\beta}^{(y)})}{1 + \sum_{z=1}^{H-1} \exp(\boldsymbol{x}'\boldsymbol{\beta}^{(z)})} \right)^{n_{y,x}} \left(\frac{1}{1 + \sum_{z=1}^{H-1} \exp(\boldsymbol{x}'\boldsymbol{\beta}^{(z)})} \right)^{n_{H,x}} \right\}$$

Here $\boldsymbol{n} = \{n_{y,x}\}$, where $n_{y,x}$ is the number of observed units with observed status $\boldsymbol{y}$ in July and corresponding predictor vector $\boldsymbol{x}$, and Ξ is the set of all distinct observed vectors $\boldsymbol{x}$. In the case of the fixed-effect model, $\boldsymbol{x}$ includes the geographical information, that is membership in a cluster k= 1, 2,...,16.

19.5.2 Fixed Effect Logistic Models for Estimating Gross Flows

We introduce four progressively simpler versions of the model ((19.7)–(19.8)) in the CPS data example leading to estimation of gross flows, with parameters naturally estimated by maximizing likelihood.

1. **Model F1** is the most elaborate model. The regressors of the logistic regression include the labour force status reported in June (status06) and age (age06), education (educ06) and state. In addition, a first-order (two-way) interaction between labour force status in June and age is included as a regressor.
2. **Model F2** drops the first order interaction between status and age from the regressors in Model F1. Otherwise it remains the same. In particular the state is retained as a predictor.
3. **Models F3, F4** are identical to models F1 and F2, respectively, except that 'state' is dropped altogether as a regressor.

SAS code for computing the MLE for the logistic version of model F1:

```
proc logistic data=CPSunitlevel;
class status07 status06 age06 edu06 state;
model status07 = age06 edu06 state status06*age06/
link=glogit;
output out=outfile pred=predicted;
```

Customised R code using the utility function **nlm** is available in the online supplementary material accompanying this book. Generalised logistic regression models can also be estimated in R using packages **mlogit** (after expressing the models in an alternative data format as econometric social-choice models) or **nnet**. A further method of fitting these models as loglinear models is to use Poisson regression in R through **glm**.

19.5.3 Equivalence between Fixed-Effect Logistic Regression and Log-Linear Models

Many authors (see e.g. Agresti 2013, p. 353) have noted the equivalence between generalised-logistic and loglinear models for broad classes of models – namely, hierarchical nominal loglinear models. This equivalence applies to the logistic models F1–F4. The loglinear version of these models would include all higher-order interactions between the predictor variables, as well as certain higher-order interactions between the dependent variable and the predictors. For example, the model F1 above includes four regressors and specifies that $\boldsymbol{y}$ (status07) depends on the regressors status06, age06, edu06, state and the interaction between age06 and status06. The equivalent loglinear model includes the four-way interaction among the regressors and in addition has the 'status07' interacting pairwise with each of the regressors and includes also the three-way interaction between 'status07', 'status06', and 'age06'. SAS code corresponding to model F1 in loglinear form is given below. The conditional probabilities of the logistic regression are retrieved from the cell inclusion probabilities of the loglinear model saved in the data set 'pred' below.

SAS code for the loglinear Equivalent of model F1:

```
proc catmod data=CPScellcount;
weight cellcount;
response /out=pred;
model status07*status06*age*edu*state =_response_  zero =
sampling;
loglin status06|age|edu|state
status07|status06 status07|age status07|edu status07|
state
status07|status06|age;
```

One attraction of loglinear models is that they support G^2likelihood-ratio goodness of fit tests, a powerful tool for model selection. Although similar tests appear in useful tabular form using the R "anova" method within ordinary logistic regression **glm**, one must specify the G^2 (or score or Wald) tests one by one within **mlogit**. In our context, the goodness of fit test associated to the equivalent loglinear model reflects only the quality of the fit of the original logistic regression, since the submodel involving only the regressors is saturated in the loglinear version and fits the data perfectly. We use this fact to classify fixed-effect models by fit in Section 19.5.

19.5.4 Weighted Estimation

The estimation of parameters within the model ((19.7)–(19.8)) has been described so far purely within a model-based framework, as though the sampled data had not been derived from a survey. The maximum likelihood estimates (MLEs) of the statistical parameter $\theta = (\beta^{(z)}, z = 1,\ldots,K\text{-}1)$ are maximisers of the likelihood or equivalently solutions of the score equation $\sum_{i\in S} \nabla_\theta \log f(\boldsymbol{Y}_i \,|\, \boldsymbol{X}_i, \theta) = \boldsymbol{0}$, where $f(y \,|\, \boldsymbol{X}_i, \theta)$ is given by expressions (19.7–19.8). This estimation method treats the sampled units as though their data vectors $(\boldsymbol{X}_i, \boldsymbol{Y}_i)$ were independent and identically distributed, and from standard large-sample theory of maximum likelihood estimators, the standard *linearization-method* variance estimator for the MLE $\hat{\theta}$ (generally, the variance estimator supplied by package statistical programs) is obtained as the inverse of the *observed information* $-\sum_{i\in S} \nabla_\theta \nabla_\theta^{tr} \log f(\boldsymbol{Y}_i \,|\, \boldsymbol{X}_i, \hat{\theta})$.

The corresponding *pseudolikelihood score equation* incorporating the survey weights is given by

$$U(\theta) = \sum_{i\in S} \boldsymbol{w}_i \; \nabla_\theta \; \log f(\boldsymbol{Y}_i \,|\, \boldsymbol{X}_i, \theta) = 0 \tag{19.9}$$

The coefficient estimates $\hat{\beta}^{(z)}$ are obtained by solving Eq. (19.9). Asymptotically correct large-sample variance estimators as given by Binder (1983, formulas (3.3)–(3.4)) are

$$\widehat{Var}(\hat{\theta}) = A^{-1} B A^{-1}, \qquad \mathrm{A} = -\sum_{i\in S} \boldsymbol{w}_i \, \nabla_\theta \nabla_\theta^{tr} \; \log f(\boldsymbol{Y}_i \,|\, \boldsymbol{X}_i, \hat{\theta}) \tag{19.10}$$

where B is an estimator assumed available and design-consistent for the variance of the pseudolikelihood score statistic $U(\theta)$ defined in (19.9). The asymptotic variance estimator (19.10) is known in statistical literature as the 'sandwich' estimator, the 'robust variance estimator', or simply as the 'linearised' variance estimator in survey statistics literature. With $f(y \,|\, \boldsymbol{X}_i, \theta)$ given by the expressions ((19.7)–(19.8)), this is the survey-weighted estimator of variance that is probably most reliable in large samples. Different valid variance estimators for the parameter MLEs are suggested by the large-sample model-based theory, when the model holds.

Binder was not explicit about the estimator B, but there are at least three different ways of constructing it (Wolter 2007). From a purely design-based point of view, the pseudolikelhood scores $U(\theta)$ are survey-weighted (vector) totals. Accordingly, they could be estimated by a method of random or paired groups. For example, if the population and sample could be partitioned into a set of F pairs M_{f1}, M_{f2} of sub-populations for which it is reasonable to assume from the complex survey structure that the sums, for $f = 1, \ldots, F$ and $g = 1,2$,

$$U_{fg}(\theta) = \sum_{i \in S \cap M_{fg}} \boldsymbol{w}_i \, \boldsymbol{\nabla}_\theta \, \log f(\boldsymbol{Y}_i \mid \boldsymbol{X}_i, \theta)$$

are independent across distinct (f, g) and identically distributed across $g = 1, 2$ for fixed f, then a variance estimator B could be constructed as

$$\boldsymbol{B}_1 = \sum_{f=1}^{F} \boldsymbol{w}_i \, (U_{f1}(\theta) - U_{f2}(\theta)) \, (U_{f1}(\theta) - U_{f2}(\theta)\,)^{tr}$$

This $\boldsymbol{B}_1$ could be approximately design-consistent only if the number G of paired groups grew with the frame and sample size (Krewski and Rao 1981).

Sometimes, in public-use data files, the decomposition into paired groups is not explicit but implicit, through sets of weight-replicates $\boldsymbol{w}_i^{(r)}$, r = 1,..,R to be used in BRR variance estimation. Such replicate-based estimation techniques are devised to reproduce (exactly for large R and approximately otherwise) quadratic-form variance estimators such as $\boldsymbol{B}_1$, so the quadratic form or BRR estimator are in that sense not really distinct estimation techniques.

If, as in the small-area estimation context of this chapter, the model $\boldsymbol{Y}_i \sim f(\boldsymbol{y} \mid \boldsymbol{X}_i, \theta)$ is assumed to hold at the unit level, independently across $i \in U$, then a simple, explicit, partially model-based consistent estimator B can be given in the form

$$\boldsymbol{B}_2 = \sum_{i \in S} \boldsymbol{w}_i \; \boldsymbol{\nabla}_\theta \; \log f(\boldsymbol{Y}_i \mid \boldsymbol{X}_i, \theta) \, \boldsymbol{\nabla}_\theta^{\,tr} \; \log f(\boldsymbol{y} \mid \boldsymbol{X}_i, \theta) \tag{19.11}$$

19.5.5 Mixed-Effect Logit Models for Gross Flows

The mixed-effect logit model is expressed similarly to the fixed-effect representation in ((19.7)–(19.8)). The geographical label K_i for sample unit i, indexes a random intercept term in the logistic regression expression for the conditional probability. Recall $\boldsymbol{X}_i = (\overset{\circ}{\boldsymbol{X}}_i, K_i)$. Let $\boldsymbol{x}^\circ$ denote the value of the fixed effect predictor $\overset{\circ}{\boldsymbol{X}}_i$, and $k = 1, \ldots, 16$ the value of K_i. Let $\boldsymbol{x}$ be an observed value for $\boldsymbol{X}_i$. Then

$\boldsymbol{x} = (\boldsymbol{x}^{\circ}, k)$, where $K_i = k$, and the vector of random effects, $\boldsymbol{\varepsilon}_k = [\varepsilon_{1,k}, \varepsilon_{2,k}]'$, is associated with unit i. We have:

$$P(\boldsymbol{Y}_i = \boldsymbol{y} \mid \boldsymbol{X}_i = \boldsymbol{x}) = \exp(\varepsilon_{y,k} + \boldsymbol{x}^{\circ\prime}\boldsymbol{\beta}^{(y)}) \Big/ \left(1 + \sum_{z=1}^{H-1} \exp(\varepsilon_{z,k} + \boldsymbol{x}^{\circ\prime}\boldsymbol{\beta}^{(z)})\right), \tag{19.12}$$

$$\boldsymbol{y} = 1, \ldots, H-1$$

$$P(\boldsymbol{Y}_i = H \mid \boldsymbol{X}_i = \boldsymbol{x}) = 1 \Big/ \left(1 + \sum_{z=1}^{H-1} \exp(\varepsilon_{z,k} + \boldsymbol{x}^{\circ\prime}\boldsymbol{\beta}^{(z)})\right) \tag{19.13}$$

In our empirical example, $H = 3$. Furthermore, the $\boldsymbol{\varepsilon}_k = [\varepsilon_{1,k}, \varepsilon_{2,k}]^{tr}$ vectors are i.i.d, with

$$\boldsymbol{\varepsilon}_k \sim N(\boldsymbol{0}, \boldsymbol{\Phi}), \qquad \text{for} \quad \boldsymbol{\Phi} = \begin{bmatrix} \phi_{11} & \phi_{12} \\ \phi_{12} & \phi_{22} \end{bmatrix}, \qquad k = 1, \ldots, 16$$

The unconditional likelihood for the model parameters after integrating out the random effects is

$$L(\boldsymbol{n}^* \mid \{\Xi_k\}, \{\boldsymbol{\beta}^{(z)}\}) = \prod_{k=1}^{16} \int \prod_{\boldsymbol{x}^{\circ} \in \Xi_k} \left\{ \prod_{y=1}^{H-1} \left(\frac{\exp(\varepsilon_{y,k} + \boldsymbol{x}^{\circ\prime}\boldsymbol{\beta}^{(y)})}{1 + \sum_{z=1}^{H-1} \exp(\varepsilon_{z,k} + \boldsymbol{x}^{\circ\prime}\boldsymbol{\beta}^{(z)})} \right)^{n_{y,k,x^{\circ}}} \times \left(\frac{1}{1 + \sum_{z=1}^{H-1} \exp(\varepsilon_{z,k} + \boldsymbol{x}^{\circ\prime}\boldsymbol{\beta}^{(z)})} \right)^{n_{3,k,x^{\circ}}} \right\} g(\varepsilon_{y,k})\, d\varepsilon_{y,k} \tag{19.14}$$

where Ξ_k denotes the set of $\boldsymbol{x}^{\circ}$ observed in the fixed-effect regression matrix involving units in state k and $\boldsymbol{n}^* = \{n_{y,k,x^{\circ}}\}$, the set of all $n_{y,kx^{\circ}}$, the number of observed units for state k, predictor $\boldsymbol{x}$ and outcome y, and $g(\varepsilon)$ denotes the density of ε incorporating the variance matrix $\boldsymbol{\Phi}$. The MLE parameter estimates maximizing the likelihood (19.14) are $\boldsymbol{\beta}^{(1)}$, $\boldsymbol{\beta}^{(2)}$, ϕ_{11}, ϕ_{12}, and ϕ_{22}. The likelihood in (19.14) no longer has a low-dimensional minimal sufficient statistic, and simple algorithms like iterative proportional fitting are not available to compute the MLE. MLEs can be computed through adaptive Gaussian quadratures (AGQ) to approximate (19.14) closely, or indirectly through or the expectation-maximization (EM) or Markov Chain Monte Carlo (MCMC) algorithms. In particular, the numerical maximization method AGQ is implemented in PROC NLMIXED in SAS, as well as in the mixed-effect logistic-regressions in packages **lme4** and **glmmML** in R. In the setting of multinomial outcomes ($H > 2$), these random-intercept logistic regression packages can successfully – but not with full efficiency – be used fit the model (19.12–19.13) in $H - 1$ separate mixed-logistic-regression stages when the off-diagonal parameters $\mathbf{\Phi}$ expressing dependence between the random intercepts and different outcome levels are all assumed to be 0.

19.5.6 Application to the Estimation of Gross Flows

As discussed in Section 19.5.4, survey-weighted estimates of the parameters in the fixed-effect models ((19.7)–(19.8)) can be obtained through maximizing the survey-weighted pseudo-log-likelihood. Application of the weighted estimates to subpopulations still relies on model assumptions, although the survey weighting results in design-based parameter estimates solving the estimating equations for the entire population under the probability design of the CPS. The weighted estimates reconstruct the weighted margins implicit in (the saturated part of) the loglinear model. These weighted margins are unbiased estimators of the population margins and so the weighted estimates are calibrated on unbiased estimates of the population margins. In general, the survey-weighting should allow *other* margins to be estimated with minimal bias versus direct Horvitz-Thompson estimates. In the CPS data, we found no substantial difference between unweighted and weighted estimates of gross flows, as would be the case if the model assumptions hold.

Models M1 and M2 revisit models F1 and F2, recasting 'state' as a categorical random effect based on the model described in (19.12, 19.13). These models are straightforward extensions of the mixed effect logistic models proposed by Jiang et al. (2002).

19.6 Results

19.6.1 Goodness of Fit Tests for Fixed Effect Models

Models F1–F4 are designed to borrow strength across the cells of the overall contingency table to produce more stable estimators for the gross flows than could be achieved by estimating the cells singly. A first attempt to validate these models is to compute the goodness of fit statistic associated with the loglinear equivalents of F1–F4.

Table 19.1 shows the goodness of fit statistics, G^2, for the loglinear version of all four fixed effect models (F1-F4), along with the corresponding residual degrees of freedom of the loglinear model and p-values (with respect to chi-square reference distributions).

Our model defines 1152 cells where there could be observed units in the longitudinal distributions. There are approximately 18 000 units in our data set, which cluster in 769 non-zero cells. For the purpose of loglinear modelling involving several hundred parameters, the zero cells are pre-treated by inserting a very small

Table 19.1 Goodness of fit for log-linear fixed-effect models, in terms of residual degrees of freedom.

Model	G^2	Degrees of freedom	*P*-value
F1	678.1	712	0.814
F2	792.9	724	0.038
F3	737.5	742	0.539
F4	854.3	754	0.006

Table 19.2 Deviance for logistic fixed-effect models, with model degrees of freedom.

Model	$-2l$	Degrees of freedom
F1	8781	54
F2	8897	42
F3	8841	24
F4	8958	12

number (0.01) in place of the zero. The zero cells are not an impediment when fitting the logistic regression. We verified that the derived conditional probabilities under logistic regression are very close to those obtained through loglinear modelling as described here.

Table 19.1 suggests that F3 may be a viable model for prediction, although we prefer the fit of F1 and its explicit parameterization of small area fixed effects. The same desirable 'state' effects enter in F2, although that model's G^2 is borderline unacceptable.

The deviances ($-2\ l$ where l is the log-likelihood of the associated fixed-effect logistic regression models F1–F4, given in Table 19.2) provide additional insight. The reductions in deviance from F3 to F1 (60) and from F4 to F2 (61) exhibit the effects of the cluster (state) on the fit of the models. These differences are highly significant ($p < 0.0001$) within standard GLM likelihood ratio tests, since their reference chi-squared distributions have 30 (54–24 and 42–12) degrees of freedom. The interactions between age (age06) and labour force status (status06) are even more significant (a difference of 116 between F2 and F1 and 117 between F4 and F3 with 12° of freedom). So F1 is clearly the preferred model, but we retain F2 because of its simplicity and near adequacy based on Table 19.1.

19.6.2 Fixed-Effect Logit-Based Estimation of Gross Flows

Our overarching goal is to use the models to estimate gross flows as counts or proportions through the estimated conditional probabilities. We fit fixed effect models using SAS PROC LOGISTIC, and the estimates $\widehat{P}(\boldsymbol{Y}_i = \boldsymbol{y} \mid \boldsymbol{X}_i = \boldsymbol{x})$ of the conditional probabilities can be expressed and computed in terms of the parameters of the generalised logistic regression. We illustrate the process of estimating the gross flows from the models by a specific example involving state, age, and education category, using model F1.

19.6.3 Mixed Effect Models

Our main interest in this section is to determine whether there are substantial differences between fixed-effect estimators of gross-flows proportions and their mixed effect counterparts. Ideally one would compare F1 to M1 and F2 to M2. We use the SAS PROC NLMIXED, which fits using AGQ (Pinheiro and Bates 1995). Fitting M1 is computationally demanding in settings, like those below, where many weight-replicated fits are needed for the estimation of variances. We were successful in replicating M2, based on the original replicate factors as explained later in this section.

Table 19.3 illustrates the computation of the estimates $\widehat{T}_{\boldsymbol{y},\boldsymbol{x}}$, $\hat{\hat{T}}_{\boldsymbol{y},\boldsymbol{x}}$, $\widetilde{T}_{\boldsymbol{y},\boldsymbol{x}}$ using the conditional probabilities $\widehat{P}(\boldsymbol{Y}_i = \boldsymbol{y} \mid \boldsymbol{X}_i = \boldsymbol{x})$, $\widetilde{P}(\boldsymbol{Y}_i = \boldsymbol{y} \mid \boldsymbol{X}_i = \boldsymbol{x})$, and totals $\widehat{T}_{\boldsymbol{x}}$ for the corresponding approaches: HT (Horvitz Thompson), fixed-effects model F1 (weighted) and F2, and mixed-effect model M2. Table 19.3 summarises the gross flows between June and July 2017 for each possible value of 'status06', the labour force status for the month of June 2017, for non-college educated individuals less than 35 years old in New Mexico.

19.6.4 Comparison of Models through BRR Variance Estimation

For the purpose of a more general analysis and comparing accuracy across states and demographic categories, it is useful to express the gross flows in terms of pairs of gross flow (employed, unemployed) proportions over all the possible values of the predictors $\boldsymbol{x} = ($ status06, age06, educ06, state $)$. We represent these estimated marginal totals by $\widehat{T}_{(+,u)}$, meaning that in $\widehat{T}_{\boldsymbol{x}}$ the June labour force status $v =$ status06 in $\boldsymbol{x} = (\boldsymbol{v}, \boldsymbol{u})$ has been summed up for each joint value of $u =$ (age06, educ06, state). The direct, or Horvitz Thompson (HT), estimator for conditional gross-flow proportions is:

$$\widehat{R}_{\boldsymbol{y},\boldsymbol{x}} = \widehat{T}_{\boldsymbol{y},\boldsymbol{x}} / \widehat{T}_{(+,u)} \tag{19.15}$$

for $\boldsymbol{x} = (\boldsymbol{v}, \boldsymbol{u})$. The model-assisted ratio is:

$$\hat{\hat{R}}_{\boldsymbol{y},\boldsymbol{x}} = \hat{\hat{T}}_{\boldsymbol{y},\boldsymbol{x}} / \widehat{T}_{(+,u)} \tag{19.16}$$

Table 19.3 Gross flow estimates for non-college educated less than 35 years old in New Mexico for HT and model F1, F2, M2.

	HT	June LFS	F1 (weighted)	F2	M2	F1	F2	M2
Gross Flows	$\widehat{T}_{y,x}$	$\widehat{T}_x$	$\widehat{P}$	$\widehat{P}$	$\widetilde{P}$	$\hat{\hat{T}}_{y,x}$	$\hat{\hat{T}}_{y,x}$	$\widetilde{T}_{y,x}$
EE	149 900	167 600	0.922	0.942	0.950	154 600	157 900	159 200
EU	6496	167 600	0.025	0.025	0.019	4341	4300	3296
EN	11 200	167 600	0.051	0.031	0.031	8599	5341	5240
UE	3402	16 860	0.167	0.14	0.188	2827	2599	3174
UU	7945	16 860	0.573	0.651	0.582	9668	10 990	9825
UN	5512	16 860	0.258	0.194	0.229	4364	3270	3859
NE	11 000	131 200	0.072	0.053	0.056	9480	7031	7391
NU	9900	131 200	0.066	0.054	0.042	8735	7125	5582
NN	110 000	131 200	0.861	0.892	0.901	112 998	117 058	118 200

Notes: Entries are rounded to four significant digits. $\widehat{P}$ and $\widetilde{P}$ estimators respectively calculated by substituting parameter estimates into ((19.7)–(19.8)) and ((19.12)–(19.13)).

Similarly, the mixed-effect ratio based on hybrid estimators as defined in (19.6) is

$$\widetilde{R}_{y,x} = \widetilde{T}_{y,x}/\widehat{T}_{(+,u)} \tag{19.17}$$

So $\widehat{R}_{y,x}$ and $\hat{\hat{R}}_{y,x}$ may represent any of the nine gross flow proportions identified by $\boldsymbol{y}$ (status07) and v = status06 for a given set of $\boldsymbol{u}$ = (age06, educ06, state) values, with $\boldsymbol{x} = (\boldsymbol{v}, \boldsymbol{u})$. We focus on comparing the frequentist properties of $\widehat{R}_{y,x}$ and $\hat{\hat{R}}_{y,x}$, model-based versus HT estimators.

To evaluate the variance and mean squared error of our estimators, we exploit the design of CPS, which enables variance estimation through BRR. This is a non-parametric variance estimation method that resembles bootstrap variance estimation except that there are only a relatively small number of 'balanced' replicates.

Within CPS, annual sampling selects groups of four households ('hits'), and then sorts these by geographic characteristics and rotation group. Within self-representing (especially large) primary sampling units (PSUs), CPS assigns replicate weights by the Fay and Train (1995) method of *Successive Difference Replication*. Within non-self-representing PSUs, CPS assigns BRR weight-factors using paired 'pseudostrata' obtained by combining pairs (and occasionally triples) of actual strata. Both replication methods use $R = 160$ weight replicates in the monthly CPS files maintained internal to the Census Bureau.

Table 19.4 Gross flow estimators (proportions) for non-college educated less than 35 years old in New Mexico – model F1.

	HT	F1 (weighted)	F2	M2	HT SE	F1 SE	F2 SE	M2 SE
Gross Flows	$\widehat{R}_{y,x}$	$\hat{\hat{R}}_{y,x}$	$\hat{\hat{R}}_{y,x}$	$\widetilde{R}_{y,x}$	$\sigma(\widehat{R}_{y,x})$	$\sigma(\hat{\hat{R}}_{y,x})$	$\sigma(\hat{\hat{R}}_{y,x})$	$\sigma(\widetilde{R}_{y,x})$
EE	0.474	0.489	0.500	0.504	0.0270	0.0254	0.0259	0.0260
EU	0.020	0.013	0.013	0.010	0.0073	0.0032	0.0030	0.0029
EN	0.035	0.027	0.016	0.016	0.0119	0.0050	0.0029	0.0029
UE	0.010	0.008	0.008	0.010	0.0053	0.0028	0.0025	0.0031
UU	0.025	0.030	0.034	0.031	0.0089	0.0078	0.0083	0.0078
UN	0.017	0.013	0.010	0.012	0.0070	0.0037	0.0029	0.0034
NE	0.034	0.030	0.022	0.023	0.0116	0.0052	0.0041	0.0042
NU	0.031	0.027	0.022	0.018	0.0058	0.0053	0.0035	0.0037
NN	0.349	0.357	0.370	0.374	0.0261	0.0251	0.0253	0.0257

Table 19.4 revisits Table 19.3 and displays the nine gross-flow proportions for the same specific example, as well as the standard errors of the estimators for these proportions. In many cases the model-assisted estimator has substantially smaller standard error than the HT. In the case of the 'employed-employed' and 'NLF-NLF' the relative error is unchanged, which can be explained partly by the large size of the populations in these cells. However, the 'Unemp-Unemp' proportion is also not improved by the model-assisted estimator showing that our approach does not always reduce small-cell variances.

Motivated by the example in Tables 19.3 and 19.4, we turn to a more general analysis. We are interested in the frequentist properties of HT, model F1 and model F2 for the small cells where savings in standard error can be very meaningful since direct HT estimates can be prohibitively noisy, as Table 19.5 shows for 77 cells of size 3. The choice of exhibiting results for these cells is partly due to privacy and confidentiality constraints. Nevertheless, cells of size 3 are similar to other small cells with respect to frequentist measurements (variance, standard error, and relative error) comparing F1 and F2 estimates to HT.

Table 19.5 suggests that the range of the estimated proportions for cells under HT, F1, F2, M2, are roughly the same. Table 19.6 shows the correlations across these cells between the HT and model-based estimators. Unsurprisingly, the model-based estimators are highly correlated, especially F2 and M2, which have the same structure. However, HT does not correlate strongly with any of the model estimators.

Table 19.5 Gross flow proportion estimates for 77 small domains represented by three data points.

Quartiles & mean	$\hat{R}_{y,x}$ HT	$\hat{\hat{R}}_{y,x}$ F1	$\hat{\hat{R}}_{y,x}$ F2	$\tilde{R}_{y,x}$ M2
Minimum	0.0076	0.0047	0.0050	0.0057
Q1	0.011	0.010	0.010	0.009
Q2	0.017	0.014	0.013	0.013
Q3	0.030	0.020	0.018	0.019
Maximum	0.123	0.218	0.230	0.234
Mean	0.026	0.019	0.019	0.020

Table 19.6 Correlations between gross flow proportion estimates for 77 small domains.

	Correlation		
	F1	F2	M2
HT	0.41	0.39	0.39
F1		0.97	0.97
F2			0.99

Table 19.7 displays the BRR estimated standard errors for the 77 gross-flow proportions under HT and the three models. The quartiles and the mean indicate very important reductions in standard errors, exceeding 50% on the average for all three models. The mixed-effect model may be the best performer, but if so its advantage is very slight, and this finding is far from definitive.

Table 19.8 displays extreme cases, instances of the largest and smallest estimates of gross-flow proportions for the models. The first case suggests that there might be not much to gain from using the models in terms of reduction of the standard error for large estimators. The second case suggests that the models particularly increase the precision of very small estimators.

19.7 Discussion

Our main point in this chapter is that estimation of conditional gross-flow transition probabilities can be done effectively using a generalised logistic (multinomial

Table 19.7 Standard errors for proportion estimates for 77 small domains.

Quartiles & mean	$\sigma(\widehat{R}_{y,x})$ HT	$\sigma(\hat{\hat{R}}_{y,x})$ F1	$\sigma(\hat{\hat{R}}_{y,x})$ F2	$\sigma(\widetilde{R}_{y,x})$ M2
Minimum	0.0035	0.0017	0.0014	0.0012
Q1	0.0067	0.0033	0.0028	0.0024
Q2	0.0106	0.0043	0.0040	0.0036
Q3	0.0180	0.0062	0.0058	0.0052
Maximum	0.071	0.070	0.075	0.076
Mean	0.015	0.0071	0.0067	0.0063

Table 19.8 Extreme standard errors for 77 small domains.

	$\widehat{R}_{y,x}$	$\hat{\hat{R}}_{y,x}$	$\hat{\hat{R}}_{y,x}$	$\widetilde{R}_{y,x}$	$\sigma(\widehat{R}_{y,x})$	$\sigma(\hat{\hat{R}}_{y,x})$	$\sigma(\hat{\hat{R}}_{y,x})$	$\sigma(\widetilde{R}_{y,x})$
GF	HT	F1	F2	M2	HT SE	F1 SE	F2 SE	M2 SE
NV 4 2 NN	0.113	0.218	0.230	0.235	0.062	0.070	0.075	0.077
GA 4 1 EN	0.0076	0.0084	0.0083	0.0083	0.0050	0.0020	0.0017	0.0013

outcome) model and then converted by the 'hybrid' approach of Thibaudeau et al. (2017) to estimates of small-area gross-flow cell counts. The benefit of such a model-based approach is that small cell counts can be estimated with substantially reduced variance by comparison with direct Horvitz-Thompson estimates. In the first instance, we implement this approach for all small cells, with fixed effects not including domain indicators. In a variant reminiscent of small area estimation, we estimate within either fixed or mixed effects models with a categorical variable indexing the small domains.

Using an extended empirical example from the US 2017 CPS viewed as a longitudinal survey embedded in a rotating panel survey, we found that it did not matter whether the model was fitted weighted or unweighted or unweighted-mixed: the estimated coefficient and small-cell predictions were virtually the same. Within that example, this data-analytic finding confirmed the statistical adequacy of log-linear or generalised-logistic models with a state intercept.

The models can be fitted within SAS by PROC LOGISTIC or (in the mixed-effect case) by PROC NLMIXED. In R, the fixed-effect unweighted generalised logistic can be fitted using the **mlogit** or **nnet** package; simple, fast customised code can be used for weighted or unweighted generalised logistic; or else one can fit multiple logistic regression models to subsets of the data (in each case, with outcomes restricted to the reference and one other category) with very little loss of efficiency, a method that in the unweighted case generalises easily (using R packages **lme4** or **glmmML**) to the random intercept case.

We view the weighted fixed-effect approach as ideal for datasets with a moderate number of small areas, and the unweighted mixed-effect (random intercept) model as suited to cases with a large number of small areas. In either case, variances can be estimated effectively using BRR. Linearization-based variance estimates are available for model coefficients, but are more problematic when coupled with direct survey variances within our hybrid approach.

Our development in this chapter highlights the need for further research in two directions. First, there is need for additional data analysis to describe state-level differences in gross-flow predictions. The apparent superiority in our example of the fixed-effect State model **F1** over random-intercept analogues suggests the need to explore models allowing for random slope effects (for age and education), reflecting the different regional and state mixtures of industrial sectors in the labour force. The second, more methodological direction in which progress is needed is the extension of gross-flow variance estimation combining linearization for estimated model parameters with design-based (BRR or other) survey variance estimates for antecedent cell-totals defined by variables $\boldsymbol{x}$ in the hybrid estimates ((19.4)–(19.6)). A similar approach to hybrid variance estimation was investigated in a general setting by Slud and Ashmead (2017) after having been implemented in small area estimation in the (non-longitudinal setting) of estimates in the Census Bureau's Voting Rights Act determinations (http://www.census.gov/rdo/pdf/3_VRA_Statistical_Methodology_Summary_V7.pdf), and research in this direction is continuing.

Acknowledgements

Any views expressed on statistical methodology, technical or operational issues are those of the authors and not necessarily those of the US Census Bureau. The authors are grateful to Khoa Dong and Tim Trudell for indispensable CPS data preparation and matching across June and July 2017 monthly data files.

References

Agresti, A. (2013). *Categorical Data Analysis*. Hoboken, NJ: Wiley.

Binder, D. (1983). On the variance of asymptotically normal estimators from complex surveys. *International Statistical Review* 51: 279–292.

Fay, R.E. and Train, G.F. (1995). Aspects of Survey and Model-Based Postcensal Estimation of Income and Poverty Characteristics for States and Counties. Joint Statistical Meetings. Section on Government Statistics.

Feder, M., Nathan, G., and Pfeffermann, D. (2000). Multilevel modeling of complex survey longitudinal data with time varying random effects. *Survey Methodology* 26: 53–65.

Fienberg, S. (1980). The measurement of crime victimization: prospects for a panel analysis of a panel survey. *Journal of the Royal Statistical Society D* 29: 313–350.

Fuller, W.A. (2009). *Sampling Statistics*. Hoboken, NJ: Wiley.

Jiang, J., Lahiri, P., and Wan, S.-M. (2002). A unified jackknife theory for empirical best prediction with M-estimation. *Annals of Statistics* 30: 1782–1810.

Korn, E. and Graubard, B. (2003). Estimating variance components by using survey data. *Journal of the Royal Statistical Society B* 65: 175–190.

Krewski, D. and Rao, J.N.K. (1981). Inference from stratified samples: properties of the lineari-zation, jackknife and balanced repeated randomization methods. *Annals of Statistics* 5: 1010–1019.

Marker, D. (1999). Organization of small area estimates using a generalized linear regression framework. *Journal of Official Statistics* 15: 1–24.

Pfeffermann, D., Skinner, C., and Humphreys, K. (1998). The estimation of gross flows in the presence of measurement error using auxiliary variables. *Journal of the Royal Statistical Society A* 161: 13–32.

Pfeffermann, D., Skinner, C., Goldstein, H. et al. (1998b). Weighting for unequal selection probabilities in multilevel models (with discussion). *Journal of the Royal Statistical Society B* 60: 23–40.

Pinheiro, J. and Bates, D. (1995). Approximations of the loglikelihood function in the nonlinear mixed effects model. *Journal of Computational and Graphical Statistics* 4: 12–35.

Purcell, N. and Kish, L. (1979). Estimates for small domains. *Biometrics* 35: 365–384.

Purcell, N. and Kish, L. (1980). Postcensal estimates from local areas (or domains). *International Statistical Review* 48: 3–18.

Rabe-Hesketh, S. and Skrondal, A. (2006). Multilevel modelling of complex survey data. *Journal of the Royal Statistical Society A*. 169: 805–827.

Rao, J.N.K. and Molina, I. (2015). *Small Area Estimation*, 2nd edition. Hoboken NJ: Wiley.

Rao, J.N.K., Verret, F., and Hidiroglou, M. (2013). A weighted estimating equations approach to inference for two-level models from survey data. *Survey Methodology* 39: 263–282.

Särndal, C.-E., Swensson, J., and Wretman, J. (1992). *Model Assisted Survey Estimation*. Springer.

Slud, E. and Ashmead, R. (2017). Hybrid BRR and parametric-bootstrap variance estimates for small domains in large surveys. In: *Proceedings of the American*

Statistical Association, Survey Research Methods Section, 1716–1730. Alexandria, VA: American Statistical Association.

Thibaudeau, Y., Slud, E., and Gottschalck, A. (2017). Modeling log-linear conditional probabilities for estimation in surveys. *Annals of Applied Statistics* 11: 680–697.

Wolter, K. (2007). *Introduction to Variance Estimation*, 2nd edition. Springer.

20

Nonparametric Estimation for Longitudinal Data with Informative Missingness

Zahoor Ahmad and Li-Chun Zhang

Department of Social Statistics and Demography, University of Southampton, Southampton, UK

20.1 Introduction

Longitudinal data analysis is of great interest in a wide array of disciplines across the medical, economic, and social sciences. Cross-sectional data can only provide a snapshot at a single point of time and does not possess the capacity to reflect change, growth, or development. Aware of the limitations in cross-sectional studies, many researchers have advanced the analytic perspective by examining data with repeated measurements. By measuring the same variable of interest repeatedly over time, the change is displayed, and constructive findings can be derived with regard to the significance of pattern revealed (Lynn, 2009a). Data with repeated measurements are referred to as longitudinal data. In many longitudinal data designs, subjects are assigned specified levels of a treatment or subjected to other risk factors over a number of time points that are separated by specified intervals.

Analysing longitudinal data poses many challenges due to several unique features inherent in such data. First, a troublesome feature of longitudinal analysis is missing data in repeated measurements. In a longitudinal survey, missing observations of the variable of interest frequently occurs. For example, in a clinical trial on the effectiveness of a new medical treatment for disease, patients may be unavailable to a follow-up investigation due to migration or health problems, or some baseline respondents may lose interest in participating at subsequent times. The missing cases may possess specific characteristics and attributes, resulting in the observed sample at later time points having a different structure to the sample initially gathered. Second, repeated measurements for the same observational unit are usually correlated, due to the fact that they are clustered within each unit. At the same time, an individual's repeated measurements may be subjected to a time-varying systematic process, resulting in serial correlation. Third,

Advances in Longitudinal Survey Methodology, First Edition. Edited by Peter Lynn.

Companion website: www.wiley.com/go/lynn/advancesinlongitudinalsurvey

longitudinal data may be ordered either in equal or unequal time intervals, where each scenario may call for a different analytic approach. Sometimes, despite an equal-spacing design, some respondents may enter a follow-up investigation after the specified survey date, which creates unequal time intervals for different individuals by chance.

To fix the scope, it is helpful to make a distinction between repeated measures in general and the type of longitudinal data considered in this chapter. Repeated measures data represent a wider concept as they sometimes involve a large number of time points and permit changing experimental or observational conditions (West et al. 2007). The longitudinal data which we consider here can be regarded as a special case of repeated measures data. They are composed of observations for the same subject ordered by a limited number of time points (i.e. *waves*) with predetermined time scale, interval, and other related conditions. This is typical of data arising from longitudinal social surveys (Lynn 2009b). In statistics and econometrics, such longitudinal data are often referred to as panel data.

As indicated above, missing data is one of the primary problems to contend with in longitudinal data analysis. Missing data can be due to units that have dropped out (unit or wave non-response) or unanswered items (item non-response). At each wave, longitudinal data are collected in a particular period of time in which the outcome and other relevant variables are recorded sequentially. Therefore, the researcher can only observe responses for those who are available within the duration of follow-up. There arise different situations of missing data. Sometimes the missing data represent a random sample of all cases. Or, missing data do not occur randomly, but the missing-ness can be controlled for with respect to observed variables such as age, gender, and health status. In these situations, missing data can cause loss of efficiency of the analysis but not necessarily bias. In many circumstances, however, the probability of data being missing is related to the missing values of the outcome variable, and failure to account for such missing data can be detrimental to the analysis of the pattern of change over time in the corresponding outcome variable. It is thus essential for the researcher to investigate and understand the nature of the missing data mechanism at hand.

Of relevance in this context is the missing data pattern, which conveys the structure of which values are observed and which values are missing in the data matrix. Some common missing data patterns are univariate, multivariate, monotone, non-monotone, and file matching (Little and Rubin 2002). A univariate missing pattern refers to the situation where missing data occur only in a single variable. As an extension, the multivariate missing pattern refers to missing data in a set of variables, either entirely for the unit or partially for particular items in a questionnaire. If a variable is missing for a particular subject not only at a specific time point but also at all subsequent waves, the missing data pattern for this individual is said to be monotone. In contrast, if a case is missing at a given time point and then returns at a later wave or waves, then the missing data pattern for this subject is referred to as nonmonotone. Sometimes, variables are not observed jointly on the same units, and such missing data are said to have a file-matching pattern. In any of the situations mentioned here, the missing data pattern mentioned can be represented by a matrix of *response indicators*, which takes value 1 if the corresponding item and unit is observed, and 0 otherwise.

Missing data mechanisms concern the relationship between the response indicators and the values of the variables in the corresponding data matrix. They are usually categorised into three classes: missing completely at random (MCAR), missing at random (MAR), and missing not at random (MNAR). If the response indicators are unrelated to both the missing outcomes and the set of observed outcomes, the observed outcomes are a random subset of the entire sample. This is referred to as MCAR. If the response indicators depend on the observed outcomes (and other auxiliary variables) but are otherwise unrelated to the missing values, the missing data are said to be MAR. MAR mechanisms are most commonly assumed in statistical analysis including longitudinal data analysis. However, in many situations, the response indicators are related to the missing values, even after controlling for all the observed values, referred to as MNAR. Ignoring the impact of the MNAR mechanism can result in serious bias of inference. Over the years, a variety of models and methods have been developed to account for MNAR mechanisms in longitudinal data analysis.

When covariates are known for every sample unit, a common way to deal with the non-response is to postulate a parametric model for the joint distribution of the outcome variable and response indicator given the covariates. Little and Rubin (2002) distinguish between selection models and pattern-mixture models, depending on how the joint distribution is factorised. For fully parametric selection models, the likelihood based on all the units, respondents or not, can be used to estimate the parameters of the model for the outcome variable as well as the model for the response probability given the outcome variable (and covariates). Qin et al. (2002) propose a semi-parametric estimation method for the case where the covariates are only known for the respondents. They assume a parametric model on the response mechanism but a non-parametric model on the distribution of the outcome variable and the covariates. Pfeffermann and Sikov (2011) propose a fully parametric estimation approach for NMAR non-response, which does not require knowledge of the covariates for the non-respondents.

There is a large literature on the use of estimating equations (EE); see, for example, Godambe (1991), Liang and Zeger (1995), and Hardin and Hilbe (2003). Robins et al. (1994) suggested a semiparametric approach based on inverse response-probability weighted EE. It is based on the assumption that the probability of non-response is either known or can be modelled parametrically. Fitz-Gerald (2002) introduced a weighting method for handling missing data using generalised estimating equations (GEEs), which also relies on the specification of a parametric non-response model.

All the aforementioned NMAR techniques require a parametric model for the response probability, regardless how the outcome variable is modelled. In this chapter, we present a *non-parametric EE* (NEE) estimation approach for longitudinal data analysis, where we neither specify a parametric model for the response probability nor the outcome variable. This can provide a useful, flexible alternative to the existing methods. The basic idea can be outlined as follows.

Let the target of estimation be given as a finite population parameter defined in terms of a population EE. For its estimation we use the *observed (respondent)* NEE, where the unknown individual response propensity is replaced by an estimate based on the response history of the same individual. For instance, one may use the observed historic response rate for a unit to estimate its individual response probability, under the assumption that the unknown response probability is 'stable' over the given period of time. There can be different assumptions of the exact nature of such stability over time, e.g. stable before the dropout for a unit with monotone missing data pattern, but over the entire history for someone with a nonmonotone pattern. To focus the idea, two different NEE-based estimators for the *change* between any two waves will be given in Section 20.2, although the NEE formulation accommodates many other types of analysis, such as estimation of regression coefficients or analysis of variance.

While the estimator of the individual response probability can be unbiased according to the given assumption, it can never be consistent due to the fact that the response history cannot be infinitely long for anyone. Moreover, the plug-in observed NEE will be somewhat biased if the 'score-term' in the population EE is correlated with the response propensity, as in the case of informative non-response. The matter will be considered in Section 20.3, including possible venues for bias adjustment. The associated variance estimation will be described in Section 20.4. We illustrate and investigate the performance of the NEE approach using a simulation study in Section 20.5. Finally, a summary of the conclusions is given in Section 20.6.

20.2 Two NEE Estimators of Change

Under the NEE approach to MNAR non-response, we do not assume a parametric model of the response probabilities that pertain to all the population units. To accommodate potentially informative missing data, we postulate an individual response probability that may depend on the longitudinal outcomes of interest and covariates specific to each observational unit. The individual response probability can be considered as a propensity of observation that accounts for the initial sample selection mechanism in addition, which may be probability sampling or non-random or informative itself. That is, the response indicator is the product of sample inclusion indicator and survey response indicator. The outcome values are also treated non-parametrically as unknown constants, just like in the design-based approach to survey sampling. Under this set-up, the observation propensity is estimated using individual-specific observation history, without involving the others in the population. The approach is applicable whenever there exist historical response/observation indicators. In other words, any unit who never responds will not be included in the estimation. To maintain the focus, we shall assume in this chapter that these never-respondents are a completely random sample from the population, without getting into the details of exploring different modelling options for them that would be necessary otherwise.

Let the population $U = \{1, \ldots, N\}$ be fixed over time points $t = 1, \ldots, T$, from the most distant ($t = 1$) to the most recent wave ($t = T$). Let y_{it} be a value associated with individual i at time t, for $i \in U$, and $\mathbf{y}_i = (y_{i1}, \ldots, y_{iT})^T$. Let δ_{it} be a 0/1 response/observation indicator for y_{it} at time t. For the finite-population change, without losing generality, let us consider

$$\Delta_t = \overline{Y}_t - \overline{Y}_{t-1} = \sum_{i \in U} y_{it}/N - \sum_{i \in U} y_{i,t-1}/N = \sum_{i \in U} (y_{it} - y_{i,t-1})/N$$

for $t \geq 2$. Let $d_{it} = y_{it} - y_{i,t-1}$. The population EE that defines Δ_t is given by

$$H(\Delta_t) = N^{-1} \sum_{i=1}^{N} S_i(\Delta_t); \qquad H(\Delta_t) = 0, \tag{20.1}$$

i.e. $S_i(\Delta_t) = S(\Delta_t; \mathbf{y}_i) = d_{it} - \Delta_t$ is the 'score', so specified that Δ_t is the solution to $H(\Delta_t) = 0$.

Let $r_{it} = \delta_{i,t-1}\delta_{it}$. Let r_{it} and $r_{it'}$ be independent, where $t \neq t'$, given $\mathbf{y}_i$ and relevant covariates $\mathbf{x}_i$. An informative non-response probability assumption of r_{it} can be given as

$$p_{it} = \Pr(r_{it} = 1 \mid \mathbf{y}_i, \mathbf{x}_i) = \Pr(\delta_{i,t-1}\delta_{it} = 1 \mid \mathbf{y}_i, \mathbf{x}_i), \tag{20.2}$$

An unbiased respondent EE for Δ_t is then given by

$$\widetilde{H}(\Delta_t) = N^{-1} \sum_{i=1}^{N} \frac{r_{it}}{p_{it}} S_i(\Delta_t); \qquad \widetilde{H}(\widetilde{\Delta}_t) = 0,$$

based on the respondents at both waves. However, $\widetilde{H}(\Delta_t)$ is not operational because p_{it} is unknown.

The observed (respondent) NEE is given by

$$\widehat{H}(\Delta_t) = N^{-1} \sum_{i=1}^{N} \frac{r_{it}}{\widehat{p}_{it}} S_i(\Delta_t); \qquad \widehat{H}(\widehat{\Delta}_t) = 0, \tag{20.3}$$

on replacing p_{it} with a suitable estimator $\hat{p}_{it}$ for each respondent with $r_{it} = 1$. The corresponding NEE-based estimator of Δ_t is simply given by

$$\hat{\Delta}_t = \sum_{i=1}^{N} \frac{r_{it}}{\hat{p}_{it}} d_{it} / \sum_{i=1}^{N} \frac{r_{it}}{\hat{p}_{it}}.$$

In its basic form, the estimator $\hat{p}_{it}$ is constructed individually for each observational unit on its own, based on the relevant historic values $r_{i2}, \ldots, r_{iT}$. For example, consider the following three estimators:

$$\hat{p}_i = \sum_{t=2}^{T} \frac{r_{it}}{T-1}, \quad \hat{p}_{1i} = \sum_{t=2}^{T_{1i}} \frac{r_{it}}{T_{1i}-1} \quad \text{and} \quad \hat{p}_{2i} = \sum_{t=2}^{T_{2i}} \frac{r_{it}}{T_{2i}-1}, \tag{20.4}$$

where $T_{1i} = \max_{t=2,\ldots,T} \delta_{it} t$ is the most recent time point of response, and $T_{2i} = \max_{t=2,\ldots,T} r_{it} t$ is that of the most recent successive responses. The three estimators are the same if $t = T$, in which case $T_{1i} = T_{2i} = T$, but they may differ if $t < T$. For instance, suppose $T = 6$ and $t = 3$. The estimator $\hat{p}_i$ uses all the five r_{it}'s, which may not be appropriate for the units observed with a monotone missing data pattern, where non-response after the dropout point can be irrelevant to the estimation of response propensity before the dropout point. The dropout is dealt with somewhat differently by $\hat{p}_{1i}$ and $\hat{p}_{2i}$. Consider a unit with $r_{i3} = 1, \delta_{i4} = \delta_{i6} = 1$, and $\delta_{i5} = 0$, such that $T_{1i} = 6$ and $T_{2i} = 4$, by which $\hat{p}_{1i}$ and $\hat{p}_{2i}$ will differ from each other, even though $\sum_{t=2}^{T_{1i}} r_{it} = \sum_{t=2}^{T_{2i}} r_{it}$ in this case.

Notice that given the form of an estimator $\hat{p}_{it}$, whether it is unbiased depends on the nature of (20.2), which generally varies from one individual to another. For instance, for someone with a non-monotone missing data pattern, $\hat{p}_i$ is unbiased if the response probability is constant over time, regardless how it depends on $\mathbf{y}_i$ and $\mathbf{x}_i$. Whereas the estimators $\hat{p}_{1i}$ and $\hat{p}_{2i}$ are biased in such a case, unless $T = t$, because the 'stopping' times T_{1i} and T_{2i} are informative otherwise. Of course, there are other situations where $\hat{p}_{1i}$ or $\hat{p}_{2i}$ may be more appropriate than $\hat{p}_i$, and so on.

Finally, there will be individuals for which all three estimators are biased, and some other estimator of p_{it} is more appropriate. For instance, one may allow $\hat{p}_{it}$ to depend on t, provided one detects a 'trend' in the r_{it}'s over time for the given individual. The flexibility of the approach here is that it allows one to *vary* the specification of $\hat{p}_{it}$ for each individual and make the assumption that is considered most appropriate given the response history of that particular individual, instead of imposing a single parametric form across the population, as under the parametric modelling approach to (20.2).

Now, the NEE (20.3) uses only the completely observed units at both t and $t-1$. This could potentially entail a loss of efficiency. For an alternative that uses the individuals who respond at either one of the two time points, consider the MNAR response probability assumption for each t:

$$\pi_{it} = \Pr(\delta_{it} = 1 \mid \mathbf{y}_i, \mathbf{x}_i). \tag{20.5}$$

The observed (respondent) NEE can then be given as

$$\begin{cases} \widehat{H}(\theta_t) = N^{-1} \sum_{i=1}^{N} \frac{\delta_{it}}{\widehat{\pi}_{it}} S_i(\theta_t) \\ \widehat{H}(\theta_{t-1}) = N^{-1} \sum_{i=1}^{N} \frac{\delta_{i,t-1}}{\widehat{\pi}_{i,t-1}} S_i(\theta_{t-1}) \end{cases} \tag{20.6}$$

where $\theta_t = \overline{Y}_t$ and $\theta_{t-1} = \overline{Y}_{t-1}$, such that $\Delta_t = \theta_t - \theta_{t-1}$, and $S_i(\theta_t) = y_{it} - \theta_t$ is such that that θ_t is the solution to the population EE, $H(\theta_t) = 0$, and similarly for $S_i(\theta_{t-1})$. Then $\widehat{\pi}_{it}$ is the response probability estimator for time t and $\widehat{\pi}_{i,t-1}$ that for time $t-1$. Below, we refer to (20.6) as $\widehat{H}(\theta_t, \theta_{t-1})$.

Having estimated the cross-sectional parameters θ_t and θ_{t-1}, one can derive the corresponding NEE-based estimate of Δ_t as the difference between the two, which is given by

$$\widehat{\Delta}_t = \sum_{i=1}^{N} \frac{\delta_{it}}{\widehat{\pi}_{it}} y_{it} \Big/ \sum_{i=1}^{N} \frac{\delta_{it}}{\widehat{\pi}_{it}} - \sum_{i=1}^{N} \frac{\delta_{i,t-1}}{\widehat{\pi}_{i,t-1}} y_{i,t-1} \Big/ \sum_{i=1}^{N} \frac{\delta_{i,t-1}}{\widehat{\pi}_{i,t-1}}.$$

The variance of this estimator is given as

$$V(\widehat{\Delta}_t) = V(\widehat{\theta}_t) - 2Cov(\widehat{\theta}_t, \widehat{\theta}_{t-1}) + V(\widehat{\theta}_{t-1}). \tag{20.7}$$

Similarly to p_{it}, one may postulate different non-parametric estimators of π_{it}. We consider the following two estimators in the simulation study later on:

$$\widehat{\pi}_i = \sum_{i=1}^{T} \frac{\delta_{it}}{T} \quad \text{and} \quad \widehat{\pi}_{1i} = \sum_{i=1}^{T_{1i}} \frac{\delta_{it}}{T_{1i}}. \tag{20.8}$$

As with the estimator $\widehat{p}_{it}$ above, one can subject the estimator $\widehat{\pi}_{it}$ to appropriate modifications, to suit different assumptions of (20.5) across the observational units.

20.3 On the Bias of NEE

The NEE $\widehat{H}(\Delta_t)$ given by (20.3) or $\widehat{H}(\theta_t, \theta_{t-1})$ given by (20.6) is not exactly unbiased, even when $\widehat{p}_{it}$ is unbiased for p_{it} or $\widehat{\pi}_{it}$ unbiased for π_{it}. Here we examine the bias and explore possible adjustments.

Let $\tau_{it} = E(\widehat{p}_{it})$ be the expectation of $\widehat{p}_{it}$, where $\tau_{it} = p_{it}$ in case $\widehat{p}_{it}$ is unbiased. Via Taylor expansion of $\widehat{p}_{it}$ around τ_{it}, provided $\widehat{p}_{it} \neq 0$, the bias of NEE (20.3) can be written as

$$B_1 = E[\widehat{H}(\Delta_t)] - H(\Delta_t) = N^{-1}\sum_{i=1}^{N} S_i(\Delta_t) E\left(\frac{r_{it}}{\widehat{p}_{it}} - 1\right)$$

$$= N^{-1}\sum_{i=1}^{N} S_i(\Delta_t)\left\{\left(\frac{2p_{it}}{\tau_{it}} - 1\right) - \frac{E(r_{it}\widehat{p}_{it})}{\tau_{it}^2} + \frac{E(r_{it}\widehat{p}_{it}^2) - 2\tau_{it}E(r_{it}\widehat{p}_{it}) + \tau_{it}^2 p_{it}}{p_i^{*3}}\right\} \tag{20.9}$$

for some p_i^* between τ_{it} and $\widehat{p}_{it}$. Further derivation of the expression (20.9) depends on the choice of $\widehat{p}_{it}$. The case of $\widehat{p}_{it} = \widehat{p}_i$ in (20.4) is given below; the other cases are similar and omitted here. We have

$$E(r_{it}\widehat{p}_i) = E\left(\frac{r_{it}}{T-1}\sum_{t'=2}^{T} r_{it'}\right) = \frac{1}{T-1}\sum_{t'=2}^{T} p_{it}p_{it'} + \frac{1}{T-1}Var(r_{it}) = p_i^2 + V(\widehat{p}_i),$$

where the last expression follows if $p_{it} \equiv p_i$, i.e. when $\widehat{p}_i$ is unbiased for p_{it}. Next, after some algebra,

$$E(r_{it}\widehat{p}_i^2) = E\left[\frac{r_{it}}{(T-1)^2}\left(\sum_{t'=2}^{T} r_{it'}\right)^2\right] = \frac{p_i\kappa_i}{(T-1)^2},$$

where $\kappa_i = 1 + 3(T-2)p_i + 2(T-2)(T-3)p_i^2$. Thus, given $\tau_{it} = p_{it}$, the bias (20.9) becomes

$$B_1 = N^{-1}\sum_{i=1}^{N} w_i S_i(\Delta_t),$$

where

$$w_i = \frac{p_i\kappa_i}{(T-1)^2 p_i^{*3}} - \frac{p_i^3}{p_i^{*3}} + \left(\frac{2p_i}{p_i^{*3}} - \frac{1}{p_i^2}\right)V(\widehat{p}_i).$$

The coefficients w_i's above are functions of p_i, such that it may depend on $\mathbf{y}_i$, even though the functional form of the dependence is unspecified under the NEE approach. In other words, the term B_1 is not zero as long as the population covariance of w_i and $S_i(\Delta_t)$ is not zero, which is given by $N^{-1}\sum_{i=1}^{N} w_i S_i$ since $N^{-1}\sum_{i=1}^{N} S_i(\Delta_t) = 0$ by definition. We refer to Ahmad (2019) for more detailed discussions of the asymptotic behaviour of the NEE in the cross-sectional setting; the basic idea is analogous if one considers $\{(d_{it}, r_{it}) : i \in U\}$ as defined 'cross-sectionally' for the pair of $(t, t-1)$.

Consider the two cross-sectional NEEs in (20.6) separately. Let $\tau_{it} = E(\widehat{\pi}_{it})$ be the expectation of $\widehat{\pi}_{it}$ for wave t. By Taylor expansion, provided $\widehat{\pi}_{it} \neq 0$, we obtain

$$B_2(t) = E[\widehat{H}(\theta_t)] - H(\theta_t) = N^{-1}\sum_{i=1}^{N} S_i(\theta_t) E\left(\frac{\delta_{it}}{\widehat{\pi}_{it}} - 1\right)$$

$$= N^{-1}\sum_{i=1}^{N} S_i(\theta_t)\left\{\left(\frac{2\pi_{it}}{\tau_{it}} - 1\right) - \frac{E(\delta_{it}\hat{\pi}_{it})}{\tau_{it}^2} + \frac{E(\delta_{it}\hat{\pi}_{it}^2) - 2\tau_{it}E(\delta_{it}\hat{\pi}_{it}) + \tau_{it}^2\pi_{it}}{\pi_i^{*3}}\right\}$$
(20.10)

for some π_i^* between τ_{it} and $\hat{\pi}_{it}$. Similarly as above, in the case of unbiased $\hat{\pi}_i$ in (20.8), we have
$E(\delta_{it}\hat{\pi}_i) = \pi_i^2 + V(\hat{\pi}_i)$ and $E(\delta_{it}\pi_i^2) = \pi_i\kappa_{i'}/T^2$, where $\kappa_{i'} = 1 + 3(T-1)\pi_i + 2(T-1)(T-2)\pi_i^2$. The corresponding bias (20.10) becomes

$$B_2(t) = N^{-1}\sum_{i=1}^{N} u_i S_i(\theta_t),$$

where

$$u_i = \frac{\pi_i\kappa_{i'}}{T^2\pi_i^{*3}} - \frac{\pi_i^3}{\pi_i^{*3}} + \left(\frac{2\pi_i}{\pi_i^{*3}} - \frac{1}{\pi_i^2}\right)V(\hat{\pi}_i)$$

One can obtain $B_2(t-1)$ as the bias of $\hat{H}(\theta_{t-1})$ similarly. Again, B_2 may not be zero if the population covariance of u_i and $S_i(\theta_t)$ is not zero.

For bias adjustment of (20.3), one can obtain an approximation to B_1 by replacing p_i^* by $\hat{p}_{it}$. Now that p_i^* lies between $\hat{p}_{it}$ and its expectation τ_{it}, its likely values can be given via the standard error (SE) of $\hat{p}_{it}$, as $p_i^* = \hat{p}_{it} \pm \alpha\widehat{SE}(\hat{p}_{it})$, for chosen α-values and subjected to the range $p_i^* \in (0,1)$. One can then estimate Δ_t based on the NEE that is adjusted by the resulting $B_1(\alpha)$. A grid of α-values will generate accordingly a set of alternative estimates of Δ_t, which provide an indication of the likely range of an unbiased estimator of Δ_t under MNAR non-response. Similarly for the NEE (20.6). The adjustment will be illustrated in the simulation study later.

20.4 Variance Estimation

The variance of $\hat{\Delta}_t$ can be obtained in a sandwich form based on Taylor expansion of the NEEs. The descriptions below are given in terms of $\hat{H}(\Delta_t)$ and $\hat{H}(\theta_t, \theta_{t-1})$, respectively; the approach is the same based on any of their bias-adjusted versions.

20.4.1 NEE (Expression 20.3)

Let $\Delta_t' = E(\hat{\Delta}_t)$. The variance of $\hat{\Delta}_t$ based on (20.3) can be approximately given as

$$Var(\hat{\Delta}_t) = G^{-1}(\Delta_t')Var[\hat{H}(\Delta_t')]G^{-T}(\Delta_t'), \tag{20.11}$$

where $\widehat{\Delta}_t - \Delta'_t \approx -G^{-1}(\Delta'_t)\widehat{H}(\Delta'_t)$ by Taylor expansion, and

$$G(\Delta'_t) = \frac{1}{N}\sum_{i=1}^{N} E\left(\frac{r_{it}}{\widehat{p}_{it}}\right)\frac{\partial S_i(\Delta'_t)}{\partial \Delta'_t}$$

$$Var[\widehat{H}(\Delta'_t)] = \frac{1}{N}\sum_{i=1}^{N} Var\left(\frac{r_{it}}{\widehat{p}_{it}}\right) S_i(\Delta'_t)S_i(\Delta'_t)^T,$$

Let $\tau_{it} = E(\widehat{p}_{it})$ as before. We have

$$E\left(\frac{r_{it}}{\widehat{p}_{it}}\right) \approx \frac{2p_{it}}{\tau_{it}} - \frac{E(r_{it}\widehat{p}_{it})}{\tau_{it}^2} + \frac{E(r_{it}\widehat{p}_{it}^2) - 2\tau_{it}E(r_{it}\widehat{p}_{it}) + \tau_{it}^2 p_{it}}{\tau_{it}^3} \stackrel{def}{=} \mu_{1i}, \quad (20.12)$$

$$E\left(\frac{r_{it}}{\widehat{p}_{it}}\right)^2 \approx \frac{3p_{it}}{\tau_{it}^2} - \frac{2E(r_{it}\widehat{p}_{it})}{\tau_{it}^3} + \frac{3(E(r_{it}\widehat{p}_{it}^2) - 2\tau_{it}E(r_{it}\widehat{p}_{it}) + \tau_{it}^2 p_{it})}{\tau_{it}^4} \stackrel{def}{=} \mu_{2i}, \quad (20.13)$$

where $E(r_{it}\widehat{p}_{it})$ and $E(r_{it}\widehat{p}_{it}^2)$ are given before in the case of $\widehat{p}_{it} = \widehat{p}_i$. To estimate the variance (20.11), we first plug in the respective estimates for each unit with $r_{it} = 1$: $\widehat{\mu}_{1i}$ for $E(r_{it}/\widehat{p}_{it})$, $\widehat{\mu}_{2i} - \widehat{\mu}_{1i}^2$ for $Var(r_{it}/\widehat{p}_{it})$ and $\widehat{\Delta}_t$ for Δ'_t. Next, the relevant terms of the respondents, i.e. $\partial S_i(\Delta'_t)/\partial\Delta'_t$ and $S_i(\Delta'_t)S_i(\Delta'_t)^T$, are weighted by $\widehat{p}_{it}^{-1}$ and summed to yield $\widehat{G}(\Delta'_t)$ and $\widehat{Var}[\widehat{H}(\Delta'_t)]$, respectively.

20.4.2 NEE (Expression 20.6)

Let $\theta'_t = E(\widehat{\theta}_t)$ and $\theta'_{t-1} = E(\widehat{\theta}_{t-1})$. The variance of $\widehat{\Delta}_t$ using the NEE (20.6) is given by (20.7), where

$$Var(\widehat{\theta}_t) = G^{-1}(\theta'_t)Var[\widehat{H}(\theta'_t)]G^{-T}(\theta'_t),$$
$$Var(\widehat{\theta}_{t-1}) = G^{-1}(\theta'_{t-1})Var[\widehat{H}(\theta'_{t-1})]G^{-T}(\theta'_{t-1}).$$

The expressions for $Var(\widehat{\theta}_t)$ can be obtained similarly as above for $Var(\widehat{\Delta}_t)$ given in (20.11) after replacing $\widehat{\Delta}_t$ with $\widehat{\theta}_t$, Δ'_t with θ'_t, r_{it} with δ_{it}, p_{it} with π_{it} and $\widehat{p}_{it}$ with $\widehat{\pi}_{it}$. Similarly for $Var(\widehat{\theta}_{t-1})$. The term $Cov(\widehat{\theta}_t, \widehat{\theta}_{t-1})$ can be approximated as follows:

$$\begin{aligned} Cov(\widehat{\theta}_t, \widehat{\theta}_{t-1}) &= Cov(\widehat{\theta}_t - \theta'_t, \widehat{\theta}_{t-1} - \theta'_{t-1}) \approx Cov[G^{-1}(\theta'_t)\widehat{H}(\theta'_t), G^{-1}(\theta'_{t-1})\widehat{H}(\theta'_{t-1})] \\ &= G^{-1}(\theta'_t)Cov(\widehat{H}(\theta'_t), \widehat{H}(\theta'_{t-1}))G^{-T}(\theta'_{t-1}), \end{aligned}$$

where

$$Cov(\widehat{H}(\theta'_t), \widehat{H}^T(\theta'_{t-1},)) = \frac{1}{N}\sum_{i=1}^{N} Cov\left(\frac{\delta_{it}}{\widehat{\pi}_{it}}, \frac{\delta_{i,t-1}}{\widehat{\pi}_{i,t-1}}\right) S_i(\theta'_t)S_i^T(\theta'_{t-1})$$

The derivation of covariance term on the right-hand side is tedious. As the result, we obtain

$$Cov\left(\frac{\delta_{it}}{\widehat{\pi}_{it}}, \frac{\delta_{i,t-1}}{\widehat{\pi}_{i,t-1}}\right) \approx (E[\widehat{\pi}_{it}\widehat{\pi}_{i,t-1} \mid \delta_{it}\delta_{i,t-1} = 1] - E\left[\widehat{\pi}_{it} \mid \delta_{it} = 1\right] E\left[\widehat{\pi}_{i,t-1} \mid \delta_{i,t-1} = 1\right])$$

$$-\pi_{it}(E[\widehat{\pi}_{i,t-1} \mid \delta_{it}\delta_{i,t-1} = 1] - E\left[\widehat{\pi}_{i,t-1} \mid \delta_{i,t-1} = 1\right])$$

$$-\pi_{i,t-1}(E[\widehat{\pi}_{it} \mid \delta_{it}\delta_{i,t-1} = 1] - E\left[\widehat{\pi}_{it} \mid \delta_{i,t} = 1\right]), \quad (20.14)$$

where the expectations involved can be calculated according to the definition of $\widehat{\pi}_{it}$ and $\widehat{\pi}_{i,t-1}$.

For plug-in estimation of this covariance, one needs to weight the units with $r_{it} = 1$, for which $S_i(\theta'_t)S_i^T(\theta'_{t-1})$ is observed, by the weights $\widehat{p}_{it}$. Whereas plug-in estimation of $Var(\widehat{\theta}_t)$ is based on the units with $\delta_{it} = 1$ and weight $\widehat{\pi}_{it}^{-1}$, and $Var(\widehat{\theta}_{t-1})$ on those with $\delta_{i,t-1} = 1$ and weight $\widehat{\pi}_{i,t-1}^{-1}$.

20.5 Simulation Study

The NEE approach provides a method for exploring informative non-response in the longitudinal setting, which is computationally easy and flexible in specification. From the outset, several factors can be expected to affect its performance in a given situation.

First, given suitable non-response assumptions for all the responding individuals, the NEE estimator should perform better given a *longer* history of response. For instance, in the case of $T = t = 3$, where there are two observations of r_{it} for each individual, there are only two possible histories for each respondent with $r_{i3} = 1$, where r_{i2} is either 1 or 0. The estimate $\widehat{p}_{it}$ that enters the NEE (20.3) either takes value 1 or 0.5, such that the estimator of Δ_t is only based on two weighting classes (whereas the *naive* estimator under the MCAR assumption is based on a single weighting class). Clearly, the ability to adjust for potentially informative non-response by the NEE approach is rather limited in this case. Thus, a factor that matters in the simulation study will be the length of response history.

Another relevant factor is the variation of y_{it}'s over time, for each given individual. Take again the estimator $\widehat{p}_t$ that is averaged over all the r_{it}'s. On the one hand, it is unbiased if the informative response propensity depends only on a scalar summary of the y_{it}'s, in which case it does not matter how volatile the y_{it}'s are over time. On the other hand, intuitively the risk of bias is heightened, when the y_{it}'s are volatile, as compared to the extreme case where $y_{it} \equiv y_i$ is completely static. Moreover, as the variance of the resulting estimator of Δ_t increases with more volatile y_{it}'s, it would be interesting to explore if this has any compounding effect together with the heightened risk of bias.

Last but not least, the non-response mechanism itself will be a critical factor to the performance of the NEE. That is, if the assumption for unbiased estimation of p_{it}'s is clearly violated, then the NEE estimator may suffer extra bias beyond the inherent bias of the NEE as explained in Section 20.3.

Table 20.1 Summary of turnover values over three waves.

Wave	Mean	Minimum	Third quartile	Maximum
First	11 000	1	10 211	2 008 585
Second	10 749	1	11 010	2 012 973
Third	11 747	1	12 295	2 026 659

Below we describe first the data used for the simulation, the models that can be used for simulating response history, and then the chosen simulation setup, including the specific response probability models, the sample size corresponding to overall response rate, and the estimators to be evaluated, before we present and discuss the simulation results.

20.5.1 Data

We have available a dataset of real turnover from 16 788 firms over three successive years. There are some (about 13%) partially missing turnovers. We impute these values using the R-package *mice*. A summary of the completed population turnover values for each of the three waves is given in Table 20.1. It can be seen that the population distributions are skewed but reasonably stable over time.

We then increase the number of waves from 3 to 10 for simulations by recycling wave 1 to be wave 4, and wave 2 to be wave 5, wave 3 to be wave 6, and so on. Below we refer to this 10-wave dataset as the *stable* population. In addition, we create a 10-wave *volatile* population where, for a given wave, we permute the turnover values among all the firms within each industrial group. In this way, the population distribution at each wave remains the same, but the y_{it}'s associated with each individual firm are perturbed quite a lot, and the individual variation of turnover over time is greatly increased.

20.5.2 Response Probability Models

One can consider a range of models, some of which are compatible with the assumptions of the NEEs, while others represent different non-response mechanisms that are used to explore the sensitivity of the NEE approach. For $t \geq 2$, let

$$\pi_{it} = \Pr(\delta_{it} = 1 \mid \delta_{i,t-1}, y_{i1}, \dots, y_{iT}, \mathbf{x}_i).$$

One can simulate monotone dropout patterns by increasing the coefficient of $\delta_{i,t-1}$ relatively, to make it the dominating predictor; one can accommodate informative non-response as long as the coefficients of $(y_{i1}, \dots, y_{iT})$ are not all zero; one can achieve informative but stable response probability if $(y_{i1}, \ldots, y_{iT})$ is replaced by $\eta_i = \eta(y_{i1}, \ldots, y_{iT})$ that is a scalar function of them.

Specifically, for the simulation study, we consider the following response probability models,

$$logit(\pi_{it}) = \gamma_0 + \gamma_1\eta_i + \gamma_2 x_i \tag{20.15}$$

$$logit(\pi_{it}) = \gamma_0 + \gamma_1\delta_{i,t-1} + \gamma_2\eta_i + \gamma_3 x_i \tag{20.16}$$

$$logit(\pi_{it}) = \gamma_0 + \gamma_1 y_{it} + \gamma_2 x_i \tag{20.17}$$

$$logit(\pi_{it}) = \gamma_0 + \gamma_1\delta_{i,t-1} + \gamma_2 y_{it} + \gamma_3 x_i \tag{20.18}$$

$$logit(\pi_{it}) = \gamma_0 + \gamma_1 x_i \tag{20.19}$$

$$logit(\pi_{it}) = \gamma_0 + \gamma_1\delta_{i,t-1} + \gamma_2 x_i. \tag{20.20}$$

The model (20.15) is congenial to the estimators (20.4) and (20.8) and the model (20.16) possibly so, the models (20.17) and (20.18) are informative mechanisms that do not necessarily have stable response probabilities for the given population of y-values, and the models (20.19) and (20.20) are MAR mechanisms. The values of $\gamma' s$ can be chosen to achieve any desired overall response rate, where relatively large coefficients for y_{it} and η_i can add more informativeness to the response mechanism.

20.5.3 Simulation Set-up

We carry out simulations separately for the stable and volatile population, as described below. The number of simulations are determined on the basis of about 1% CV of the Monte Carlo errors.

We experiment with the different response probability models (20.15)–(20.20). The results using different estimators $\widehat{p}_{it}$ in (20.4) and $\widehat{\pi}_{it}$ in (20.8) are largely similar under models (20.16), (20.18), and (20.20), compared to those under models (20.15), (20.17), and (20.19), respectively, as long as the coefficient of $\delta_{i,\,t-1}$ under the former group is not large enough to induce monotone response patterns. Alternatively, when clear monotone response patterns are simulated, the estimators $\widehat{p}_{1i}$ and $\widehat{p}_{2i}$ would outperform $\widehat{p}_{it}$, as can be expected; similarly for $\widehat{\pi}_{1i}$ compared to $\widehat{\pi}_i$. Moreover, the model (20.19) yields stable MAR response probability, so that the NEE estimators are nearly unbiassed. Due to limited space, below we focus on simulation under the models (20.15) and (20.17), where η_i is set to the mean of $y_{i1}, \ldots, y_{iT}$, and the additional covariate x_i is a random variable from $LogN(2, 2)$.

For each response probability model, we set the γ-coefficients such that the overall response rate is about 60% or 80%, to be referred to as the *low* or *high* response setting, respectively. As explained in Section (20.3), the bias of the NEE estimators depends on the correlation between w_i and $S_i(\Delta_t)$ or, similarly, that between u_i and $S_i(\theta_t)$. We can vary the correlation by changing the coefficient γ_1 in (20.15) and (20.17), relative to the other γ-terms, while holding the overall response rate at the required setting. We simulate a *low* correlation scenario, where the population correlation between w_i and $S_i(\Delta_t)$ is e.g. −0.0536 at wave 6, and a *high* correlation scenario, where the correlation between w_i and $S_i(\Delta_t)$ is −0.1363 at the same wave. The high correlation induces non-negligible bias of the NEE estimators.

Given each response probability model, the response indicators δ_{it} are generated independently for all the 10 waves, based on which the NEE estimator $\widehat{\Delta}_t$ from (20.3) is calculated using $\widehat{p}_i$ in (20.4), denoted by **EE**($\widehat{p}_i$), for various combinations of (T, t). Given that we have full knowledge of the simulated data, we can calculate the hypothetical estimator using the NEE based on the individual mean of the true response probabilities over time. Including this hypothetical estimator, denoted by **EE_**h, allows us to understand when a result is due to the empirical property of the observed NEE, and when it is caused by the misspecified non-response assumption, i.e. when the data are simulated under the model (20.17) but the estimator is ideal under the model (20.15). Finally, the naïve estimator under the MCAR assumption is included as the baseline estimator for comparison. Similarly for the NEE estimator $\widehat{\Delta}_t$ from (20.6), calculated using $\widehat{\pi}_i$ in (20.8) and denoted by **EE**($\widehat{\pi}_i$), together the corresponding hypothetical estimator **EE_**h and naïve estimator under the MCAR assumption.

To explore bias adjustment of the NEE (20.3), discussed at the end of Section 20.3, we consider the following options: **EE_**i, simply dropping the term B_1 in (20.9) involving p_i^*; **EE_**ii, replacing p_i^* by $\widehat{p}_i$; **EE**(α), let $p_i^* = \widehat{p}_i \pm \alpha\widehat{\text{SE}}(\widehat{p}_i)$, for various α-values, and $p_i^* \in (0, 1]$. Similarly for the NEE (20.6).

20.5.4 Results

Below we report for each estimator its absolute percent relative bias (APRB), its standard error (SE), and the expectation of the SE estimator (ESEE) given as the square root of the variance estimator.

Table 20.2 shows the results under the model (20.15), in the high-response setting and high-correlation scenario, for $T = t = 4, 7$ and 10. Due to the cyclic pattern of the population data, the target parameter is the same on all these three occasions, where $\Delta_4 = \Delta_7 = \Delta_{10} = -746.55$, so that the differences in the results are chiefly caused by the length of response history. Notice that one may e.g. consider the NEE estimator by (20.3) to be based on 3, 6, and 9 weighting classes (of possible values of $\widehat{p}_i$), respectively, for $T = 4, 7$, and 10. We observe the following:

- The hypothetical estimator EE_h is unbiased for both NEEs under the model (20.15). Both the NEE estimators are biased in this high-correlation scenario. The bias is nevertheless greatly reduced compared to the naïve estimator, and it decreases as T increases. The reason for the latter is that the bias has two causes: informative non-response and the non-linear term $1/\hat{p}_i$. With large T and smaller variance of $\hat{p}_i$, the contribution of non-linearity to the bias decreases. As explained before, increased volatility of the individual y_{it}'s does not affect the bias of the NEE estimators here.
- Due to differential weighting, the SE of the hypothetical estimator can still be higher than the naïve estimator. The NEE estimators have even higher SEs, as can be expected. The bias-variance trade-off compared to the naïve estimator is clearly affected by the volatility of the individual y_{it}'s, although in these simulations it is still in favour of the NEE estimators for the volatile population.

While the variance of the NEE estimator by (20.3) is increased dramatically from the stable to the volatile population, that from the NEE (20.6) is about the same for both. As the δ_{it}'s are independent over time, the covariance term in (20.7) is close to zero, so that the NEE (20.6) actually loses efficiency compared to (20.3) for the stable population, even though it ostensibly uses more observations. Moreover, its variance remains about the same for the volatile population, because the population distribution is about the same over time according to how this volatile population is generated.

- When it comes to bias adjustment, dropping the indeterminable term (involving p_i^* or π_i^*) is not appealing *a priori*. Replacing p_i^* by $\hat{p}_i$ (or π_i^* by $\hat{\pi}_i$) can be problematic because of the variance of $\hat{p}_i$. For instance, in the case of $T = t = 10$, the bias is actually increased with **EE**_h compared to **EE**($\hat{p}_i$) because of this. Similarly with ad hoc choices of **EE**(α), illustrated for $\alpha = 0.5$ here.
- An interesting effect of these bias adjustments is that they clearly improve the variance estimation, e.g. when p_i^* is replaced by $\hat{p}_t$. The sandwich variance estimator based on the direct plug-in observed NEE tends to overestimate the variance, sometimes considerably, in these simulations. The sandwich variance estimator derived from the various bias-adjusted observed NEEs performs much better, and the improvement seems not related to the effect on point estimation.

Table 20.3 shows the results under the model (20.17), also in the high response setting and high correlation scenario, for $T = t = 7$ only, and the message is the same for the other choices of (T, t). The results show clearly that the underlying non-response assumption needs to be fairly close to the truth, in order for the NEE estimators to perform well. The risk of using the individual average of response probabilities over time is heightened with increasing volatility of the individual y_{it}'s, as can be seen from the bias of **EE**_h for the volatile population; in contrast, **EE**_h remains nearly unbiased for the stable population. Since lack of mean

Table 20.2 Results under model (20.15), high-response, high-correlation. population: stable (left), volatile (right, italic).

	$T = t = 4, \Delta_4 = -746.55$											
	naïve	**EE_h**	**EE($\widehat{p}_i$)**	**naïve**	**EE_h**	**EE($\widehat{\pi}_i$)**	**naïve**	**EE_h**	**EE($\widehat{p}_i$)**	**naïve**	**EE_h**	**EE($\widehat{\pi}_i$)**
APRB	48.85	0.03	12.62	24.16	0.25	22.00	*64.43*	*0.03*	*14.83*	*31.54*	*0.17*	*15.77*
SE	10.36	10.44	14.57	56.96	58.63	73.50	*50.25*	*45.48*	*69.36*	*52.76*	*52.49*	*65.86*
ESEE	258.05	10.61	21.95	517.51	59.01	102.39	*631.49*	*45.91*	*105.75*	*524.17*	*52.54*	*89.77*
	EE_i	**EE_ii**	**EE(0.5)**	**EE_i**	**EE_ii**	**EE(0.5)**	**EE_i**	**EE_ii**	**EE(0.5)**	**EE_i**	**EE_ii**	**EE(0.5)**
APRB	8.21	20.58	0.70	14.19	26.17	3.60	*12.29*	*25.40*	*0.80*	*17.79*	*20.93*	*6.79*
SE	19.24	12.85	17.16	88.23	66.33	81.46	*93.48*	*60.65*	*82.71*	*79.51*	*60.87*	*73.00*
ESEE	14.63	13.97	14.65	67.57	73.92	68.92	*71.50*	*67.56*	*71.37*	*62.60*	*98.15*	*63.04*
	$T = t = 7, \Delta_4 = -746.55$											
	naïve	**EE_h**	**EE($\widehat{p}_i$)**	**naïve**	**EE_h**	**EE($\widehat{\pi}_i$)**	***naïve***	**EE_h**	**EE($\widehat{p}_i$)**	***naïve***	**EE_h**	**EE($\widehat{\pi}_i$)**
APRB	51.82	0.00	1.36	25.99	0.09	1.54	*68.18*	*0.11*	*0.06*	*32.78*	*0.06*	*0.51*
SE	10.70	10.84	14.31	57.20	58.54	68.98	*71.19*	*64.43*	*85.07*	*56.98*	*55.77*	*64.18*
ESEE	258.26	10.73	20.14	517.76	58.86	90.02	*633.52*	*64.87*	*116.89*	*523.33*	*56.87*	*78.22*
	EE_i	**EE_ii**	**EE(0.5)**	**EE_i**	**EE_ii**	**EE(0.5)**	**EE_i**	**EE_ii**	**EE(0.5)**	**EE_i**	**EE_ii**	**EE(0.5)**
APRB	15.42	10.06	6.89	18.23	6.84	8.88	*21.12*	*9.99*	*10.38*	*15.15*	*3.33*	*7.28*
SE	17.52	12.85	15.86	78.93	65.22	73.59	*102.52*	*79.38*	*93.08*	*70.66*	*62.34*	*67.01*
ESEE	13.91	13.66	14.12	68.70	84.57	69.08	*86.21*	*85.89*	*87.21*	*64.87*	*83.51*	*64.98*
	$T = t = 10, \Delta_4 = -746.55$											
	naïve	**EE_h**	**EE($\widehat{p}_i$)**	**naïve**	**EE_h**	**EE($\widehat{\pi}_i$)**	***naïve***	**EE_h**	**EE($\widehat{p}_i$)**	***naïve***	**EE_h**	**EE($\widehat{\pi}_i$)**
APRB	53.00	0.01	1.46	26.57	0.11	0.23	*62.80*	*0.02*	*2.36*	*30.62*	*0.06*	*0.16*
SE	10.63	10.76	13.39	57.33	58.86	65.79	*86.59*	*77.52*	*94.64*	*63.00*	*61.24*	*66.54*
ESEE	258.35	10.80	18.74	518.02	58.84	76.09	*630.71*	*77.13*	*118.73*	*520.87*	*61.68*	*73.09*
	EE_i	**EE_ii**	**EE(0.5)**	**EE_i**	**EE_ii**	**EE(0.5)**	**EE_i**	**EE_ii**	**EE(0.5)**	**EE_i**	**EE_ii**	**EE(0.5)**
APRB	14.52	4.58	7.32	9.91	1.44	4.30	*17.08*	*3.50*	*8.87*	*7.34*	*0.48*	*2.92*
SE	15.89	12.47	14.50	70.24	64.56	67.71	*107.52*	*91.04*	*99.93*	*68.59*	*66.27*	*67.33*
ESEE	13.59	13.21	13.75	64.99	91.05	65.53	*94.78*	*94.67*	*95.73*	*66.35*	*71.52*	*66.72*

heterogeneity is a potential shortcoming for any parametric estimation approach in the presence of NMAR mechanisms, more empirical research is worthwhile regarding how to sensibly tailor the individual specification of $\widehat{p}_{it}$(or $\widehat{\pi}_{it}$) under the NEE approach.

In Table 20.4 the different response-rate settings and correlation scenarios are contrasted with each other, for $T = t = 6$. We observe the following:

- The effects of low response setting on the variance is clear and as expected, where all the SEs are increased, which is more dramatic for the volatile population when holding the correlation scenario fixed. Bias adjustment of the NEEs greatly improves the variance estimation again.
- The NEE estimators yield useful bias reduction compared to the naïve estimator in all the cases, even in the low response setting where the NEE estimator $\mathbf{EE}(\widehat{p}_i)$ from (20.3) has large bias itself.
- It is intriguing to observe how the bias of the NEE estimator $\mathbf{EE}(\widehat{p}_i)$ from (20.3) varies. The bias is increased in the high correlation scenario, but more so under the low-response setting. Moreover, the bias is much higher absolutely in the low-response setting, where $SE(\widehat{p}_i)$ is larger and the contribution of non-linear $1/\widehat{p}_i$ is relatively greater. Bias adjustment is also affected, especially in the low-response and low-correlation combination, where the bias is almost entirely due to the non-linearity, and simply dropping the interminable term p_i^* is more beneficial than replacing it with $\widehat{p}_i$.

Finally, the results for $t < T$ are omitted here, as these provide few new insights, since estimation for $t < T$ is based on the same $\widehat{p}_i$ and $\widehat{\pi}_i$ as $t = T$. One can remove the arbitrary difference in Δ_t and Δ_T by e.g. letting $t = T - 3$, given the cyclic data here. The differences in the results would be entirely due to Monte Carlo variation in the simulated δ_{it}'s and δ_{iT}'s.

20.6 Conclusions

In this chapter we propose a new NEE approach to estimation based on longitudinal data subjected to informative missing mechanisms. The following conclusions emerge from the theoretical investigation and the simulation study.

The NEE approach is easy to compute and flexible in specification of the estimators of individual response probabilities. This makes it a widely applicable technique for exploratory data analysis of longitudinal missing data mechanisms, based on the observed response history. The results can provide a basis for deciding whether more sophisticated modelling is needed in a given situation.

The NEE estimators are clearly better than the naïve estimator, provided sensible choices of the response probability estimator, which can easily accommodate

Table 20.3 Results under model (20.17), high response, high correlation. Population: stable (Left), volatile (Right, italic).

	$T = t = 7, \Delta_4 = -746.55$											
	naïve	**EE_h**	**EE($\hat{p}_i$)**	**naïve**	**EE_h**	**EE($\hat{\pi}_i$)**	**naïve**	**EE_h**	**EE($\hat{p}_i$)**	**naïve**	**EE_h**	**EE($\hat{\pi}_i$)**
APRB	61.35	0.79	5.66	41.32	10.21	13.18	*43.07*	*26.99*	*36.76*	*41.27*	*22.24*	*35.49*
SE	22.07	21.40	27.55	59.68	60.40	69.96	*245.66*	*246.55*	*376.42*	*59.16*	*59.41*	*121.79*
ESEE	260.95	21.51	39.04	521.45	60.31	91.51	*587.46*	*325.30*	*665.94*	*526.65*	*216.53*	*302.23*
	EE_i	**EE_ii**	**EE(0.5)**	**EE_i**	**EE_ii**	**EE(0.5)**	**EE_i**	**EE_ii**	**EE(0.5)**	**EE_i**	**EE_ii**	**EE(0.5)**
APRB	12.80	15.74	3.46	7.69	18.80	2.24	*36.16*	*35.63*	*36.48*	*33.30*	*35.77*	*34.43*
SE	32.92	25.88	30.05	79.29	66.59	74.27	*509.44*	*314.02*	*440.79*	*155.88*	*116.73*	*136.78*
ESEE	28.06	27.93	28.42	69.97	68.46	70.47	*472.62*	*437.18*	*472.23*	*265.52*	*263.32*	*264.86*

Table 20.4 Results under model (20.15), by response and correlation. population: stable (left), volatile (right, italic).

	$T = t6, \Delta_6 = 998.03$, low response and low correlation											
	naïve	**EE_h**	**EE($\hat{p}_i$)**	**naïve**	**EE_h**	**EE($\hat{\pi}_i$)**	**naïve**	**EE_h**	**EE($\hat{p}_i$)**	**naïve**	**EE_h**	**EE($\hat{\pi}_i$)**
APRB	61.84	0.13	11.61	26.09	0.31	4.70	*55.35*	*0.82*	*10.06*	*24.50*	*0.02*	*0.12*
SE	90.27	65.06	94.08	157.73	131.99	156.8	*489.66*	*534.93*	*663.13*	*325.30*	*334.46*	*401.58*
ESEE	317.10	65.68	129.10	690.10	129.91	196.6	*756.76*	*545.83*	*932.20*	*581.22*	*329.65*	*542.28*
	EE_i	**EE_ii**	**EE(0.5)**	**EE_i**	**EE_ii**	**EE(0.5)**	**EE_i**	**EE_ii**	**EE(0.5)**	**EE_i**	**EE_ii**	**EE(0.5)**
APRB	1.77	23.66	4.14	3.25	9.07	2.39	*3.34*	*22.58*	*2.58*	*6.51*	*1.87*	*3.21*
SE	106.59	89.15	100.46	168.66	153.22	162.1	*789.03*	*569.02*	*730.74*	*482.44*	*368.63*	*440.11*
ESEE	83.16	99.04	88.74	144.96	151.90	149.6	*589.31*	*644.55*	*620.15*	*391.34*	*376.64*	*394.60*
	$T = t = 6, \Delta_6 = 998.03$, low response and high correlation											
	naïve	**EE_h**	**EE($\hat{p}_i$)**	**naïve**	**EE_h**	**EE($\hat{\pi}_i$)**	***naïve***	**EE_h**	**EE($\hat{p}_i$)**	***naïve***	**EE_h**	**EE($\hat{\pi}_i$)**
APRB	108.81	0.02	16.24	45.13	0.08	6.52	*92.23*	*0.25*	*14.64*	*38.32*	*0.09*	*3.31*
SE	44.89	29.59	43.17	105.61	82.30	99.32	*297.06*	*194.46*	*291.02*	*160.29*	*131.30*	*156.83*
ESEE	329.72	30.11	58.88	695.22	82.67	137.9	*1075.6*	*195.87*	*386.22*	*685.68*	*133.02*	*197.21*
	EE_i	**EE_ii**	**EE(0.5)**	**EE_i**	**EE_ii**	**EE(0.5)**	**EE_i**	**EE_ii**	**EE(0.5)**	**EE_i**	**EE_ii**	**EE(0.5)**
APRB	7.80	37.08	2.85	1.08	14.69	0.86	*5.33*	*32.41*	*3.48*	*4.71*	*6.75*	*2.47*
SE	47.98	41.75	45.56	107.03	93.45	103.3	*325.36*	*279.46*	*308.30*	*171.15*	*153.09*	*163.23*
ESEE	37.68	46.55	40.52	89.11	100.53	94.48	*250.13*	*303.36*	*267.97*	*147.84*	*156.45*	*152.56*

Table 20.4 (Continued)

	$T = t = 6, \Delta_6 = 998.03$, high response and low correlation											
	naïve	**EE_h**	**EE($\widehat{p}_i$)**	**naïve**	**EE_h**	**EE($\widehat{\pi}_i$)**	**naïve**	**EE_h**	**EE($\widehat{p}_i$)**	**naïve**	**EE_h**	**EE($\widehat{\pi}_i$)**
APRB	18.67	0.03	3.35	10.08	0.03	1.43	27.76	0.02	3.14	14.30	0.03	0.49
SE	33.51	40.03	53.15	78.51	83.93	97.06	187.56	229.10	295.80	139.33	152.11	173.98
ESEE	183.43	40.81	79.95	524.35	85.21	126.3	586.81	229.78	447.84	513.86	148.15	221.40
	EE_i	**EE_ii**	**EE(0.5)**	**EE_i**	**EE_ii**	**EE(0.5)**	**EE_i**	**EE_ii**	**EE(0.5)**	**EE_i**	**EE_ii**	
APRB	3.62	7.62	0.23	0.25	3.28	0.12	6.54	8.15	1.80	2.72	0.97	1.18
SE	70.49	45.10	61.70	111.27	93.65	103.3	393.26	250.82	343.83	204.33	167.90	187.23
ESEE	55.67	50.16	55.13	101.79	97.22	100.6	311.19	277.75	307.65	179.36	167.16	176.51

	$T = t = 6, \Delta_6 = 998.03$, high response and high correlation											
	naïve	**EE_h**	**EE($\widehat{p}_i$)**	**naïve**	**EE_h**	**EE($\widehat{\pi}_i$)**	***naïve***	**EE_h**	**EE($\widehat{p}_i$)**	***naïve***	**EE_h**	**EE($\widehat{\pi}_i$)**
APRB	43.65	0.01	3.50	22.01	0.05	2.00	45.49	0.12	1.95	22.11	0.00	1.05
SE	9.36	9.62	12.48	57.42	58.44	70.36	62.99	56.77	78.80	54.60	53.79	63.04
ESEE	190.44	9.49	18.29	528.27	58.76	95.38	645.15	56.65	111.04	533.44	54.40	80.88
	EE_i	**EE_ii**	**EE(0.5)**	**EE_i**	**EE_ii**		**EE_i**	**EE_ii**	**EE(0.5)**	**EE_i**	**EE_ii**	**EE(0.5)**
APRB	12.19	11.96	0.36	5.01	6.16	0.50	14.11	10.24	1.34	1.76	3.52	0.14
SE	15.31	10.91	12.86	82.11	65.59	71.75	98.42	71.09	81.55	71.76	60.73	64.26
ESEE	11.96	11.92	12.28	69.47	68.97	69.09	79.00	77.29	79.40	63.81	61.81	63.20

NMAR mechanisms. This is especially the case given low volatility of the individual outcome variables over time, even though considerable variation of the same outcome variable may exist across the population. The ability to vary the non-response assumption for different individuals makes it potentially a flexible alternative to standard parametric modelling approach, where the same model parameters are assumed to apply across the population.

The NEE estimator is not exactly unbiased, even when the response probability estimator is unbiased. The bias has two sources: the correlation between e.g. p_{it} and y_{it}, and the non-linearity of $1/\hat{p}_{it}$, or the variance of $\hat{p}_{it}$. The variance of $\hat{p}_{it}$ is naturally reduced given longer response history. To reduce the bias caused by the non-linearity in situations of short history, one could possibly improve the efficiency of $\hat{p}_{it}$ by grouping the units with a similar response pattern *and* similar observed y_{it}'s. Therefore, a topic for the future is to study empirically how to vary $\hat{p}_{it}$ according to the different historic response patterns, while improving its efficiency based on similar individuals in both senses. The same consideration applies to the choice between the different NEEs, such as (20.3) vs. (20.6).

Another, more difficult topic that requires further theoretical development is how to adjust the bias caused by the correlation e.g. between p_{it} and y_{it}. The simulation study shows that this is desirable, despite the large reduction of bias compared to the naïve estimator even without such adjustment, also because it can improve the associated variance estimation. But it is not an easy task, because simply plugging in $\hat{p}_{it}$(or $\hat{\pi}_{it}$) is problematic due to its variance.

References

Ahmad, Z. (2019). Nonignorable Nonresponse Adjustment using Fully Nonparametric Approach, Unpublished PhD Thesis, University of Southampton, UK.

FitzGerald, E.B. (2002). Extended generalized estimating equations for binary familial data with incomplete families. *Biometrics* 58: 718–726.

Godambe, V.P. (1991). *Estimating Functions.* New York: Oxford Science Publications.

Hardin, J. and Hilbe, J.M. (2003). *Generalized Estimating Equations.* Boca Raton: Chapman and Hall.

Liang, K.Y. and Zeger, S.L. (1995). Inference based on estimating functions in the presence of nuisance parameters (with discussion). *Statistical Science* 10: 158–172.

Little, R.J.A. and Rubin, D.B. (2002). *Statistical Analysis with Missing Data*. New York: Wiley.

Lynn, P. (2009a). Methods for longitudinal surveys. In: *Methodology of Longitudinal Surveys* (ed. P. Lynn), 1–19. Chichester: Wiley.

Lynn, P. (2009b). *Methodology of Longitudinal Surveys.* New York: Wiley.

Pfeffermann, D. and Sikov, A. (2011). Imputation and estimation under nonignorable nonresponse in household surveys with missing covariate information. *Journal of Official Statistics* 27: 181–209.

Qin, J., Leung, D., and Shao, J. (2002). Estimation with survey data under nonignorable nonresponse or informative sampling. *Journal of the American Statistical Association* 97: 193–200.

Robins, J.M., Rotnitzky, A., and Zhao, L.P. (1994). Estimation of regression coefficients when some regressors are not always observed. *Journal of the American Statistical Association* 89: 846–866.

West, P., Sweeting, H., and Young, R. (2007). Smoking in Scottish youths: personal income, parental social class and the cost of smoking. *Tobacco Control* 16: 329–333.

Index

Advances in Longitudinal Survey Methodology, First Edition. Edited by Peter Lynn.

Companion website: www.wiley.com/go/lynn/advancesinlongitudinalsurvey